Selected Topics in Synthetic Biology

Selected Topics in Synthetic Biology

Editor: Daniel McGuire

R CALLISTO
REFERENCE

www.callistoreference.com

Callisto Reference,
118-35 Queens Blvd., Suite 400,
Forest Hills, NY 11375, USA

Visit us on the World Wide Web at:
www.callistoreference.com

© Callisto Reference, 2017

ISBN: 978-1-63239-859-8 (Hardback)

The publisher's policy is to use permanent paper from mills that operate a sustainable forestry policy. Furthermore, the publisher ensures that the text paper and cover boards used have met acceptable environmental accreditation standards.

Trademark Notice: Registered trademark of products or corporate names are used only for explanation and identification without intent to infringe.

Printed in the United States of America.

Cataloging-in-publication Data

Selected topics in synthetic biology / edited by Daniel McGuire.
   p. cm.
Includes bibliographical references and index.
ISBN 978-1-63239-859-8
1. Synthetic biology. 2. Bioengineering. I. McGuire, Daniel.
TA164 .S45 2017
660.6--dc23

Table of Contents

Preface

This book outlines the processes and applications of synthetic biology in detail. It provides comprehensive insights into this field. Synthetic biology is the amalgamation of biology and engineering. It refers to the study of designing and producing biological systems and modules. It uses the elements of molecular biology, computer engineering, biophysics and genetic engineering, among other subjects. This text is compiled in such a manner that it will provide in-depth knowledge about this field to the readers. The various sub-fields of synthetic biology along with technological process that have future implications are glanced at. Most of the topics introduced in this book cover new techniques and applications of synthetic biology. All those who are interested in this subject area will benefit alike from it. It will serve as a resource guide for students and experts and contribute to the growth of the discipline.

This book was inspired by the evolution of our times; to answer the curiosity of inquisitive minds. Many developments have occurred across the globe in the recent past which has transformed the progress in the field.

This book was developed from a mere concept to drafts to chapters and finally compiled together as a complete text to benefit the readers across all nations. To ensure the quality of the content we instilled two significant steps in our procedure. The first was to appoint an editorial team that would verify the data and statistics provided in the book and also select the most appropriate and valuable contributions from the plentiful contributions we received from authors worldwide. The next step was to appoint an expert of the topic as the Editor-in-Chief, who would head the project and finally make the necessary amendments and modifications to make the text reader-friendly. I was then commissioned to examine all the material to present the topics in the most comprehensible and productive format.

I would like to take this opportunity to thank all the contributing authors who were supportive enough to contribute their time and knowledge to this project. I also wish to convey my regards to my family who have been extremely supportive during the entire project.

Editor

1

Protective Epitopes of the *Plasmodium falciparum* SERA5 Malaria Vaccine Reside in Intrinsically Unstructured N-Terminal Repetitive Sequences

Masanori Yagi[1]ⓢ, Gilles Bang[2]ⓢ, Takahiro Tougan[1], Nirianne M. Q. Palacpac[1], Nobuko Arisue[1], Taiki Aoshi[3,4], Yoshitsugu Matsumoto[5], Ken J. Ishii[3,4], Thomas G. Egwang[6], Pierre Druilhe[2]¤, Toshihiro Horii[1]*

1 Department of Molecular Protozoology, Research Institute for Microbial Diseases, Osaka University, Suita, Osaka, Japan, **2** Laboratoire de Parasitologie Bio-Médicale, Institut Pasteur, Paris, France, **3** Laboratory of Adjuvant Innovation, National Institute of Biomedical Innovation, Ibaraki, Osaka, Japan, **4** Laboratory of Vaccine Science, Immunology Frontier Research Center, Osaka University, Suita, Osaka, Japan, **5** Laboratory of Molecular Immunology, School of Agriculture and Life Sciences, The University of Tokyo, Tokyo, Japan, **6** Med Biotech Laboratories, Kampala, Uganda

Abstract

The malaria vaccine candidate antigen, SE36, is based on the N-terminal 47 kDa domain of *Plasmodium falciparum* serine repeat antigen 5 (SERA5). In epidemiological studies, we have previously shown the inhibitory effects of SE36 specific antibodies on *in vitro* parasite growth and the negative correlation between antibody level and malaria symptoms. A phase 1 b trial of the BK-SE36 vaccine in Uganda elicited 72% protective efficacy against symptomatic malaria in children aged 6–20 years during the follow-up period 130–365 days post-second vaccination. Here, we performed epitope mapping with synthetic peptides covering the whole sequence of SE36 to identify and map dominant epitopes in Ugandan adult serum presumed to have clinical immunity to *P. falciparum* malaria. High titer sera from the Ugandan adults predominantly reacted with peptides corresponding to two successive N-terminal regions of SERA5 containing octamer repeats and serine rich sequences, regions of SERA5 that were previously reported to have limited polymorphism. Affinity purified antibodies specifically recognizing the octamer repeats and serine rich sequences exhibited a high antibody-dependent cellular inhibition (ADCI) activity that inhibited parasite growth. Furthermore, protein structure predictions and structural analysis of SE36 using spectroscopic methods indicated that N-terminal regions possessing inhibitory epitopes are intrinsically unstructured. Collectively, these results suggest that strict tertiary structure of SE36 epitopes is not required to elicit protective antibodies in naturally immune Ugandan adults.

Editor: Georges Snounou, Université Pierre et Marie Curie, France

Funding: This work was supported by Grant-in-Aid for Young Scientists (B) (22770151) and Grant-in-Aid for Global COE (Centers of Excellence) Program to MY; Grant-in-Aid for Scientific Research (A) (24249024), Program for the Promotion of International Policy Dialogues Contributing to the Development of Science and Technology Diplomacy to TH from the Japanese Ministry of Education, Science, Sports, Culture and Technology. The authors would like to acknowledge the funding support of Global Health Innovative Technology Fund (GHIT RFP 2013-001) to TH under the project Clinical development of BK-SE36/CpG malaria vaccine. The funders had no role in the study design, data collection and analysis, decision to publish, or preparation of the manuscript.

Competing Interests: The authors have declared that no competing interests exist.

* E-mail: horii@biken.osaka-u.ac.jp

ⓢ These authors contributed equally to this work.

¤ Current address: The Vac4all initiative, Paris, France

Introduction

Despite the vast malaria burden no effective malaria vaccine exists [1,2]. The development of malaria vaccines has mainly focused on *Plasmodium falciparum*, the most deadly of five *Plasmodium* species that infect humans. Malaria vaccine development strategies vary depending on the target stages of the parasite life cycle, i.e. sporozoite, intra-hepatocytic stage, asexual erythrocyte stages, gametocyte, and mosquito midgut stages. Asexual erythrocyte stage antigens are thought to elicit antibodies which reduce blood parasitemia and lessen the severity of malaria symptoms. However, sequence polymorphism of many antigens, as observed in several vaccine candidates such as merozoite surface protein (MSP)-1, MSP-2 [3] and apical membrane antigen-1 (AMA-1) [4], hamper the systematic vaccine development strategy based on host immune responses against malaria parasites.

P. falciparum serine repeat antigen 5 (*Pf*SERA5) is one of the candidate vaccines in human trial [5–7]. Abundantly expressed in the parasitophorous vacuole and on the merozoite surface, and belonging to the SERA protein family, the 120 kDa protein is processed into 47, 50, 6 and 18 kDa domains at the time of schizont rupture. While the 50 and 6 kDa domains are secreted, the 47 and 18 kDa domains are covalently linked by disulfide bond(s) and remain on the merozoite surface [7–9]. *Pf*SERA5 was the first physiological substrate identified for *P. falciparum* subtilisin-like serine protease (*Pf*SUB1) [10].

Sequence analysis of 445 *P. falciparum* field isolates from nine countries worldwide revealed that sequence polymorphism of

*Pf*SERA5 is remarkably limited unlike other malaria vaccine candidates [11]. Moreover, high antibody level against the N-terminal 47 kDa domain correlated with the absence of fever or low parasitemia [5,12,13]. Under *in vitro* conditions, antibodies against the N-terminal domain were also suggested to correlate with antiparasitic effects through several mechanisms. At high antibody concentration, inhibition of parasite growth was found to be associated with merozoite agglutination [14] or complement mediated cell lysis of segmented schizont [15]. At low antibody concentrations, monocyte-mediated antibody dependent cellular inhibition (ADCI) activity has been demonstrated [16].

SE36 is based on the N-terminal 47 kDa domain constructed by removing the polyserine region located in the middle of the domain [5]. A recent phase 1 b clinical trial and follow-up study 365 days post-second vaccination elicited 72% protective efficacy against symptomatic malaria in Ugandan children aged 6–20 years [6]. Although the exact function of *Pf*SERA5 remains unknown, a parasite inhibitory epitope defined by a murine monoclonal antibody was mapped onto amino acids 17–73 of the Honduras-1 strain, a well conserved region in diverse geographical isolates of *P. falciparum* [17,18].

Intrinsically unstructured proteins (IUPs), also called intrinsically disordered proteins or natively unfolded proteins, have for the past 10–20 years generated interest because of their unusual way to carry out molecular recognition different from traditional protein structure-function paradigms. IUPs contain polypeptide chains lacking stable tertiary structure when they exist alone, however, some of them are known to switch to more ordered conformation upon recognition of their binding partners and play their biological roles [19,20].

*Pf*SERA5 (17–73) has an octamer repeat (OR) region and a serine rich (SR) region. The repeat number of octamer motifs varies depending on strains but the basic motif sequences are well conserved in *P. falciparum* [11]. These OR and SR regions have biased amino acid composition and are low complexity regions with little diversity in their amino acids [21]. Low complexity regions are often found in *Plasmodium* species and, due to lack of hydrophobic amino acids, such regions are expected to be intrinsically unstructured [22]. *P. falciparum* proteins, such as MSP-2 and trophozoite exported protein 1 (Tex1), are reported to have intrinsically unstructured regions (IURs) [23–25]. P27A, an IUR found in Tex-1, was well recognized by sera from individuals living in malaria endemic areas. Moreover, murine antibodies purified from P27A immunized mice showed high ADCI activities [24]. Immunogenicity of IUR in naturally infected humans was also reported for *P. vivax* AMA-1 [26].

In the present study, we performed epitope mapping with overlapping synthetic peptides covering the whole sequence of SE36 utilizing serum from Ugandan volunteers, and serum from previously vaccinated mice and squirrel monkeys. We identified the N-terminal repetitive sequence regions of SE36 as immunodominant IgG epitopes in Ugandan individuals presumed to have clinical immunity to *P. falciparum* malaria. We have demonstrated previously that antibodies raised against N-terminal region of *Pf*SERA5 strongly inhibit *in vitro* parasite growth by ADCI at concentrations which do not show any detectable direct inhibition of growth [16], thus we used this assay as a screen for functional inhibition activity of anti-SE36 IgG. Affinity purified antibodies against N-terminal repetitive sequence regions of SE36 showed high ADCI activities suggesting that the regions are protective epitopes. Additionally, the OR and SR regions are revealed to be intrinsically unstructured by spectroscopic methods and protein structure predictions. These results show that the N-terminal repetitive sequences have characteristics of an intrinsically

unstructured region and are highly immunogenic in Ugandan adults eliciting protective antibodies against malaria.

Materials and Methods

Ethics Statement

Serum samples from pool of individuals living in endemic areas and individual Ugandan serum samples were from participants in an earlier epidemiological study [9]. Briefly, the study utilizes residual samples from a cross-sectional study of 40 (37 sera are available for this study) healthy Ugandan adults living in Atopi Parish, a malaria holoendemic area, located 5 km west of Apac Town, 300 km north of Kampala. Ethical clearance for sampling and consent was obtained and approved by the Uganda National Council for Science and Technology under the 1997 Guidelines for Health Research Involving Human Subjects in Uganda [9]. In agreement with the local community leadership, a process of dialogue was done. Information about the study was given to the head of the community, household and study participants. Verbal consent was obtained for voluntary participation and for blood samples to be taken and stored for use in future studies. It was deemed culturally-sensitive in this community that experienced recent government conflict that verbal informed consent be sought (written consent were not practiced, disliked and viewed as mistrust). Being a cross-sectional study, signatures will also be the only record of their participation and risk of privacy is minimized if their signature is not recorded. No other records exist for their participation. Blood samples were coded during blood collection, processed within a few hours after collection and separated into sera, which was stored at $-20°C$ and $-70°C$ until analyses.

Animal housing, care and handling of squirrel monkeys were done in strict compliance with "The guidelines for the care and use of laboratory animals" by the University of Tokyo [5]. Briefly, male squirrel monkeys (*Saimiri sciureus*) of Guyana phenotype were bred in captivity. The monkeys were quarantined and conditioned for at least a month prior to the commencement of the study. Thorough medical examinations revealed that the animals were free of all intestinal and any blood stage infections including malaria, and they were declared to be in a general good health by a veterinarian. They were housed in the Amami Laboratory of Injurious Animals, Institute of Medical Science, University of Tokyo in individual safety cabinets with an exercise bar at a controlled environment of $24\pm2°C$ and $50\pm10\%$ humidity. Monkeys were fed with new world monkey chow (Clea Japan Inc., Tokyo, Japan) and allowed free access to water. Lighting was automatically regulated on a 12 hours light-dark cycle. The monkeys, weighing between 680 and 760 g at the beginning of the experiment, were divided into two treatment groups that received their intramuscular injection on the left thigh 5 and 3 weeks before challenge infection. All procedures were performed under anesthesia and all efforts were made to minimize suffering. All experimental procedures were approved by the School of Agriculture and Life Sciences, the University of Tokyo. Additional details of animal welfare/care and steps taken to ameliorate suffering were in accordance with the recommendations of the Weatherall report, "The use of non-human primates in research". During the study no monkey died or was sacrificed.

Animal experiments using mice were approved by the Animal Care and Use Committee of the Research Institute for Microbial Diseases, Osaka University, Japan. Mice care and steps to ameliorate suffering was conducted in accordance with the guidelines of the committee and immunization experiments were in accordance with the GERBU adjuvant protocol described below (GERBU Biotechnik GmbH, Heidelberg, Germany). During the study, no mice were sacrificed.

Animal Blood Samples

Residual serum samples from squirrel monkeys (*Saimiri sciureus*) that received 50 µg SE36 protein with 500 µg aluminum hydroxide gel in 0.5 ml of PBS and those in the control group that received the same volume of PBS by intra-muscular injection were utilized [5]. In brief, after 2 or 3 immunizations, these monkeys were followed through after *P. falciparum* challenge infection [5]. For mouse immunization, 30 ddY mice were purchased from Japan SLC, Inc. (Hamamatsu, Japan). Each mouse was subcutaneously immunized with 50 µl of 1 mg/ml SE36 protein and 50 µl GERBU adjuvant (100 µl in total) 4 times at 2-week intervals. Two weeks after last immunization, blood draw was performed from the mouse tail and blood samples from the mice were pooled for the experiments.

Recombinant SE36 Protein

GMP grade SE36 protein was expressed in *E. coli* using a codon optimized synthetic gene and purified as previously described [5].

Synthetic Peptides

Fifteen synthetic peptides of 40–42 residues covering the whole sequence of SE36 protein were synthesized by Operon Biotechnology Inc. (Tokyo, Japan) (Fig. 1 and Table S1, series I). Each peptide was designed to overlap with two adjacent peptides at its N- and C-terminal halves, respectively.

Enzyme-Linked Immunosorbent Assay

Enzyme-linked immunosorbent assay (ELISA) was performed using flat-bottomed 96-well Nunc-Immuno plates (Nunc, Roskilde, Denmark). SE36 protein or the synthetic peptides were dissolved in carbonate buffer (pH 9.6) as coating buffer at a concentration of 0.1 µg/ml. For ELISA assays of synthetic peptides, each plate was coated with the whole peptide series. The plates were coated overnight at 4°C with 100 µl of the protein or peptide solutions, washed three times with PBS containing 0.05% Tween 20 (PBS/T) and blocked for an hour with 5% skim milk in PBS at 37°C. The plates were again washed three times with PBS/T prior to addition of serum samples or purified IgG prepared in 5% skim milk in PBS/T. Test samples were added to wells at optimized concentration and incubated for an hour at 37°C. After washing with PBS/T, peroxidase-conjugated goat IgG fraction to human IgG (whole molecule) (55220; Cappel ICN Pharmaceuticals Inc, Aurora, OH) diluted 1:2000; or horseradish peroxidase-conjugated rabbit anti-human IgG antibody (A8792; Sigma-Aldrich Corp., St. Louis, MO) diluted 1:2000; or peroxidase conjugated affiniPure goat anti-mouse IgG antibody (H+L) (115-035-166; Jackson ImmunoResearch Laboratories, Inc., West Grove, PA) diluted 1:5000 in 5% skim milk in PBS/T was added to the plates and incubated at 37°C for 1 hour. The plates were washed and

incubated with 100 µl freshly prepared citrate-phosphate buffer (pH 5.0) containing 0.2% hydrogen peroxide and OPD tablet (154-01673; Sigma-Aldrich Corp., St. Louis, MO) for 15 minutes. The reaction was stopped with 100 µl of 2 M sulfuric acid and optical density was read at 492 nm.

Purification of Antibodies

To prepare anti-OR and anti-SR antibodies for ADCI experiments, the antibodies were purified from Ugandan high antibody titer serum pool (SE36-positive serum pool) [9]. Prior to the purification of antibodies specific to the OR or SR region, whole antibodies were purified from the serum pool with HiTrap Protein G HP columns (GE Healthcare UK Ltd, Buckinghamshire, UK). Antibodies purified by Protein G columns were then loaded to either OR or SR-specific peptide columns. The OR/SR peptide columns were prepared with SulfoLink Immobilization Kit for Peptides (Thermo Fisher Scientific, Waltham, MA), so that OR/SR peptides with cysteine residues at the N-termini (series II peptides 1 and 3, Table S2) were immobilized to the columns via thiol groups, following recommendations of the manufacturer. Antibodies bound to the columns were eluted with 0.1 M Gly-HCl at pH 2.7 and immediately neutralized with 1 M Tris-HCl at pH 8.5.

Murine serum pool from SE36-vaccinated mice was used to purify the antibody against the whole SE36 molecule. However, due to limited sample volume, after applying to Protein G column no further purification was done. All antibodies were dialyzed against RPMI 1640 (Nakalai Tesque, Kyoto, Japan) prior to ADCI assays.

Antibody-Dependent Cellular Inhibition Assay (ADCI) and Assessment of Parasitemia by Flow Cytometry

The ADCI assay was as previously described [27,28] and carried out with either (i) human IgG purified with protein G and then affinity purified for specific peptides; or (ii) protein G purified IgG from mice immunized with SE36. Final concentration of IgG was at 0.3 mg/ml. As a positive control, a pool of hyperimmune African adults IgG (PIAG) [27] was used at 2 mg/ml to assess reproducibility between each assay. Monocytes (MN) from peripheral blood mononuclear cells were further enriched using EasySep Human Monocyte Enrichment Kit Without CD16 Depletion according to manufacturer's instruction (StemCell Technologies Inc., Vancouver, BC, Canada). Monocyte monolayer was obtained after incubation of 2×10^5 MN for 30 minutes at 37°C in 5% CO_2 atmosphere. A synchronized asexual blood stage parasite culture (K1 clone) with very mature schizonts (0.5% parasitemia, 2.5% haematocrit) was added on the MN monolayer in addition to murine and human IgG to be tested. Intrinsic anti-parasitic effect of control and test IgG sera was assessed in wells containing the blood stage parasites without MN. Prior to ADCI assay, only MN with non-significant phagocytosis effect against *in vitro* growth of asexual blood stage parasites were selected. Samples were tested in duplicate wells. Plates were incubated in a candle jar

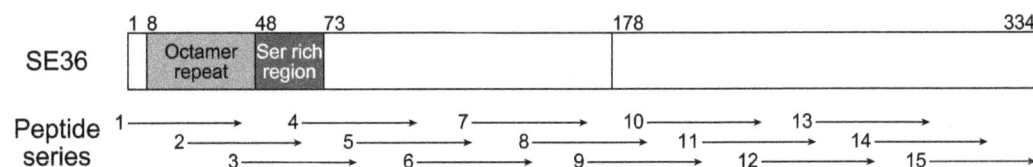

Figure 1. Schematic representation of synthetic peptide series I covering the whole sequence of SE36 protein. The number 178 denotes the position of polyserine sequence present in *Pf*SERA5 but deleted in SE36 [5].

at $37°C$, in a 5% CO_2 incubator. At 48 and 72 hours, 50 µl of complete medium was added to each well. At 96 hours the assay was stopped and the parasitemia determined by flow cytometry (FACSCalibur, BD Biosciences, CA). Assays were performed at the same day by two independent researchers.

Flow cytometry enumeration of infected erythrocytes with viable malaria parasites was performed by double staining of DNA and RNA using hydroethidine (HE) and thiazole orange (TO) (Sigma-Aldrich Corp.). Briefly, erythrocytes were incubated for 20 min at $37°C$ in the dark with 20 µg/ml of HE diluted in PBS-1% FCS (FACS buffer), washed three times in FACS buffer, followed by another incubation for 30 min at room temperature in the dark with TO diluted at 1:15000 in FACS buffer. Analysis was performed on 1×10^5 erythrocytes with the CellQuest Pro software. Parasitemia was determined as the percentage of double stained infected erythrocytes among the whole erythrocyte population. The specific growth inhibitory index (SGI) was calculated according to the following formula: SGI = $100 \times$ [1-(percent parasitemia with MN and test IgG/percent parasitemia with test IgG)/(percent parasitemia with MN and naïve IgG/ percent parasitemia with naïve IgG)]. An SGI effect was considered as significant if yielding a value >30% [29].

Protein Structure Prediction

The amino acid sequences of the N-terminal domains of PfSERA1-9 (3D7) were aligned with Multiple Sequence Alignment Tool version 1.1 (http://cib.cf.ocha.ac.jp/KYG/onlyalign.html) [30] after deletion of N-terminal signal sequences predicted by SignalP 3.0 [31]. Since PfSERA8 has no corresponding region, amino acid sequence identity and similarity of relatively conserved regions among PfSERA1-7 and 9 were calculated based on Clustal W alignment and similarity classification in NPS@server (http://npsa-pbil.ibcp.fr/) [32,33]. Percentage "identity" and "strong similarity" were calculated from the sum of the number of amino acids (Table 1). For prediction of disordered/ordered structure, the sequences were applied to Consensus Disorder Prediction (http://protease.burnham.org/www/tools/html/disorder.html) [34]. Secondary structure prediction of SE36 was performed by Consensus secondary structure prediction in NPS@ server [33].

Circular Dichroism

Circular dichroism (CD) spectra were acquired with a J-820 spectropolarimeter (Jasco, Tokyo, Japan) at $5–37°C$. Samples were prepared at 0.2 mg/ml for SE36 or 0.1 mg/ml for peptides in 50 mM sodium phosphate and 150 mM NaCl with or without 40% 2,2,2-trifluoroethanol (TFE). Each spectrum is an average of 20–40 times measurements. The obtained data were converted into mean residue ellipticity, $[\theta]$. Peptide concentrations were determined using BCA Protein Assay Kit (Thermo Fisher Scientific). For SE36, the concentration was determined from absorbance at 280 nm using an extinction coefficient calculated as reported by Gill and von Hippel [5,35].

Tryptophan Fluorescence

Tryptophan fluorescence spectra were acquired with a F-7000 fluorescence spectrophotometer (Hitachi, Tokyo, Japan) at $25°C$ with excitation wavelength at 295 nm and emission detection wavelength between 300 to 450 nm. Samples were prepared at 0.2 mg/ml in 50 mM sodium phosphate and 150 mM NaCl with or without 8 M urea.

Table 1. Amino acid sequence identity and similarity of the relatively conserved regions (brown bars in Fig. 5) in the N-terminal domain between SE36 (Honduras-1 SERA5) and 3D7 SERA proteins.

3D7 SERA	1	2	3	4	5	6	7	9
Identity (%)	57.2	50.7	54.6	51.3	98.0	53.9	50.7	52.0
Similarity (%)	80.9	77.0	80.9	77.0	99.3	74.3	78.3	80.3

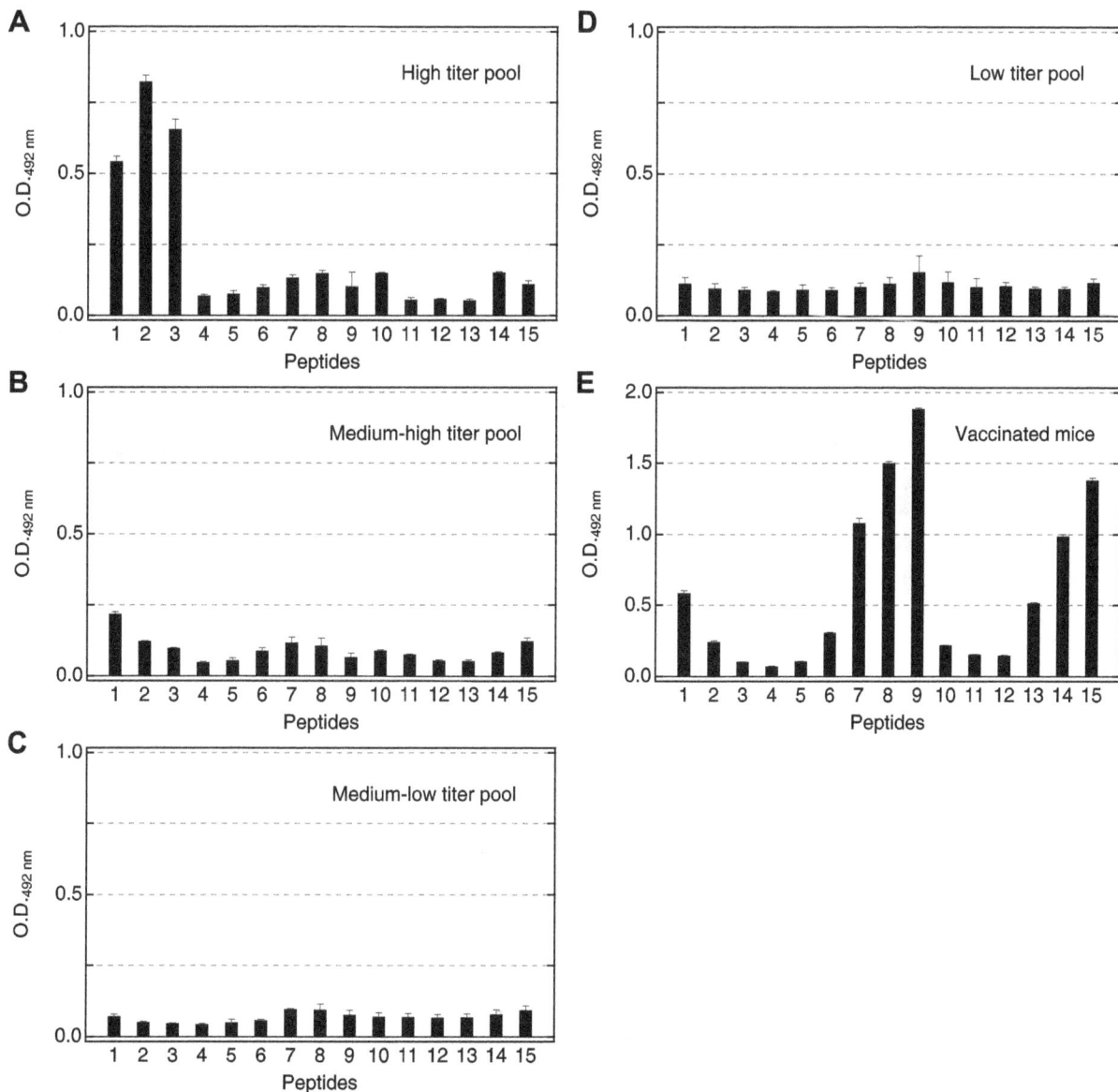

Figure 2. Reactivity of pooled Ugandan serum samples and a vaccinated mouse serum pool against synthetic peptide series I. (A) High, **(B)** medium-high, **(C)** medium-low, **(D)** low titer sera pool. Each pool consists of equal aliquots of 9–10 individual sera. The geometric mean anti-SE36 IgG titers of the individual samples are in Table S2. The patterns of reactivity for 18 individual sera are shown in Fig. S2. Serum samples were diluted 800-fold. Secondary antibody was peroxidase-conjugated goat IgG fraction to human IgG (whole molecule) (55220; Cappel ICN Pharmaceuticals Inc, Aurora, OH) diluted 1:2000. **(E)** Pooled serum from five mice used at 1:1,600. Secondary antibody was peroxidase conjugated affiniPure goat anti-mouse IgG antibody (H+L) (115-035-166; Jackson ImmunoResearch Laboratories, Inc., West Grove, PA) diluted 1:5000. All sera were tested for ELISA at least four times. Error bars reflect standard deviation. Reactivity of malaria naïve Japanese serum and naïve mouse serum are shown in Fig. S2.

Results

Reactivity of Anti-SE36 Positive Ugandan Serum against the Synthetic Peptides

To determine antigenic regions of SE36, overlapping synthetic peptides corresponding to the N-terminal domain of *Pf*SERA5 (Honduras-1) were prepared (Fig. 1). Based on anti-SE36 IgG levels (Table S2), four pools of high (9 individuals), medium-high (9 individuals), medium-low (9 individuals) and low (10 individuals) titer sera were made and tested for reactivity with peptide series I (Details are in Table S1). As shown in Fig. 2A, pooled high titer sera predominantly reacted with peptides 1, 2 and 3. Peptides 1–3

correspond to the OR and SR regions. Looking at individual sera, more than half of the individuals in the high titer group (or those presumed to have clinical immunity to *P. falciparum* infection) predominantly reacted with peptides 1, 2 and/or 3 (Fig. S2A–D, F, G, I). Medium and low titer pooled sera (Fig. 2B–D), as well as the Japanese naïve control serum (Fig. S2S), reacted poorly to these peptides. In addition, two randomly chosen high titer Ugandan serum samples (PRI and T69) were also examined with another overlapping peptide set (Table S1, series II) which had a different span in the SE36 protein (Fig. S1A, a set of 26 peptides with 20–40 residues). Again, both high titer sera predominantly reacted with peptides 1 and 2, with Ugandan T69 serum also

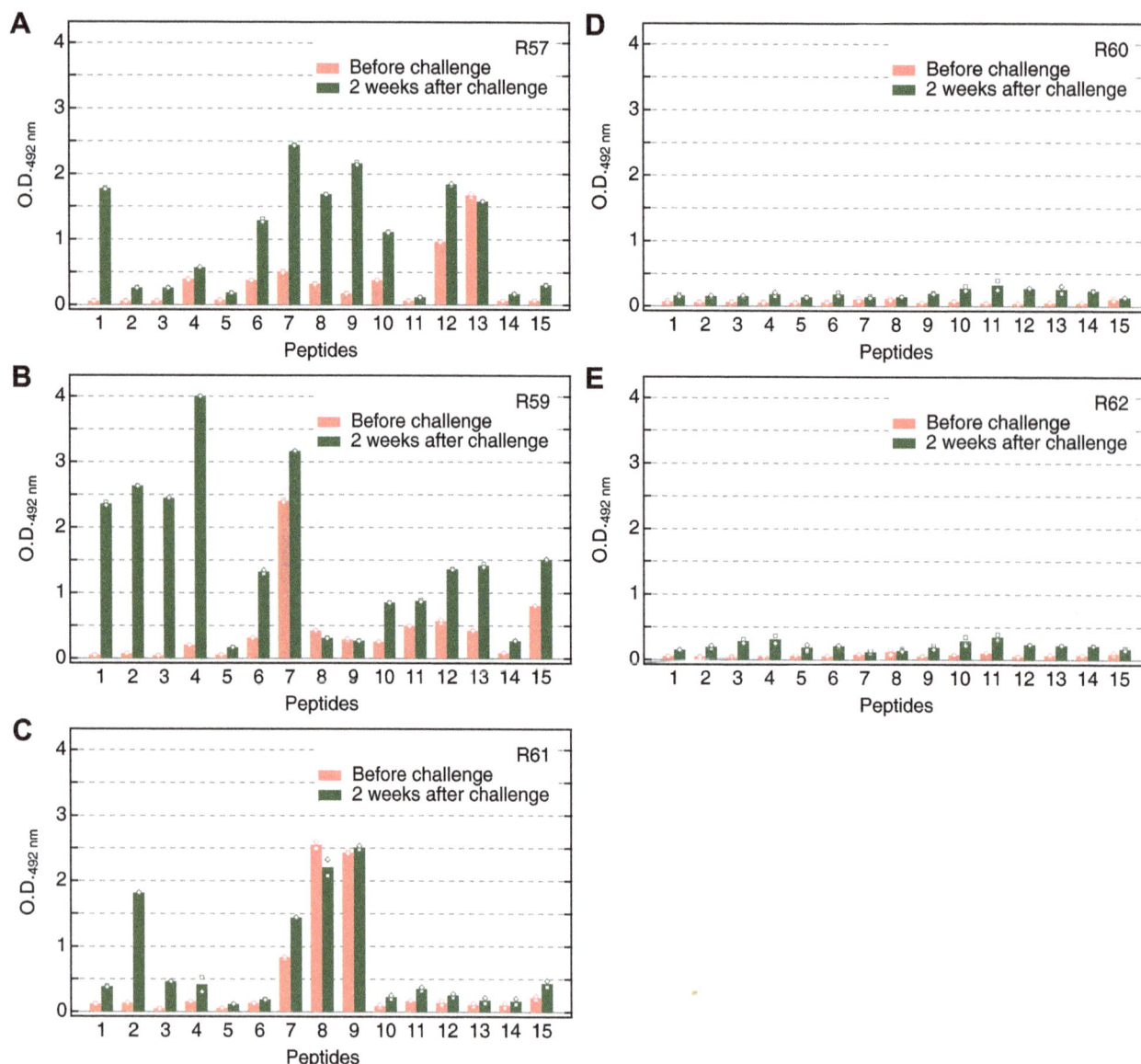

Figure 3. Reactivity in squirrel monkeys. The reactivity of the serum samples from vaccinated (**A–C**) and non-vaccinated (**D, E**) squirrel monkeys against the different peptides in Fig. 1 and Table S1 series I. Red and green bars represent samples before challenge infection and two weeks after challenge infection, respectively. R57, R59-62 are subject codes for squirrel monkeys. Monkey serum were diluted 1:400; secondary antibody was horseradish peroxidase-conjugated rabbit anti-human IgG antibody (A8792; Sigma-Aldrich Corp., St. Louis, MO) diluted 1:2000. Mean values from duplicate ELISA with individual data points are shown.

showing reactivity to peptide 3 (Fig. S1B–1 and B–2). This is in contrast to an SE36 vaccinated mouse serum pool that reacted broadly with a number of regions/peptides as shown in Fig. 2E. Reactivity to naïve mouse serum is shown in Fig. S2T.

From Ugandan serum reactivity, it appears that the OR and SR repetitive sequences in the N-terminal regions are highly antigenic in Ugandan adults with high anti-SE36 antibody titers.

Reactivity of Vaccinated Squirrel Monkey Serum against the Synthetic Peptides

We performed epitope mapping using serum samples obtained previously from SE36 vaccinated squirrel monkeys. Sera from those monkeys vaccinated with SE36 before and after *P. falciparum* challenge infection were used. Before challenge infection, the spectra of reactivity against the peptides were broad, similar to

vaccinated mouse serum pool. After challenge infection, some reactivity was observed to the peptides corresponding to OR and SR sequences (Fig. 3A–C). In contrast, monkeys in the control group (without SE36 vaccination) did not show marked response to the synthetic peptides even after challenge infection (Fig. 3D, E). Thus, as earlier reported [5], without SE36 vaccination, no anti-SE36 antibody can be induced despite challenge infection. However, priming the host with SE36 vaccination resulted in a boosting of immune response at the OR and SR sequences by challenge infection. We cannot exclude the presence of other protective epitopes outside the OR and SR sequences (located downstream of the OR and SR sequences) since we did not observe dominant reactivity against OR and/or SR sequences yet all three vaccinated squirrel monkeys were protected from high parasitemia [5].

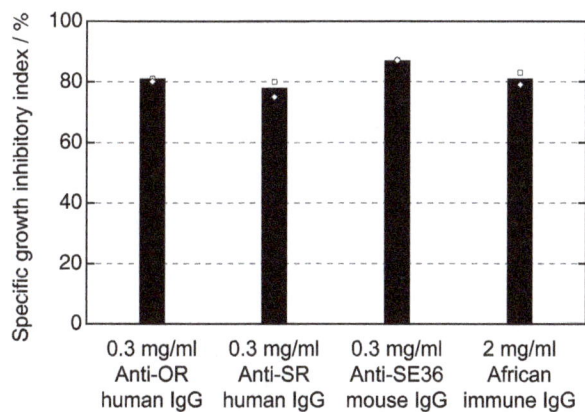

Figure 4. Antibody-dependent cellular inhibition activity with affinity purified antibodies. *In vitro* parasite specific growth inhibition (SGI) in the presence of human monocytes and human affinity-purified anti-OR or anti-SR IgG or murine anti-SE36 IgG at 0.3 mg/ml. A pool of polyclonal immune African IgG from individuals living in endemic areas was used as positive control at a final concentration of 2 mg/ml. IgG purified from malaria naïve human sera was used as negative control and included in the formula for SGI calculation as described in Materials and Methods. Mean values from duplicate ELISA with individual data points are shown.

Antibody-Dependent Cellular Inhibition (ADCI) Assay with Affinity Purified Antibodies

To examine whether antibodies specific to OR and SR sequences can exert any parasite growth inhibitory effect, we conducted *in vitro* parasite growth inhibition assays with either murine IgG induced by SE36 or affinity purified human natural IgG specific to OR or SR regions in both direct and human monocyte dependent ADCI assays (Fig. 4). The purified human antibodies were tested for their selectivity and reactivity by ELISA (Fig. S1C, D). No significant direct inhibitory effect was observed in all tested IgG. In contrast, anti-parasitic ADCI activity was strong using either induced murine anti-SE36 IgG or human IgG affinity purified against the two synthetic peptides (OR or SR peptide) (Fig. 4). It is noteworthy that these specific IgG preparation used at final concentration of 0.3 mg/ml have a similar growth inhibitory activity to a 2 mg/ml pool of polyclonal immune African IgG previously used in passive transfer experi-

ments of IgG to malaria patients [27,36]. These results indicate that OR and SR regions are protective epitopes.

Structure Prediction and Physicochemical Characterization of the N-Terminal Domain of SERA5

The structural features of both OR and SR sequences as well as the other parts of SE36 protein (based on the SERA5 N-terminal domain of the Honduras-1 strain) were examined using several structure prediction servers. Consensus Disorder Prediction [34] discriminated ordered and disordered regions in the sequence (Fig. 5 and Fig. S3). The region from the N-terminal end to Asp-76 (Fig. 5) was predicted to be predominantly disordered. This region corresponds to OR and SR sequences which were revealed as protective epitopes inducing antibodies capable of parasite growth inhibition. Other disordered regions identified were the polyserine sequence and its N- and C-terminal adjacent regions (matching series I peptides 7–9). All other parts in SE36 were predicted to be ordered. Using the secondary structure prediction program in NPS@ server [33], the assigned secondary structures matched the ordered regions identified with Consensus Disorder Prediction (Fig. 5). The predictions of ordered regions were further confirmed by the fact that all cysteine residues could be aligned at the same positions and high sequence identity and similarity could be obtained in the relatively conserved region shown in Fig. 5 in all of *Pf*SERA family, suggesting a common tertiary structure among the SERA family proteins (Table 1 and Fig. S3).

The synthetic peptides corresponding to OR and SR regions were subjected to CD experiments to define their structural characteristics. The peptide for the OR region did not show any ability to form rigid, typical secondary structure even in the presence of 40% TFE, an inducer and stabilizer for secondary structure (Fig. 6A–C). The behavior of the SR peptide was similar to OR peptide in the absence of TFE. However, with 40% TFE and at lower temperature, the SR peptide showed spectral change distinct from the OR peptide (Fig. 6D–F). Considering the low complexity due to biased amino acid composition of the SR region and the non-significant spectral change, an intrinsically unstructured nature of the region can be suggested. However, the structure predictions also assigned short ordered structure on a cluster of valines (VSTVSVSQ) in the SR region. These results may suggest a possible role of this region for hydrophobic interaction with other molecule(s).

The CD spectrum of the whole SE36 protein suggests an ordered structure (Fig. 7A). The shoulders near 208 and 222 nm suggest the existence of an α-helical structure. In the above

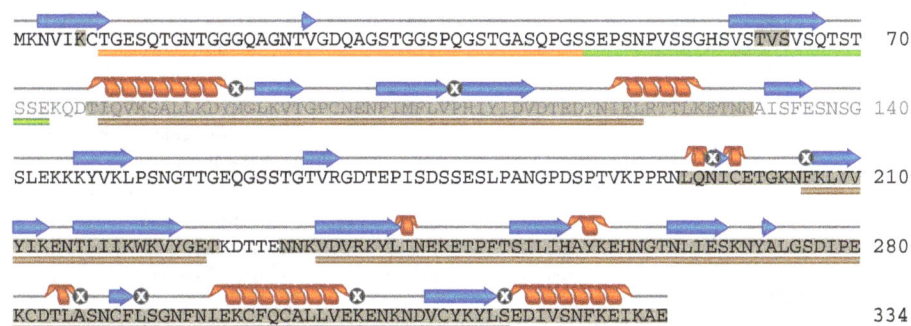

Figure 5. Schematic summary of the sequence analyses of SE36 by structure prediction servers. The shaded amino acids were predicted as ordered residues by the program, Consensus Disorder Prediction. The regions predicted as α-helix and β-structure by NPS@ are represented as helices and arrows above the sequence, respectively. Amino acid denoted by a symbol "⊗"did not reach consensus. The bars in orange, green and brown below the sequence denote OR region, SR region and relatively conserved regions among SERA family genes 1–7 and 9, respectively.

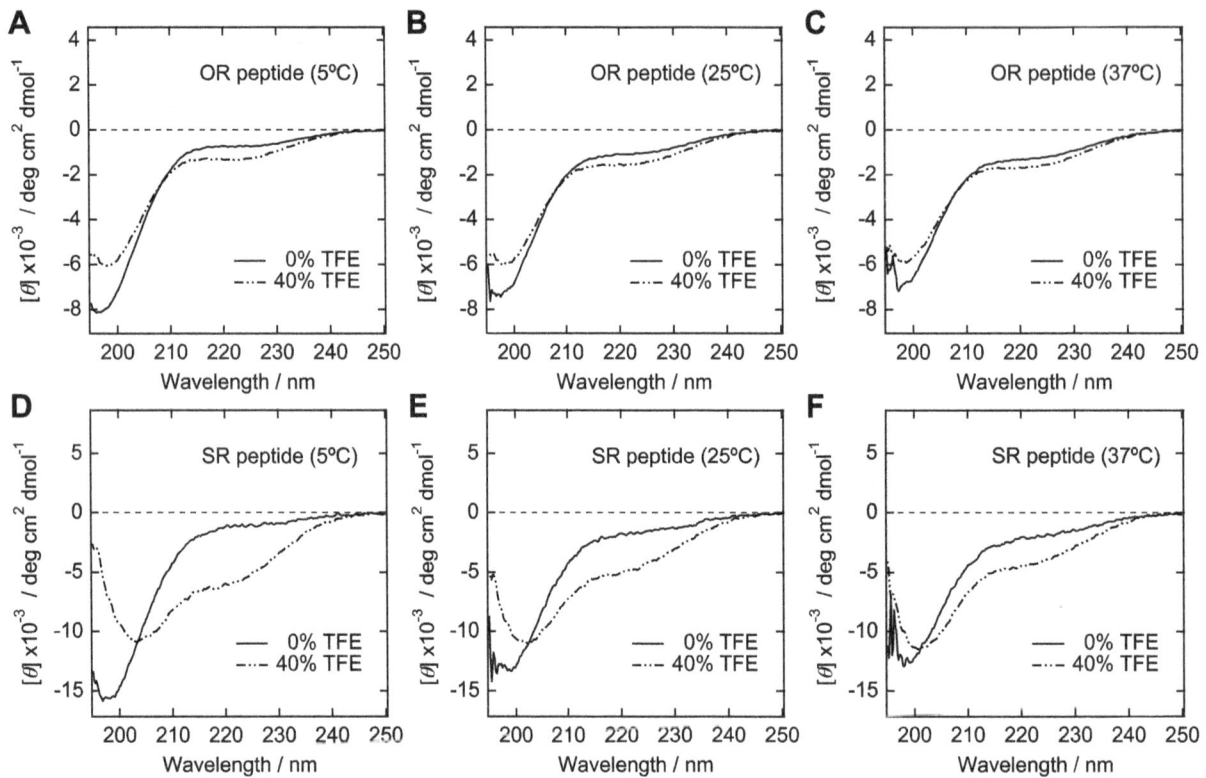

Figure 6. CD spectra of the synthetic peptides. Upper panel (**A–C**) corresponds to OR peptide region. Lower panel (**D–F**) corresponds to SR peptide region. The sequences of the corresponding peptides are in Table S1, series II (peptide 1 and 3, respectively). The measurements were performed at 5, 25 and 37°C with or without 40% TFE.

analysis, N-terminal domain of *Pf*SERA5, including SE36, contains only one tryptophan residue that was predicted in an ordered region (Fig. 5). Therefore, tryptophan fluorescence spectra provide a clue to assess the extent of formation of hydrophobic cluster in recombinant SE36 protein. The spectrum of SE36 protein in the absence of denaturant showed a peak at 330–340 nm, suggesting that the region around the tryptophan residue was located in relatively hydrophobic condition (Fig. 7B). The peak of the spectrum shifted to ~350 nm and the intensity increased upon the addition of 8 M urea. This spectral change implies that fluorescence quenching component(s) such as a polar residue and/or a disulfide bond existed near the tryptophan residue in a non-denatured state. Results from CD and tryptophan fluorescence experiments support that there are some driving forces to form compact tertiary structure in the sequence of SE36. The structure prediction and spectroscopic studies showed the existence of both structured and unstructured parts in SE36 with OR and SR epitopes belonging to unstructured parts.

Discussion

In this study, we first investigated which sequences in our SE36 vaccine candidate are predominantly recognized by sera from malaria endemic areas. From the results of the epitope mapping using the synthetic peptides, we identified N-terminal repetitive sequences, OR and SR regions, as the dominant epitopes in SE36 protein in Ugandan high titer adult sera. In contrast, vaccination of SE36 to mice or squirrel monkeys did not result in specific induction of OR- and SR-biased antibodies. However, it is interesting that the vaccinated monkeys increased the antibody

reactivity against OR and SR regions after challenge infection, while non-vaccinated monkeys did not show any significant response to SE36 even after challenge infection. In malaria endemic areas, as previously reported [5], the frequency of individuals seropositive (or having high antibody titers) to SE36 is not high even in adults. Although the mechanism is unclear, vaccination by SE36 led to induction in immune response to SE36, including OR and SR regions after challenge infection in squirrel monkeys. To examine the inhibitory effect of antibodies against OR and SR dominant epitopes, we performed an ADCI assay. The assay showed that the antibodies specific to both OR and SR regions inhibited in vitro growth of asexual blood stage parasites in cooperation with blood monocytes, suggesting that these regions are indeed protective epitopes in SE36. ADCI has been shown to require for specific inhibition of infected erythrocytes cooperation between the Fc domain of cytophilic IgG and the Fc-γ receptors of monocytes [28,37]. The recognition and subsequent activation likely induce the release of cytotoxic mediators leading to parasite killing at the intra-erythrocytic level. Although some parasites are indeed phagocytosed [28], this indirect intra-erythrocytic effect is the main mode of action of ADCI. What is generally observed in blood thin smears is the presence of many picnotic or crisis forms of parasites at the end of the assay [37].

The protective epitopes identified in this study are consistent with our previous results using mouse antibodies, i.e., glutathione-S-transferase-fused proteins containing the N-terminal regions corresponding to OR and SR regions were recognized by parasite-inhibitory antibodies [17,18]. However, the OR and SR regions may not be the sole protective epitopes, since murine SE36 specific

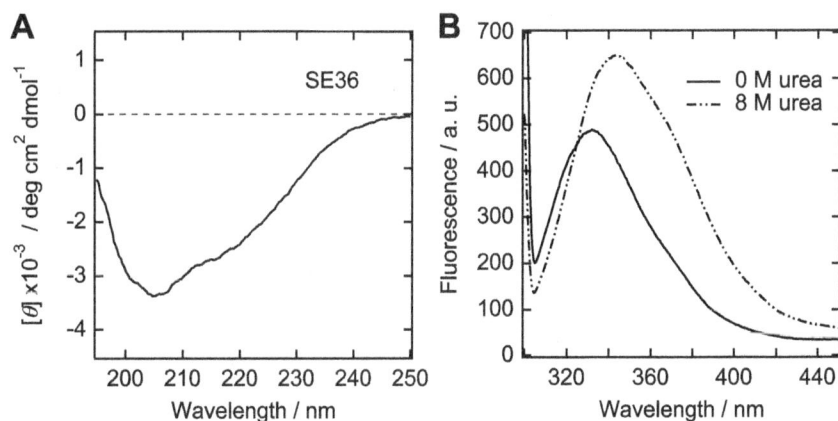

Figure 7. CD (A) and tryptophan fluorescence (B) spectra of SE36. Tryptophan fluorescence measurements were done with or without 8 M urea at 25°C.

antibodies showed broad reactivity against different regions of SE36 and correspondingly have comparable parasite inhibitory effects at similar antibody concentrations (Fig. 2E and Fig. 4). Squirrel monkeys vaccinated with the SE36 protein, likewise, gained measurable protection without showing dominant OR and SR specific antibodies [5]. Additionally, there are also some individual samples (T65, TO28, T64, T68 and TO08) that showed reactivity to other peptide regions (Fig. S2I, L, M, N, P).

We further characterized the physicochemical properties of OR and SR protective epitopes. Structure prediction programs and spectroscopic experiments indicated that the OR and SR regions are predominantly disordered, referred to as IURs. IURs are found in many eukaryotes, especially in apicomplexan parasites including *P. falciparum* [22]. The OR region consists of octamer repeats with closely related sequence motifs and the number of repeats largely differs among 445 field isolates [11]. All of the octamer repeat sequences are similar and lack bulky hydrophobic residues, suggesting that they all are intrinsically unstructured. The sequence motif corresponding to the SR region, on the other hand, is highly conserved although with slight variation in the number of repeats.

One example of the biological function of IURs is interaction of transcription factors with nucleic acids [38,39]. The fly-casting mechanism has been suggested as an advantage, allowing the flexibility of IURs [40,41]. Since the OR region has strain-specific octamer repeat numbers [11], the OR region may function to interact with other molecule(s) without strict structural requirement which can be observed in a traditional enzyme-substrate interaction. In contrast to the OR region, the tendency to form a secondary structure at lower temperature with TFE (Fig. 6) and the higher sequence conservation of SR region may reflect rigid structure formation either by binding-coupled folding or more strictly controlled interaction to other molecule(s).

There are a few reports which refer to the immunogenicity of intrinsically unstructured regions [24,26,42]. Here, we found strong antigenicity of OR and SR regions of *Pf*SERA5 protein that were found to be intrinsically unstructured. The antibodies recognizing the unstructured peptides have strong antiparasitic effect in an ADCI assay. The intrinsically unstructured characteristic of the protective epitope(s) is an advantage for a vaccine candidate. Some malaria antigens such as PfCP-2.9 (a fusion protein consisting of AMA-1 domain III and AMA-1 19 kDa C-terminal domain fragment) are known to have conformational epitopes which require strict stereo structure of antigens [43].

However, the protective epitopes of SE36 do not require strict stereo structure. Even though the epitopes are unstructured and flexible in free states, upon interacting with an antibody, unstructured peptides would be fixed for antibody specific interactions as shown in Fig. S1C.

Conclusions

We identified the epitopes targeted by biologically active antibodies in the malaria vaccine candidate SE36 using sera from people living in an endemic area, Uganda. The epitopes have repetitive sequences and have characteristics of intrinsically unstructured region. The polymorphism of the epitope regions is limited and they do not require strict stereo structure to elicit functional antibodies inhibiting *in vitro* growth of asexual blood stage parasites. These results support SE36 as a promising malaria vaccine antigen.

Supporting Information

Figure S1   Reactivity studies with peptide series II.

Figure S2   Reactivity studies with peptide series I.

Figure S3   Sequence alignment of *Pf*SERA1-9.

Table S1   The sequences of synthetic peptides used in this study.

Table S2   Anti-SE36 antibody titers of Ugandan individuals.

Acknowledgments

We would like to acknowledge Prof. Yuji Goto (Institute for Protein Research, Osaka University) for the use of spectrometers for the CD and fluorescence measurements.

Author Contributions

Conceived and designed the experiments: TH KJI PD MY. Performed the experiments: MY GB TT NMQP NA TA YM TGE. Analyzed the data: MY TH TT NMQP GB PD. Contributed reagents/materials/analysis tools: TGE TT PD YM. Wrote the paper: MY TH NMQP GB PD.

References

1. WHO World malaria report 2012.
2. Murray CJ, Rosenfeld LC, Lim SS, Andrews KG, Foreman KJ, et al. (2012) Global malaria mortality between 1980 and 2010: a systematic analysis. Lancet 379: 413–431.
3. Chauhan VS, Yazdani SS, Gaur D (2010) Malaria vaccine development based on merozoite surface proteins of *Plasmodium falciparum*. Hum Vaccin 6(9): 757–762.
4. Healer J, Murphy V, Hodder AN, Masciantonio R, Gemmill AW, et al. (2004) Allelic polymorphisms in apical membrane antigen-1 are responsible for evasion of antibody-mediated inhibition in *Plasmodium falciparum*. Mol Microbiol 52(1): 159–168.
5. Horii T, Shirai H, Jie L, Ishii KJ, Palacpac NMQ, et al. (2010) Evidences of protection against blood-stage infection of *Plasmodium falciparum* by the novel protein vaccine SE36. Parasitol Int 59(3): 380–386.
6. Palacpac NMQ, Ntege E, Yeka A, Balikagala B, Suzuki N, et al. (2013) Phase 1b randomized trial and follow-up study in Uganda of the blood-stage malaria vaccine candidate BK-SE36. PLoS ONE 8(5), e64073.
7. Palacpac NMQ, Arisue N, Tougan T, Ishii KJ, Horii T (2011) *Plasmodium falciparum* serine repeat antigen 5 (SE36) as a malaria vaccine candidate. Vaccine 29(35): 5837–5845.
8. Li J, Mitamura T, Fox BA, Bzik DJ, Horii T (2002) Differential localization of processed fragments of *Plasmodium falciparum* serine repeat antigen and further processing of its N-terminal 47 kDa fragment. Parasitol Int 51(4): 343–352.
9. Aoki S, Li J, Itagaki S, Okech BA, Egwang TG, et al. (2002) Serine repeat antigen (SERA5) is predominantly expressed among the SERA multigene family of *Plasmodium falciparum*, and the acquired antibody titers correlate with serum inhibition of the parasite growth. J Biol Chem 277(49): 47533–47540.
10. Yeoh S, O'Donnell RA, Koussis K, Dluzewski AR, Ansell KH, et al. (2007) Subcellular discharge of a serine protease mediates release of invasive malaria parasites from host erythrocytes. Cell 131(6): 1072–1083.
11. Tanabe K, Arisue N, Palacpac NM, Yagi M, Tougan T, et al. (2012) Geographic differentiation of polymorphism in the *Plasmodium falciparum* malaria vaccine candidate gene SERA5. Vaccine 30(9): 1583–1593.
12. Okech BA, Nalunkuma A, Okello D, Pang XL, Suzue K, et al. (2001) Natural human immunoglobulin G subclass responses to *Plasmodium falciparum* serine repeat antigen in Uganda. Am J Trop Med Hyg 65(6): 912–917.
13. Okech B, Mujuzi G, Ogwal A, Shirai H, Horii T, et al. (2006) High titers of IgG antibodies against *Plasmodium falciparum* serine repeat antigen 5 (SERA5) are associated with protection against severe malaria in Ugandan children. Am J Trop Med Hyg 74(2): 191–197.
14. Pang XL, Mitamura T, Horii T (1999) Antibodies reactive with the N-terminal domain of *Plasmodium falciparum* serine repeat antigen inhibit cell proliferation by agglutinating merozoites and schizonts. Infect Immun 67(4): 1821–1827.
15. Pang XL, Horii T (1998) Complement-mediated killing of *Plasmodium falciparum* erythrocytic schizont with antibodies to the recombinant serine repeat antigen (SERA). Vaccine 16(13): 1299–1305.
16. Soe S, Singh S, Camus D, Horii T, Druilhe P (2002) *Plasmodium falciparum* serine repeat protein, a new target of monocyte-dependent antibody-mediated parasite killing. Infect Immun 70(12): 7182–7184.
17. Fox BA, Pang XL, Suzue K, Horii T, Bzik DJ (1997) *Plasmodium falciparum*: an epitope within a highly conserved region of the 47-kDa amino-terminal domain of the serine repeat antigen is a target of parasite-inhibitory antibodies. Exp Parasitol 85(2): 121–134.
18. Fox BA, Horii T, Bzik DJ (2002) *Plasmodium falciparum*: fine-mapping of an epitope of the serine repeat antigen that is a target of parasite-inhibitory antibodies. Exp Parasitol 101(1): 69–72.
19. Uversky VN (2011) Intrinsically disordered proteins from A to Z. Int J Biochem Cell Biol 43(8): 1090–1103.
20. Tompa P (2012) Intrinsically disordered proteins: a 10-year recap. Trends Biochem Sci 37(12): 509–516.
21. Zilversmit MM, Volkman SK, DePristo MA, Wirth DF, Awadalla P, et al. (2010) Low-complexity regions in *Plasmodium falciparum*: missing links in the evolution of an extreme genome. Mol Biol Evol 27(9): 2198–2209.
22. Mohan A, Sullivan WJ Jr, Radivojac P, Dunker AK, Uversky VN (2008) Intrinsic disorder in pathogenic and non-pathogenic microbes: discovering and analyzing the unfoldomes of early-branching eukaryotes. Mol Biosyst 4(4): 328–340.
23. Zhang X, Perugini MA, Yao S, Adda CG, Murphy VJ, et al. (2008) Solution conformation, backbone dynamics and lipid interactions of the intrinsically unstructured malaria surface protein MSP2. J Mol Biol 379(1): 105–121.
24. Olugbile S, Kulangara C, Bang G, Bertholet S, Suzarte E, et al. (2009) Vaccine potentials of an intrinsically unstructured fragment derived from the blood stage-associated *Plasmodium falciparum* protein PFF0165c. Infect Immun 77(12): 5701–5709.
25. Kulangara C, Luedin S, Dietz O, Rusch S, Frank G, et al. (2012) Cell biological characterization of the malaria vaccine candidate trophozoite exported protein 1. PLoS One. 7(10): e46112.
26. Bueno LL, Lobo FP, Morais CG, Mourão LC, de Ávila RA, et al. (2011) Identification of a highly antigenic linear B cell epitope within *Plasmodium vivax* apical membrane antigen 1 (AMA-1). PLoS One. 6(6): e21289.
27. Bouharoun-Tayoun H, Attanath P, Sabchareon A, Chongsuphajaisiddhi T, Druilhe P (1990) Antibodies that protect humans against *Plasmodium falciparum* blood stages do not on their own inhibit parasite growth and invasion in vitro, but act in cooperation with monocytes. J Exp Med 172(6): 1633–1641.
28. Jafarshad A, Dziegiel MH, Lundquist R, Nielsen LK, Singh S, et al. (2007) A novel antibody-dependent cellular cytotoxicity mechanism involved in defense against malaria requires costimulation of monocytes FcγRII and FcγRIII. J Immunol 178(5): 3099–3106.
29. Druilhe P, Spertini F, Soesoe D, Corradin G, Mejia P, et al. (2005) A malaria vaccine that elicits in humans antibodies able to kill *Plasmodium falciparum*. PLoS Med 2(11): e344.
30. Kim OTP, Yura K, Go N (2006) Amino acid residue doublet propensity in the protein-RNA interface and its application to RNA interface prediction. Nucleic Acids Res 34(22): 6450–6460.
31. Emanuelsson O, Brunak S, von Heijne G, Nielsen H (2007) Locating proteins in the cell using TargetP, SignalP, and related tools. Nat Protoc 2(4). 953–971.
32. Thompson JD, Higgins DG, Gibson TJ (1994) CLUSTAL W: improving the sensitivity of progressive multiple sequence alignment through sequence weighting, position-specific gap penalties and weight matrix choice. Nucleic Acids Res 22(22): 4673–4680.
33. Combet C, Blanchet C, Geourjon C, Deléage G (2000) NPS@: network protein sequence analysis. Trends Biochem Sci 25(3): 147–150.
34. Igarashi Y, Heureux E, Doctor KS, Talwar P, Gramatikova S, et al. (2009) PMAP: databases for analyzing proteolytic events and pathways. Nucleic Acids Res 37:D611–618.
35. Gill SC, von Hippel PH (1989) Calculation of protein extinction coefficients from amino acid sequence data. Anal Biochem 182(2): 319–326.
36. Sabchareon A, Burnouf T, Ouattara D, Attanath P, Bouharoun-Tayoun H, et al. (1991) Parasitologic and clinical human response to immunoglobulin administration in *falciparum* malaria. Am J Trop Med Hyg 45(3): 297–308.
37. Bouharoun-Tayoun H, Oeuvray C, Lunel F, Druilhe P (1995) Mechanisms underlying the monocyte-mediated antibody-dependent killing of *Plasmodium falciparum* asexual blood stages. J Exp Med 182(2): 409–418.
38. Liu J, Perumal NB, Oldfield CJ, Su EW, Uversky VN, et al. (2006) Intrinsic disorder in transcription factors. Biochemistry 45(22): 6873–6888.
39. Hilser VJ, Thompson EB (2011) Structural dynamics, intrinsic disorder, and allostery in nuclear receptors as transcription factors. J Biol Chem 286(46): 39675–39682.
40. Shoemaker BA, Portman JJ, Wolynes PG (2000) Speeding molecular recognition by using the folding funnel: the fly-casting mechanism. Proc Natl Acad Sci U S A 97(16): 8868–8873.
41. Levy Y, Onuchic JN, Wolynes PG (2007) Fly-casting in protein-DNA binding: frustration between protein folding and electrostatics facilitates target recognition. J Am Chem Soc 129(4): 738–739.
42. Adda CG, MacRaild CA, Reiling L, Wycherley K, Boyle MJ, et al. (2012) Antigenic characterization of an intrinsically unstructured protein, *Plasmodium falciparum* merozoite surface protein 2. Infect Immun 80(12): 4177–4185.
43. Pan W, Huang D, Zhang Q, Qu L, Zhang D, et al. (2004) Fusion of two malaria vaccine candidate antigens enhances product yield, immunogenicity, and antibody-mediated inhibition of parasite growth *in vitro*. J Immunol 172(10): 6167–6174.

2

Acetylation of Human TCF4 (TCF7L2) Proteins Attenuates Inhibition by the HBP1 Repressor and Induces a Conformational Change in the TCF4::DNA Complex

Susanne Elfert[1]**, Andreas Weise**[1¤]**, Katja Bruser**[1]**, Martin L. Biniossek**[1]**, Sabine Jägle**[1,2]**, Niklas Senghaas**[1,2]**, Andreas Hecht**[1,2,3]*

1 Institute of Molecular Medicine and Cell Research, Albert-Ludwigs-University Freiburg, Freiburg, Germany, 2 Faculty of Biology, Albert-Ludwigs-University Freiburg, Freiburg, Germany, 3 BIOSS Centre for Biological Signalling Studies, Albert-Ludwigs-University Freiburg, Freiburg, Germany

Abstract

The members of the TCF/LEF family of DNA-binding proteins are components of diverse gene regulatory networks. As nuclear effectors of Wnt/β-catenin signaling they act as assembly platforms for multimeric transcription complexes that either repress or activate gene expression. Previously, it was shown that several aspects of TCF/LEF protein function are regulated by post-translational modification. The association of TCF/LEF family members with acetyltransferases and deacetylases prompted us to investigate whether vertebrate TCF/LEF proteins are subject to acetylation. Through co-expression with p300 and CBP and subsequent analyses using mass spectrometry and immunodetection with anti-acetyl-lysine antibodies we show that TCF4 can be acetylated at lysine K_{150} by CBP. K_{150} acetylation is restricted to TCF4E splice variants and requires the simultaneous presence of β-catenin and the unique TCF4E C-terminus. To examine the functional consequences of K_{150} acetylation we substituted K_{150} with amino acids representing the non-acetylated and acetylated states. Reporter gene assays based on Wnt/β-catenin-responsive promoter regions did not indicate a general role of K_{150} acetylation in transactivation by TCF4E. However, in the presence of CBP, non-acetylatable TCF4E with a $K_{150}R$ substitution was more susceptible to inhibition by the HBP-1 repressor protein compared to wild-type TCF4E. Acetylation of K_{150} using a bacterial expression system or amino acid substitutions at K_{150} alter the electrophoretic properties of TCF4E::DNA complexes. This result suggests that K_{150} acetylation leads to a conformational change that may also represent the mechanism whereby acetylated TCF4E acquires resistance against HBP1. In summary, TCF4 not only recruits acetyltransferases but is also a substrate for these enzymes. The fact that acetylation affects only a subset of TCF4 splice variants and is mediated preferentially by CBP suggests that the conditional acetylation of TCF4E is a novel regulatory mechanism that diversifies the transcriptional output of Wnt/β-catenin signaling in response to changing intracellular signaling milieus.

Editor: Michael Klymkowsky, University of Colorado, Boulder, United States of America

Funding: This work was supported by the German Research Foundation (Deutsche Forschungsgemeinschaft, DFG; http://www.dfg.de) grant He2004/8-1 and CRC-850/B5 to AH. Publication fees for this article were covered by funding from the DFG and the Albert-Ludwigs-University Freiburg (Open Access Publishing program; http://www.ub.uni-freiburg.de/go/oapf). The funders had no role in study design, data collection and analysis, decision to publish, or preparation of the manuscript.

Competing Interests: The authors have declared that no competing interests exist.

* E-mail: andreas.hecht@mol-med.uni-freiburg.de

¤ Current address: Institute of Biology II, Cell Biology, Albert-Ludwigs-University Freiburg, Freiburg, Germany

Introduction

Proper embryonic development and postnatal tissue homeostasis critically depend on precisely controlled gene expression. T-cell factors and lymphoid enhancer factor (TCF/LEF) constitute a family of highly conserved transcriptional regulators that is comprised of four members in humans: TCF1, LEF1, TCF3 and TCF4 (gene symbols *TCF7*, *LEF1*, *TCF7L1* and *TCF7L2*, respectively) [1,2]. Loss-of-function studies in different organisms revealed that TCF/LEF proteins perform essential functions in various aspects of embryogenesis and adult life by acting as nuclear effectors of growth factor and mitogenic signaling cascades including the Wnt/β-catenin pathway [1–14].

Structural features common to all members of the TCF/LEF protein family include an N-terminal β-catenin-binding domain, interaction sites for Groucho-related-gene (Grg)/transducin-like enhancer of split (TLE) transcriptional corepressors, a high-mobility-group-box (HMG-box) DNA-binding domain and an adjacent nuclear localization signal (NLS) [1,2]. Outside of these domains, however, the sequence similarity drops considerably. Structural divergence among TCF/LEF family members is further enhanced by alternative splicing. This divergence is most pronounced for the mouse and human *TCF7* and *TCF7L2* genes [15–18]. Thus, *TCF7L2* may give rise to more than one hundred transcript variants which can be grouped into three main categories of M-, S- and E-types based on their capacity to generate protein products (denominated TCF4) with structural similarities in their C-termini [15,16,18,19]. TCF4E isoforms contain binding motifs for the carboxy-terminal binding protein (CtBP) [20], the lysine acetyl-transferase (KAT) p300 [21] and a so-called C-clamp, which represents a second DNA-binding domain in addition to the HMG-box [22,23]. None of these features are found in the TCF4M and TCF4S variants. Differences in domain composition are likely to be the major cause of differences in promoter-specific transactivation potential between TCF4 protein isoforms and also within the TCF/LEF family [18,21].

TCF/LEF proteins engage in transcriptional control in at least two different but not mutually exclusive ways. The HMG-box of TCF/LEF proteins belongs to the HMGB-domain subtype that recognizes and bends specific DNA sequences [24,25]. Accordingly, TCF/LEF proteins can facilitate the juxtaposition of non-adjacent transcription factor binding sites by deformation of the DNA helix and thereby aid in the assembly of transcriptional complexes [24]. As an example of a second mode of action TCF/LEF proteins can occupy the promoters of Wnt/β-catenin target genes and support the recruitment of multi-component transcription complexes through direct interactions with specific protein-binding partners. Repressive complexes that form in the absence of a Wnt stimulus can contain Grg/TLE proteins, histone deacetylases (HDACs) and the carboxy-terminal binding protein (CtBP) [26]. Upon activation of the Wnt/β-catenin pathway, β-catenin translocates into the nucleus where it interacts with TCF/LEF proteins to displace or inactivate their co-repressors. Additionally, β-catenin-mediated recruitment of histone-modifying enzymes, chromatin remodelers and factors at the interface of the basal transcription machinery ultimately results in the activation of target gene transcription [26].

Interactions with their binding partners and other functions of TCF/LEF proteins are subject to regulation by diverse post-translational modifications. The Nemo-like kinase (NLK) phosphorylates human TCF4 at threonine residues T_{178} and T_{189}; these modifications inhibit DNA-binding and promote TCF4 degradation [27,28]. TCF4 is also phosphorylated by the TRAF2-and-NCK-interacting-protein-kinase (TNIK) at serine S_{154}, which results in the activation of Wnt/β-catenin target gene expression through an unknown mechanism [29,30]. Phosphorylation and sumoylation can also affect the intracellular distribution of TCF/LEF proteins [31–33].

Another modification that plays important roles in Wnt/β-catenin signaling is the reversible acetylation of lysine residues. HDACs that remove acetyl moieties and thereby antagonize KATs, compete with β-catenin for binding to TCF/LEF proteins [34–36] and are integral components of TCF/LEF-based transcriptional repression complexes [34]. Creb-binding protein (CBP) and the closely related p300 are two KATs that modify both histones and non-histone proteins [37,38]. In *Drosophila melanogaster*, dCBP acts as transcriptional coactivator of Armadillo (the fly orthologue of β-catenin) [39]. Additionally, dCBP negatively regulates the interactions between Armadillo and LEF1 by acetylating lysine 25, which is located in the β-catenin binding domain of LEF1 [40]. Pop-1, a *C. elegans* TCF/LEF family member, can also be acetylated by CBP. In this case, however, acetylation affects the cytoplasmic/nuclear distribution of Pop-1 [41]. In vertebrates, the CBP-catalyzed acetylation of β-catenin increases its affinity for TCF4 [42,43]. Furthermore, as in *Drosophila*, CBP and p300 function as crucial transcriptional cofactors of β-catenin [44–47]. However, depending upon the promoter-context, CBP and p300 may interchangeably function as coactivators or they can exhibit contrasting stimulatory and inhibitory activities, respectively [39,48,49].

HMG-box protein 1 (HBP1) is a sequence-specific DNA-binding protein and multifunctional transcriptional regulator involved in the coordination of differentiation and cellular proliferation in a large number of cellular backgrounds [50–58]. HBP1 has the properties of a tumor suppressor; this tumor suppressor activity is likely due to the ability of HBP1 to transcriptionally repress cell cycle regulators and the pro-inflammatory macrophage migration inhibitory factor [50,51,55,58]. Moreover, HBP1 induces cellular senescence in response to oncogenic stress by upregulating CDKN2A/P16INK4A [54]. The expression and activity of HBP1 itself are under the control of oncogenic pathways involving miR-17-5p and different post-translational modifications [53,59]. Additionally, a link between oncogenic Wnt/β-catenin signaling and HBP1 has been established by the finding that HBP1 physically interacts with two distinct domains in TCF4 to prevent its DNA binding and the resultant Wnt/β-catenin target gene expression in tumor cells [55]. HBP1 exerts a similar effect on c-MYC, a downstream target of Wnt/β-catenin signaling [52]. Thus, HBP1 is an important modulator and negative regulator of Wnt/β-catenin pathway activity. However, it is not known whether the inhibition of Wnt/β-catenin signaling by HBP1 occurs constitutively or can be regulated somehow.

While KATs and HDACs have been functionally linked to Wnt/β-catenin signaling and β-catenin/TCF-mediated transcriptional control, mechanistic insights and knowledge about the precise targets of these enzymes are still scarce. In particular, it would be interesting to know whether vertebrate TCF/LEF proteins not only participate in the recruitment of KATs and HDACs but are also subject to conditional acetylation themselves. Here, we show that CBP acetylates human TCF4 E-type splice variants at lysine residue K_{150}. Acetylation of K_{150} requires β-catenin and the C-terminal tail of TCF4E2, raising the possibility that acetylation occurs in a triple protein complex. Although K_{150} is remote from the HMG-box and C-clamp DNA-binding domains of TCF4E isoforms, acetylation of K_{150} alters the mobility of TCF4::DNA complexes and changes their structure. Substitutions that mimic the acetylated or non-acetylated states of K_{150} have no impact on the transactivation capacity of TCF4E2 in general. However, they differentially affect the ability of the HBP1 repressor protein to attenuate TCF4E2 transcriptional activity. Thus, acetylation of K_{150} may have a regulatory function in context-dependent transcriptional control by TCF4 and protect against the inhibitory influence of HBP1.

Results

TCF4E2 is preferentially acetylated by CBP

To investigate the potential acetylation of different vertebrate TCF/LEF family members, LEF1, the TCF1E splice form, TCF3 and the TCF4E2 splice form were co-expressed with the acetyltransferases p300 and CBP in HEK293 cells. All TCF/LEF factors harbored C-terminal HA-epitope tags that allowed for their immunoprecipitation from whole cell lysates. Immunoprecipitates were analyzed for the presence of acetylation by SDS-polyacrylamide gel electrophoresis (SDS-PAGE) and western blotting using a pan-acetyl-lysine antibody (Fig. 1, upper panel). For LEF1 and TCF1E we observed weak but clearly detectable acetylation at levels that were comparable in the presence of both p300 and CBP. TCF4E2 was also weakly acetylated by p300 but the signal was much more pronounced in the presence of CBP (Fig. 1, compare lanes 10 and 15). No acetyl-lysine immunoreactivity was observed for TCF3 neither upon co-expression with p300 nor with CBP. Comparable expression and similar efficiency of immunoprecipitation of the TCF/LEF family members was confirmed in the input material and immunoprecipitates (Fig. 1, middle panels). Moreover, p300 and CBP were expressed at similar levels and

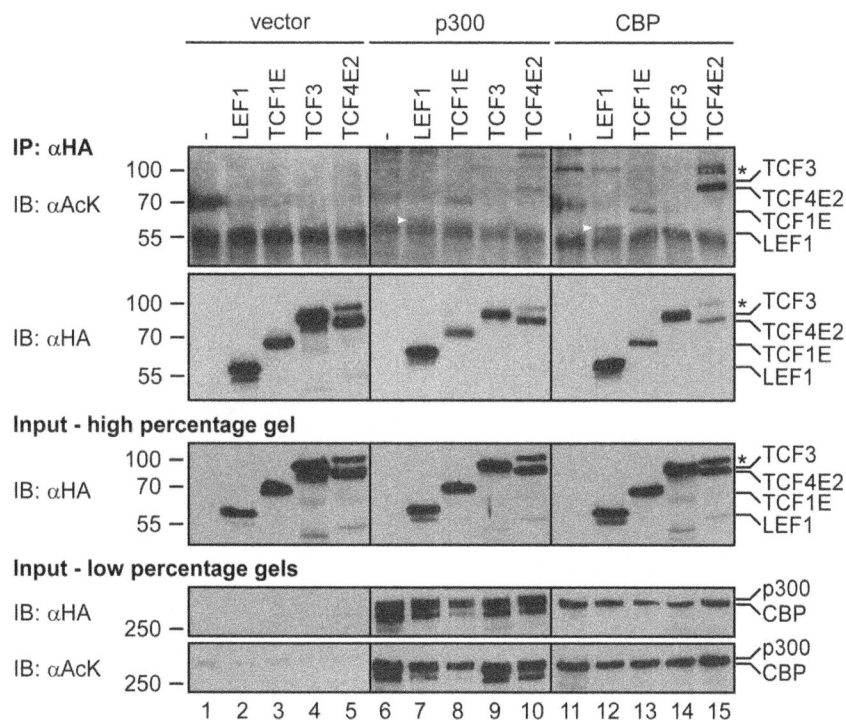

Figure 1. Acetylation of TCF/LEF family members by p300 and CBP. Expression vectors for HA-tagged versions of LEF1, TCF1E, TCF3 and TCF4E2 were transfected into HEK293 cells in the presence or absence of expression vectors for the acetyl-transferases p300 or CBP as indicated. To be able to detect p300 and CBP and to demonstrate comparable expression levels of the two proteins we used HA-tagged versions of p300 and CBP in this series of experiments. To detect acetylation of TCF/LEF proteins after immunoprecipitation with an antibody against the HA-tag, the samples were analyzed by SDS-PAGE and western blotting with a pan-anti-acetyl-lysine antibody (αAcK, top panel). The expression and immunoprecipitation of TCF/LEF proteins was controlled with the anti-HA antibody (middle panels). High percentage gels were used for the analyses of TCF/LEF family members. The expression of p300 and CBP in input material was also monitored by SDS-PAGE and western blotting with the anti-HA antibody. In addition, autoacetylation of p300 and CBP was detected with the pan-anti-acetyl-lysine antibody to demonstrate their enzymatic activity. Two distinct low percentage gels were used for the analyses of p300 and CBP expression and acetylation. The asterisks denote a species of TCF4E2 running at higher than expected molecular weight presumably due to an unknown post-translational modification. White arrowheads mark the location of LEF1. Molecular weight standards are shown on the left of the panels. IP: immunoprecipitation; IB: immunoblot.

both proteins are enzymatically active as shown by their autoacetylation (Fig. 1, lower panels) [60,61]. Thus, TCF/LEF family members appear to be differentially acetylated by p300 and CBP, with TCF4E2 being a preferential target of CBP.

Acetylation of TCF4E2 at K_{150}

Because acetylation of LEF1 was previously described [40] and we were intrigued by the differential effects of p300 and CBP, we focused on the acetylation of TCF4E2. To identify potential acetyl acceptor lysines, TCF4E2 was co-expressed with CBP, immunoprecipitated as before and then subjected to peptide analyses by mass spectrometry. Although TCF4E2 was not easily accessible by a typical proteomic approach because of its high proline content [62], nanoflow-HPLC-MS/MS (high performance liquid chromatography tandem mass spectrometry) measurement followed by Mascot search gave the first hint of acetylation at lysine residue K_{150} of TCF4E2 (Figures S1 and S2). The twofold charged parent ion of the TCF4E2-derived peptide DVQAGSLQSRQALK$_{150}$ with K_{150} being acetylated was assigned a Mascot score value of 87 for the MS/MS spectrum and a mass deviation of 14 ppm for the parent ion (Figure S1). The total ion current signal was close to the detection limit. We additionally compared the MS/MS spectrum of the sample measurement with the spectrum of an acetylated synthetic reference peptide with the same amino acid sequence. Both showed good similarity. The threefold charged

parent ion was also assigned by Mascot with a mass deviation of 7 ppm but a lower score value of 40 (Figure S2). The MS/MS spectrum also showed similarity to the reference peptide but contained a base peak which could not be assigned. Therefore, the evidence for acetylation of K_{150} is mainly supported by the spectrum of the twofold charged ion. The peptide was also found in its non-acetylated form in the sample suggesting that the TCF4E2 sample was only partially acetylated.

To independently confirm the acetylation of TCF4E2 at K_{150}, a polyclonal peptide antibody against acetylated K_{150} (αK_{150}ac) was produced. To test and verify the specificity of this peptide antibody, we generated bacterially expressed recombinant TCF4E2 proteins that harbored either lysine, alanine (TCF4E2K$_{150}$A) or arginine (TCF4E2K$_{150}$R) at amino acid position 150. In addition, we obtained TCF4E2 that was specifically acetylated at K_{150} (TCF4E2K$_{150}$ac) by introducing a stop codon into the TCF4E2 cDNA at the position corresponding to K_{150} to exploit the site-specific acetylation system described by Neumann and coworkers [63]. All TCF4E2 variants were purified, separated by SDS-PAGE and analyzed by western blotting either with the K_{150}ac antibody or with anti-c-Myc antibodies recognizing a C-terminal epitope tag common to all of them. While the anti-c-Myc antibodies detected all of the TCF4E2 variants, the K_{150}ac antibody recognized only TCF4E2K$_{150}$ac but not mutant or non-acetylated TCF4E2 (Fig. 2 A), which confirms the specificity of this antibody.

A

B

C

D

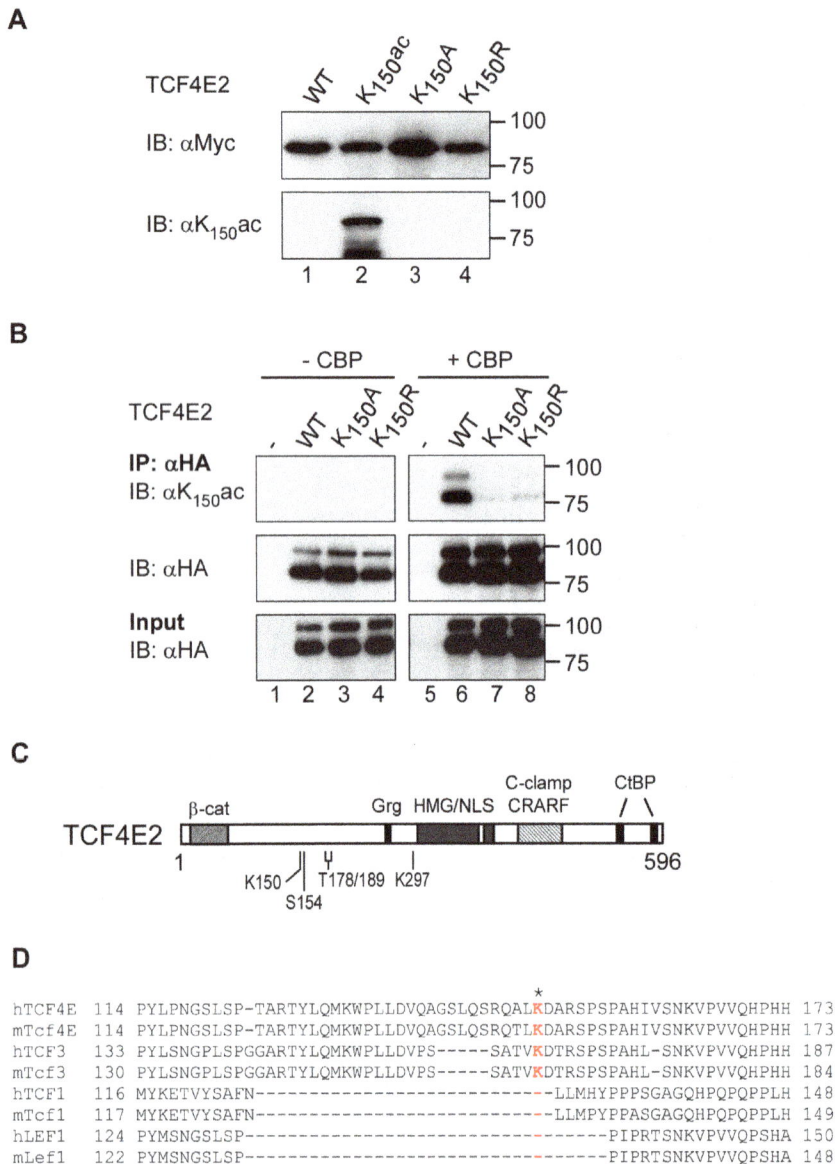

Figure 2. Acetylation of TCF4E2 at lysine 150. (**A**) Recombinant TCF4E2 variants expressed in *E. coli* were affinity-purified and analyzed by SDS-PAGE and western blotting with anti-c-Myc (top) and anti-K_{150}ac antibodies (bottom). Molecular weight standards are indicated on the right. (**B**) HEK293 cells were transfected with expression vectors for HA-tagged TCF4E2 and TCF4E2 K_{150}A or K_{150}R mutants in the presence or absence of an expression vector for CBP (without HA-tag) as indicated. TCF4E2 was immunoprecipitated with anti-HA antibodies and precipitates were analyzed by SDS-PAGE and western blotting with anti-K_{150}ac antibodies (upper panel) and anti-HA antibodies (middle panel). The presence of TCF4E2 proteins in the input material was also monitored by SDS-PAGE and western blotting using anti-HA antibodies (bottom panel). Molecular weight standards are shown on the right. (**C**) Schematic representation of TCF4E2 domain structure. β-Cat: β-catenin binding domain; Grg: Groucho/TLE binding motif; HMG: HMG-box; NLS: nuclear localization signal; CtBP: CtBP binding sites. Amino acid residues targeted by acetylation (K150), phosphorylation (S154, T178, T189) and sumoylation (K297) are marked. (**D**) Amino acid sequence conservation around TCF4E2 K_{150} (colored red and marked with an asterisk) in human (h) and mouse (m) TCF/LEF family members. The numbers reflect the positions of terminal amino acids shown for each of the factors. IP: immunoprecipitation; IB: immunoblot.

The αK_{150}ac antibodies were then used to verify the acetylation of TCF4E2 at K_{150} in mammalian cells as indicated by the MS/MS analysis. TCF4E2, TCF4E2K_{150}A and TCF4E2K_{150}R were co-expressed with CBP in HEK293 cells, immunoprecipitated with anti-HA antibodies and analyzed by SDS-PAGE and western blotting (Fig. 2 B). In the absence of CBP, no acetylation could be detected. In the presence of CBP, a strong signal was detected for TCF4E2; this signal was significantly reduced for TCF4E2K_{150}A and TCF4E2K_{150}R (Fig. 2 B, upper panel), arguing that TCF4E2 is acetylated by CBP at K_{150}.

Inspection of the location of K_{150} relative to the known domain structure of TCF/LEF family members and other protein features reveals that K_{150} resides within a part of TCF4E2 located between the β-catenin binding domain and the Grg/TLE interaction site (Fig. 2 C). Interestingly, potential phosphorylation sites for the serine/threonine kinases TNIK, homeodomain-interacting protein kinase 2 (HIPK2), and NLK [27,30,64] are in the near vicinity of K_{150}, raising the possibility of cross-talk among these modifications. Amino acid sequence comparison within the TCF/LEF family and across different species shows that K_{150} is conserved in mouse and human TCF4E2 but not in mouse or human TCF1, LEF1, *Drosophila* dTCF or *C. elegans* POP1 (Fig. 2 D, data not shown). Mouse and human TCF3 proteins contain a stretch of amino acids related to the TCF4 sequence around K_{150} (Fig. 2 D). However, the absence of a GSLQS motif from TCF3 and other amino acid sequence deviations immediately preceding K_{150} may explain why TCF3, unlike TCF4, was not found to be acetylated in our initial analysis.

A dual requirement of the β-catenin binding domain and C-terminal sequences for efficient acetylation of TCF4E splice variants at K_{150}

We and others have previously shown that p300 and CBP interact with β-catenin and that p300 binds to the C-terminal portion of TCF4E2 downstream of the HMG-box [21,45–47]. To gain insight into the potential interplay of TCF4E2, CBP and β-catenin with respect to acetylation of K_{150}, we performed co-immunoprecipitation experiments with TCF4E2, TCF4E2K_{150}A, TCF4E2K_{150}R and TCF4E2 deletion mutants lacking either the β-catenin binding domain (TCF4E2ΔN) or the C-terminus (TCF4E2ΔC) (Fig. 3 A). All TCF4E2 variants were present at comparable levels in whole cell extracts prepared from transfected HEK293 cells except for TCF4E2ΔN which was somewhat underrepresented (Fig. 3 B). This underrepresentation, however, was largely equalized in the immunoprecipitates. Furthermore, the presence of CBP led to a general increase in TCF4E2 levels. Regardless of these complications, the analyses of β-catenin co-immunoprecipitation and TCF4E2K_{150} acetylation yielded un-equivocal results. All TCF4E2 variants except TCF4E2ΔN were able to precipitate β-catenin in the absence or presence of CBP (Fig. 3 B), suggesting that K_{150} and its acetylation do not impact on the interaction of TCF4E2 with β-catenin. In contrast, acetylation of TCF4E2, which was dependent on the co-expression of CBP (Fig. 3 B, compare lanes 2 and 10), was strongly reduced in the absence of β-catenin co-immunoprecipitation (Fig. 3 B, lane 11). The pronounced decrease of the acetylation signal seen with TCF4E2K_{150} mutants further confirms that K_{150} is targeted by CBP (Fig. 3 B, compare lanes 10, 12, 13). Deletion of the TCF4E2 C-terminus also impaired the CBP-mediated acetylation of K_{150}, even though β-catenin still co-precipitated with TCF4E2ΔC (Fig. 3 B, lane 14). Thus, efficient acetylation of TCF4E2K_{150} shows a dual requirement for the N- and C-terminal protein sequences.

The finding that the C-terminus of TCF4E2 is required to promote the acetylation of K_{150} is intriguing because there are numerous naturally occurring TCF4 splice variants with C-terminal amino acid sequences that diverge from TCF4E2 to varying extents [15,16,18,19,65]. We therefore analyzed whether the TCF4E3 isoform and representatives of the M- and S-types of TCF4 splice variants (for structures see Fig. 3 A) also undergo acetylation at K_{150}. TCF4E3 lacks amino acid residues derived from exon 13 and harbors a CRALF-type C-clamp. TCF4S2 has an incomplete C-clamp and resembles TCF4E2ΔC. The C-terminus of TCF4M1 differs completely and has none of the C-clamp elements at all. When tested in our combined co-expression/co-immunoprecipitation approach all TCF4 isoforms co-immunoprecipitated β-catenin (Fig. 4). However, only TCF4E2 and TCF4E3 were efficiently acetylated by CBP at K_{150}. In contrast, neither TCF4S2 nor TCF4M1 exhibited K_{150} acetyla-tion (Fig. 4), which is in agreement with the behavior of the TCF4E2ΔC deletion mutant. Thus, K_{150} acetylation appears to be a special feature of TCF4 E-type isoforms.

K_{150} acetylation does not alter the transactivation capacity of TCF4E2

Next, we were interested in determining the potential impact of K_{150} and its acetylation on the protein function of TCF4E2. However, the acetylation of K_{150} does not appear to influence the intracellular localization of TCF4E2, nor does it affect TCF4E2 phosphorylation, ubiquitination or protein stability, as revealed by immunofluorescence analyses, phosphatase treatment and protea-some inhibitor studies (Figures S3 and S4). Finally, to determine whether acetylation at K_{150} modulates the transactivation capacity of TCF4E2, we performed reporter gene assays in HEK293 and U-2 OS cells and determined the luciferase activities driven by the β-catenin/TCF-inducible *Cdx1*, *Axin2* or *c-MYC* promoters (Fig-ures S5 and S6). However, wild-type TCF4E2, and the TCF4E2K_{150}A and TCF4E2K_{150}R mutants that mimic the acetylated or non-acetylated states of K_{150}, respectively, were indistinguishable in their ability to stimulate *Cdx1*, *Axin2* and *c-MYC* promoter activity upon co-expression with a constitutively active form of β-catenin in both cell lines. Thus, under the conditions of these experiments, K_{150} and its potential acetylation do not seem to affect the TCF4E2 transactivation capacity.

K_{150} is involved in HBP1-mediated repression of TCF4E2 transactivation

Even though the substitution of K_{150} did not alter the transcriptional activation by TCF4E2 *per se*, it is still possible that there are conditions in which K_{150} has an influence on the activity of TCF4E2. HBP1 is a transcriptional repressor that has been shown to interact with two domains in TCF4; these domains encompass amino acid residues 53–171 and 327–400 and thus include K_{150} and the TCF4 HMG-box, respectively [55]. HBP1 impairs transactivation by TCF4 by apparently preventing TCF4 DNA-binding. To test whether K_{150} plays a role in HBP1-mediated repression of TCF4 activity, we performed luciferase reporter gene assays in human HEK293 cells. Specifically, we asked whether the co-expression of HBP1 can affect the activation of a mouse *Cdx1* promoter construct by wild-type TCF4E2, TCF4E2K_{150}Q, or TCF4E2K_{150}R in the presence of constitu-tively active β-catenin. Additionally, CBP was co-expressed to force TCF4E2 acetylation. As expected, co-transfection of increasing amounts of HBP1 expression plasmid gradually decreased reporter gene activation by β-catenin and TCF4E2 (Fig. 5 A). In the absence of CBP, no significant difference between wild-type TCF4E2, TCF4E2K_{150}Q, or TCF4E2K_{150}R was observed. However, in the presence of CBP, wild-type TCF4E2 and TCF4E2K_{150}Q proved to be less susceptible to inhibition by HBP1. Importantly, the protein levels of wild-type and mutant TCF4E2 were not reduced in the presence of HBP1 (Fig. 5 B). Overall, the observations that differences between wild-type TCF4E2, TCF4E2K_{150}Q, and TCF4E2K_{150}R arise only in the presence of CBP and that the TCF4E2K_{150}Q mutant which mimics the acetylated state of K_{150}, behaves like wild-type TCF4E2 suggest that non-acetylated TCF4E2 is the preferred target of HBP1.

In a first attempt to determine how K_{150} acetylation affects the inhibition of TCF4E2 by HBP1 we analyzed their interaction by co-immunoprecipitation. However, regardless of the presence or absence of CBP, HBP1 co-precipitated wild-type TCF4E2, TCF4E2K_{150}Q, and TCF4E2K_{150}R with similar efficiencies from whole cell extracts (data not shown). Apparently, the disruption of the TCF4::HBP1 complex does not underlie the protective effect of K_{150} acetylation.

A   TCF4 variants:

B

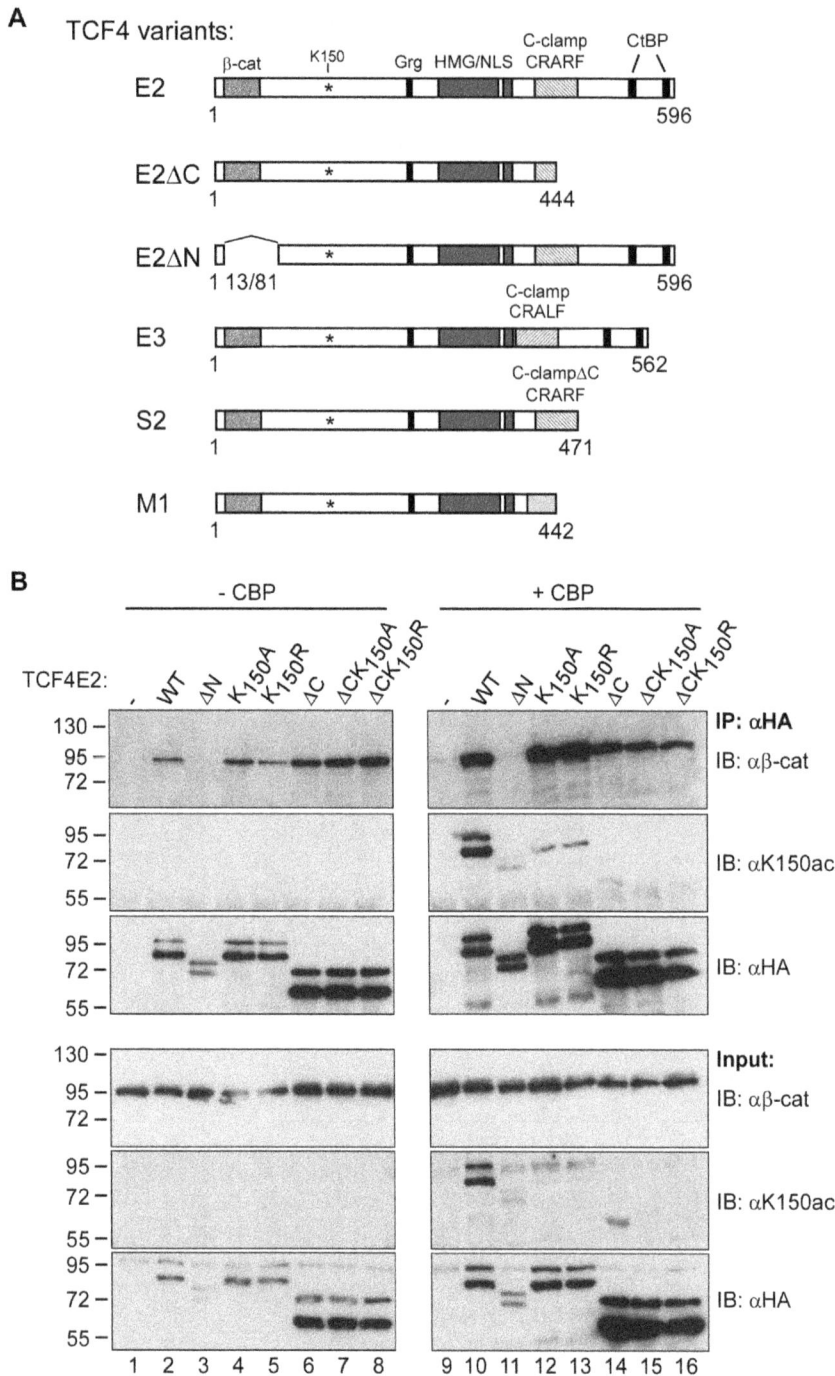

Figure 3. A dual requirement of the β-catenin binding domain and C-terminal sequences for efficient acetylation of TCF4E2 K$_{150}$. (**A**) Schematic representation of protein structures of TCF4 splice variants and deletion mutants. β-Cat: β-catenin binding domain; Grg: Groucho/TLE binding motif; HMG: HMG-box; NLS: nuclear localization signal; CtBP: CtBP binding sites. CRARF/CRALF: amino acid signature motifs within the C-clamps of TCF4E2 and TCF4E3. K$_{150}$ and coordinates of terminal amino acids are shown. (**B**) Expression constructs for TCF4E2, TCF4E2ΔN and TCF4E2ΔC (either wild-type or mutated at K$_{150}$ to alanine or arginine as indicated) were transfected into HEK293 cells in the presence or absence of an expression vector for CBP. The TCF4E2 variants were immunoprecipitated from whole cell lysates with anti-HA antibodies. The presence of endogenous β-catenin in the precipitates or cell lysates was analyzed by SDS-PAGE and western blotting using anti-β-catenin antibodies (αβ-cat). The acetylation of TCF4E2 variants was analyzed by anti-K$_{150}$ac antibodies (αK150ac) and the presence of TCF4E2 variants in the cell lysates and immunoprecipitates was analyzed by anti-HA antibodies (αHA). Molecular weight standards are indicated on the right. IP: immunoprecipitation; IB: immunoblot.

Acetylation of K$_{150}$ alters DNA binding by TCF4E2

HBP1 interacts with amino acid residues 53–171 of TCF4 and affects DNA binding by TCF4. By analogy, K$_{150}$ and its acetylation may also have an impact on DNA binding by TCF4E2 and could thus indirectly interfere with HBP1 repressor function. To test this hypothesis, we performed electrophoretic mobility shift

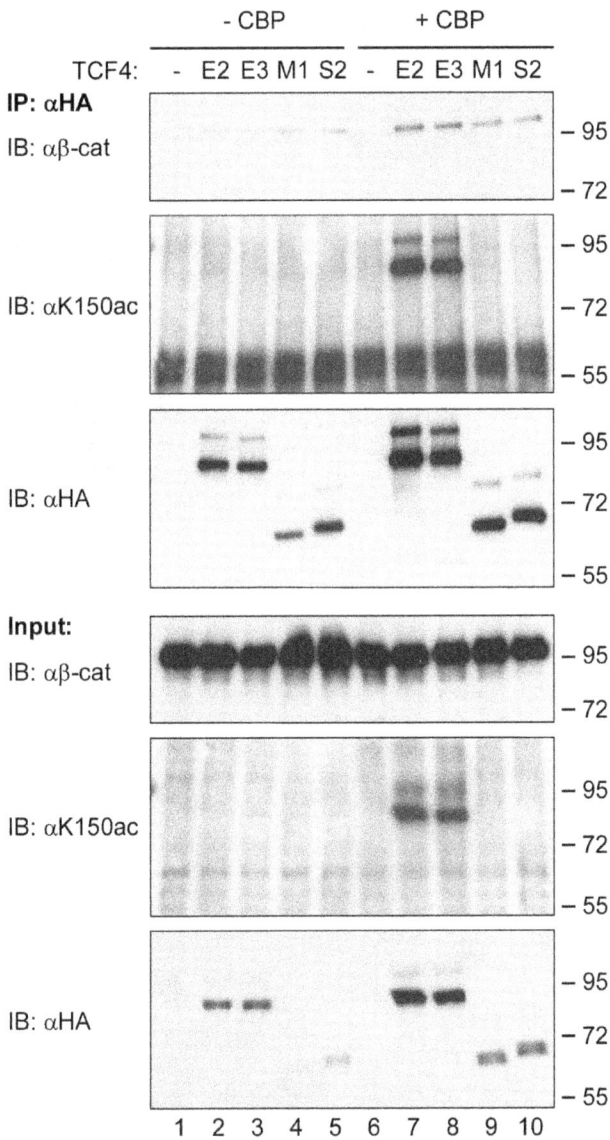

Figure 4. Acetylation of K$_{150}$ is a special feature of TCF4E splice variants. Expression constructs for TCF4E2, TCF4E3, TCF4S2 and TCF4M1 were transfected into HEK293 cells in the presence or absence of an expression vector for CBP. TCF4 variants were immunoprecipitated from whole cell lysates with anti-HA antibodies. The presence of endogenous β-catenin in precipitates or cell lysates was analyzed by SDS-PAGE and western blotting using anti-β-catenin antibodies (αβ-cat). The acetylation of TCF4 isoforms was analyzed by anti-K$_{150}$ac antibodies (αK150ac), and the presence of TCF4 isoforms in cell lysates and immunoprecipitates was analyzed by anti-HA antibodies (αHA). Molecular weight standards are indicated on the right. IP: immunoprecipitation; IB: immunoblot.

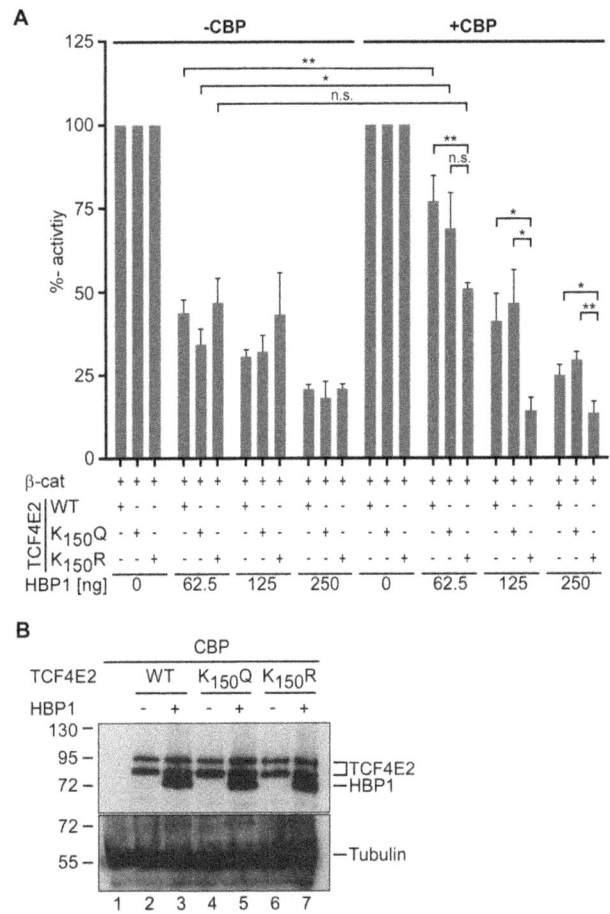

Figure 5. Mutation of TCF4E2 K$_{150}$ attenuates the inhibitory influence of HBP1 in a CBP-dependent manner. (A) HEK293 cells were cotransfected with combinations of firefly and Renilla luciferase reporter genes and expression vectors for a constitutively active form of β-catenin, TCF4E2 variants, CBP and HBP1 as indicated. Firefly luciferase expression was driven by the promoter of the Wnt/β-catenin target gene *Cdx1*. Reporter gene activities were determined 40 h post transfection. Bars represent percentages of luciferase activity. Reporter activities obtained with TCF4E2 WT and the TCF4E2 K$_{150}$ mutants in the absence of HBP1 were each set to 100 per cent both in the absence and presence of CBP. The average values and standard errors of the mean are shown. The asterisks mark statistically significant differences (**: p<0.01; *: p<0.05; n=4; unpaired Student's t-test) **(B)** A relevant subset of the whole cell extracts used for luciferase measurements were also employed to monitor the expression levels of TCF4E2 variants and HBP1 by SDS-PAGE and western blotting with anti-HA antibodies. Note that the TCF4E2 variants and HBP1 are all marked by HA-epitope tags. The samples analyzed had been transfected with expression vectors for β-catenin, the TCF4E2 variants, CBP and the highest amount of HBP1 expression vector. Antibodies against α-tubulin were used to monitor equal loading. Molecular weight standards are shown on the left.

assays (EMSAs) with oligonucleotides containing *Cdx1* TCF-binding element (TBE) 4 and the adjacent 5'-RCCG-3' motif as probes (Fig. 6 A) [18]. First, we analyzed DNA binding by full-length TCF4E2, TCF4E2K$_{150}$A, TCF4E2K$_{150}$R and an additional TCF4E2K$_{150}$Q mutant obtained by transcription and translation *in vitro*. TCF4E2K$_{150}$Q was included in the assay because it shares structural similarity with acetylated K$_{150}$. All TCF4E2 variants were present in similar amounts (Fig. 6 B) and bound to the wild-type *Cdx1* probe with equal efficiency (Fig. 6 C). As expected, mutation of the core TBE abolished the interaction

with TCF4E2 proteins. Mutation of the 5'-RCCG-3' motif also reduced the formation of TCF4E2::DNA complexes which is consistent with the idea that the 5'-RCCG-3' motif provides an essential contact for the C-clamp specifically found in the TCF4E splice forms [18,22,23]. Interestingly, the protein::DNA complexes formed by TCF4E2 and TCF4E2K$_{150}$R migrated slightly faster compared to those containing TCF4E2K$_{150}$A and TCF4E2K$_{150}$Q. The difference in mobility of the TCF4E2::DNA complexes was preserved with the probe containing a mutated 5'-RCCG-3' motif (Fig. 6 C), suggesting that the effect of the amino acids at position 150 is independent of the C-clamp. To verify this hypothesis, we used TCF4E2 mutants lacking the C-clamp (TCF4E2ΔC) for the EMSA (Fig. 6 D). As with the full-length TCF4E2 forms, the shortened TCF4E2ΔC proteins were expressed at similar levels and bound the *Cdx1* TBE4 with equal efficiencies (Fig. 6 D). They also exhibit the same requirements for an intact TBE core motif. However, the absence of the C-clamp alleviates the necessity of the 5′-RCCG-3′ motif. Importantly, the wild-type and mutant TCF4E2ΔC::DNA complexes showed differences in mobility analogous to those observed with the full-length TCF4E2 variants but in an even more pronounced manner.

Differences in the mobility of the protein::DNA complexes formed by TCF4E2, TCF4E2K$_{150}$A, TCF4E2K$_{150}$R, and TCF4E2K$_{150}$Q and the corresponding ΔC versions may arise from the presence or absence of a positively charged amino acid at position 150. To investigate this possibility, we used acidic urea PAGE to separate the TCF4E2ΔC variants according to their

native charge. However, all TCF4E2ΔC variants migrate at the same position in acidic urea gels (Fig. 6 E). This result makes it highly unlikely that the retarded migration of the DNA complexes containing TCF4E2ΔCK$_{150}$A and TCF4E2ΔCK$_{150}$Q results from the lack of a single positively charged amino acid and instead argues in favor of differences in complex structure as the underlying cause.

TCF4E2 expressed *in vitro* is not acetylated (data not shown) and alanine, glutamine or arginine substitutions are only imperfect proxies for the acetylated and non-acetylated states of lysine residues, respectively. Therefore, the DNA binding properties of TCF4E2 when acetylated at K$_{150}$ needed to be determined. For this determination, we used the bacterial expression system allowing site-specific acetylation of TCF4E2ΔC mentioned above [63]. TCF4E2ΔC, TCF4E2ΔCK$_{150}$A, TCF4E2ΔCK$_{150}$R, TCF4E2ΔCK$_{150}$Q and TCF4E2ΔCK$_{150}$ac were affinity purified from bacterial lysates and the presence of comparable protein

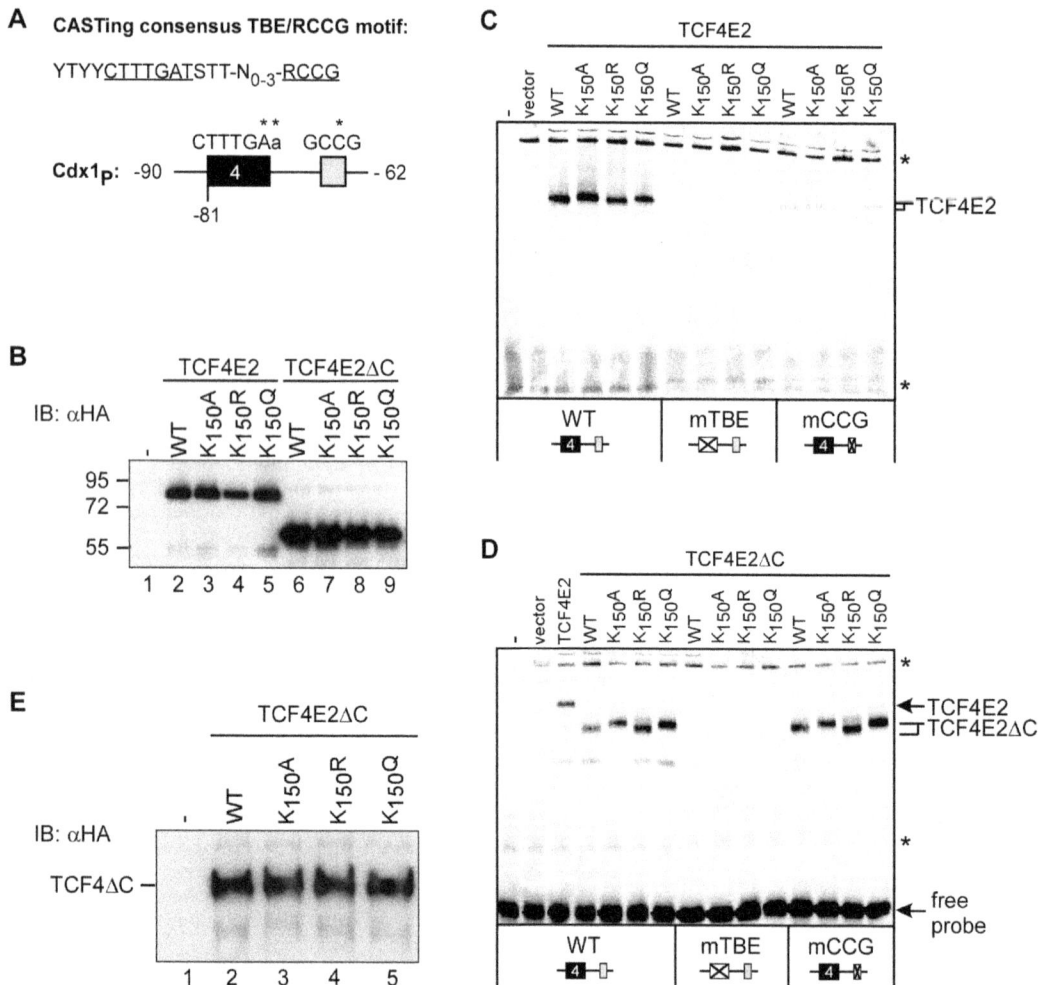

Figure 6. Mutations of K$_{150}$ induce charge-independent changes in the migration of TCF4E2::DNA complexes *in vitro*. (**A**) Top: Extended consensus TCF-binding element (TBE) and adjacent 5′-RCCG-3′ motif derived from CASTing experiments. TBE core element and 5′-RCCG-3′ motif are underlined. Bottom: schematic depiction of the *Cdx1* TBE4 probe used for EMSAs. Numbers refer to coordinates of the *Cdx1* promoter (Cdx1$_P$). Positions where mutations were inserted in the *Cdx1* TBE or 5′-RCCG-3′ motif are indicated by asterisks. (**B**) Wild-type and mutant TCF4E2 and TCF4E2ΔC proteins expressed *in vitro* were analyzed for equal abundance by SDS-PAGE and western blotting using anti-HA antibodies. Molecular weight standards are indicated on the left. Lane 1: control without protein (-). (**C, D**) EMSAs with TCF4E2 (C) or TCF4E2ΔC (D) using *Cdx1* TBE4 probes as depicted. Positions of protein::DNA complexes containing TCF4E2 and its derivatives are marked by arrows or brackets. Asterisks indicate non-specific bands. The control binding reaction received no protein (-) or wheat germ lysate mock-programmed with empty expression vector (vector). (**E**) Acidic urea PAGE of wild-type and mutant TCF4E2ΔC proteins expressed *in vitro*. TCF4E2ΔC variants were visualized upon electrophoresis by western blotting using anti-HA antibodies. Lane 1: control without protein (-).

amounts and K_{150} acetylation was determined by western blot (Fig. 7 A). When used in DNA-binding studies, the bacterial proteins showed a similar behavior to their counterparts expressed *in vitro*. As observed previously, DNA complexes formed by the non-acetylated wild-type protein and the $K_{150}R$ mutant migrated faster than those formed by the $K_{150}A$ and the $K_{150}Q$ variants. Strikingly, the DNA complex formed by acetylated TCF4E2ΔC migrated even more slowly (Fig. 7 B). No DNA binding was observed with a mutant *Cdx1* TBE4 or in the absence of bacterially expressed TCF4E2ΔC proteins. Taken together, the results of our DNA-binding studies suggest that acetylation of TCF4E2 at K_{150}

has a strong influence on the structure of the TCF4E2::DNA complex.

Discussion

Members of the TCF/LEF family of transcription factors perform critical functions as components of diverse gene regulatory networks in a variety of cell types and tissues [66]. Accordingly, their activities have to be tightly controlled. Previously, it was shown that the DNA binding, intracellular localization, and protein stability of TCF/LEF proteins can be regulated by phosphorylation and sumoylation [27–33,64,67]. The cytoplasmic-nuclear trafficking of invertebrate family members and the interaction with Armadillo, the fly orthologue of β-catenin, are controlled by lysine acetylation [40,41]. Here, we have investigated whether vertebrate TCF/LEF proteins are also subject to acetylation. By co-expression with KATs and subsequent analyses using immunodetection with anti-acetyl-lysine antibodies we provide evidence that LEF1, TCF1E and TCF4E variants indeed can be acetylated. LEF1 and TCF1E are similarly modified by p300 and CBP whereas acetylation of TCF4E variants is much more pronounced in the presence of CBP. Acetylation of LEF1 within its β-catenin binding domain by *Drosophila* CBP had been previously reported [40] and was not further investigated. The acetylation of TCF1E and the TCF4E isoforms, however, is a novel finding and shows that TCF/LEF family members not only help to recruit KATs and HDACs to TCF/LEF target genes but also appear to be substrates for these enzymes themselves.

In many cases, CBP and p300 appear to be functionally interchangeable as coactivators of a large number of transcription factors but some differences between the two enzymes concerning their range of target genes exists [37,38]. For example, both p300 and CBP cooperate with β-catenin in the regulation of certain Wnt target genes, but the two factors have opposing functions in Wnt/β-catenin mediated expression of the *BIRC5/Survivin* and *EPHB* receptor genes [48,68,69]. The molecular basis for these differences is unclear. Our finding that CBP can acetylate TCF4E isoforms much more efficiently than p300 further supports the idea that CBP and p300 can differentially contribute to Wnt/β-catenin signaling processes. It would also be interesting to explore whether differential acetylation of TCF4E plays a role in the contrasting activities of CBP and p300 at *BIRC5/Survivin* and similar genes.

Mass spectrometry provided first the evidence that K_{150} is an acetyl acceptor site in TCF4E2. This result was confirmed by mutagenesis and site specific anti-K_{150}ac antibodies. Efficient acetylation of K_{150} required the simultaneous presence of the β-catenin binding site and the extended C-terminus of the TCF4E splice variants. This result is consistent with our previous identification of a p300/CBP interaction site at the C-terminus of TCF4E2 and raises the possibility that acetylation of TCF4E isoforms occurs in triple complexes containing TCF4E variants, CBP and β-catenin, although other scenarios cannot be excluded at this point. On the other hand, because of the involvement of β-catenin, one could hypothesize that K_{150} acetylation occurs conditionally and is regulated by Wnt signaling. However, this hypothesis awaits further investigation.

The *TCF7L2* gene, which codes for TCF4 proteins, generates a large number of structurally diverse splice variants [15,16,18,19]. The TCF4M and TCF4S isoforms differ from TCF4E isoforms with respect to the amino acid composition of their C-termini. Generally, the transactivation capacity of the TCF4M and TCF4S isoforms is lower compared to TCF4E and they are likely to act upon different sets of target genes [18]. Here, we have shown that

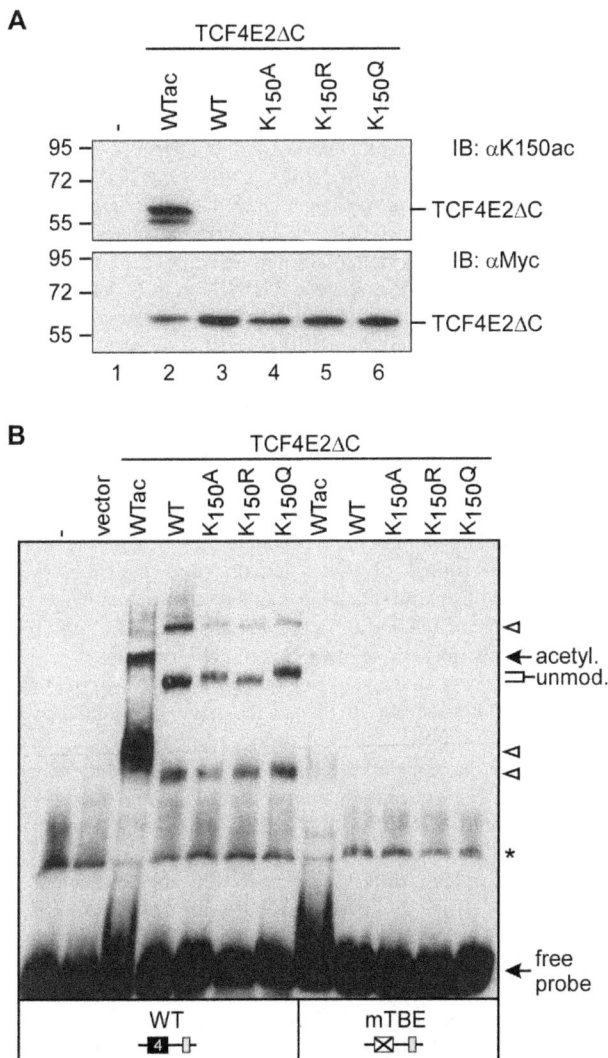

Figure 7. Acetylation of K_{150} affects the mobility of the TCF4E2::DNA complex more strongly than amino acid substitutions. (A) Recombinant TCF4E2ΔC variants expressed in *E. coli* were affinity-purified and analyzed by SDS-PAGE and western blotting with anti-K_{150}ac (top) and anti-c-Myc antibodies (bottom). Molecular weight standards are indicated on the left. Lane 1: control without protein (-). (B) EMSA with TCF4E2ΔC using *Cdx1* TBE4 probes as depicted. Positions of protein::DNA complexes containing the unmodified wild-type and mutant TCF4E2ΔC, or the acetylated wild-type protein (WTac) are marked. Arrow-heads: TCF4::DNA complex generated by partially degraded TCF4 proteins or multimers. Asterisk: non-specific band. Control binding reaction received no protein (-) or mock protein purifications (vector) of bacteria transformed with empty expression vector.

the TCF4M and TCF4S isoforms are not acetylated by CBP. This result further underscores the idea that TCF4 splice variants are functionally different and confirms that the C-terminus of TCF4 proteins is a critical determinant that confers unique properties upon TCF4E isoforms with respect to DNA binding, protein-protein interactions and post-translational modifications [18,20–22]. Thus, the phenotypic effects of K_{150} acetylation on TCF4 splice variant function may be restricted to target genes that are specifically regulated by TCF4E isoforms.

K_{150} is located downstream of the β-catenin binding domain in TCF4 proteins and close to several serine and threonine residues that can be phosphorylated by TNIK, HIPK2 and NLK [27,30,64,67]. These phosphorylation events affect transactivation by TCF4 proteins, DNA binding and protein stability. The location of K_{150} therefore raises the possibility of crosstalk between acetylation of K_{150} and phosphorylation of S_{154} and $T_{178/189}$ in TCF4E variants. Corresponding interdependencies among these types of post-translational modifications have been reported for p53, histones and other factors [70,71]. However, in our co-immunoprecipitation experiments, the acetylation of K_{150} or the mutation of this residue did not appear to influence the interaction with β-catenin. Thus, while complex formation with β-catenin is critical for K_{150} acetylation, there appears to be no reverse effect through positive or negative feedback. Similarly, our experiments with wild-type TCF4E2 and TCF4E2 mutants with alanine, arginine or glutamine substitutions at K_{150} involving phosphatase treatment and proteasome inhibition did not provide evidence that acetylation of K_{150} by CBP has an impact on TCF4E2 phosphorylation and turnover. Still, it may be premature to rigorously exclude the possibility of crosstalk between acetylation and phosphorylation of TCF4E isoforms. Overexpression of TCF4E2 and CBP may have biased our experimental system by exhausting the potentially limiting amounts of endogenous kinases in HEK293 cells. Future experiments involving TCF4 variants with additional substitutions at phospho-acceptor sites and co-expression of TNIK, NLK or HIPK2 should help to clarify whether there are any reciprocal effects between acetylation and phosphorylation of TCF4 proteins.

CBP has been shown to acetylate *C. elegans* Pop-1, altering intracellular distribution [41]. In our study, neither the co-expression of CBP nor mutation of K_{150} affected the intracellular localization of TCF4E2 which we found to reside exclusively in the nucleus. The apparent difference concerning changes in the localization of Pop-1 and TCF4E2 in response to CBP-mediated acetylation may be explained by the fact that CBP targets different protein domains in Pop-1 and TCF4E2 (this study and [41]). Furthermore, pairwise alignment of TCF4E2 and Pop-1 shows no or only very little amino acid sequence similarity in the regions corresponding to K_{150} in TCF4E2 and $K_{185,187,188}$ in Pop-1. Thus, although acetylation affects several members of the TCF/LEF family, we hypothesize that this post-translational modification has been acquired independently during evolution of the TCF/LEF family to fulfill different regulatory purposes.

To date, our analyses concerning the acetylation of TCF4E2 have been performed under conditions of transient overexpression and with recombinant proteins *in vitro*. Therefore, it is an open question whether acetylation of TCF4E2 at K_{150} occurs under physiological conditions. Preliminary attempts to detect K_{150} acetylation after immunoprecipitation of TCF4 from colorectal cancer cells were not successful (A. W., S. E. and A.H., unpublished). We ascribe this result to the fact that K_{150} acetylation appears to be restricted to TCF4E isoforms and likely requires the formation of triple complexes between TCF4E isoforms, CBP and β-catenin. Currently available TCF4 antibod-

ies and other analytical tools may not be suitable or sensitive enough to sufficiently enrich a subpopulation of TCF4E proteins that carry the K_{150} acetylation. The pronounced context-dependence of K_{150} acetylation also presents a challenge for the detection of gene regulatory phenotypes of TCF4E isoforms with amino acid substitutions at K_{150}. Overcoming these hurdles may require the development of novel, isoform-specific TCF4 antibodies for selective immunoprecipitation and gene-replacement strategies to substitute K_{150} mutants for wild-type TCF4 proteins, thereby allowing the study of their functions at endogenous Wnt/β-catenin target genes.

When assessed in reporter gene assays using the *Cdx1*, *Axin2* and *c-MYC* promoters we found that the transactivation by TCF4E2 was neither impaired nor augmented by K_{150} mutation, suggesting that acetylation of K_{150} is not generally involved in transcriptional regulation by TCF4E variants. However, we cannot exclude that the functional importance of K_{150} and its acetylation presents differently depending upon experimental conditions, for example upon forced expression of NLK, TNIK, HIPK2 or when examined with other target gene promoters. Nonetheless, we favor the idea that acetylation of K_{150} exerts its effects only in a specific context. In agreement with this assumption we found that in the presence of CBP, wild-type TCF4E2 and TCF4E2K_{150}Q which mimics acetylation, were partially protected against the inhibitory influence of HBP1. HBP1 is widely expressed and has a strong repressive effect on Wnt/β-catenin signaling [55,58]. Therefore, it is easily conceivable that the interaction between HBP1 and TCF4E2 is subject to some form of regulation that could be provided by the acetylation of TCF4E2K_{150}. Intriguingly, the two binding domains for HBP1 in TCF4E2 comprise amino acids 53–171 and 327–400, the latter corresponding to the HMG-box DNA binding domain [55]. Acetylation of TCF4E2 occurs at the more N-terminal of these domains and the phenotypic consequences of this modification manifest themselves in altered DNA binding of TCF4E2. This parallel and the observed functional consequences of the K_{150} mutation in reporter gene assays suggests to us that acetylation of TCF4E2 at K_{150} could be a relevant regulatory event to thwart the repressive influence by HBP1.

How could the acetylation of K_{150} in TCF4E2 interfere with HBP1-mediated repression? Mutation of K_{150} does not prevent the interaction of TCF4E2 with HBP1 in co-immunoprecipitation experiments (K. B. and A. H., unpublished) arguing that K_{150} acetylation does not directly interfere with the interaction of TCF4E2 and HBP1. Alternatively, the protective effect of K_{150} acetylation might manifest itself only under conditions in which HBP1 also exerts its inhibitory influence, namely DNA binding by TCF4E2. In support of this idea, we observed differences in the migratory behavior of protein::DNA complexes formed by wild-type and mutant TCF4E2. Migration of TCF4E2::DNA complexes was retarded by the substitution of K_{150} with alanine and glutamine and was slowed down even further upon acetylation of K_{150}. We ruled out charge-dependent effects as an underlying cause because wild-type TCF4E2 and all mutant versions of TCF4E2 migrated equally fast in acid-urea gels. Moreover, the electrophoretic properties of TCF4E2K_{150}A, TCF4E2K_{150}Q and TCF4E2K_{150}ac differed in EMSAs even though all three proteins show the same loss of a single positive charge. From this result, we conclude that charge neutralization and acetylation of K_{150} induce a structural change in the TCF4E2::DNA complex similar to what has been proposed for thymine DNA glycosylase [71]. Accordingly, the acetylation of K_{150} may result in a conformation of the TCF4E2::DNA complex that is at least partially resistant to the action of HBP1.

Conformational changes induced by acetylation may have additional implications aside from their potential impact on HBP1-mediated repression. TCF/LEF proteins are known to bend DNA and can function as architectural factors [24,25]. Thus, one could speculate that different structures of TCF4E2::DNA complexes brought about by the acetylation of K_{150} contribute to the assembly or disassembly of multimeric transcription factor complexes at regulatory regions of genes regulated by TCF4E isoforms.

Materials and Methods

Plasmids and mutagenesis

Expression constructs for TCF1E, Lef1, TCF3, TCF4E2, TCF4E2ΔN, TCF4E2ΔC, p300, p300-HA, CBP, CBP-HA, HBP1, a constitutively active form of β-catenin and the luciferase reporter constructs pGL3b-Axin2, pGL3b-Cdx1 and pBV-luc c-MYC 4×TBE2 were described previously [21,44,55,72–75]. Expression constructs for TCF4E3, TCF4S2 and TCF4M1 were derived from the TCF4E2 expression vector following a PCR-based cloning strategy as described [18]. Details are available upon request. The Renilla luciferase expression vector pRL-CMV was purchased from Promega, Heidelberg, Germany. TCF4E2 mutants TCF4E2 K_{150}A, K_{150}R and K_{150}Q in pCS2+ [76] were generated by site-directed mutagenesis according to the Stratagene QuickChange mutagenesis protocol using the following primers: For K_{150}A: 5′-GTAGACAAGCCCTCGCG-GATGCCCGGTCC-3′; K_{150}R: 5′-GTAGACAAGCCCT-CAGGGATGCCCGGTC-3′; K_{150}Q: 5′-GTAGA-CAAGCCCTCCAGGATGCCCGGTC-3′ (mutated codons are underlined). The TCF4E2 K_{150}A, K_{150}R and K_{150}Q mutations were also introduced into the TCF4E2ΔC expression construct using the same protocol and primers. For site-specific acetylation, TCF4E2 K_{150}TAG was generated in the pCal-c (Stratagene, Heidelberg, Germany) vector background with the same method using the primer 5′-GTAGACAAGCCCTCTAG-GATGCCCGGTC-3′. All mutants were sequence verified by DNA sequencing using the BigDye® Terminator v1.1 cycle sequencing kit (Applied Biosystems, Darmstadt, Germany) and a MegaBACE 500 capillary sequencer (GE healthcare, Munich, Germany) operated by the sequencing core facility of the University Medical Center Freiburg. For bacterial expression, TCF4E2 coding sequences were transferred into the pCal-c expression vector using BamHI/BstBI restriction enzymes. Wild-type and mutant forms of TCF4E2ΔC were derived from these constructs by conventional cloning techniques (details available on request). To facilitate the detection of TCF4E2 proteins, constructs based on pCS2+ added a C-terminal HA-tag while constructs based on pCal-c added a C-terminal Myc-His-tag to TCF4E2 sequences.

Cell culture, transient transfection and luciferase reporter assays

HEK293 and U-2 OS cells (ATCC # CRL-1573 and ATCC # HTB-96) were cultured in DMEM (PAN-biotech, Aidenbach, Germany) containing 10% fetal calf serum (PAN-Biotech, Aidenbach, Germany), 10 mM HEPES buffer (PAN-Biotech, Aidenbach, Germany), 1% MEM-non essential amino acids (Invitrogen, Karlsruhe, Germany) and 100 units/ml penicillin/streptomycin (Invitrogen, Karlsruhe, Germany). For western blotting and immunoprecipitations, 3×10^6 cells in 10 cm culture dishes were transfected with 10 μg plasmid DNA for TCF/LEF proteins and 15 μg of expression vectors for p300 or CBP in 1 ml calcium-phosphate precipitate 6 hours after seeding. The DNA amounts were equalized with empty expression vector. For luciferase reporter assays, 1×10^5 HEK293 cells were seeded per well of a 24-well plate. Cells were transfected with FuGENE6 reagent (Roche Applied Science, Mannheim, Germany) 4 h after plating. DNA mixtures of 100 ng of the pGL3b-Cdx1 firefly luciferase reporter, 10 ng of the Renilla luciferase expression vector pRL-CMV, 100 ng of plasmid DNA for expression of a constitutively active form of β-catenin, 50 ng of the expression constructs for WT and mutant TCF4E2, 250 ng of an expression vector for human CBP and three different amounts (62.5 ng, 125 ng or 250 ng) of an expression construct for rat HBP1 were used for the transfections. The total amounts of DNA transfected were kept constant by the addition of empty pCS2+ expression vector [76]. For the transfection experiments shown in Figures S5 and S6, 1×10^5 HEK293 cells or 5×10^4 U-2 OS cells seeded per well of a 24-well plate were transfected with FuGENE6 reagent (Roche Applied Science, Mannheim, Germany) 4 h after plating. DNA mixtures of 100 ng firefly luciferase reporter plasmids, 10 ng of the Renilla luciferase expression vector pRL-CMV, 100 ng of plasmid DNA for expression of a constitutively active form of β-catenin and three different amounts (75 ng, 150 ng or 300 ng) of each expression construct for the TCF4E2 variants were used for the transfections. The total amounts of DNA transfected were kept constant by addition of empty pCS2+ expression vector. Luciferase assays were performed 40 h after transfection as described [18,21]. All results represent the average values of at least three independent experiments including standard deviations.

Whole cell extracts and immunoprecipitation

To prevent enzymatic deacetylation, 1 μM trichostatin A (TSA) was added to the cells and to the cell lysis buffer 2 hours prior to protein preparation. Whole cell extracts were prepared 48 hours after transfection by lysing the cells in IPN$_{150}$ buffer [50 mM Tris/HCl (pH 7.6), 150 mM NaCl, 5 mM MgCl$_2$, 0.1% NP40, Complete^{TM} protease inhibitor (Roche Applied Science, Mannheim, Germany), 1 μM TSA] for 30 minutes on ice. Cell lysates were cleared by centrifugation at $20000 \times g$ and 4°C for 10 min. The protein concentrations in the cell lysates were determined using the DC protein assay kit (BioRad, Munich Germany). For immunoprecipitation, whole cell lysate with a protein content of 1 mg was combined with 1 μg of anti-HA antibody (clone 3F10, Roche Applied Science, Mannheim, Germany) and 40 μl 50% (v/v) protein-G-sepharose (GE healthcare, Munich, Germany) and incubated over-night at 4°C under constant rotation. Afterwards, the probes were washed three times with IPN$_{150}$ buffer and eluted by boiling in 2× SDS-loading buffer for 5 minutes at 95°C followed by SDS-PAGE analysis. For the purification of TCF4E2 for MS/MS analysis, HEK293 cells were transfected and lysed as described above. Cleared lysates from six 10 cm culture dishes were pooled and TCF4E2 was enriched by immunoprecipitation with 6 μg of anti-HA antibody (clone 3F10, Roche Applied Science, Mannheim, Germany) and 160 μl 50% (v/v) protein-G-sepharose (Roche Applied Science, Mannheim, Germany) in IPN$_{150}$ over-night at 4°C. Following three washing steps with IPN$_{150}$ lysis buffer, TCF4E2 was eluted from the protein-G-sepharose beads with 160 μl of 0.1 M glycine (pH 2.5) containing 1 μM TSA and Complete^{TM} protease inhibitors (Roche Applied Science, Mannheim, Germany). After incubation at 4°C under gentle agitation the supernatant was removed from the protein-G-sepharose beads and neutralized by addition of 8 μl 1 M Na$_2$HPO$_4$ (pH 8.0). For further enrichment of the acetylated TCF4E2, a second immunoprecipitation step using a polyclonal rabbit pan-acetyl-lysine antibody (α-AcK; AB 3879, Millipore, Schwalbach, Germany) was performed. For this assay, the neutralized supernatant was added to 800 µl of 10 mM sodium phosphate buffer (pH 6.8) containing 2 µg of the anti-AcK antibody and 40 µl of 50% (v/v) protein-G-sepharose. After incubation for 3 h at 4°C, the protein-G-sepharose beads were washed three times with 10 mM sodium phosphate buffer (pH 6.8) and TCF4E2 was eluted with 50 µl 0.1 M TFA for 30 min at 4°C under gentle agitation. The supernatant was removed from the beads and neutralized by the addition of 2.5 µl of 1 M Na_2HPO_4 (pH 8.0). Fractions of the lysates and eluates from the different immunoprecipitation steps were used in western blots to monitor the presence and/or enrichment of non-acetylated or acetylated forms of TCF4E2.

Silver staining of preparative gels and in gel endoproteinase digests

For preparative SDS-PAGE, subsequent silver staining and endoproteinase digests all the buffers and solutions were produced with HPLC grade water and all the equipment was thoroughly cleaned and treated with 5% formic acid prior to usage. In addition, all the steps were performed in a fume hood to prevent keratin contamination of the samples. After silver staining [77], TCF4E2 protein bands derived either from co-transfections with CBP expression construct or from transfections with TCF4E2 expression construct only were excised and used for in gel endoproteinase digests. As background controls for the MS/MS analysis, the gel pieces from empty neighboring lanes in the SDS-PAGE were excised and treated the same way as the TCF4E2 protein bands. Asp-N (Roche Applied Science, Mannheim, Germany) in gel endoproteinase digests were performed as follows: gel pieces were cut into small cubes with a scalpel and transferred to a 0.5 ml reaction tube. After the addition of 100 µl of 50 mM $(NH_4)HCO_3$, gel pieces were sonicated for 10 min in an ice water bath; this treatment was repeated for 20 min after the addition of 100 µl acetonitrile to the reaction. The supernatant was discarded and gel pieces were incubated for 10 min in water at room temperature. Supernatant was discarded following 10 min incubation at room temperature in 100 µl acetonitrile. The supernatant was removed and gel pieces were dried in a SpeedVac for 10 min. The dried gel pieces were incubated in 10 mM DTT in 50 mM $(NH_4)HCO_3$ for 30 min at 56°C in a thermo mixer at 500 rpm. The supernatant was removed, and gel pieces were incubated in 100 µl of 55 mM iodoacetamide in 50 mM $(NH_4)HCO_3$ for 30 min in the dark. Next, 100 µl of 50 mM $(NH_4)HCO_3$ was added, and after a 15 min incubation at room temperature, the supernatant was replaced with 100 µl of acetonitrile. After a 10 min incubation at room temperature, the supernatant was removed and gel pieces were dried in a SpeedVac as before. In each reaction tube 50 µl of sodium phosphate buffer (pH 8.0) containing 4 ng of Asp-N endoproteinase was added to the gel pieces following incubation at 37°C for 18 hours. The supernatant was removed and put into a new 0.5 ml reaction tube. The gel pieces were treated by another 15 min sonication step in 50 µl of acetonitrile, and the supernatant was added to the same new reaction tube. The supernatant was completely dried in a SpeedVac. Samples were kept at −20°C until MS/MS analysis.

Nanoflow-HPLC- MS/MS analysis

Mass spectrometric measurements of protease-digested protein samples and synthetic control peptides were performed on a LTQ-FT mass spectrometer (Thermo Fisher Scientific, Bremen, Germany) coupled to an Ultimate3000 micro pump (Dionex, Idstein, Germany). Synthetic control peptides DVQAGSLQSR-QALK and DVQAGSLQSRQALK(ac) were obtained from GenScript Corporation Piscataway, NJ, USA. HPLC-column tips (fused silica) with 75 µm inner diameter (New Objective Inc., Woburn, USA) were self packed [78] with Reprosil-Pur 120 ODS-3 (Dr. Maisch, Ammerbuch, Germany). For LCMS, peptide separation was performed using a gradient of 0.5% acetic acid (ACS Reagent, Sigma-Aldrich, Taufkirchen, Germany) in water to 0.5% acetic acid in 80% ACN (HPLC gradient grade, SDS, Peypin, France). Water and ACN were at least gradient-grade quality. The mass spectrometer was operated in the data-dependent mode and switched automatically between MS and MS/MS with a normalized collision energy setting of 35. Each MS scan was followed by a maximum of five MS/MS scans. The MASCOT-Software [79] (Matrixscience, London, United Kingdom) in combination with the NCBInr Database (National Center for Biotechnology Information, Bethesda, MD, USA) was used for protein identification (Search criteria: monoisotopic m/z, mass accuracy for MS: 15 ppm or better, for MS/MS 0.8 Da or better, up to three missed cleavages, variable modifications: Acetyl (K, N-term, Protein N-Term), Carbamidomethyl (C), Propionamide (C), Oxidation (M), Gln→pyro-Glu (N-term Q), Glu→pyro-Glu (N-term E).

Western blotting and antibodies

For western blotting, proteins were transferred onto nitrocellulose membranes using a semi dry blotting apparatus after SDS-PAGE. As a blocking reagent, 2% skim milk powder in TBS-T [20 mM Tris/HCl (pH 7.6), 150 mM NaCl, 0.1% Tween-20] buffer was used. HBP1 and TCF4E2 were detected by a monoclonal rat anti-HA antibody (3F10, Roche Applied Science, Mannheim, Germany; 1:2000). The acetylation of TCF/LEF proteins was analyzed with a polyclonal rabbit pan-acetyl-lysine antibody (α-AcK, AB 3879, Millipore, Schwalbach, Germany; 1:1000). The acetylation of TCF4E2 K_{150} was detected with a polyclonal antibody to K_{150}ac (Pineda Antibody Service, Berlin, Germany; 1:500). This antibody was generated by immunizing rabbits with the acetylated peptide NH2-CRQALKacDAR-$CONH_2$ derived from TCF4E2. Anti-K_{150}ac specific polyclonal antibodies were purified from the resulting antisera by dual affinity purification. In the first step, a sepharose column charged with the modified peptide used for immunization was employed. From the eluate of this column antibodies reactive towards to non-acetylated TCF4E2 were removed through adsorption to a column charged with the unmodified peptide. Polyclonal rabbit antibodies were used to detect p300 and CBP (p300: C-20, Santa Cruz, Heidelberg, Germany; 1:200; CBP: 06-294, Chemicon, Hofheim, Germany; 1:1000). For detection of β-catenin and α-tubulin, mouse monoclonal antibodies (610154, BD Transduction Laboratories, Heidelberg, Germany; 1:1000; T9026, Sigma-Aldrich, Taufkirchen, Germany; 1:10000) were used. For the detection of recombinant His-purified TCF4E2 proteins expressed in E. coli a monoclonal mouse anti-c-Myc antibody (9E10, Roche Applied Science, Mannheim, Germany; 1:2000) was used.

Bacterial expression, site-directed acetylation and affinity-purification

For bacterial expression, TCF4E2 variants in the pCal-c expression vector were transformed into E. coli BL-21 and cultivated in LB-medium with ampicillin (150 µg/ml) to $OD_{600} = 0.7–0.8$ at 37°C. Protein expression was induced with 0.5 mM IPTG. After 3 to 5 hours of induction, cells were harvested by centrifugation at $10000 \times g$ for 10 minutes at 4°C and lysed in His-lysis buffer [50 mM NaH_2PO_4 (pH 8.0), 300 mM NaCl, 10 mM imidazole, 1 mg/ml lysozyme, 0.5% protease inhibitor cocktail for His-tagged proteins (Sigma-Aldrich, Taufkirchen, Germany)] for 30 minutes on ice followed by 6×10 seconds of sonication on ice. Lysates were cleared by centrifugation at $10000 \times g$ for 10 minutes at $4°C$ and stored on ice. For site-specific acetylation of TCF4E2 K_{150}, the mutated expression vector containing the cDNA for TCF4E2 K_{150}TAG was co-transformed into E. coli BL-21 together with pCDF_PylT-1 and pBK_AcRS3 encoding a $tRNA_{CUA}$ and the according acetyl-lysyl-tRNA synthetase [63]. Cultures were grown in LB-medium supplemented with ampicillin (150 µg/ml), spectinomycin (50 µg/ml) and kanamycin (50 µg/ml) at $37°C$. At $OD_{600} = 0.7–0.8$, 10 mM Nε-acetyl-lysine (Sigma-Aldrich, Taufkirchen, Germany) and 20 mM nicotinamide (Sigma-Aldrich, Taufkirchen, Germany) were added for additional 30 minutes before induction with 0.5 mM IPTG. After 5 hours cultivation, cells were collected by centrifugation at 10000 g for 10 minutes at $4°C$ and lysed as described above. For NiNTA purification, cleared lysates were incubated with NiNTA-Agarose (Qiagen, Hilden, Germany) over-night at $4°C$ with constant rotation. The agarose was collected by centrifugation at $600 \times g$ for 3 minutes and the unbound fraction was removed. The agarose was washed twice with His-wash buffer [50 mM NaH_2PO_4 (pH 8.0), 300 mM NaCl, 10 mM imidazole] for 10 minutes followed by centrifugation at 600 g for 3 minutes. After the second washing step, the agarose was resuspended in 1 bed volume of His-wash buffer and applied to a chromatography column. 10 bed volumes His-wash buffer were added to the column for the third washing step. His-tagged proteins were eluted with His-elution buffer [50 mM NaH_2PO_4 (pH 8.0), 300 mM NaCl, 250 mM Imidazol, 0.5% protease inhibitor cocktail for His-tagged proteins (Sigma-Aldrich, Taufkirchen, Germany)], and 1 ml fractions were collected and analyzed by SDS-PAGE. Elution fractions that contained the highest amounts of the His-tagged protein were pooled, dialyzed twice against 20 mM HEPES (pH 7.9), 75 mM NaCl, 2 mM $MgCl_2$, 10% glycerol over-night at $4°C$, concentrated 10-fold using Amicon Ultra centrifugal units (Millipore, Schwalbach, Germany) and stored at $-80°C$.

Immunofluorescence

For indirect immunofluorescence, 1×10^5 U-2 OS cells seeded onto glass cover slips in 24-well plates were transfected with 100 ng of plasmid DNA using FuGENE6 (Roche Applied Science, Mannheim, Germany) according to the recommendations of the manufacturer. After 48 hours, the cells were fixed with 4% paraformaldehyde, permeabilised with 0.5% Triton X-100 and stained with anti-HA antibody (3F10, Roche Applied Science, Mannheim, Germany; 1:200) over-night. As a secondary antibody, goat anti-rat Alexa-555 was used at a 1:200 dilution (Molecular Probes, Darmstadt, Germany). Nuclei were counterstained with DAPI (1:1000).

Treatment with λ-phosphatase and proteasome inhibitor

To determine the phosphorylation status of proteins, 20 µg of whole cell lysate was treated with 400 units of λ-phosphatase (NEB, Frankfurt, Germany) for 20 minutes at $30°C$ and analyzed by SDS-PAGE and western blotting. For analysis of proteasomal degradation, cells were treated with 20 µM MG132 in DMSO or a solvent control for two hours prior to cell lysis. Whole cell lysates were prepared as described above and further analyzed by SDS-PAGE and western blotting.

Transcription and translation in vitro and electrophoretic mobility shift assays (EMSAs)

Proteins for use in DNA binding assays were either expressed in E. coli and purified as described above, or they were produced by transcription and translation in vitro using the TNT SP6 high-yield

wheat germ protein expression system (Promega, Heidelberg, Germany) with 6 µg of plasmid DNA in 50-µl reactions. Equal expression levels of the different TCF4E2 variants obtained either way were controlled by western blotting. Binding reactions for EMSAs including biotinylated oligonucleotides representing Cdx1 TCF-binding element 4 (TBE4) were composed as before and the samples were further processed as described [18].

Acidic Urea PAGE

To determine if there are any potential influences of protein charge on the migratory behavior of TCF4E2, material transcribed and translated in vitro was separated by acidic urea PAGE [3M urea, 5% acetic acid, 10% polyacrylamide:bisacrylamide] and blotted towards the cathode onto a PVDF membrane in 5% acetic acid using a semi-dry blotting apparatus. After blotting, the membranes were equilibrated in TBS-T for at least 20 minutes and blocked with 2% skim milk in TBS-T buffer. Proteins were detected with a monoclonal rat anti-HA antibody (3F10, Roche Applied Sciences, Mannheim, Germany; 1:2000).

Supporting Information

Figure S1 Evidence for acetylation of TCF4 at K_{150}. Comparison of the MS/MS spectra for peptides found in the AspN-digested TCF4 sample and assigned by MASCOT (**A, C**) versus the MS/MS spectra of acetylated and non-acetylated versions of the synthetic peptide **DVQAGSLQSRQALK (B, D)**. The pictures show the MS/MS spectra of the doubly charged precursor ions as assigned by Mascot. (**A**) Sample spectrum assigned to the peptide acetylated at the lysine with a Mascot score value of 87. (**B**) Spectrum of acetylated synthetic control peptide. (**C**) Sample spectrum assigned to the non-acetylated peptide with a Mascot score value of 65. (**D**) Spectrum of non-acetylated synthetic control peptide.

Figure S2 Evidence for acetylation of TCF4 at K_{150}. Comparison of the MS/MS spectra for peptides found in the AspN-digested TCF4 sample and assigned by MASCOT (**A, C**) versus the MS/MS spectra of acetylated and non-acetylated versions of the synthetic peptide **DVQAGSLQSRQALK (B, D)**. The pictures show the MS/MS spectra of the triply charged precursor ions as assigned by Mascot. (**A**) Sample spectrum assigned to the peptide acetylated at the lysine with a Mascot score value of 40. (**B**) Spectrum of acetylated synthetic control peptide. (**C**) Sample spectrum assigned to the non-acetylated peptide with a Mascot score value of 26, which is below the Mascot threshold value for indication of identity at p<0.05. (**D**) Spectrum of non-acetylated synthetic control peptide.

Figure S3 Intracellular localization of TCF4E2 WT, K_{150}A and K_{150}R. U-2 OS cells transfected with expression constructs for the TCF4E2 variants with or without CBP were stained with anti-HA antibodies and secondary antibodies coupled to Alexa-555 to visualize the TCF4E2 variants (TCF4, red). Nuclei were counterstained with DAPI. Phase contrast and overlay of TCF4E2 and DAPI staining are shown. Bar: 20 µm.

Figure S4 Acetylation of K_{150} does not affect phosphorylation or proteasomal degradation of TCF4E2. (**A**) To analyze if K_{150} acetylation influences the phosphorylation of TCF4E2, extracts of HEK293 cells transfected with the expression constructs for TCF4E2, TCF4E2K_{150}A or TCF4E2K_{150}R in the absence or presence of CBP were treated with λ-phosphatase (λ-

PPase) and analyzed by SDS-PAGE and western blotting with anti-HA antibodies. Prior to λ-phosphatase treatment, all TCF4 variants showed comparable migration patterns by SDS-PAGE, suggesting that they are equally phosphorylated. λ-phosphatase treatment resulted in a mobility shift and faster migration of the TCF4 variants due to dephosphorylation. Again, wild-type TCF4E2 and TCF4E2 mutants showed uniform behavior. (**B**) HEK293 cells transfected with expression constructs for TCF4E2, TCF4E2K$_{150}$A or TCF4E2K$_{150}$R with or without CBP were treated with 20 µM MG132 for two hours prior to cell lysis and whole cell extracts were analyzed by SDS-PAGE and western blotting with anti-HA and anti-β-catenin (αβ-cat) antibodies. For TCF4E2 no change in protein amount and no additional protein bands that would indicate polyubiquitination were detected. The presence or absence of CBP made no difference. In contrast, for β-catenin, MG132 treatment resulted in the appearance of additional protein bands and a stronger signal, suggesting polyubiquitination of the protein. Molecular weight standards are indicated on the right of the panels.

Figure S5 Mutation of K$_{150}$ has no influence on the transactivation capacity of TCF4E2 at different promoters in HEK293 cells.

HEK293 cells were cotransfected with combinations of firefly and Renilla luciferase reporter genes, control vector, expression vector for a constitutively active form of β-catenin and increasing amounts of TCF4E2 variants as indicated. Firefly expression was driven by promoters from the Wnt/β-catenin target genes *Cdx1* (**A**), *Axin2* (**B**) and *c-MYC* (**C**). Reporter gene activities were determined 40 h post transfection. Bars represent relative luciferase activity compared to values obtained with the lowest amount of TCF4E2 WT expression vector and β-catenin (set to 100%). The average values and standard deviations from at least three independent experiments are shown. (**D**) Expression levels of all TCF4E2 variants in whole

cell extracts of cells transfected with 150 ng DNA as used in the luciferase reporter assays was controlled by SDS-PAGE and western blotting using anti-HA antibodies. Molecular weight standards are shown on the left.

Figure S6 Mutation of K$_{150}$ has no influence on the transactivation capacity of TCF4E2 at different promoters in U-2 OS cells.

U-2 OS cells were cotransfected with combinations of firefly and Renilla luciferase reporter genes, control vector, expression vector for a constitutively active form of β-catenin and increasing amounts of TCF4E2 variants as indicated. Firefly expression was driven by promoters from the Wnt/β-catenin target genes *Cdx1* (**A**), *Axin2* (**B**) and *c-MYC* (**C**). Reporter gene activities were determined 40 h post transfection. Bars represent relative luciferase activity compared to values obtained with the lowest amount of TCF4E2 WT expression vector and β-catenin (set to 100%). The average values and standard deviations from at least three independent experiments are shown.

Acknowledgments

We thank Amy S. Yee for providing the pEF-Bos HBP1 expression vector. We thank Michael M. Hoffmann and members of the sequencing core facility of the University Medical Center Freiburg for sequencing of plasmids.

Author Contributions

Conceived and designed the experiments: SE AW NS AH. Performed the experiments: SE AW KB MLB SJ NS. Analyzed the data: SE AW KB MLB SJ NS AH. Contributed reagents/materials/analysis tools: SE AW MLB SJ NS AH. Wrote the paper: SE MLB AH.

References

1. Arce L, Yokoyama NN, Waterman ML (2006) Diversity of LEF/TCF action in development and disease. Oncogene 25: 7492–7504.
2. Hoppler S, Kavanagh CL (2007) Wnt signalling: variety at the core. J Cell Science 120: 385–393.
3. Nguyen H, Merrill BJ, Polak L, Nikolova M, Rendl M, et al. (2009) Tcf3 and Tcf4 are essential for long-term homeostasis of skin epithelia. Nat Genet 41: 1068–1075.
4. Brunner E, Peter O, Schweizer L, Basler K (1997) pangolin encodes a Lef-1 homologue that acts downstream of Armadillo to transduce the Wingless signal in Drosophila. Nature 385: 829–833.
5. van de Wetering M, Cavallo R, Dooijes D, van Beest M, van Es J, et al. (1997) Armadillo coactivates transcription driven by the product of the Drosophila segment polarity gene dTCF. Cell 88: 789–799.
6. Galceran J, Farinas I, Depew MJ, Clevers H, Grosschedl R (1999) Wnt3a-/-like phenotype and limb deficiency in Lef1(-/-)Tcf1(-/-) mice. Genes Dev 13: 709–717.
7. Gregorieff A, Grosschedl R, Clevers H (2004) Hindgut defects and transformation of the gastro-intestinal tract in Tcf4(-/-)/Tcf1(-/-) embryos. EMBO J 23: 1825–1833.
8. Korinek V, Barker N, Moerer P, van Donselaar E, Huls G, et al. (1998) Depletion of epithelial stem-cell compartments in the small intestine of mice lacking Tcf-4. Nat Genet 19: 379–383.
9. Lin R, Thompson S, Priess JR (1995) pop-1 encodes an HMG box protein required for the specification of a mesoderm precursor in early C. elegans embryos. Cell 83: 599–609.
10. Liu F, van den Broek O, Destree O, Hoppler S (2005) Distinct roles for Xenopus Tcf/Lef genes in mediating specific responses to Wnt/beta-catenin signalling in mesoderm development. Development 132: 5375–5385.
11. Merrill BJ, Pasolli HA, Polak L, Rendl M, Garcia-Garcia MJ, et al. (2004) Tcf3: a transcriptional regulator of axis induction in the early embryo. Development 131: 263–274.
12. Roel G, Hamilton FS, Gent Y, Bain AA, Destree O, et al. (2002) Lef-1 and Tcf-3 transcription factors mediate tissue-specific Wnt signaling during Xenopus development. Curr Biol 12: 1941–1945.
13. van Genderen C, Okamura RM, Farinas I, Quo RG, Parslow TG, et al. (1994) Development of several organs that require inductive epithelial-mesenchymal interactions is impaired in LEF-1-deficient mice. Genes Dev 8: 2691–2703.
14. Verbeek S, Izon D, Hofhuis F, Robanus-Maandag E, te Riele H, et al. (1995) An HMG-box-containing T-cell factor required for thymocyte differentiation. Nature 374: 70–74.
15. Duval A, Rolland S, Tubacher E, Bui H, Thomas G, et al. (2000) The human T-cell transcription factor-4 gene: structure, extensive characterization of alternative splicings, and mutational analysis in colorectal cancer cell lines. Cancer Res 60: 3872–3879.
16. Howng SL, Huang FH, Hwang SL, Lieu AS, Sy WD, et al. (2004) Differential expression and splicing isoform analysis of human Tcf-4 transcription factor in brain tumors. Int J Oncol 25: 1685–1692.
17. Van de Wetering M, Castrop J, Korinek V, Clevers H (1996) Extensive alternative splicing and dual promoter usage generate Tcf-1 protein isoforms with differential transcription control properties. Mol Cell Biol 16: 745–752.
18. Weise A, Bruser K, Elfert S, Wallmen B, Wittel Y, et al. (2010) Alternative splicing of Tcf7l2 transcripts generates protein variants with differential promoter-binding and transcriptional activation properties at Wnt/beta-catenin targets. Nucleic Acids Res 38: 1964–1981.
19. Shiina H, Igawa M, Breault J, Ribeiro-Filho L, Pookot D, et al. (2003) The human T-cell factor-4 gene splicing isoforms, Wnt signal pathway, and apoptosis in renal cell carcinoma. Clin Cancer Res 9: 2121–2132.
20. Valenta T, Lukas J, Korinek V (2003) HMG box transcription factor TCF-4's interaction with CtBP1 controls the expression of the Wnt target Axin2/Conductin in human embryonic kidney cells. Nucleic Acids Res 31: 2369–2380.
21. Hecht A, Stemmler MP (2003) Identification of a promoter-specific transcriptional activation domain at the C terminus of the Wnt effector protein T-cell factor 4. J Biol Chem 278: 3776–3785.
22. Atcha FA, Syed A, Wu B, Hoverter NP, Yokoyama NN, et al. (2007) A unique DNA binding domain converts T-cell factors into strong Wnt effectors. Mol Cell Biol 27: 8352–8363.
23. Chang MV, Chang JL, Gangopadhyay A, Shearer A, Cadigan KM (2008) Activation of wingless targets requires bipartite recognition of DNA by TCF. Curr Biol 18: 1877–1881.

24. Grosschedl R, Giese K, Pagel J (1994) HMG domain proteins: architectural elements in the assembly of nucleoprotein structures. Trends Genet 10: 94–100.

25. Love JJ, Li X, Chung J, Dyson HJ, Wright PE (2004) The LEF-1 high-mobility group domain undergoes a disorder-to-order transition upon formation of a complex with cognate DNA. Biochemistry 43: 8725–8734.

26. Mosimann C, Hausmann G, Basler K (2009) Beta-catenin hits chromatin: regulation of Wnt target gene activation. Nature Rev Mol Cell Biol 10: 276–286.

27. Ishitani T, Ninomiya-Tsuji J, Matsumoto K (2003) Regulation of lymphoid enhancer factor 1/T-cell factor by mitogen-activated protein kinase-related Nemo-like kinase-dependent phosphorylation in Wnt/beta-catenin signaling. Mol Cell Biol 23: 1379–1389.

28. Yamada M, Ohnishi J, Ohkawara B, Iemura S, Satoh K, et al. (2006) NARF, an nemo-like kinase (NLK)-associated ring finger protein regulates the ubiquityla-tion and degradation of T cell factor/lymphoid enhancer factor (TCF/LEF). J Biol Chem 281: 20749–20760.

29. Mahmoudi T, Li VS, Ng SS, Taouatas N, Vries RG, et al. (2009) The kinase TNIK is an essential activator of Wnt target genes. EMBO J 28: 3329–3340.

30. Shitashige M, Satow R, Jigami T, Aoki K, Honda K, et al. (2010) Traf2- and Nck-interacting kinase is essential for Wnt signaling and colorectal cancer growth. Cancer Res 70: 5024–5033.

31. Najdi R, Syed A, Arce L, Theisen H, Ting JH, et al. (2009) A Wnt kinase network alters nuclear localization of TCF-1 in colon cancer. Oncogene 28: 4133–4146.

32. Yamamoto H, Ihara M, Matsuura Y, Kikuchi A (2003) Sumoylation is involved in beta-catenin-dependent activation of Tcf-4. EMBO J 22: 2047–2059.

33. Sachdev S, Bruhn L, Sieber H, Pichler A, Melchior F, et al. (2001) PIASy, a nuclear matrix-associated SUMO E3 ligase, represses LEF1 activity by sequestration into nuclear bodies. Genes Dev 15: 3088–3103.

34. Billin AN, Thirlwell H, Ayer DE (2000) Beta-catenin-histone deacetylase interactions regulate the transition of LEF1 from a transcriptional repressor to an activator. Mol Cell Biol 20: 6882–6890.

35. Henderson BR, Galea M, Schuechner S, Leung L (2002) Lymphoid enhancer factor-1 blocks adenomatous polyposis coli-mediated nuclear export and degradation of beta-catenin. Regulation by histone deacetylase 1. J Biol Chem 277: 24258–24264.

36. Ye F, Chen Y, Hoang T, Montgomery RL, Zhao XH, et al. (2009) HDAC1 and HDAC2 regulate oligodendrocyte differentiation by disrupting the beta-catenin-TCF interaction. Nature Neurosci 12: 829–838.

37. Bedford DC, Kasper LH, Fukuyama T, Brindle PK (2010) Target gene context influences the transcriptional requirement for the KAT3 family of CBP and p300 histone acetyltransferases. Epigenetics 5: 9–15.

38. Kalkhoven E (2004) CBP and p300: HATs for different occasions. Biochemical Pharmacology 68: 1145–1155.

39. Li J, Sutter C, Parker DS, Blauwkamp T, Fang M, et al. (2007) CBP/p300 are bimodal regulators of Wnt signaling. EMBO J 26: 2284–2294.

40. Waltzer L, Bienz M (1998) Drosophila CBP represses the transcription factor TCF to antagonize Wingless signalling. Nature 395: 521–525.

41. Gay F, Calvo D, Lo MC, Ceron J, Maduro M, et al. (2003) Acetylation regulates subcellular localization of the Wnt signaling nuclear effector POP-1. Genes Dev 17: 717–722.

42. Levy L, Wei Y, Labalette C, Wu Y, Renard CA, et al. (2004) Acetylation of beta-catenin by p300 regulates beta-catenin-Tcf4 interaction. Mol Cell Biol 24: 3404–3414.

43. Wolf D, Rodova M, Miska EA, Calvet JP, Kouzarides T (2002) Acetylation of beta-catenin by CREB-binding protein (CBP). J Biol Chem 277: 25562–25567.

44. Hecht A, Vleminckx K, Stemmler MP, van Roy F, Kemler R (2000) The p300/CBP acetyltransferases function as transcriptional coactivators of beta-catenin in vertebrates. EMBO J 19: 1839–1850.

45. Miyagishi M, Fujii R, Hatta M, Yoshida E, Araya N, et al. (2000) Regulation of Lef-mediated transcription and p53-dependent pathway by associating beta-catenin with CBP/p300. J Biol Chem 275: 35170–35175.

46. Sun Y, Kolligs FT, Hottiger MO, Mosavin R, Fearon ER, et al. (2000) Regulation of beta -catenin transformation by the p300 transcriptional coactivator. Proc Natl Acad Sci U S A 97: 12613–12618.

47. Takemaru KI, Moon RT (2000) The transcriptional coactivator CBP interacts with beta-catenin to activate gene expression. J Cell Biol 149: 249–254.

48. Ma H, Nguyen C, Lee KS, Kahn M (2005) Differential roles for the coactivators CBP and p300 on TCF/beta-catenin-mediated survivin gene expression. Oncogene 24: 3619–3631.

49. Miyabayashi T, Teo JL, Yamamoto M, McMillan M, Nguyen C, et al. (2007) Wnt/beta-catenin/CBP signaling maintains long-term murine embryonic stem cell pluripotency. Proc Natl Acad Sci U S A 104: 5668–5673.

50. Berasi SP, Xiu M, Yee AS, Paulson KE (2004) HBP1 repression of the p47phox gene: cell cycle regulation via the NADPH oxidase. Mol Cell Biol 24: 3011–3024.

51. Chen YC, Zhang XW, Niu XH, Xin DQ, Zhao WP, et al. (2010) Macrophage migration inhibitory factor is a direct target of HBP1-mediated transcriptional repression that is overexpressed in prostate cancer. Oncogene 29: 3067–3078.

52. Escamilla-Powers JR, Daniel CJ, Farrell A, Taylor K, Zhang X, et al. (2010) The tumor suppressor protein HBP1 is a novel c-myc-binding protein that negatively regulates c-myc transcriptional activity. J Biol Chem 285: 4847–4858.

53. Li H, Bian C, Liao L, Li J, Zhao RC (2011) miR-17-5p promotes human breast cancer cell migration and invasion through suppression of HBP1. Breast Cancer Res Treat 126: 565–575.

54. Li H, Wang W, Liu X, Paulson KE, Yee AS, et al. (2010) Transcriptional factor HBP1 targets P16(INK4A), upregulating its expression and consequently is involved in Ras-induced premature senescence. Oncogene 29: 5083–5094.

55. Sampson EM, Haque ZK, Ku MC, Tevosian SG, Albanese C, et al. (2001) Negative regulation of the Wnt-beta-catenin pathway by the transcriptional repressor HBP1. EMBO J 20: 4500–4511.

56. Yao CJ, Works K, Romagnoli PA, Austin GE (2005) Effects of overexpression of HBP1 upon growth and differentiation of leukemic myeloid cells. Leukemia 19: 1958–1968.

57. Zhuma T, Tyrrell R, Sekkali B, Skavdis G, Saveliev A, et al. (1999) Human HMG box transcription factor HBP1: a role in hCD2 LCR function. EMBO J 18: 6396–6406.

58. Tevosian SG, Shih HH, Mendelson KG, Sheppard KA, Paulson KE, et al. (1997) HBP1: a HMG box transcriptional repressor that is targeted by the retinoblastoma family. Genes Dev 11: 383–396.

59. Wang W, Pan K, Chen Y, Huang C, Zhang X (2012) The acetylation of transcription factor HBP1 by p300/CBP enhances p16INK4A expression. Nucleic Acids Res 40: 981–995.

60. Black JC, Choi JE, Lombardo SR, Carey M (2006) A mechanism for coordinating chromatin modification and preinitiation complex assembly. Mol Cell 23: 809–818.

61. Ceschin DG, Walia M, Wenk SS, Duboe C, Gaudon C, et al. (2011) Methylation specifies distinct estrogen-induced binding site repertoires of CBP to chromatin. Genes Dev 25: 1132–1146.

62. Leymarie N, Berg EA, McComb ME, O'Connor PB, Grogan J, et al. (2002) Tandem mass spectrometry for structural characterization of proline-rich proteins: application to salivary PRP-3. Anal Chem 74: 4124–4132.

63. Neumann H, Hancock SM, Buning R, Routh A, Chapman L, et al. (2009) A method for genetically installing site-specific acetylation in recombinant histones defines the effects of H3 K56 acetylation. Mol Cell 36: 153–163.

64. Hikasa H, Sokol SY (2011) Phosphorylation of TCF proteins by homeodomain-interacting protein kinase 2. J Biol Chem 286: 12093–12100.

65. Nazwar TA, Glassmann A, Schilling K (2009) Expression and molecular diversity of Tcf7l2 in the developing murine cerebellum and brain. J Neurosci Res 87: 1532–1546.

66. Archbold HC, Yang YX, Chen L, Cadigan KM (2012) How do they do Wnt they do?: regulation of transcription by the Wnt/beta-catenin pathway. Acta Physiologica 204: 74–109.

67. Hikasa H, Ezan J, Itoh K, Li X, Klymkowsky MW, et al. (2010) Regulation of TCF3 by Wnt-dependent phosphorylation during vertebrate axis specification. Dev Cell 19: 521–532.

68. Teo JL, Kahn M (2010) The Wnt signaling pathway in cellular proliferation and differentiation: A tale of two coactivators. Adv Drug Delivery Rev 62: 1149–1155.

69. Kumar SR, Scehnet JS, Ley EJ, Singh J, Krasnoperov V, et al. (2009) Preferential induction of EphB4 over EphB2 and its implication in colorectal cancer progression. Cancer Res 69: 3736–3745.

70. Yang XJ, Seto E (2008) Lysine acetylation: codified crosstalk with other posttranslational modifications. Mol Cell 31: 449–461.

71. Mohan RD, Litchfield DW, Torchia J, Tini M (2010) Opposing regulatory roles of phosphorylation and acetylation in DNA mispair processing by thymine DNA glycosylase. Nucleic Acids Res 38: 1135–1148.

72. Eckner R, Ewen ME, Newsome D, Gerdes M, DeCaprio JA, et al. (1994) Molecular cloning and functional analysis of the adenovirus E1A-associated 300-kD protein (p300) reveals a protein with properties of a transcriptional adaptor. Genes Dev 8: 869–884.

73. Aulehla A, Wehrle C, Brand-Saberi B, Kemler R, Gossler A, et al. (2003) Wnt3a plays a major role in the segmentation clock controlling somitogenesis. Dev Cell 4: 395–406.

74. Lickert H, Domon C, Huls G, Wehrle C, Duluc I, et al. (2000) Wnt/(beta)-catenin signaling regulates the expression of the homeobox gene Cdx1 in embryonic intestine. Development 127: 3805–3813.

75. He TC, Sparks AB, Rago C, Hermeking H, Zawel L, et al. (1998) Identification of c-MYC as a target of the APC pathway. Science 281: 1509–1512.

76. Turner DL, Weintraub H (1994) Expression of achaete-scute homolog 3 in Xenopus embryos converts ectodermal cells to a neural fate. Genes Dev 8: 1434–1447.

77. Blum H, Beier H, Gross HJ (1987) Improved Silver Staining of Plant-Proteins, Rna and DNA in Polyacrylamide Gels. Electrophoresis 8: 93–99.

78. Olsen JV, Ong SE, Mann M (2004) Trypsin cleaves exclusively C-terminal to arginine and lysine residues. Mol Cell Proteomics 3: 608–614.

79. Perkins DN, Pappin DJ, Creasy DM, Cottrell JS (1999) Probability-based protein identification by searching sequence databases using mass spectrometry data. Electrophoresis 20: 3551–3567.

3

Expression of the Genetic Suppressor Element 24.2 (GSE24.2) Decreases DNA Damage and Oxidative Stress in X-Linked Dyskeratosis Congenita Cells

Cristina Manguan-Garcia[1,2]**, Laura Pintado-Berninches**[1]**, Jaime Carrillo**[1]**, Rosario Machado-Pinilla**[1,2]**, Leandro Sastre**[1,2]**, Carme Pérez-Quilis**[3,4]**, Isabel Esmoris**[3,4]**, Amparo Gimeno**[3,4]**, Jose Luis García-Giménez**[2,3,4]**, Federico V. Pallardó**[2,3,4]**, Rosario Perona**[1,2]*

1 Instituto de Investigaciones Biomédicas CSIC/UAM, Madrid, Spain, **2** CIBER de Enfermedades Raras, Valencia, Spain, **3** Biomedical Research Institute INCLIVA, Valencia, Spain, **4** Department of Physiology, Faculty of Medicine and Dentistry, University of Valencia, Valencia, Spain

Abstract

The predominant X-linked form of Dyskeratosis congenita results from mutations in *DKC1*, which encodes dyskerin, a protein required for ribosomal RNA modification that is also a component of the telomerase complex. We have previously found that expression of an internal fragment of dyskerin (GSE24.2) rescues telomerase activity in X-linked dyskeratosis congenita (X-DC) patient cells. Here we have found that an increased basal and induced DNA damage response occurred in X-DC cells in comparison with normal cells. DNA damage that is also localized in telomeres results in increased heterochromatin formation and senescence. Expression of a cDNA coding for GSE24.2 rescues both global and telomeric DNA damage. Furthermore, transfection of bacterial purified or a chemically synthesized GSE24.2 peptide is able to rescue basal DNA damage in X-DC cells. We have also observed an increase in oxidative stress in X-DC cells and expression of GSE24.2 was able to diminish it. Altogether our data indicated that supplying GSE24.2, either from a cDNA vector or as a peptide reduces the pathogenic effects of Dkc1 mutations and suggests a novel therapeutic approach.

Editor: Gabriele Saretzki, University of Newcastle, United Kingdom

Funding: This work was supported by grants: PI11-0949 and PI12/02263 FIS, SAF2008-01338 from the Ministerio de Ciencia e Innovación. Grants, PROMETEO2010/074 from Generalitat Valenciana and Fundación Salud 2000 to FVP. CMG, JLGG and CPQ are supported by CIBER de Enfermedades Raras. The funders had no role in study design, data collection and analysis, decision to publish, or preparation of the manuscript.

Competing Interests: The authors have declared that no competing interests exist.

* Email: RPerona@iib.uam.es

Introduction

Telomeres are nucleoprotein complexes located at the ends of linear chromosomes and consist of tandem repeats of simple DNA sequences (TTAGGG in humans) and proteins that interact directly or indirectly with these sequences [1]. Sequence erosion of terminal repeats is inherent to each round of genome replication. The replenishment of the telomeric repeats is accomplished by the extension of their 3′ ends, through a reaction mediated by the telomerase complex [2]. In humans, the active telomerase complex consists of a minimum of three essential components: hTERT, hTR and dyskerin [3]. Besides forming part of the telomerase complex, dyskerin is a pseudouridine synthase component of H/ACA small nuclear RNPs [4], complexes that mediate the conversion of specific uridines (U) to pseudouridine in newly synthesized ribosomal RNAs [5] [6] [7]. Point mutations in dyskerin cause a rare disease named X-linked dyskeratosis congenita (X-DC) [8]. Individuals with X-DC display features of premature ageing, as well as nail dystrophy, mucosal leukoplakia, interstitial fibrosis of the lung and increased susceptibility to cancer [9]. The tissues affected by X-DC, such as bone marrow and skin, are characterized by the high rate of turnover of their progenitor cells.

Telomere shortening prevents the formation of the loop-like structure maintained by a nucleoprotein structure consisting of telomeric DNA and 6 proteins that are together known as shelterin [1]. This capping structure prevents the otherwise exposed ends of different chromosomes from being recognized as double strand breaks (DSBs) by the cell's DNA repair machinery which would result in telomere fusion. When telomeres become critically short or unprotected because of shelterin deficiency, they trigger a DNA damage response (DDR), leading to the activation of an ataxia telangiectasia mutated (ATM) or ataxia telangiectasia and Rad3 related (ATR)-dependent DNA damage response at chromosome ends [10] [11] [12] [13]. 53BP1 is a C-non-homologous-end-joining (C-NHEJ) component and an ATM target that accumulates at DSBs and uncapped telomeres [14] [15]. The binding of 53BP1 close to DNA breaks impacts the dynamic behavior of the local chromatin and facilitates the non-homologous-end-joining (NHEJ) repair reactions that involve distant sites [16]. ATM phosphorylates Chk2 leading to activation of cell cycle checkpoints. Chk2 acts as a signal distributor, dispersing checkpoint signal to downstream targets such as p53, Cdc25A, Cdc25C, BRCA1 and E2F1 [17].

Senescence, initially described as stable cell proliferation arrest, can be induced by telomere shortening and also by activated oncogenes, DNA damage and drug-like inhibitors of specific enzymatic activities [18]. Senescent cells are typically characterized by a large flat morphology and the expression of a senescence-associated-β-galactosidase (SA-β-gal) activity of unknown function. In the nucleus of senescent cells chromatin undergoes dramatic remodeling through the formation of domains of heterochromatin called senescence-associated heterochromatin-foci (SAHF). SAHF contain histone modifications and proteins characteristic of silent heterochromatin such as methylated lysine 9 of histone H3 (H3K9me), heterochromatin protein 1 (HP1), and the histone H2A variant macroH2A.1 [18] [19]. Proliferation-promoting genes such as E2F-target genes (e.g. cyclin A) are recruited into SAHF, dependent on the pRB suppressor protein, thereby irreversibly silencing expression of those genes.

In cultured cells and animal models, telomere erosion promotes chromosomal instability via breakage-fusion-bridge cycles, contributing to the early stages of tumorigenesis. Telomere shortening in Dyskeratosis congenita is associated with a higher risk of some types of cancer such as head and neck squamous cell carcinoma (HNSCC) (mostly tongue), skin squamous cell carcinoma (SCC), anogenital, stomach, esophagus, and lymphomas, as well as myelodysplastic syndrome (MDS) [20] [21] [22] [23]. Altogether, these findings provide direct clinical evidence that short telomeres in hematopoietic cells are dysfunctional, mediate chromosomal instability and predispose to malignant transformation in a human disease.

We previously isolated the peptide GSE24.2, in a screen of cDNAs for those that confer survival ability on cells treated with cisplatin [24]. Intriguingly GSE24.2 turned out to be a short dyskerin fragment containing two highly conserved motifs implicated in pseudouridine synthase catalytic activity. GSE24.2 prevents telomerase inhibition mediated by different chemotherapeutic agents, including cisplatin and telomerase inhibitors. In X-DC cells and WI-38-VA13 cells, GSE24.2 induces an increase in hTERT mRNA levels and the recovery of telomerase activity [24]. Mutations in DKC1 lead to severe destabilization of telomerase RNA (TR), a reduction in telomerase activity and a significant continuous loss of telomere length during growth [25]. When a peptide encoding GSE24.2, was introduced into mutant cells, it rescued telomerase activity and prevented the decrease in TR levels induced by the Dkc1 mutation [26] GSE24.2 was recently approved as an orphan drug for the treatment of Dyskeratosis congenita (EU/3/12/1070 - EMA/OD/136/11).

To obtain more information on the biological activity of this dyskerin fragment we studied its effect on the DNA damage pathway in patient derived X-DC cells and a mouse F9 cell line carrying the A353V mutation in the Dkc1 gene [26]. This is the mutation most frequently found in patients with X-DC (about 40% of patients) [27] [28] and is localized in the PUA RNA binding domain, the putative site for interaction with hTR. Recently, it has been described that mouse ES cells expressing a small dyskerin deletion, removing exon 15 of Dkc1, additionally showed decreased proliferative rate and increased sensitivity to DNA damage [29] that was independent on telomere length suggesting that decreased telomerase activity induced by the mutation in Dkc1 resulted in induction of DNA damage probably by extratelomeric activity of Dkc1 gene. Therefore the use of a mouse F9A353V model would allow study the effect of GSE24.2 directly on DNA damage, independently of telomeric elongation. Here we show that human X-DC cells showed both basal DNA damage foci and phosphorylation of ATM and CHK2 together with increased content of heterochromatin. Expression of the GSE24.2 was able to reduce DNA damage in X-DC patient and F9 X-DC mouse cell line models, by decreasing the formation of DNA damage foci. Finally, we also report that expression of

GSE24.2 decreases oxidative stress in X-DC patient cells and that may result in reduced DNA damage. These data support the contention that expression of GSE24.2, or related products, could prolong the lifespan of dyskeratosis congenita cells.

Materials and Methods

Cell lines and constructs

Dermal fibroblasts from a control proband (X-DC-1787-C) and two X-DC patients (X-DC-1774-P and X-DC3) were obtained from Coriell Cell Repository. GSE24.2, DKC, motif I and motif II were cloned as previously described in the pLXCN vector [24]. PGATEV protein expression plasmid [30] was obtained from Dr. G. Montoya. PGATEV-GSE24.2 was obtained by subcloning the GSE24.2 fragment into the NdeI/XhoI sites of the pGATEV plasmid as previously described [24].

F9 cells and F9 cells transfected with A353V targeting vector were previously described [31] [26]. F9A353V cells were cultured in Dulbecco modified Eagle medium (DMEM) 10% fetal bovine serum, 2 mM glutamine (Gibco) and Sodium bicarbonate (1,5 gr/ml).

Cell transfection and analysis of gene expression

F9 cells were transfected with 16 μg of DNA/10^6 cells, using lipofectamine plus (Invitrogen, Carlsbad, USA), according to the manufacturer's instructions. Peptides transfection was performed by using the Transport Protein Delivery Reagent (50568; Lonza, Walkersville, USA) transfection kit. Routinely from 6 to 15 μg were used per 30 mm dish.

Antibodies. The source of antibodies was as follow: phospho-Histone H2A.X Ser139 (2577; Cell Signaling), phospho-Histone H2A.X Ser139 clone JBW301 (05-636; Millipore), macroH2A.1 (ab37264; abcam), 53BP1 (4937; Cell Signaling), anti-ATM Protein Kinase S1981P (200-301-400; Rockland), phospho-Chk2-Thr68 (2661; Cell Signaling), Monoclonal Anti-α-tubulin (T9026; Sigma-Aldrich), Anti-8-Oxoguanine Antibody, clone 483.15 (MAB3560, Merck-Millipore). Fluorescent antibodies were conjugated with Alexa fluor 488 (A11029 and A11034, Molecular Probes) and Alexa fluor 647 (A21236, Molecular Probes, Carlsbad, USA)).

Immunofluorescence and Fluorescence in situ hybridization (FISH) for telomeres

Protein localization was carried out by fluorescence microscopy. For this purpose, cells were grown on coverslips, transfected and fixed in 3.7% formaldehyde solution (47608; Fluka, Sigma, St. Louis, USA) at room temperature for 15 min. After washing with 1x PBS, cells were permeabilized with 0.2% Triton X-100 in PBS and blocked with 10% horse serum before overnight incubation with γ-H2A.X, 53BP1, p-ATM, p-CHK2 antibodies. Finally, cells were washed and incubated with secondary antibodies coupled to fluorescent dyes (alexa fluor 488 or/and alexa fluor 647).

For immuno-FISH, immunostaining of 53BP1 was performed as described above and followed by incubation in PBS 0,1% Triton X-100, fixation 5 min in 2% paraformaldehyde (PFA), dehydration with ethanol and air-dried. Cells were hybridized with the telomeric PNA-Cy3 probe (PNA Bio) using standard PNA-FISH procedures. Imaging was carried out at room temperature in Vectashield, mounting medium for fluorescence (Vector Laboratories, Burlingame, USA). Images were acquired with a Confocal Spectral Leica TCS SP5. Using a HCX PL APO Lambda blue 63×1.40 OIL UV, zoom 2.3 lens. Images were acquired using LAS-AF 1.8.1 Leica software and processed using LAS-AF 1.8.1

Leica software and Adobe Photoshop CS. Colocalization of 53BP1 foci and the PNA FISH probe was quantified in at least 200 cells.

Telomeric repeat amplification protocol (TRAP) assay

Telomerase activity was measured using the TRAPeze kit [32] (Millipore, Billerica, MA USA) according to the manufacturer's recommendations. TRAP assay activity was normalized with the internal control [24].

Real-time quantitative PCR

RNA isolation and cDNA synthesis. Total cellular RNA was extracted using Trizol (Invitrogen, Carlsbad, USA) according to the manufacturer's instructions. For reverse transcription reactions (RT), 1 µg of the purified RNA was reverse transcribed using random hexamers with the High-Capacity cDNA Archive kit (Applied Biosystems, P/N: 4322171; Foster City, CA) according to the manufacturer's instructions. RT conditions comprised an initial incubation step at 25°C for 10 min. to allow random hexamers annealing, followed by cDNA synthesis at 37°C for 120 min, and a final inactivation step for 5 min. at 95°C.

Measurement of mRNA Levels. The mRNA levels were determined by quantitative real-time PCR analysis using an ABI Prism 7900 HT Fast Real-Time PCR System (Applied Biosystems, Foster City, CA). Gene-specific primer pairs and probes for *SOD1* (*SOD Cu/Zn*), *SOD2* (*SOD Mn*), *GPX1 (Glutathione peroxidase 1)* and *CAT (Catalase)* (Assay-on-demand, Applied Biosystems), were used together with TaqMan Universal PCR Master Mix (Applied Biosystems, Foster City, USA) and 2 µl of reverse transcribed sample RNA in 20 µl reaction volumes. PCR conditions were 10 min. at 95°C for enzyme activation, followed by 40 two-step cycles (15 sec at 95°C; 1 min at 60°C). The levels of glyceraldehyde-3-phosphate dehydrogenase (*GAPDH*) expression were measured in all samples to normalize gene expression for sample-to-sample differences in RNA input, RNA quality and reverse transcription efficiency. Each sample was analyzed in triplicate, and the expression was calculated according to the $2^{-\Delta\Delta Ct}$ method.

GSE24.2 peptide production and purification

E. Coli DH5a cells were transformed with pGATEV GSE24.2 and lysates prepared as described [30]. The fusion protein was purified with glutathione-sepharose and purity analyzed by gel electrophoresis. GSE24.2 was obtained from the purified fusion protein by TEV protease digestion according to the manufacturer's recommendations. Typically, over 90% of the fusion protein was cleaved, as determined by SDS-PAGE. The protein was passed twice over a 5 ml Hi-Trap Ni-NTA column to remove the polyhistidine tags, un-cleaved protein, TEV protease and impurities. Synthetic GSE24.2 was obtained from Peptide 2.0 Inc (Chantilly, USA) and purified by HPLC

Western Blot

Whole-cell extracts were prepared essentially as described previously [33]. Nuclear extracts were obtained as previously reported [24]. Western blotting was performed using standard methods [33]. Protein concentration was measured by using the Bio-Rad protein assay.

Senescence analysis

Control and X-DC fibroblasts (1×10^4 cells) were plated onto 6 well plates and fixed after four days to assay the SA-β-gal (Senescence Detection Kit, BioVision, Milpitas, USA). The percentage of senescent cells was calculated in 6 images per sample taken in the bright field microscopy at 100× magnification (Nikon Eclipse TS100 Microscopy, Melville, NY, USA).

Determination of reactive oxygen species (ROS) content with dihydroethidium

Cells were cultured in 12 chamber plates for 4 days (at confluence). Afterwards cells were washed 2 times with pre-warmed PBS medium, 2 µL/mL of diluted dihydroethidium (Dihydroethidium, D7008-Sigma, St. Louis, USA) was added to the plate. Cells were incubated at 37°C for 20 min. After washing the plate with PBS, medium was replaced, and cells cultured for an additional 1 hour at 37°C. The fluorescence was measured using spectraMAX GEMINIS (Molecular Device, Sunnyvale, USA), with 530 nm of excitation wavelength and 610 nm of emission wavelength. Mean fluorescence intensity (MFI) for each cell line, was normalized by the cellular protein content.

Measurement of CuZnSOD and MnSOD activity

To determine MnSOD and CuZnSOD activity the cells were treated as described in the Cayman "Superoxide Dismutase Assay kit" (Ann Arbor, USA). After centrifugation at 10,000 g for 10 min, supernatant was used to measure CuZnSOD activity. The mitochondrial pellet was lysed using a lysis buffer compatible with the manufacturer's instructions (10 mM HEPES, pH7.9, 420 mM NaCl, 1,5 mM MgCl$_2$, 0,5 mM EDTA, 0.1% Triton X-100) for 20 min on ice. After centrifugation at 12,000 g for 5 min, the supernatant was collected for MnSOD activity assay. Measurements of CuZnSOD and MnSOD activities were performed in a 96 well plate prepared using 3–4 replicates from different cellular extracts for each sample. The final absorbance was measured at 450 nm using a spectrophotometer spectraMAXPLUS 384 (Molecular Devices, Sunnyvale, USA).

Measurement of catalase activity

The method for measuring the catalase enzymatic activity was based on the reaction of the enzyme with methanol in the presence of hydrogen peroxide to produce formaldehyde. Cells were lysed using freeze (liquid N$_2$, 10 s) and thaw (ice, 15 min) procedure repeated three times. After centrifugation of the cell lysate at 13,000 g, for 10 min. at 4°C, supernatants were recovered and quantified using Lowry method. A 96 well plate was prepared using at least 4 replicates for each sample, obtained from different cellular extracts.

Assay reaction consisted in mixing on a 96 well plate: 100 µL of phosphate buffer 100 mM pH 7.0; 30 µL methanol and 20 µL of the sample with the same protein concentration. Then, the reaction was started with 20 µL of 85 mM H$_2$O$_2$, maintained during 20 min at room temperature and finally stopped using 30 µL of KOH 10 M. The formaldehyde produced reacts with 35 mM purpald reagent dissolved in 0,5 M HCl during 10 min at room temperature. Finally, 10 µL of 0.5% KIO$_4$ in KOH 0.5 M were added and the absorbance at the wavelength of 540 nm was measured with spectrophotometer spectra MAXPLUS 384 (Molecular Devices, Sunnyvale, USA).

Measurement of glutathione peroxidase activity

Gpx activity was measured by using a glutathione peroxidase assay kit (Cayman (Ann Arbor, USA). Briefly, cells were collected and lysed using cold buffer (50 mM Tris-HCl, pH 7.5, 5 mM EDTA and 1 mM DTT) and two freeze-thaw cycles as described above. The lysates were centrifuged at 10,000 g for 15 min at 4°C and the supernatants recovered in fresh tubes. A 96 well plate was prepared using at least 3 replicates for each sample from different

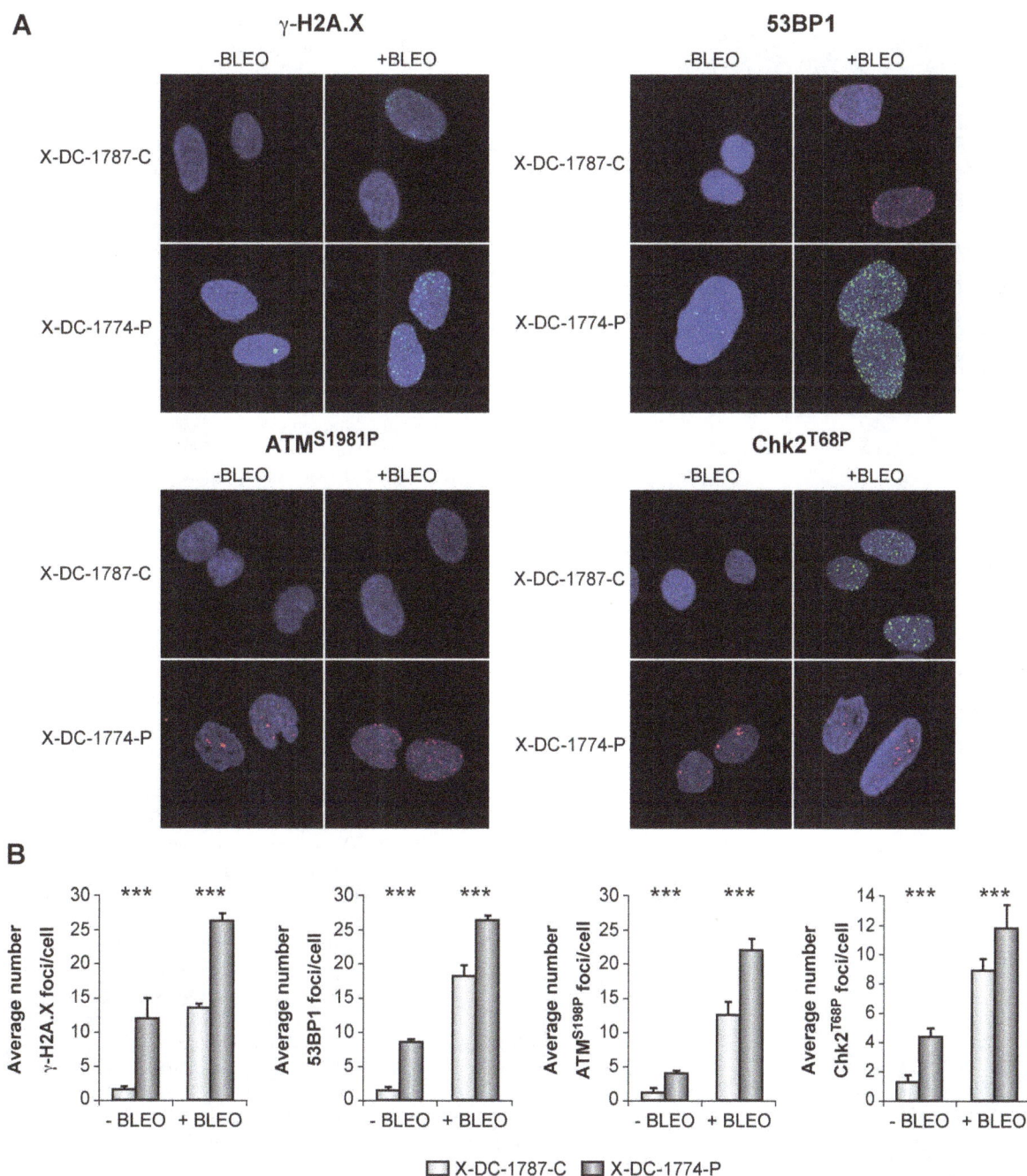

Figure 1. DNA damage signaling in X-DC patient cells. (A) Immunofluorescence staining of DNA damage proteins. Control X-DC-1787-C and patient X-DC-1774-P cells were, either not treated (-Bleo) or treated (+Bleo) with bleomycin (10 μg/ml) for 24 hours, fixed and incubated with antibodies against γ-H2AX, 53BP1, p-ATM or p-CHK1 and secondary fluorescent antibodies. Nuclear DNA was counterstained with DAPI (blue). (B). Quantification of γ-H2A.X foci, pATM, 53BP1 and pCHK2 associated foci in X-DC-1787-C and X-DC-1774-P cells. More than 200 cells were analyzed in each cell line and indicated as the average number of foci/cell. Asterisks indicate significant differences in relation to control cells lines or to untreated cells. Average values and standard deviations of two independent experiments are shown. Experiments were repeated 3 times with similar results.

cellular extracts. After protein quantification by Lowry method, samples containing 20 μg of total proteins were added to the 96 well plate containing a solution with 1 mM GSH, 0.4 U/mL of glutathione reductase, 0.2 mM NADPH. The reaction was initiated by adding 0.22 mM of cumene hydroperoxide and the reduction of the absorbance was recorded at 340 nm each 1 min during 8 min. The Gpx activity was determined by the rate of decrease in absorbance at 340 nm (1 mU/mL Gpx). Molar coefficient extinction for NADPH was 0.00622 $mM^{-1} cm^{-1}$.

Statistical analysis

For the statistical analysis of the results, the mean was taken as the measurement of the main tendency, while standard deviation was taken as the dispersion measurement. T-Student was performed. The significance has been considered at *p<0.05, **

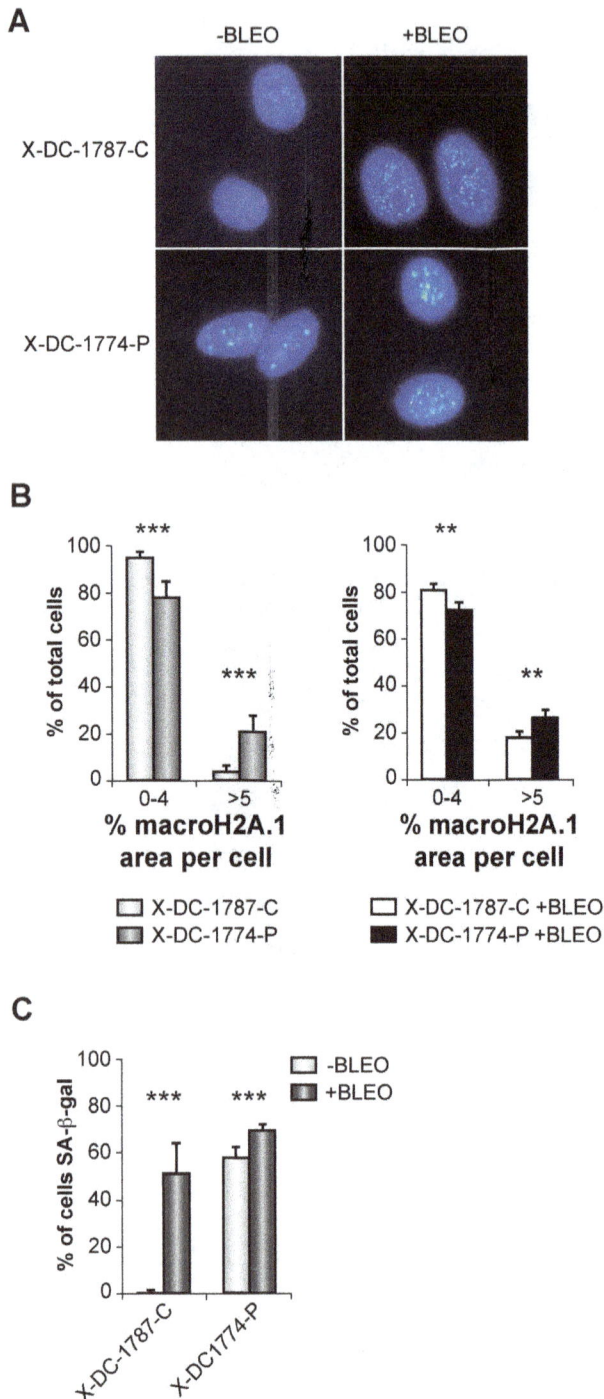

Figure 2. Determination of Histone-macroH2A.1-associated heterochromatin and senescence in X-DC cells. (A) Histone-macroH2A.1-associated heterochromatin detection in X-DC cells. X-DC-1787-C and X-DC-1774-P cells were either not treated (-Bleo) or treated (+Bleo) with bleomycin (10 mg/ml) for 24 hours, fixed and incubated with an antibody against Histone-macroH2A.1 followed by a secondary fluorescence labeled antibody. (B) Quantification of Histone-macroH2A.1-associated heterochromatin. More than 200 cells were analyzed in each cell line and grouped to the area presenting Macro H2A.1 foci per cell. Asterisks indicate significant differences between cells lines. Average values and standard deviations of two independent experiments are shown. (C) SA-β-gal activity in X-DC-1787-C and X-DC-1774-P cells either untreated (-Bleo) or treated (+Bleo) with bleomycin (10 µg/ml). Senescent cells were quantified in 6 images of random regions. Experiments were repeated 3 times with similar results. Asterisks indicate significant differences in response to bleomycin.

for p<0.01 and *** for p<0.001. GraphPad Software v5.0 was used for statistical analysis and graphic representations.

Results

1-Basal and induced DNA damage response in X-DC cells involves 53BP1, ATM and CHK2 and results in increased heterochromatin formation and senescence

It has been previously demonstrated [29] that a pathogenic mutation in murine *Dkc1* causes growth impairment and the enhancement of DNA damage responses after treatment with the chemotherapeutic agent etoposide. In the context of telomeres of normal length, cells with the dyskerin mutation $Dkc1^{\Delta15}$ (deletion of exon 15) showed increased number of DNA damage foci as observed by detection of p-H2A.X^{Ser139} (γ-H2A.X) foci and activation of the ATM/p53 pathway.

We have used paired human cell lines (heterozygous carrier and patient) harboring the same mutation in *DKC* gene, responsible for X-DC and studied the DNA damage response pathway. Telomere length of the control cell line (healthy carrier grandmother from X-DC-1774-P patient) was the right length for the age of this control (60 year old and 10.7 kpb). Both basal DNA damage and that produced in response to the DNA damaging agent bleomycin were studied. Our results show that the number of γ-H2A.X-associated foci/cell was dramatically higher in cells obtained from the X-DC-1774-P patient than in the carrier cell line X-DC-1787-C (Fig. 1A and B). When cells were treated with bleomycin, which induces double strand breaks, we found an increase in the number of γ-H2A.X associated foci/cell in both X-DC-1774-P and X-DC-1787-C cells. Although basal DNA damage in X-DC-1774-P was already much higher than that of control cells the increase was similar or even lower to that observed in control cells. We also investigated the presence of 53BP1 foci in these cell lines, since 53BP1 is recruited to DNA-damage associated foci. We found the average number of foci/cell was similar to that observed for γ-H2A.X, higher to that observed in control cells but even if there is an increase after bleomycin treatment, the increase in the number of foci/cell was smaller than control cells ATM protein is also recruited to DNA-damage sites at the chromatin and phosphorylated, we found that X-DC-1774-P cells showed higher number of foci/cell with phosphorylated ATM compared to carrier cells. In bleomycin treated cells both patient and carrier cells showed increased response to DNA damage although similar to what happen with the other indicators of DNA damage the increase observed in X-DC-1774-P was lower than control cells. CHK2 is a protein, substrate of ATM-kinase. We studied the number of cells with phosphorylated CHK2 at Thr68 and found a higher number of foci/cells in X-DC-1774-P in untreated cells, which increases after bleomycin treatment, but such increase is lower than control cells. Altogether these results indicate that basal DNA damage is higher in X-DC patient cells that in mutation carrier cells, in response to bleomycin this increase is not higher in X-DC cells probably due to the high basal damage observed in these cells. Furthermore we have found that the signaling pathway associated with this DNA damage, include at least 53BP1, ATM and CHK2, although we cannot exclude the participation of other proteins.

In order to verify if X-DC cells harbor an increased heterochromatin content we studied the nuclear distribution of histone-macroH2A.1-associated heterochromatin in X-DC-1774-P and X-DC-1787-C cells, both in basal conditions and after bleomycin treatment (Fig. 2A). X-DC-1774-P cells already showed an average of 20% of the nuclear area with positive expression for macroH2A.1, and after bleomycin treatment we detected an increase up to 30% (Fig. 2B). X-DC-1787-C cells showed a very

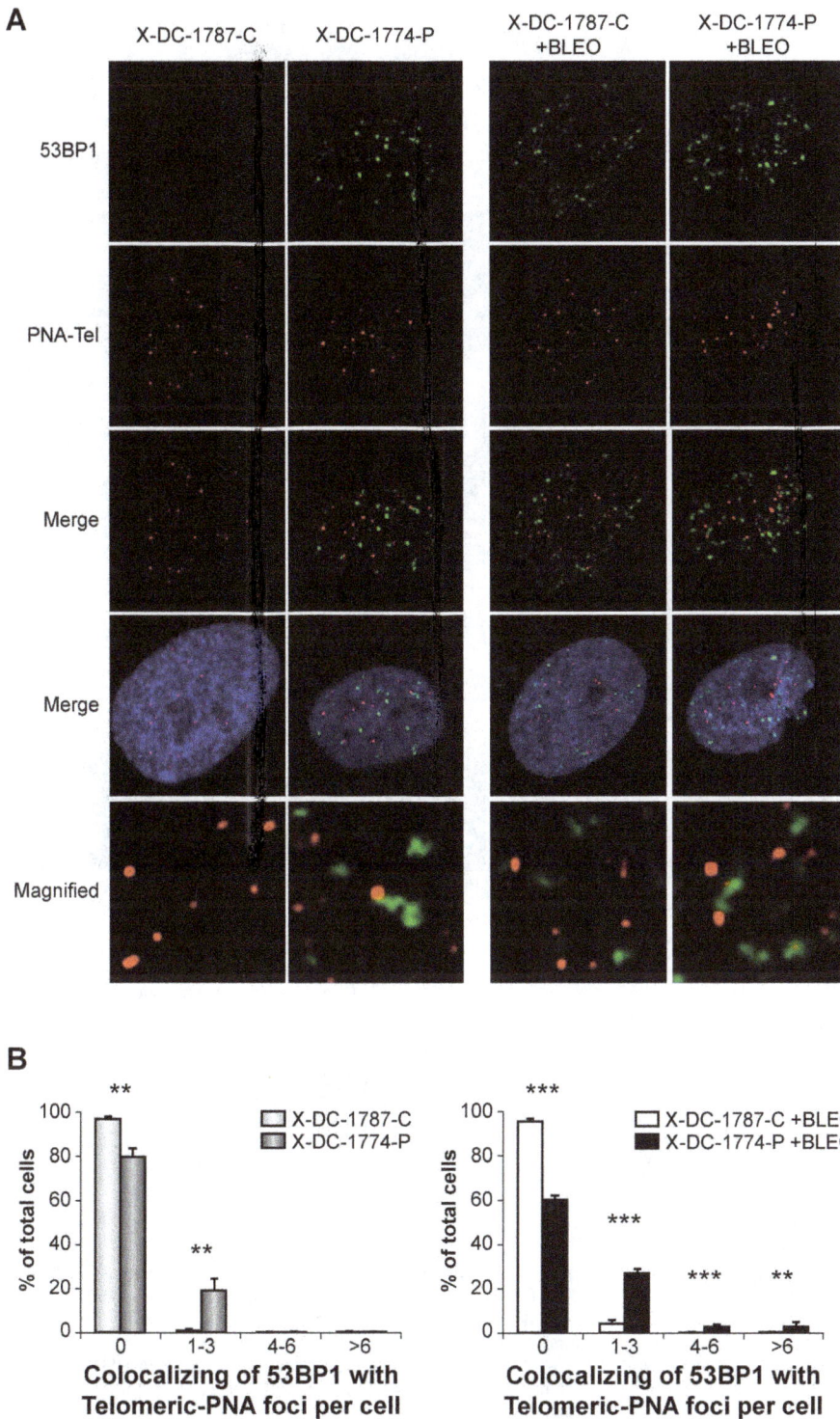

Figure 3. Localization of 53BP1 foci to telomeres in X-DC patient cells. X-DC-1787-C and X-DC-1774-P cells untreated (-Bleo) or treated (+ Bleo) with bleomycin (10 µg/ml) and incubated with γ-H2A.X and PNA-FISH probe. (A) Colocalization of 53BP1 foci (green) and telomeres as identified by hybridizing with a PNA-FISH probe (red). DNA was counterstained with DAPI (blue). Magnified views of merged images showing details of the colocalization are shown in the two lower series of panels (B) Colocalized 53BP1 foci and PNA-FISH probe at telomeres was quantified. More than 200 cells were analyzed in each cell line in an experiment performed three times with similar results. Asterisks indicate significant differences in relation to different cell lines.

low expression in basal conditions that increases to almost 20% after bleomycin treatment (Fig. 2C). These data indicated that X-DC patient cells show extensive areas of heterochromatin that

further increased in response to bleomycin. Thus, both basal and induced DNA damage may trigger a relevant silencing of gene expression in these cells. Almost 60% of X-DC-1774-P cells were

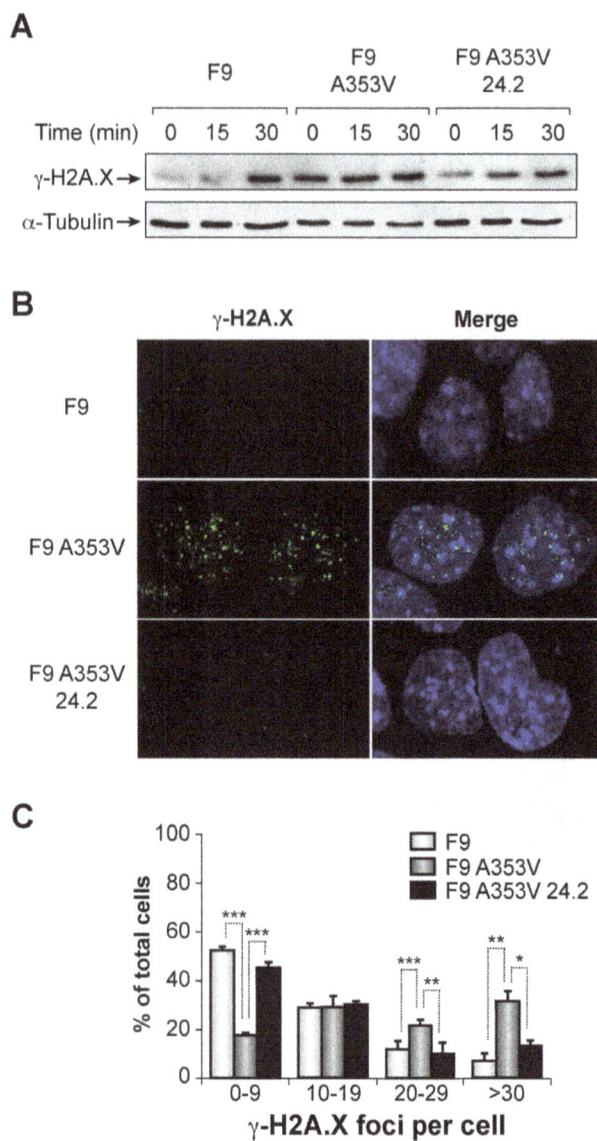

Figure 4. F9A353V cells show enhanced, basal and bleomycin induced, DNA damage response. (A) F9, F9A353V and F9A353V cells transfected with GSE24.2 (10 μg DNA per million cells). F9A353V 24.2 cells were treated with bleomycin (10 μg/ml). After 0, 15 or 30 minutes of treatment cells were lysed and the experiment analyzed by western blot with antibodies against γ-H2A.X or α-tubulin as a loading control. (B) Immunofluorescence staining of γ-H2A.X (green) in F9, F9A353V and F9A353V 24.2 cells (10 μg DNA per million cells). Nuclear DNA was counterstained with DAPI (blue). (C) Quantification of γ-H2AX foci in F9, F9A353V or F9A353V 24.2 cells. More than 200 cells were analyzed in each cell line and grouped to the number of γ-H2A.X foci observed per cell. Experiments were repeated 3 times with similar results. Asterisks indicate significant differences in relation to different cell lines.

positive for the senescence SA-β-gal activity that increases to almost 70% after bleomycin treatment. X-DC-1787-C cells showed low expression of SA-β-gal that also increases further after bleomycin treatment.

2- DNA damage is localized in telomeres in X-DC cells

Since telomere length is greatly diminished in X-DC patient cells we investigated if DNA damage was enriched at telomeres,

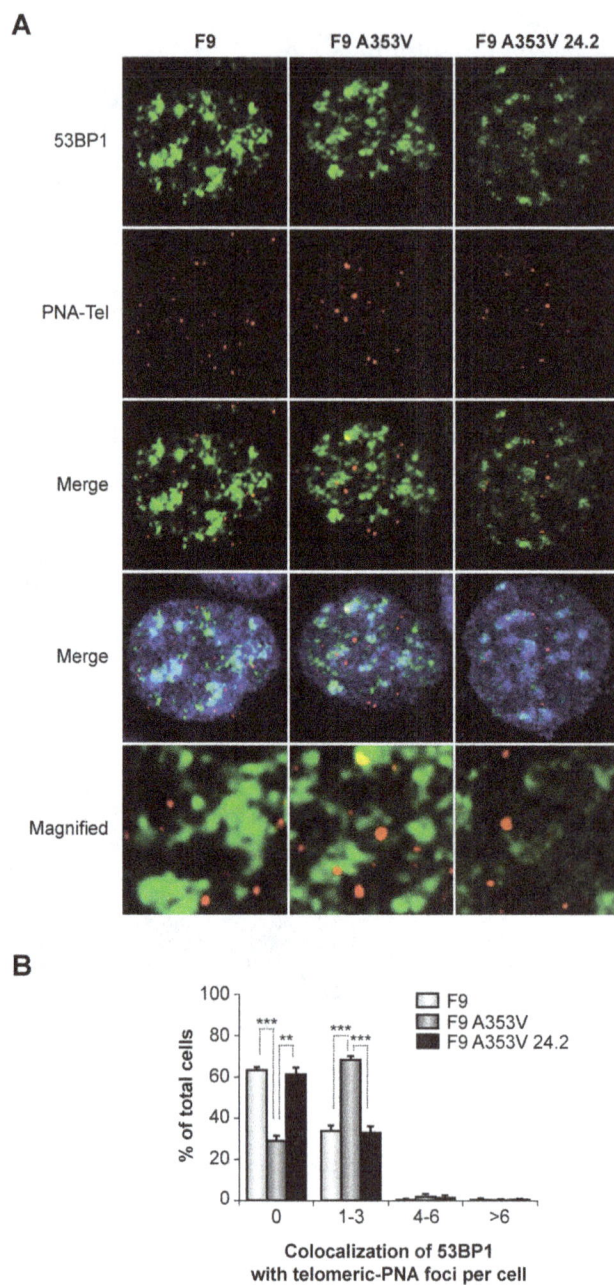

Figure 5. Localization of 53BP1 foci to telomeres in F9A353V cells. F9, F9A353V and F9A353V cells transfected with GSE24.2 (F9A353V 24.2) (F9 cells were treated with bleomycin,10 μg/ml for 24 hours) and incubated with 53BP1 antibodies and with a PNA-FISH probe. (A) Colocalization of 53BP1 foci (green) and PNA-FISH probe that identified telomeres (PNA-Tel, red). DNA was counterstained with DAPI (blue). Magnified views of merged images showing details of the colocalization are shown in the lower panels. (B) Quantification of the colocalization of 53BP1 foci and telomere signals shown in panel A. More than 200 cells were analyzed in each cell line and grouped to the number of 53BP1 foci associated to telomeres (PNA-Tel) per cell. Experiments were repeated 3 times with similar results. Asterisks indicate significant differences in relation to different cell lines.

both in basal conditions and after DNA damage induction. In order to investigate this, we combined a PNA FISH probe as a telomere marker, and 53BP1 for DNA damage detection. The results showed that there was a high association of damaged DNA

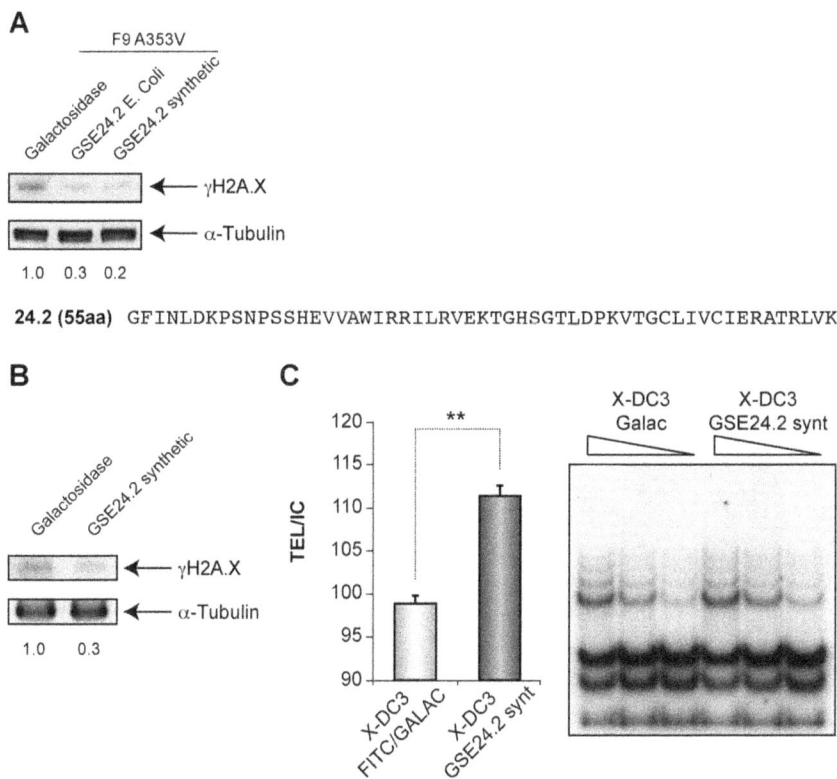

24.2 (55aa) GFINLDKPSNPSSHEVVAWIRRILRVEKTGHSGTLDPKVTGCLIVCIERATRLVK

Figure 6. Activity of the GSE24.2 peptide expressed in bacteria or chemically-synthesized. (A) F9A353V cells were transfected with 15 μg of β-galactosidase as a control (galactosidase), or GSE24.2 purified from E. Coli (GSE24.2E.coli) or obtained by chemical synthesis (GSE24.2 synthetic). After 24 hours cells were lysed and the levels of γ-H2AX and α-tubulin determined by western blot. The values at the bottom were obtained after quantification of the blot and show the ration between expression levels of γ-H2AX and α-tubulin in each line and referred to those found in β-galactosidase transfected cells. (B) Same experiment described in A, performed in X-DC3 cells transfected with β-galactosidase or chemically synthesized GSE24.2. (C) Reactivation of telomerase activity by chemically synthesized GSE24.2. X-DC3cells were transfected with β-galactosidase or chemically synthesized GSE24.2 and telomerase activity determined by TRAP assay (right). Different amounts extract were used for each TRAP assay as indicated. The activity was quantified by evaluating the intensity of the bands in relation with the internal control (TEL/IC) (left panel). The values for GSE24.2 transfected cells were referred to the β-galactosidase transfected cells. The experiments were repeated at least three times with similar results. Asterisks indicate significant differences between the two different transfected peptides.

at the telomeres in X-DC-1774-P cells that was not found in carrier X-DC-1787-C cells (Fig. 3A). Furthermore the increase in DNA damage observed after bleomycin treatment (Fig. 1B) was strongly associated with telomeres in X-DC-1774-P cells in contrast to X-DC-1787-C cells (Fig. 3) indicating the relevance of telomere shortening in the response to DNA damage in X-DC patient cells.

3-Expression of GSE24.2 impairs the induction of γ-H2A.X foci after DNA damage

Since F9 cells represent a good model system to study DNA damage responses as previously demonstrated (29), we used them in order to investigate if the expression of GSE24.2 could modify the activation of the DNA damage response. Therefore, we transfected F9A353V and control F9 cells [26] with the GSE24.2 expression plasmid and treated them either with bleomycin or etoposide, a topoisomerase inhibitor known to induce DNA double-stranded breaks. We found that, as expected, bleomycin treatment induced γ-H2A.X in both cell lines (Fig. 4A). However, the basal level of γ-H2A.X was much higher in F9A353V cells than in F9 cells expressing the WT dyskerin, indicating that the mutation renders the cells more susceptible to DNA damage. In the presence of the GSE24.2 F9A353V, γ-H2A.X decreased to values very similar to those observed in F9 cells in both, basal and

bleomycin-induced levels. Similar results were obtained in etoposide-treated cells (data not shown). We next investigated the presence of γ-H2A.X containing foci in basal conditions and the results confirmed those obtained in the western blot studies (Fig. 4B and 4C). Most F9 cells showed very few foci; the number increased in F9A353V cells but was reduced at similar level to those of F9 cells when the mutant cells were transfected with GSE24.2. Altogether, the results indicated that the expression of GSE24.2 decreases the DNA damage produced by the dyskerin mutation.

Afterwards, we investigated if the increased DNA damage in F9A353V cells was enriched at the telomeres (as already found in X-DC patient cells, Fig. 3) and also whether the protection from DNA damage induced by GSE24.2 also applies to damage at the telomeres. We use combined immunological detection of 53BP1 and PNA-FISH probe. The results (Fig. 5A and 5B) indicated that F9A353V cells have a stronger association of 53BP1 to the telomeres than in F9 cells treated with bleomycin, up to 60%. However in F9A353V cells transfected with the GSE24.2 there is little association of 53BP1 foci at the telomeres (30% 1–3 53BP1 foci per cell). These results indicate that the elevated DNA damage response found in F9A353V cells is probably caused by defects at the telomeres induced by the *Dkc1* mutation, in agreement with the results obtained in *Dkc1*$^{\Delta 15}$ MEF cells. Interestingly, expression

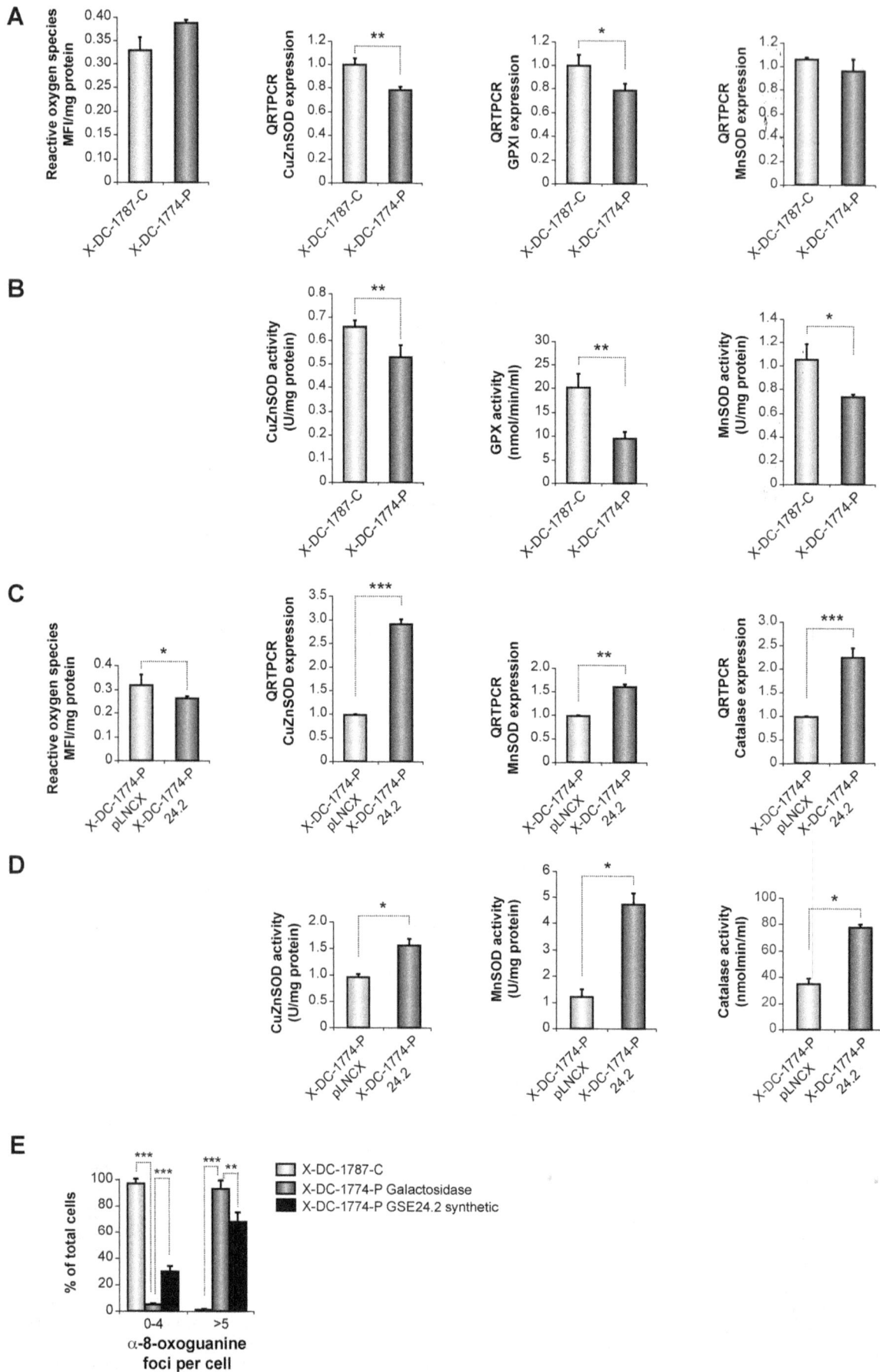

Figure 7. Oxidative stress analysis in X-DC fibroblasts after GSE24.2 transfection. (A) ROS levels were determined in fibroblasts from the carrier DC1787, and fibroblasts from the patient X-DC1774-P. Levels were determined using the fluorescent probe dihydroethidium in confluent cells (left panel). RNA expression was determined for CuZnSOD, MnSOD, and GPX1 by qRT-PCR (A right panels). B) Enzymatic activities of CuZnSOD,MnSOD, and Glutathione peroxidase 1 were also determined. C) ROS levels were studied in X-DC1774-P fibroblasts (expressing pLNCX vector) and X-DC-1774-P cells expressing GSE24.2 (X-DC1774-PGSE24.2, left panel). Cu/ZnSOD, MnSOD, and catalase expression levels were determined by qRT-PCR. D) Cu/ZnSOD, MnSOD, and catalase activities in confluent pLNCX and 24.2 cells are shown in left panels. E) X-DC1774-P and X-DC1774-PGSE24.2, cells were transfected with GSE24.2 synthetic peptide and levels of 8-oxoguanine studied by immunofluorescence. The 8-oxoguanine foci signal was expressed as the average number of foci/cell in 200 cells. Results are expressed as mean ± standard deviation from three independent experiments. Statistical significance is expressed as (*) p<0.05.

of GSE24.2 reverted the telomere damage in F9A353V cells, indicating its biological importance in the reversion of the mutant *Dkc1* phenotype.

4-Treatment of X-DC cells with GSE24.2 peptide rescues DNA damage

We have previously reported that the GSE24.2 peptide purified from bacteria was able to increase telomerase activity in F9A353V cells [26] therefore we next tested if the activity of the GSE24.2 peptide either purified from E-coli or chemically synthesized reduced the DNA damage. We found that the levels of γ-H2A.X in F9A353V cells decreased after transfecting this peptide (Fig. 6A) either obtained from bacteria or chemically synthesized to 30 and 20%, respectively. Moreover the synthetic peptide also decreased the DNA damage in X-DC3 cells (DKC1 mutated lymphocytes) by 30% (Fig. 6B). This decrease in DNA damage correlated well with the ability of the synthetic peptide to increase telomerase activity in these cells (Fig. 6C).

5- Oxidative stress in X-DC cells is decreased by expression of GSE24.2

Oxidative stress is one of the causes of DNA damage producing both single-strand breaks (SSBs) and double-strand breaks (DSBs). SSBs are the result from the interaction of hydroxyl radicals with deoxyribose and subsequent generation of peroxyl-radicals. These reactive oxygen species (ROS) are then responsible for nicking phosphodiester bonds that form the backbone of each helical strand of DNA [34]. To clarify the presence of higher oxidation levels in X-DC cells we have studied ROS levels, and the expression of antioxidant enzymes CuZn (SOD1) and Mn (SOD2) superoxide dismutase, glutathione peroxidase 1 (GPX1) and their corresponding enzymatic activities in X-DC-1787-C and X-DC-1774-P cell lines. Levels of ROS were elevated in X-DC-1774-P cells compared with X-DC-1787-C carrier cells and also higher than in GM03348, an age-matched cell line from a healthy subject (data not shown). In agreement with this result we found a decrease in gene expression levels of the antioxidant enzymes CuZnSOD and MnSOD and GPX1 when compared the X-DC-1774-P to the carrier cell line (Fig. 7A). We also determined the activity of the three enzymes with decreased expression in the X-DC-1774-P cells that also showed decreased activity in agreement with the gene expression data (Fig. 7B).

In order to investigate if expression of GSE24.2 was able to overcome the increased oxidative stress found in X-DC-1774-P cells, we expressed in this cell line either pLNCX-GSE24.2 or the empty vector (pLNCX). The results indicated that X-DC-1774-P cells expressing GSE24.2 showed lower levels of ROS. We also studied the expression levels of CuZnSOD, MnSOD and catalase in both cell lines and found that expression levels of these antioxidant enzymes were higher in X-DC-1774-P-24.2. When the corresponding protein activities were analyzed, we observed an increase in CuZnSOD, MnSOD and catalase activities (Fig. 7C) in X-DC-1774-P-24.2 when compared with the empty vector

transfected cells. Altogether, the data indicated that the observed decrease in oxidative stress in X-DC cells expressing GSE24.2 should contribute to protect these cells from DNA damage. We finally investigated if treatment with the GSE24.2 synthetic peptide was also able to induce a decrease in oxidative DNA damage. We transfected X-DC-1774-P cells with the GSE24.2 synthetic peptide and evaluated the levels of 8-oxoguanine by immunofluorescence (Fig 7E). The results showed that indeed the synthetic GSE24.4 reduced the signal obtained with 8-oxoguanine-antibody.

Discussion

We have previously reported that expression of a *dyskerin* internal peptide (GSE24.2) reactivates telomerase activity in cells that are deficient in this activity by increasing TERT and TERC levels [24] We have also reported that expression of GSE24.2 increases TR levels by stabilizing this RNA [26]. Because of this activity GSE24.2 has been recently approved as an orphan drug by EMA for the treatment of Dyskeratosis congenita. We have now studied the role of GSE24.2 in the DNA damage response of X-DC patient cells in an effort to better understand the mechanism of GSE24.2 action in X-DC. We studied several proteins involved in the DNA-damage response and found, as other authors have [35] [36], that X-DC patient cells presented higher levels of DNA damage associated foci detected by γ-H2A.X and 53BP1 and to a lesser extent p-ATM and p-CHK2. We also found increased levels of DNA damage in response to bleomycin that was more evident when we studied -H2A.X, p-ATM and p-CHK2 associated foci as previously described in mice (29) but this increase was not higher than that obtained in control cells probably because X-DC cells already have massive damage in basal conditions. Previous reports described increased levels of DNA damage in DC cells harboring mutations in *DKC1, TERC* or *TERT*. However in fibroblasts and lymphocytes from these patients the response to induced DNA-damage was not increased [35] in contrast to another study [29]. We have here used X-DC patient cells which exhibited short telomeres, p53 activation and senescence [37]. Indeed, a high level of DNA damage, both at basal and induced by bleomycin, was observed at telomeres suggesting that the shortening of telomeres in these cells induces further damage by preventing repair. Dysfunctional telomeres trigger a DNA damage response most likely because they are too short to adopt the normal t-loop structure needed to form the telomere with correctly ordered shelterin components. Recruitment of histone-macroH2A.1 has been associated to heterochromatin and senescent associated foci (SAHF) [18] [19]. We found that both senescence and macroH2A.1 associated-foci are increased in X-DC patient cells and also that bleomycin treatment increases these values, suggesting that the impairment in the repair of DNA lesions in X-DC cells likely contributes to the senescent phenotype.

Using the *in vitro* generated *Dkc1* mutant F9A353V cells we have found, in agreement with our previous results (and also [29]), that

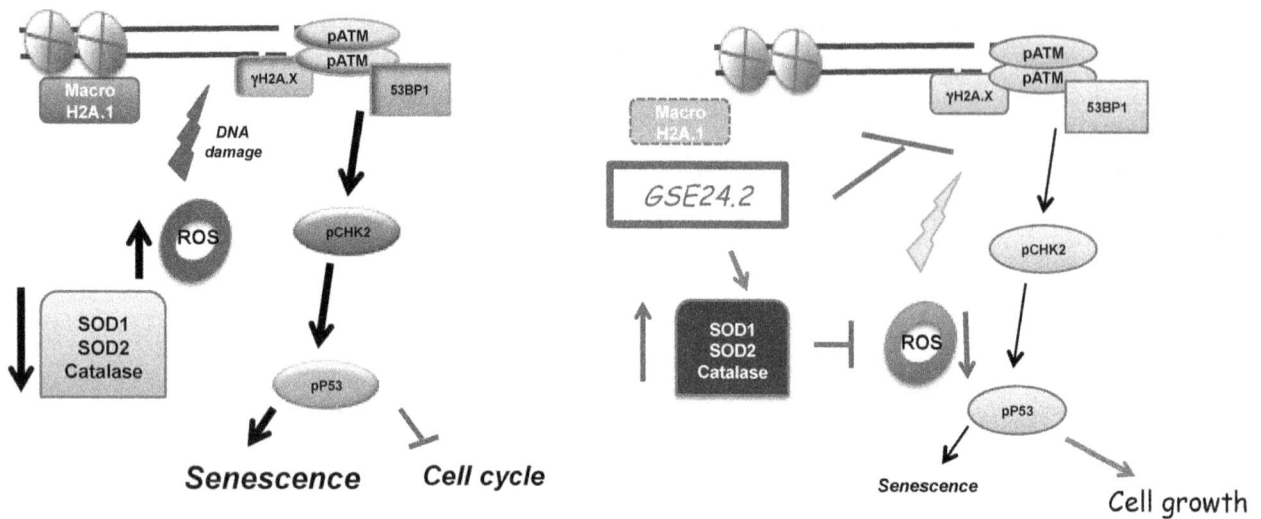

Figure 8. Proposed biological activity of GSE24.2 on DNA damage and oxidative stress. Dyskeratosis congenita cells display high basal DNA damage detected by increased γH2AX, p-ATM, p-CHK2 and 53BP1 foci. Additionally there are increased levels of ROS, and decreased expression and activity of antioxidant enzymes resulting in higher oxidative damage and senescence (left panel). GSE24.2 peptide increases expression of antioxidant enzymes and as a consequence decreased ROS levels (right panel). In parallel there is increased telomerase activity that may help to decrease global and telomeric DNA damage. Globally these two activities of GSE24.2 might result in increased viability and growth of DC cells (26).

these cells showed increased DDR compared with F9 cells, both in the steady state and when treated with bleomycin or etoposide. Other *Dkc1* mutations such as *Dkc1*$^{\Delta15}$ have been shown to accumulate DNA damage indicating that DC cells have cellular defects even in the context of long telomeres [29]. We previously reported that an internal fragment of Dyskerin, the peptide GSE24.2 induces an increase in telomerase activity in X-DC cells [24]. Now we are showing that expression of GSE24.2 is able to induce protection against DNA damage. Furthermore, the repair of pre-existing DNA lesions should also take place at telomeres in F9A353V cells as shown by the decrease in 53BP1 and PNA-FISH telomeric colocalization (Fig. 5B). Interestingly, the observed decrease in DNA damage mediated by GSE24.2 expression in F9A353V cells, also occurs when we used either bacterially produced or chemically synthesized peptide, reinforcing the idea that GSE24.2 reactivates telomerase activity, by acting directly at the telomeric DNA [26] and/or changing telomere folding. According with these results the transfection of the GSE24.2 synthetic peptide into X-DC3 human patient lymphocytes resulted in both increased telomerase activity and decreased DNA damage. On the other hand the consequences of A353V-X-DC mutation on DNA damage resemble to those found in cells with mutations in *Tin2* and *Pot1*, which are structural components of telomeres [38] [39].

Diseases with telomerase deficiency are linked to oxidative stress. Elevated levels of the lipoperoxide malondialdehide (MDA) [40], and MDA-DNA adducts have been reported in rare degenerative diseases [41] and in aging [42]. In addition, oxidative stress conditions caused by H_2O_2 increased the rate of telomere shortening in fibroblasts from ataxia-telangiectasia patients [43]. Furthermore, increased accumulation of ROS is involved in decreased cell growth in a *DKC*$^{\Delta15}$ mouse model [36], though there is very little information about oxidative stress in human X-DC cells. Interestingly, the existence of oxidative stress in lymphocytes from patients with an autosomal dominant form of DC with mutations in *TERC* has been recently reported [44]. To further increase the characterization of the oxidative stress profile in X-DC we characterize the levels of ROS and the expression

and activity of the main antioxidant enzymes. We found in X-DC-1774-P an increase in ROS levels and a decrease in the expression and activity of antioxidant enzymes in patient cells when compared to carrier cells. Interestingly expression of GSE24.2 results in an increase in SOD1, SOD2 and catalase expression that might decrease ROS levels in X-DC-1774-P 24.2 cells. Different groups [45] [46] [47] [48] have reported decreased cellular ROS levels in stressed hTERT over-expressing cells, demonstrating that telomerase re-expression contributes to decrease oxidative stress [49]. The work by Westin et al. demonstrated that the reduction of the levels of superoxide in DC cells was not dependent of the localization of TERT in the mitochondria, but also p53/p21$^{WAF/CIP}$-dependent process in the context of telomere shortening in cells from DC patients. Therefore, our findings reinforce the notion that increased telomerase activity [24] and repair of DNA damage at telomeres induced by GSE24.2 is concomitant with a decrease in oxidative stress in X-DC cells. Alternatively the decreased DNA damage detected by γ-H2A.X, might corresponds to decreased oxidative damage, in agreement of our results evaluating the levels of 8-oxoguanine that decreased after transfection of the GSE24.2 synthetic peptide.

In summary our results show that, GSE24.2 attenuates the impact of the *DKC1* mutations on DNA damage and its incidence on telomeres (Fig. 8). Furthermore, oxidative stress decreases in GSE24.2 expressing cells, and this should contribute to decrease the rate of DNA damage and therefore enable restoration of cell cycle progression. Indeed, we have previously shown that expression of GSE24.2 X-DC fibroblasts restores proliferation [26].

Since GSE24.2 has been approved as an orphan drug for the treatment of DC, the results presented here indicate that expression of GSE24.2 may form the basis of a useful and safe therapeutic strategy for X-DC patients either by using it as a permanent or as a temporal telomerase activator. These results indicate that GSE24.2 expression has a broad effect on DC cells, reducing oxidative stress and DNA damage in addition to reactivating telomerase activity. All these protective effects could cooperatively contribute to increase DC cells survival and proliferation [26] and give further support to the recent approval of GSE24.2 as an orphan drug for DC treatment.

Acknowledgments

We thank to P. Mason for the F9 cell mouse model and for the critical review of the manuscript.

Author Contributions

Conceived and designed the experiments: CMG LPB JC RMP LS CPQ IE AG JLGG FVP RP. Performed the experiments: CMG LPB JC RMP LS CPQ IE AG JLGG FVP RP. Analyzed the data: CMG LPB JC RMP LS CPQ IE AG JLGG FVP RP. Contributed reagents/materials/analysis tools: CMG LPB JC RMP LS CPQ IE AG JLGG FVP RP. Wrote the paper: CMG RMP LS JLGG FVP RP.

References

1. Palm W, de Lange T (2008) How shelterin protects mammalian telomeres. Annu Rev Genet 42: 301–334.
2. Osterhage JL, Friedman KL (2009) Chromosome end maintenance by telomerase. J Biol Chem 284: 16061–16065.
3. Cohen SB, Graham ME, Lovrecz GO, Bache N, Robinson PJ, et al. (2007) Protein composition of catalytically active human telomerase from immortal cells. Science 315: 1850–1853.
4. Meier UT, Blobel G (1994) NAP57, a mammalian nucleolar protein with a putative homolog in yeast and bacteria. J Cell Biol 127: 1505–1514.
5. Ni J, Tien AL, Fournier MJ (1997) Small nucleolar RNAs direct site-specific synthesis of pseudouridine in ribosomal RNA. Cell 89: 565–573.
6. Yang Y, Isaac C, Wang C, Dragon F, Pogacic V, et al. (2000) Conserved composition of mammalian box H/ACA and box C/D small nucleolar ribonucleoprotein particles and their interaction with the common factor Nopp140. Mol Biol Cell 11: 567–577.
7. Decatur WA, Fournier MJ (2002) rRNA modifications and ribosome function. Trends Biochem Sci 27: 344–351.
8. Heiss NS, Knight SW, Vulliamy TJ, Klauck SM, Wiemann S, et al. (1998) X-linked dyskeratosis congenita is caused by mutations in a highly conserved gene with putative nucleolar functions. Nat Genet 19: 32–38.
9. Kirwan M, Dokal I (2008) Dyskeratosis congenita: a genetic disorder of many faces. Clin Genet 73: 103–112.
10. de Lange T (2009) How telomeres solve the end-protection problem. Science 326: 948–952.
11. Martinez P, Blasco MA (2010). Role of shelterin in cancer and aging. Aging Cell 9: 653–666.
12. Tejera AM, Stagno d'Alcontres M, Thanasoula M, Marion RM, Martinez P, et al. (2010).TPP1 is required for TERT recruitment, telomere elongation during nuclear reprogramming, and normal skin development in mice. Dev Cell 18: 775–789.
13. Martinez P, Flores JM, Blasco MA (2012) 53BP1 deficiency combined with telomere dysfunction activates ATR-dependent DNA damage response. J Cell Biol 197: 283–300.
14. Rappold I, Iwabuchi K, Date T, Chen J (2001) Tumor suppressor p53 binding protein 1 (53BP1) is involved in DNA damage-signaling pathways. J Cell Biol 153: 613–620.
15. Fernandez-Capetillo O, Chen HT, Celeste A, Ward I, Romanienko PJ, et al. (2002) DNA damage-induced G2-M checkpoint activation by histone H2AX and 53BP1. Nat Cell Biol 4: 993–997.
16. Dimitrova N, Chen YC, Spector DL, de Lange T (2008) 53BP1 promotes non-homologous end joining of telomeres by increasing chromatin mobility. Nature 456: 524–528.
17. Perona R, Moncho-Amor V, Machado-Pinilla R, Belda-Iniesta C, Sanchez Perez I (2008) Role of CHK2 in cancer development. Clin Transl Oncol 10: 538–542.
18. Zhang R, Chen W, Adams PD (2007) Molecular dissection of formation of senescence-associated heterochromatin foci. Mol Cell Biol 27: 2343–2358.
19. Xu C, Xu Y, Gursoy-Yuzugullu O, Price BD (2012).The histone variant macroH2A1.1 is recruited to DSBs through a mechanism involving PARP1. FEBS Lett 586: 3920–3925.
20. Hartwig FP, Collares T (2013). Telomere dysfunction and tumor suppression responses in dyskeratosis congenita: balancing cancer and tissue renewal impairment. Ageing Res Rev 12: 642–652.
21. Young NS (2012). Bone marrow failure and the new telomere diseases: practice and research. Hematology 17 Suppl 1: S18–21.
22. Stewart JA, Chaiken MF, Wang F, Price CM (2012). Maintaining the end: roles of telomere proteins in end-protection, telomere replication and length regulation. Mutat Res 730: 12–19.
23. Alter BP, Giri N, Savage SA, Rosenberg PS (2009) Cancer in dyskeratosis congenita. Blood 113: 6549–6557.
24. Machado-Pinilla R, Sanchez-Perez I, Murguia JR, Sastre L, Perona R (2008) A dyskerin motif reactivates telomerase activity in X-linked dyskeratosis congenita and in telomerase-deficient human cells. Blood 111: 2606–2614.
25. Zeng XL, Thumati NR, Fleisig HB, Hukezalie KR, Savage SA, et al. (2012).The accumulation and not the specific activity of telomerase ribonucleoprotein determines telomere maintenance deficiency in X-linked dyskeratosis congenita. Hum Mol Genet 21: 721–729.
26. Machado-Pinilla R, Carrillo J, Manguan-Garcia C, Sastre L, Mentzer A, et al. (2012). Defects in mTR stability and telomerase activity produced by the Dkc1 A353V mutation in dyskeratosis congenita are rescued by a peptide from the dyskerin TruB domain. Clin Transl Oncol 14: 755–763.

27. Knight SW, Heiss NS, Vulliamy TJ, Greschner S, Stavrides G, et al. (1999) X-linked dyskeratosis congenita is predominantly caused by missense mutations in the DKC1 gene. Am J Hum Genet 65: 50–58.
28. Vulliamy TJ, Marrone A, Knight SW, Walne A, Mason PJ, et al. (2006) Mutations in dyskeratosis congenita: their impact on telomere length and the diversity of clinical presentation. Blood 107: 2680–2685.
29. Gu BW, Bessler M, Mason PJ (2008) A pathogenic dyskerin mutation impairs proliferation and activates a DNA damage response independent of telomere length in mice. Proc Natl Acad Sci U S A 105: 10173–10178.
30. Kalinin A, Thoma NH, Iakovenko A, Heinemann I, Rostkova E, et al. (2001) Expression of mammalian geranylgeranyltransferase type-II in Escherichia coli and its application for in vitro prenylation of Rab proteins. Protein Expr Purif 22: 84–91.
31. Mochizuki Y, He J, Kulkarni S, Bessler M, Mason PJ (2004) Mouse dyskerin mutations affect accumulation of telomerase RNA and small nucleolar RNA, telomerase activity, and ribosomal RNA processing. Proc Natl Acad Sci U S A 101: 10756–10761.
32. Wright WE, Shay JW, Piatyszek MA (1995) Modifications of a telomeric repeat amplification protocol (TRAP) result in increased reliability, linearity and sensitivity. Nucleic Acids Res 23: 3794–3795.
33. Sanchez-Perez I, Murguia JR, Perona R (1998) Cisplatin induces a persistent activation of JNK that is related to cell death. Oncogene 16: 533–540.
34. Taghizadeh K, McFaline JL, Pang B, Sullivan M, Dong M, et al. (2008) Quantification of DNA damage products resulting from deamination, oxidation and reaction with products of lipid peroxidation by liquid chromatography isotope dilution tandem mass spectrometry. Nat Protoc 3: 1287–1298.
35. Kirwan M, Beswick R, Walne AJ, Hossain U, Casimir C, et al. (2011). Dyskeratosis congenita and the DNA damage response. Br J Haematol 153: 634–643.
36. Gu BW, Fan JM, Bessler M, Mason PJ (2011) Accelerated hematopoietic stem cell aging in a mouse model of dyskeratosis congenita responds to antioxidant treatment. Aging Cell 10: 338–348.
37. Carrillo J, Gonzalez A, Manguan-Garcia C, Pintado-Berninches L, Perona R (2013). p53 pathway activation by telomere attrition in X-DC primary fibroblasts occurs in the absence of ribosome biogenesis failure and as a consequence of DNA damage. Clin Transl Oncol. [Epub ahead of print]
38. Walne AJ, Vulliamy T, Beswick R, Kirwan M, Dokal I (2008) TINF2 mutations result in very short telomeres: analysis of a large cohort of patients with dyskeratosis congenita and related bone marrow failure syndromes. Blood 112: 3594–3600.
39. Hockemeyer D, Palm W, Wang RC, Couto SS, de Lange T (2008) Engineered telomere degradation models dyskeratosis congenita. Genes Dev 22: 1773–1785.
40. Ahamed M, Kumar A, Siddiqui MK (2006) Lipid peroxidation and antioxidant status in the blood of children with aplastic anemia. Clin Chim Acta 374: 176–177.
41. Patel KJ, Joenje H (2007) Fanconi anemia and DNA replication repair. DNA Repair (Amst) 6: 885–890.
42. Voss P, Siems W (2006) Free Radic Res. 40(12):1339–49
43. Tchirkov A, Lansdorp PM (2003) Role of oxidative stress in telomere shortening in cultured fibroblasts from normal individuals and patients with ataxia-telangiectasia. Hum Mol Genet 12: 227–232.
44. Pereboeva L, Westin E, Patel T, Flaniken I, Lamb L, et al. (2013). DNA damage responses and oxidative stress in dyskeratosis congenita. PLoS One 8: e76473.
45. Ahmed S, Passos JF, Birket MJ, Beckmann T, Brings S, et al. (2008) Telomerase does not counteract telomere shortening but protects mitochondrial function under oxidative stress. J Cell Sci 121: 1046–1053.
46. Kang HJ, Choi YS, Hong SB, Kim KW, Woo RS, et al. (2004) Ectopic expression of the catalytic subunit of telomerase protects against brain injury resulting from ischemia and NMDA-induced neurotoxicity. J Neurosci 24: 1280–1287.
47. Saretzki G (2009) Telomerase, mitochondria and oxidative stress. Exp Gerontol 44: 485–492.
48. Saretzki G, Murphy MP, von Zglinicki T (2003) MitoQ counteracts telomere shortening and elongates lifespan of fibroblasts under mild oxidative stress. Aging Cell 2: 141–143.
49. Westin ER, Aykin-Burns N, Buckingham EM, Spitz DR, Goldman FD, et al. (2011) The p53/p21(WAF/CIP) pathway mediates oxidative stress and senescence in dyskeratosis congenita cells with telomerase insufficiency. Antioxid Redox Signal;14(6):985–97.

4

Mechanism of Intramembrane Cleavage of Alcadeins by γ-Secretase

Yi Piao[1], Ayano Kimura[1], Satomi Urano[1], Yuhki Saito[1], Hidenori Taru[2,3], Tohru Yamamoto[1], Saori Hata[1], Toshiharu Suzuki[1]*

1 Laboratory of Neuroscience, Graduate School of Pharmaceutical Sciences, Hokkaido University, Sapporo, Japan, **2** Laboratory of Neural Cell Biology, Graduate School of Pharmaceutical Sciences, Hokkaido University, Sapporo, Japan, **3** Creative Research Institute, Hokkaido University, Sapporo, Japan

Abstract

Background: Alcadein proteins (Alcs; Alcα, Alcβand Alcγ) are predominantly expressed in neurons, as is Alzheimer's β-amyloid (Aβ) precursor protein (APP). Both Alcs and APP are cleaved by primary α- or β-secretase to generate membrane-associated C-terminal fragments (CTFs). Alc CTFs are further cleaved by γ-secretase to secrete p3-Alc peptide along with the release of intracellular domain fragment (Alc ICD) from the membrane. In the case of APP, APP CTFβ is initially cleaved at the ε-site to release the intracellular domain fragment (AICD) and consequently the γ-site is determined, by which Aβ generates. The initial ε-site is thought to define the final γ-site position, which determines whether Aβ40/43 or Aβ42 is generated. However, initial intracellular ε-cleavage sites of Alc CTF to generate Alc ICD and the molecular mechanism that final γ-site position is determined remains unclear in Alcs.

Methodology: Using HEK293 cells expressing Alcs plus presenilin 1 (PS1, a catalytic unit of γ-secretase) and the membrane fractions of these cells, the generation of p3-Alc possessing C-terminal γ-cleavage site and Alc ICD possessing N-terminal ε-cleavage site were analysed with MALDI-TOF/MS. We determined the initial ε-site position of all Alcα, Alcβ and Alcγ, and analyzed the relationship between the initially determined ε-site position and the final γ-cleavage position.

Conclusions: The initial ε-site position does not always determine the final γ-cleavage position in Alcs, which differed from APP. No additional γ-cleavage sites are generated from artificial/non-physiological positions of ε-cleavage for Alcs, while the artificial ε-cleavage positions can influence in selection of physiological γ-site positions. Because alteration of γ-secretase activity is thought to be a pathogenesis of sporadic Alzheimer's disease, Alcs are useful and sensitive substrate to detect the altered cleavage of substrates by γ-secretase, which may be induced by malfunction of γ-secretase itself or changes of membrane environment for enzymatic reaction.

Editor: Stephen D. Ginsberg, Nathan Kline Institute and New York University School of Medicine, United States of America

Funding: This work was supported in part by Grants-in-Aid for Scientific Research (23390017 to TS, 24790062 to SH) from the Ministry of Education, Culture, Sports, Science and Technology (MEXT), and a grant from the New Energy and Industrial Technology Development Organization (NEDO) in Japan. The funders had no role in study design, data collection and analysis, decision to publish, or preparation of the manuscript.

Competing Interests: The authors have declared that no competing interests exist.

* E-mail: tsuzuki@pharm.hokudai.ac.jp

Introduction

The γ-secretase is comprised of four membrane proteins, presenilin 1 (PS1) or 2 (PS2), nicastrin (NCT), anterior pharynx defective 1 (APH-1), and presenilin enhancer 2 (PEN-2) [1]. PS functions as the catalytic unit of this aspartyl protease complex [2]. Prior to intramembrane cleavage of type I membrane proteins by γ-secretase, the substrate membrane proteins are subject to primary extracellular/intraluminal cleavage at the juxtamembrane region by a sheddase such as a disintegrin and metalloproteinase (ADAM) [3]. This primary cleavage is essential for the subsequent intramembrane γ-cleavage, although the exact regulation of intramembrane cleavage by γ-secretase remains unclear.

There are no distinct consensus amino acid sequences of γ-cleavage sites among over 60 different proteins reported as substrates of γ-secretase [1]. However, the molecular mechanisms of γ-cleavage of the Alzheimer's disease (AD)-related β-amyloid precursor protein (APP) and Notch have been well characterized [4,5]. APP is cleaved by β-secretase (BACE) in addition to α-secretase (ADAM 10 and ADAM 17), and retains the C-terminal fragments, amyloidogenic CTFβ and amyloidolytic CTFα, in the membrane while secreting the large extracellular N-terminal fragments [3]. When CTFβ is further cleaved by γ-secretase, the AD-related amyloid β protein (Aβ) is generated, while metabolically labile p3 peptide is generated from CTFα by γ-cleavage. This intramembrane γ-cleavage of APP CTF occurs initially at ε-cleavage sites. ε-cleavage between Leu645 and Val646 of APP695 generates Aβ49, which is processed to generate Aβ46 and Aβ43, and γ-cleavage at a site between Val636 and Ile637 generates Aβ40, a major Aβ species. When alternative ε-cleavage occurs between Thr644 and Leu645, Aβ48, Aβ45, and Aβ42 are sequentially generated, and cleavage at Gly634 generates Aβ38. These Aβ peptides are generated by processing of every three to four amino acids from the initial ε-site by γ-secretase [6–10].

A

B

C

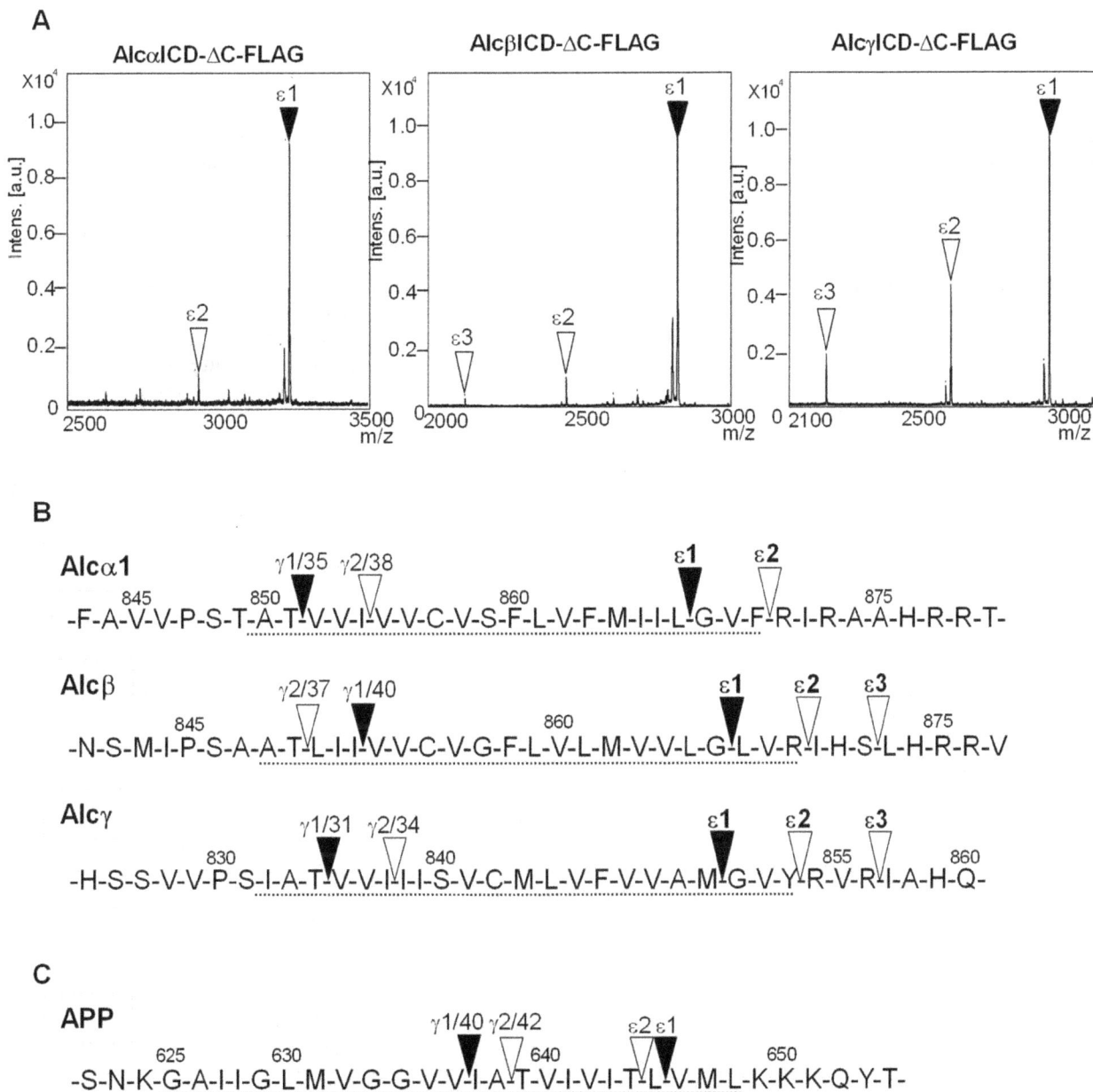

Figure 1. Determination of intramembrane ε-cleavage sites of Alcadeins. Representative mass spectra of Alc ICD-ΔC-FLAG generated by *in vitro* γ-secretase assay with membranes from HEK293 cells expressing Alc-ΔC-FLAG (**A**), and localization of ε-cleavage sites on the amino acid sequence (**B**), along with the comparison to γ- and ε-cleavage sites of APP (**C**). **A.** AlcαICD-ΔC-FLAG generated from Alcα-ΔC-FLAG (left), Alcβ ICD-ΔC-FLAG generated from Alcβ-ΔC-FLAG (middle), and Alcγ ICD-ΔC-FLAG generated from Alcγ-ΔC-FLAG (right). Closed arrowheads indicate the major product cleaved at the ε1 site, and open arrowheads indicate the minor products cleaved at the ε2 and ε3 sites. Amino acid sequence of Alc ICD-ΔC-FLAG was determined (**Fig. S3**). Other peaks, which are not indicated with arrowheads, are not products derived from Alc-ΔC-FLAG, because they are detectable in cells expressing an inactive/dominant-negative PS1 D385A mutant (**Fig. S2**). **B.** Amino acid sequence of human Alcα1, Alcβ and Alcγ (numbers indicate amino acid position, and the broken underline indicates putative transmembrane region). The major (ε1) and minor (ε2 and ε3) ε-cleavage sites are indicated, along with the previously identified major (γ1) and minor (γ2) γ-cleavage sites (13). In Alcα (upper), cleavage at γ1 generates p3-Alcα35, while cleavage at γ2 generates p3-Alcα38. Therefore, the γ1 cleavage site "γ1/35" and the γ2 cleavage site "γ2/38" are shown. Cultured cell lines generate p3-Alcα2N+35 and p3-Alcα2N+38, which possess two extra amino acids at the N terminal but have identical γ-cleavage sites to p3-Alcα35 and p3-Alcα38 in human CSF and are therefore considered the products cleaved at γ1 and γ2, respectively. In Alcβ (middle), cleavage at γ1 (γ1/40) generates p3-Alcβ40, while cleavage at γ2 (γ2/37) generates p3-Alcβ37. In cultured cell lines, "γ1/40" is the major γ-cleavage site, but "γ2/37" is the major site in human CSF. In Alcγ (lower), cleavage at the major cleavage siteγ1 (γ1/31) generates p3-Alcγ31, while cleavage at the minor site γ2 (γ2/34) generates p3-Alcγ34. Schematic pictures of protein constructs used in this study are shown in **Fig. S1. C.** Major and minor γ- and ε-cleavage sites of APP. The major γ1 (γ1/40) and minor γ2 (γ2/42) cleavage sites are shown. In APP, the major ε1 site largely defines the γ1 site to generate Aβ40, and the minor ε2 site promotes the γ2 site to generate Aβ42 or Aβ38.

Therefore, based on these observations, the initial ε-cleavage site nearly defines the position of the final γ-cleavage site in APP.

If this procedure is basically common among type I membrane protein substrates, the first-determined ε-cleavage site is prerequisite to the generation of alternative products toward a specific γ-cleavage site. For example, the familial AD (FAD)-linked PS mutations are thought to alter the position of the first ε-site, which in turn contributes to changing the production ratio of γ-site cleaved products [11,12]. In fact, FAD-linked mutations of PS1 increase the Aβ42/Aβ40 ratio in comparison to wild-type PS1 by increasing the alternative ε-site cleavage of APP CTFβ [13]. Alternatively, an alteration of γ-secretase function may not reach to major γ-site position and increase minor γ-cleavage, by which production of Aβ42 increases while Aβ38 generation decreases in APP [14,15]. In this study, we explored whether the correlation between γ- and ε-cleavage sites is also common to the Alcadein family of proteins (Alcs): Alcα, Alcβ and Alcγ.

Alcs are encoded by independent genes and expressed largely in neurons [16]. Alcs are primarily cleaved by APP α-secretase to generate Alc CTFs, which are consequently cleaved by γ-secretase, like APP, to secrete a short peptide p3-Alc into cell media or cerebrospinal fluid (CSF), along with liberation of the intracellular cytoplasmic domain fragment Alc ICD [17]. Thus, the C-terminal amino acid residue of p3-Alc contains a γ-cleavage site of Alc, and the N-terminal amino acid residue of Alc ICD demonstrates an initial ε-cleavage site. We previously showed that cells expressing FAD-linked PS1 mutation demonstrated an altered ratio of γ-cleavage products, with the increase of minor γ-cleaved products (minor p3-Alc species) to major p3-Alc species in these cells reflected by the increase of Aβ42 (the minor species) to Aβ40 (the major species) [17]. Because the magnitude of alternative γ-cleavage of Alcs and APP to generate p3-Alcα, p3-Alcβ, p3-Alcγ, and Aβ was not equivalent in cells expressing FAD-linked PS1 mutants [17], we speculated that their mechanisms of intramembrane substrate cleavage by γ-secretase may differ, or their sensitivity to altered γ-secretase activity may differ. Furthermore, we reported the increase of a minor species, p3-Alcα38, in the CSF of sporadic AD (SAD) patients, suggesting that γ-secretase dysfunction occurs in some populations of SAD patients [18]. Thus, understanding the mechanism by which Alcs are cleaved by γ-secretase is important to gaining deeper insight into the pathogenesis of AD.

The γ-secretase was found to cleave Alcs (Alcα, Alcβ and Alcγ) [19], and their γ-cleavage sites were determined as the C-termini of p3-Alcα, p3-Alcβ, and p3-Alcγ. These γ-cleavage sites were also demonstrated in human CSF [17,18]. We previously reported that in human, the major p3-Alcα species is p3-Alcα35 with C-terminal cleavage at Thr851 (numbering for Alcα1 isoform), and the minor species is p3-Alcα38 with C-terminal cleavage at Ile854. The major p3-Alcβ species in human CSF is p3-Alcβ37 with C-terminal cleavage at Thr849, while the minor species is p3-Alcβ40 with C-terminal cleavage at Ile852. However, in cells expressing Alcβ, the major species is p3-Alcβ40, and the minor species is p3-Alcβ37. In both human CSF and cells expressing Alcγ, the major p3-Alcγ species is p3-Alcγ31 with C-terminal cleavage at Thr834, and the minor species is p3-Alcγ34 with C-terminal cleavage at Ile837 [17]. Determination of the Alcs γ-site was somewhat equivocal, because the major γ-site of p3-Alcβ differed between CSF and cultured cells [17]. Therefore, to explore the mechanism for intramembrane cleavage of Alcs in this study, we first determined the ε-cleavage sites of Alcs. We then analyzed the relationship between initial determination of ε-cleavage site and final γ-cleaved site.

Results

Determination of Intramembrane ε-cleaving Site of Alcadeins

To identify the ε-cleavage sites of Alcs, we determined the N-terminal amino acid of Alc ICD. HEK293 cells expressing AlcΔC-FLAG (**Fig. S1**) were subjected to treatment with the γ-secretase inhibitor DAPT (3,5-(Difuorophenyl)acetyl-L-alanyl-L-2-phenyl-glycine t-butyl ester) to accumulate membrane-associated Alc CTFΔC-FLAG, which is the C-terminal product of AlcΔC-FLAG cleaved by primary α-secretase. Membranes were prepared from the cells and then subjected to in vitro γ-secretase assay to facilitate the cleavage of Alc CTFΔC-FLAG. The generated intracellular Alc ICDΔC-FLAG was isolated by immunoprecipitation, and representative MS spectra are shown (**Fig. 1A**). In Fig. 1A, the signals indicated with arrowheads are γ-secretase-dependent products because cells expressing a dominant-negative PS1 mutant carrying D385A substitution did not generate these Alc ICD species (**Fig. S2**).

The molecular masses observed by TOF/MS analysis (**Fig. 1A**) were compared with expected values (**Table S1**), and the amino acid sequence was determined with matrix-assisted laser desorption ionization time-of-flight tandem mass spectrometry (MALDI-MS/MS) analysis (**Fig. S3**). The γ-cleavage site was also confirmed by MALDI-MS/MS analysis of p3-Alc secreted by cells expressing AlcΔC-FLAG as described [17]. The ε- and γ-cleavages observed upon in vitro membrane incubation were consistent with the cleavages observed in cells. The minor/major γ-cleavage ratio of p3-Alc generated by the in vitro γ-secretase assay with membrane coincided with the minor/major γ-cleavage ratio of p3-Alc secreted by cells, and the production levels of p3-Alc were very similar between the in vitro γ-secretase assay and cultured cells (**Fig. S4**).

In the in vitro γ-secretase assay with membrane prepared from cells expressing AlcαΔC-FLAG, Alcα truncated at Gly886 with C-terminal FLAG (**Fig. S1,** and amino acid sequence around the ε-sites shown in **Fig. 1B**) generated two Alcα ICD species, MW3238 (major ε1) and MW2934 (minor ε2), including the FLAG-tag. Therefore, the major Alcα ICD species possessed Gly868 and minor Alc ICD species possessed Arg871 at their N-termini (**Fig. 1A, left; Fig. S3A**). Cells expressing AlcαΔC-FLAG secreted major p3-Alcα2N+35 (γ1) and minor p3-Alcα2N+38 (γ2) species (**Fig. S4A**), indicating that the γ-cleavage of Alcα was not affected by truncation of the cytoplasmic region and addition of the FLAG tag (See **ref. 13,** for "2N+"species that possess the same C-terminal γ-site to p3-Alcα35 (γ1) and p3-Alcα38 (γ2), respectively; however, "2N+" species were predominantly generated from cultured cells.).

Identical analyses were performed with AlcβΔC-FLAG truncated at Ile881 and AlcγΔC-FLAG truncated at Ile866 (**Fig. S1**). The major Alcβ ICD species demonstrated MW 2824 (ε1) with two minor species of MW 2455 (ε2) and MW 2118 (ε3) including the FLAG tag. Therefore, the major Alcβ ICD species possessed Leu867, and the minor Alcβ ICD species possessed Ile870 and Leu873, respectively, at their N-termini (**Fig. 1A, middle; Fig. S3B**). The major Alcγ ICD demonstrated MW 2946 (ε1) with two minor species of MW 2627 (ε2) and MW 2216 (ε3) including the FLAG tag. Therefore the major Alcγ ICD species has an N-terminal Gly851 residue, and the minor Alcγ ICD species possess Arg854 and Ile857, respectively, at their N termini (**Fig. 1A, right; Fig. S3C**). The γ-cleavage sites of AlcβΔC-FLAG and AlcγΔC-FLAG were confirmed by determination of the amino acid sequences of p3-Alcβ and p3-Alcγ (**Fig. S4B and C**).

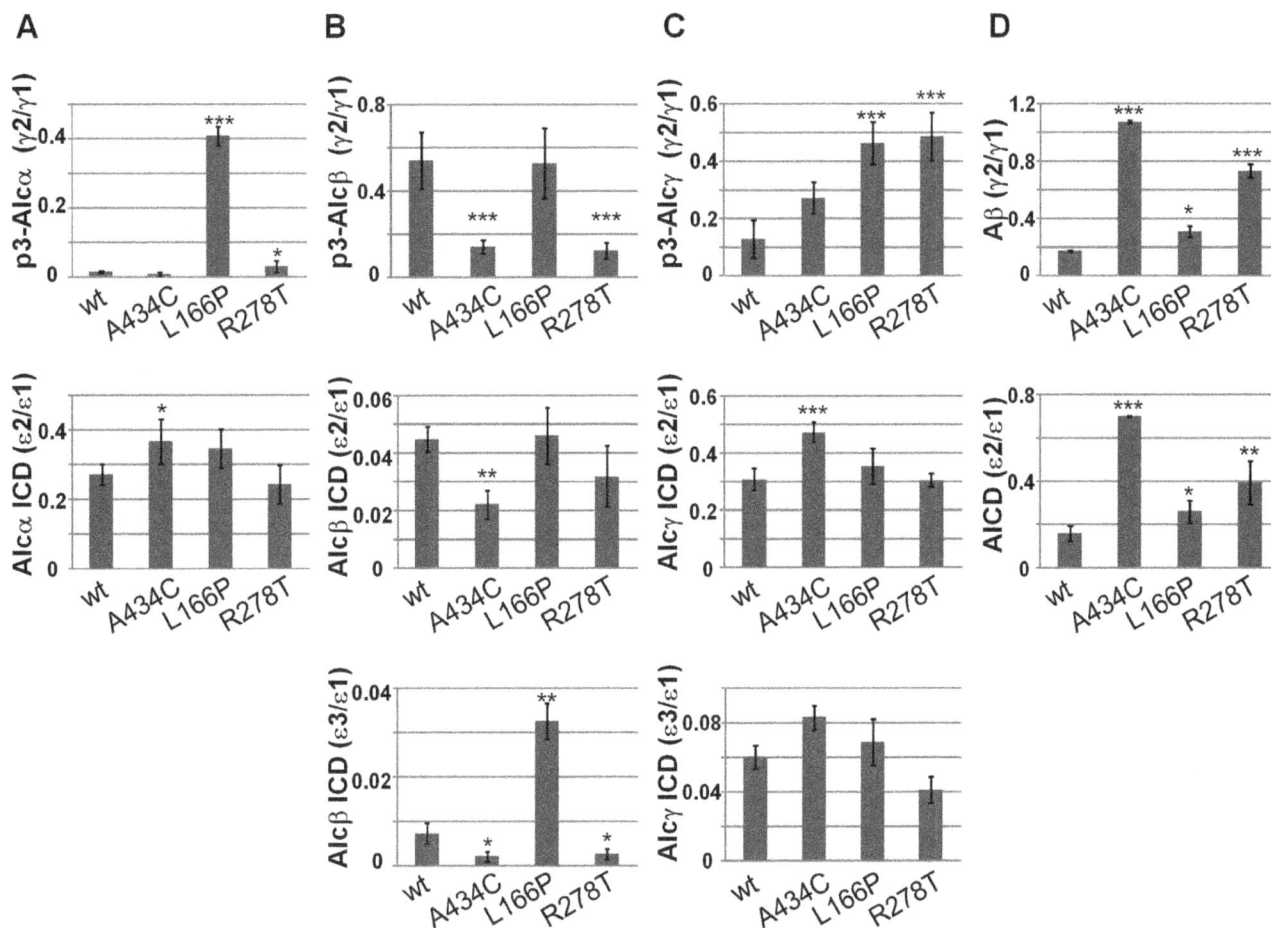

Figure 2. Magnitudes of the minor to major ratios of γ-cleavage and ε-cleavage in Alcs and APP from cells expressing FAD-linked PS1 mutants. Minor to major p3-Alc ratio (γ2/γ1, first rows) was determined by quantitative MS analysis of p3-Alc secreted by HEK293 cells expressing Alc-ΔC-FLAG with either wild-type PS1 or PS1 carrying a FAD-linked mutation (A434C, L166P, or R278T) (**Fig. S5**). Minor to major Alc ICD ratios (ε2/ε1, second rows; ε3/ε1, third rows) were also determined by MS analysis of Alc ICD-ΔC-FLAG generated by *in vitro* γ-secretase assay with membranes from the same cells (**Fig. S5**). **A. Comparison of Alcα γ-cleavage ratio to ε-cleavage ratio.** The ratio (γ2/γ1) of p3-Alcα2N+38 (minor) to p3-Alcα2N+35(major) (first row) and the ratio (ε2/ε1) of the Alcα ICD-ΔC-FLAG product cleaved at ε2 (minor) to the product cleaved at ε1 (major) (second row) are shown. **B. Comparison of Alcβ γ-cleavage ratio to ε-cleavage ratio.** The ratio (γ2/γ1) of p3-Alcβ37 (minor) to p3-Alcβ40 (major) (first row) and the ratio (ε2/ε1 or ε3/ε1) of the Alcβ ICD-ΔC-FLAG product cleaved at ε2 or ε3 (minor) to the product cleaved at ε1 (major) (second row, ε2/ε1; third row, ε3/ε1) are shown. **C. Comparison of Alcγ γ-cleavage ratio to ε-cleavage ratio.** The ratio (γ2/γ1) of p3-Alcγ34 (minor) to p3-Alcγ31 (major) (first row) and the ratio (ε2/ε1 or ε3/ε1) of the Alcγ ICD-ΔC-FLAG product cleaved at ε2 or ε3 (minor) to the product cleaved at ε1 (major) (second row, ε2/ε1; third row, ε3/ε1) are shown. **D. Comparison of APP γ-cleavage ratio to ε-cleavage ratio.** The ratio (γ2/γ1) of Aβ42 (minor) to Aβ40 (major) (first row) and the minor to major ratio of AICD (second row, ε2/ε1) are shown. Net Aβ values are shown in **Table S2**, and representative mass spectra of AICD-FLAG are shown in **Fig. S5D**. Statistical analysis was performed using Dunnett's multiple comparison test. Significance is indicated relative to the ratio of wild-type PS1 (wt) (mean ± S.E., n = 4, *P<0.05, **P<0.001, ***P<0.0001).

These results show that γ-secretase first cleaves Alcα CTF between Leu867 and Gly868 (ε1) for a major product and between Phe870 and Arg871 (ε2) for a minor product. Alcβ CTF is first cleaved between Gly866 and Leu867 (ε1) for a major product and between Arg869 and 870Ile (ε2) or Ser872 and Leu873 (ε3) for minor products, and Alcγ CTF is first cleaved between Met850 and Gly851 (ε1) for a major product and between Tyr853 and Arg854 (ε2) or Arg856 and Ile857 (ε3) for minor products. Some minor ε-cleavage sites may be located outside of putative Alcs transmembrane domains. However, again, the ε-cleavages are due to γ-secretase because PS1 dominant negative mutant cannot cleave Alc ICD at these ε-sites (**Fig. S2**). The locations of Alcα, Alcβ, and Alcγ, γ- and ε-cleavage sites were then compared to those of APP (**Fig. 1B and C**).

Relationship between Altered Cleavage at γ-sites and at ε-sites in Alcs and APP

Previous observation revealed that FAD-linked PS1 mutations altered the ratio of γ-cleavages of Alcα, Alcβ and Alcγ, as observed in APP, although the magnitudes of altered γ-cleavage varied across these substrates [17]. We next examined whether the altered γ-cleavage of Alcs in cells expressing a FAD-linked PS1 mutation relates to the alteration of first ε-cleavage position. Alcs-ΔC-FLAG proteins were coexpressed in cells together with wild-type PS1 or FAD-linked PS1 mutants. The p3-Alcs secreted by cells were analyzed with MALDI-TOF/MS (**Fig. S5, upper rows in panels A to C**), and the minor/major ratios of γ-cleavage products were determined (**Fig. 2A to C, upper panels**). Alcs ICD-FLAG was generated by *in vitro* γ-secretase assay with membrane and identified with MALDI-TOF/MS (**Fig.**

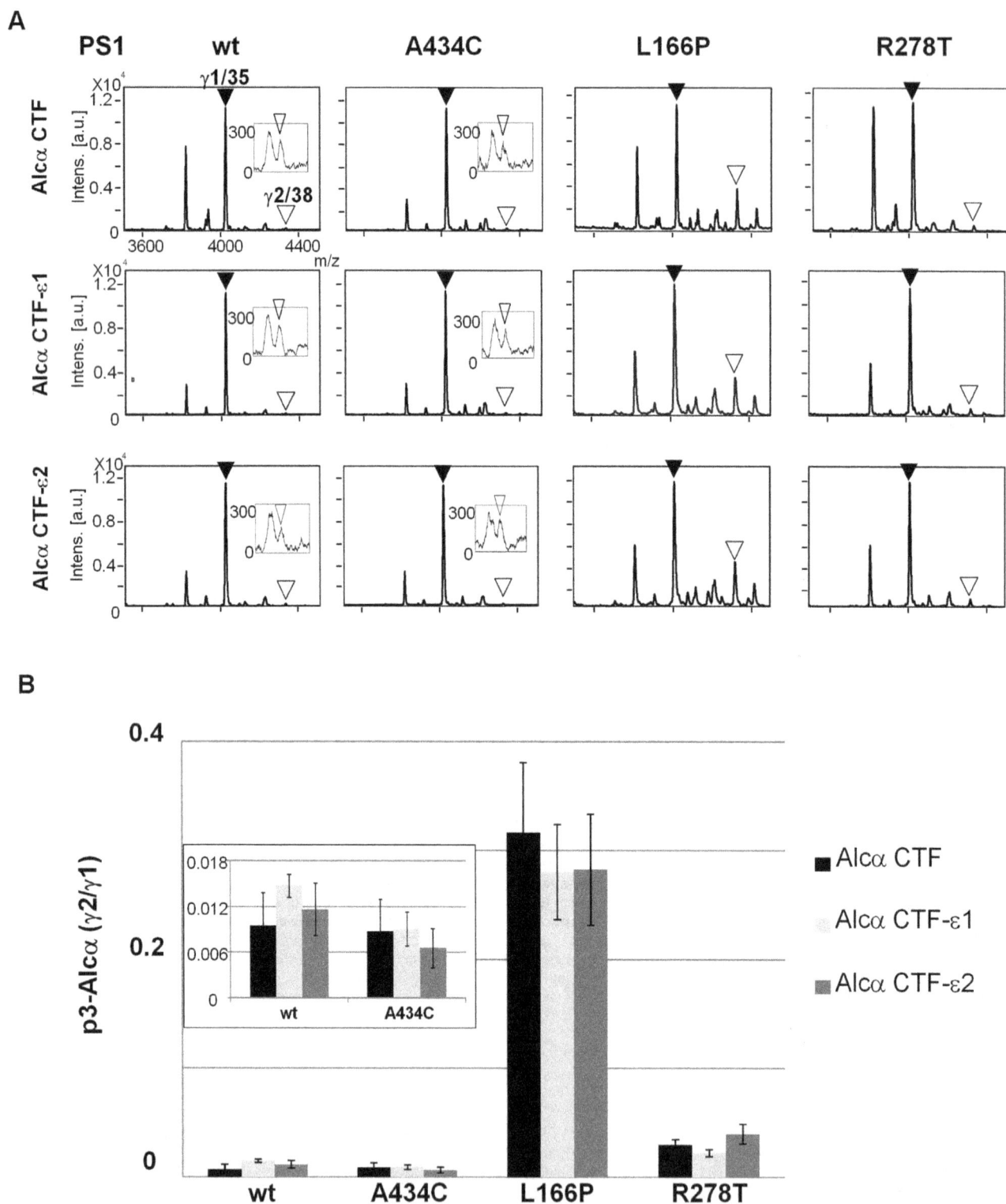

Figure 3. Alteration of ε-cleavage is not necessarily prerequisite to determine a specific γ-cleavage site in Alcα. A. Representative MS spectra of p3-Alcα secreted by HEK293 cells expressing Alcα CTF, Alcα CTF-ε1, or Alcα CTF-ε2 with either wild-type PS1 (wt) or a FAD-linked PS1 mutant (A434C, L166P, or R278T). The p3-Alcα species in cell culture media were immunoprecipitated and subjected to MALDI-TOF/MS analysis. Closed arrowheads indicate the major product with γ1 site (p3-Alcα2N+35, "γ1/35"), while open arrowheads indicate the minor product with γ2 site (p3-Alcα2N+38, "γ2/38"). The spectra of the minor p3-Alcα38 product are enlarged in windows in which intensity of 300 on the y-axis corresponds to 0.03 in the original panels. **B.** The peak area of p3-Alcα2N+38 (minor species) was compared with that of p3-Alcα2N+35 (major species), and the minor to major ratios (p3-Alcα2N+38/p3-Alcα2N+35) are indicated as γ2/γ1. Statistical analysis was performed using one-way analysis of variance followed by the Tukey-Kramer multiple comparison test (means ± S.E., n = 4). Significance in comparison to the ratio of Alcα CTF was not observed in cells expressing wild-type PS1 (wt) or FAD-linked PS1 mutants. The columns of "wt" and "A434C" are enlarged in window.

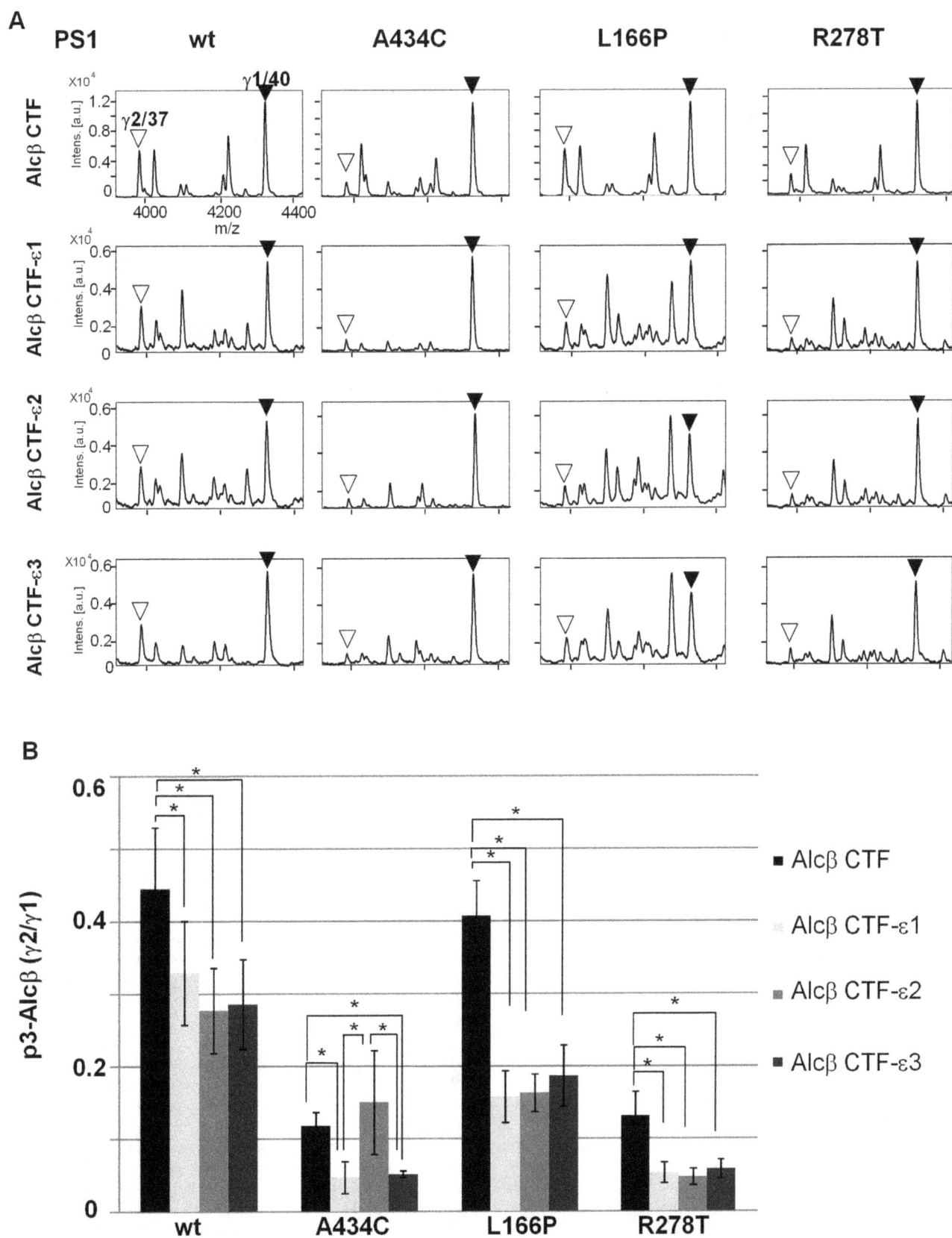

Figure 4. Alteration of ε-cleavage is not necessarily prerequisite to determine a specific γ-cleavage site in Alcβ. A. Representative MS spectra of p3-Alcβ species secreted by HEK293 cells expressing Alcβ CTF, Alcβ CTF-ε1, Alcβ CTF-ε2, or Alcβ CTF-ε3 with either wild-type PS1 (wt) or a FAD-linked PS1 mutant (A434C, L166P, or R278T). The p3-Alcβ species in cell culture media were immunoprecipitated and subjected to MALDI-TOF/MS analysis. Closed arrowheads indicate the major product with γ1 site (p3-Alcβ40, "γ1/40"), while open arrowheads indicate the minor product with

γ2 site (p3-Alcβ37, "γ2/37"). **B.** The peak area of p3-Alcβ37 (minor species) was compared with that of p3-Alcβ40 (major species), and the minor to major ratios (p3-Alcβ37/p3-Alcβ40) are indicated as γ2/γ1. Statistical analysis was performed using one-way analysis of variance followed by the Tukey-Kramer multiple comparison test (mean ± S.E., n = 4, *$P < 0.05$).

S5, lower rows in panels A to C), and the minor/major ratios of ε-cleavages were compared to those of γ-cleavages (**Fig. 2A to C, middle and lower panels**). APP CTFβ/C99-FLAG was also analyzed for major (Aβ40) and minor (Aβ42) γ-cleavage products by sandwich ELISA (sELISA; **Table S2**) along with major and minor ε-cleavage products, APP intracellular domain fragment (AICD-FLAG; **Fig. S5D**). The minor/major ratio of γ-cleavage was compared to that of ε-cleavage (**Fig. 2D**). Positions of APP γ- and ε-cleavage sites are indicated for comparison with the positions in Alcs (**Fig. 1C**).

In APP, as expected, the magnitudes of altered ε-cleavage demonstrated minor/major (ε2/ε1) ratios well correlated with the altered γ-cleavage minor/major (Aβ42/Aβ40) ratios in cells expressing the wild type and FAD-linked mutants of PS1 ($R^2 = 0.7356$; **Fig. S6D and Fig. 2D**). In contrast to APP, Alcs tended to show that the first-determined ε-cleavage position is not necessarily prerequisite to determine a specific γ-cleavage position (compare upper panels with middle and lower panels in **Fig. 2A–C**). In Alcα, minor γ-cleavage to generate p3-Alcα38 increased remarkably in cells expressing PS1 L166P and less significantly in cells expressing PS1 R278T, while no significant effect was observed in cells expressing PS1 A434C mutant, when compared to cells expressing wild-type PS1. Minor ε-cleavage (ε2) increased slightly in cells expressing PS1 A434C mutant, while no significant alternation was detected in cells expressing PS1 L166P or R278T mutants when compared to cells expressing wild-type PS1 (**Fig. S5A**). The comparison of the minor/major (38/35) ratio of γ-cleavage with minor/major (ε2/ε1) ratio of ε-cleavage in cells expressing the respective PS1 mutants suggests that covariance between the magnitude of ε-cleavage and γ-cleavage positions alteration in Alcα was low ($R^2 = 0.1597$; **Fig. S6A and Fig. 2A**).

We performed identical analyses for Alcβ and Alcγ (**Fig. 2B and C**). In Alcβ, the minor/major ratio of γ-cleavage is indicated as the p3-Alcβ37/p3-Alcβ40 (γ2/γ1) ratio. Because Alcβ demonstrated at least three ε-cleavage sites, a major ε1 site with minor ε2 and ε3 sites (**Fig. 1B**), we examined both ε2/ε1 and ε3/ε1 ratios to determine the minor/major ratio of ε-cleavage. Minor γ-cleavage was significantly reduced in cells expressing PS1 A434C and R278T mutants (**Fig. S5B**). Decreased minor ε-cleavage seemed to occur in cells expressing PS1 A434C, as reflected in both the ε2/ε1 and ε3/ε1 ratios (**Fig. 2B**). PS1 R278T decreased the ε3/ε1 ratio significantly and tended to decrease the ε2/ε1 ratio, but not significantly. PS1 L166P did not affect the ε2/ε1 ratio, but the ε3/ε1 ratio was greatly increased. These results may suggest that alteration of ε-cleavage tends to reflect the position of the γ-cleavage site in Alcβ ($R^2 = 0.8518$ for ε2/ε1 ratio versus γ2/γ1 ratio and $R^2 = 0.4675$ for ε3/ε1 ratio versus γ2/γ1; **Fig. S6B**). However, upon careful analysis of **Fig. 2B**, the ratios of ε2/ε1 and ε3/ε1 are very low (<0.05), while the ratio of γ2/γ1 is approximately 0.5 in wild-type PS1 and L166P mutant, and 0.15 in A434C and R278T mutants. This observation indicates that both γ1 and γ2 cleavages are largely derived from ε1 cleavage position in Alcβ, or in other words, the ε1 site is exclusively dominant among Alcβ ε-cleavage sites. Therefore, one dominant ε-site is likely to determine two γ-sites, and the alteration of γ-site is not affected by a small magnitude of alternation at the ε-site position in Alcβ.

In Alcγ, the minor/major ratio of γ-cleavage is indicated as the p3-Alcγ34/p3-Alcγ31 (γ2/γ1) ratio. Alcγ demonstrated three ε-cleavage sites, a major ε1 site with minor ε2 and ε3 sites (**Fig. 1C**). Thus, as in the case of Alcβ, we examined both ε2/ε1 and ε3/ε1 ratios to determine the minor/major ratio of ε-cleavage. Cells expressing FAD-linked PS1 L166P and R278T mutants demonstrated significantly increased minor γ-cleavage, while only the PS1 A434C mutant demonstrated increased ε2/ε1 ratio, and no PS1 mutants demonstrated a significant change in ε3/ε1 ratio compared to wild-type PS1 (**Fig. 2C and Fig. S5C**). These results also suggest that alteration of ε-cleavage positions does not largely correlate with the alteration of Alcγ γ-cleavage sites ($R^2 = 0.0201$ for ε2/ε1 ratio versus γ2/γ1 ratio and $R^2 = 0.1372$ for ε3/ε1 ratio versus γ2/γ1 ratio; **Fig. S6C and Fig. 2C**).

Overall, these findings indicate that alteration of ε-cleavage sites in Alcs does not influence the determination of γ-cleavage site, unlike APP, which demonstrates a significant covariance of changes in magnitude between ε- and γ-cleavage products.

A specific ε-cleavage Site is not Necessarily Prerequisite to Determine a Specific γ-cleavage Positions in Alcs

To further examine whether one ε-cleavage position can determine a specific γ-cleavage position, we expressed Alcs truncated at ε-cleavage sites in cells expressing wild-type and FAD-linked mutants of PS1, and analyzed the alteration of γ-cleavage sites (**Fig. 3 Fig. 4 Fig. 5**). Alcα CTF-ε1 (truncated at site ε1) and Alcα CTF-ε2 (truncated at site ε2), along with Alcα CTF, were expressed in cells (**Fig. 3 and Fig. S7A**). The p3-Alcα in the culture media was analyzed with MALDI-TOF/MS (**Fig. 3A**), and the minor/major (p3-Alcα 2N+38/p3-Alcα 2N+35 or γ2/γ1) ratios were determined (**Fig. 3B**). Production levels of p3-Alcα were highly similar between Alc CTF and Alc CTF-ε. The minor/major (γ2/γ1) ratios of p3-Alcα from cells expressing wild-type (wt) and FAD-linked mutants of PS1 (A434C, L166P, and R278T) did not significantly differ between Alcα CTF and Alcα CTF-ε. These results indicate that both ε1 and ε2 cleavage generate ratios of the major γ-cleavage product that are identical to those of Alcα CTF with an intact cytoplasmic region. This analysis clearly indicates that the first-determined ε-cleavage is not necessarily prerequisite to determine γ-cleavage position; that is, both ε1 and ε2-sites predominantly reach the γ1 site as the major γ-cleavage position. We confirmed that FAD-linked PS1 mutation L166P demonstrated the greatest effect on increasing the generation of minor γ2-cleaved product [13] (**Fig. S5A**), but the effect of this mutation may be not due to the position of initial ε-cleavage site.

The same analysis was performed for Alcβ (**Fig. 4 and Fig. S7B**) and Alcγ (**Fig. 5 and Fig. S7C**). Alcβ CTF-ε1, Alcβ CTF-ε2, Alcβ CTF-ε3, and Alcβ CTF were expressed in cells along with different forms of PS1, and the secreted p3-Alcβ was analyzed (**Fig. 4A**). Alcβ CTF truncated at ε1, ε2, orε3 generated almost identical levels of major γ1 (p3-Alcβ40) and minor γ2 (p3-Alcβ37) products, but their production levels decreased to 50–60% of those of Alcβ CTF including an intact cytoplasmic region (note "*Intens*" shown in **Fig. 4A**). The γ2/γ1 ratios of the three truncated Alcβ CTF-ε species were not largely affected by the FAD-linked PS1 mutation, except for the A434C mutant, in which Alcβ CTF-ε1 and Alcβ CTF-ε3 decreased the γ2/γ1 ratio.

The results largely demonstrated identical minor/major (γ2/γ1) ratios for γ-cleavage products among AlcβCTF-ε species truncated at the ε1, ε2, and ε3 sites (**Fig. 4B**). However, Alc CTFβ-ε

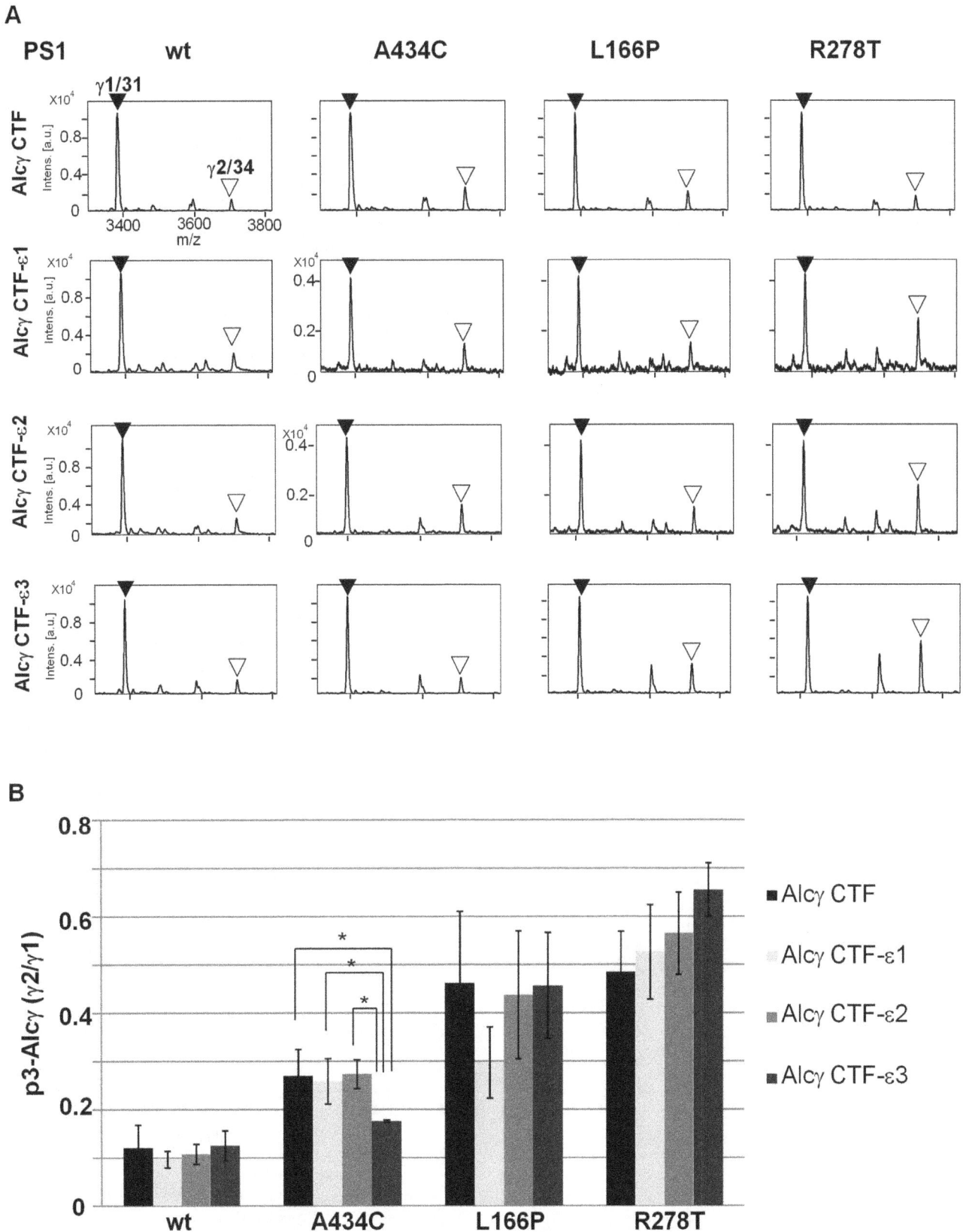

Figure 5. Alteration of ε-cleavage is not necessarily prerequisite to determine a specific γ-cleavage site in Alcγ. A. Representative MS spectra of p3-Alcγ secreted by HEK293 cells expressing Alcγ CTF, Alcγ CTF-ε1, Alcγ CTF-ε2, or Alcγ CTF-ε3 with either wild-type PS1 (wt) or FAD-linked PS1 mutants (A434C, L166P, R278T). The p3-Alcγ species in cell culture media were immunoprecipitated and subjected to MALDI-TOF/MS analysis. Closed arrowheads indicate the major product with γ1 site (p3-Alcγ31, "γ1/31"), while open arrowheads indicate the minor product with γ2 site (p3-Alcγ34, "γ2/34"). **B.** The peak area of p3-Alcγ34 (minor species) was compared with that of p3-Alcγ31 (major species), and the ratios (p3-Alcγ34/p3-

Alcγ31) are indicated as γ2/γ1. Statistical analysis was performed using one-way analysis of variance followed by the Tukey-Kramer multiple comparison test (mean ± S.E., n = 4, *$P < 0.05$).

significantly decreased the minor γ2-cleaved product (p3-Alcβ37), resulting in the decrease of the minor/major (γ2/γ1) ratio compared to that of Alcβ CTF, which seems to differ slightly from the case of Alcα and Alcγ (compare **Fig. 4B** with **Figs. 3B and 5B**). However, overall, the results indicate that one ε-cleavage position has no determining effect on the dominance of a specific γ-cleavage position, and that the effect of FAD-linked PS1 mutations on the ratio of γ-cleaved products is not largely due to the alteration of ε-cleavage position.

Alcγ CTF-ε truncated at ε1, ε2, and ε3 sites also secreted p3-Alcγ with a fixed minor/major (p3-Alcγ34/p3-Alcγ31 or γ2/γ1) ratio that was highly similar to the ratio of Alcγ CTF, with the exception of Alcγ CTF-ε3 in cells expressing A434C PS1 mutant (**Fig. 5 and Fig. S7C**). The production levels of p3-Alcγ derived from Alcγ CTF-ε positions were highly similar overall to that of p3-Alcγ derived from Alcγ CTF with the entire C-terminal in cells expressing wild-type PS1, but decreased by 40% for Alcγ CTF-ε1 and Alcγ CTF-ε2 in cells expressing FAD-linked PS1 mutants (note "*Intens*" shown in **Fig. 5A**). This analysis indicates that the alteration of ε-cleavage position does not affect the determination of γ-cleavage position of Alcγ.

Taken together, these findings indicate that the position of physiological ε-site is not necessarily prerequisite to determine a specific γ-cleavage position in Alcs. Moreover, the alteration of initial ε-cleavage site does not contribute to the changes of their minor/major ratio of γ-cleaved products in cells expressing FAD-linked mutation of PS1. The observation with Alcs may differ from the conclusion obtained from APP CTFs truncated at ε-sites. In APP, cells express CTF 1–49 (ε1 site) secreted predominantly Aβ40 while those expressing CTF 1–48 (ε2 site) secreted preferentially Aβ42 [**8**].

Alteration of γ-cleavage upon Disturbance of Physiological ε-cleavage Sites in Alcs

We next asked whether the γ-cleavage sites are altered when Alc CTFs are initially truncated at artificial ε-cleavage positions (**Fig. 6**). Pseudo-ε-cleavage sites were designated toward the N termini of the physiological major and minor ε-cleavage sites. Alcα CTFs with physiological ε-cleavage sites, Alcα CTF-ε1 and Alcα CTF-ε2, or Alcα CTFs with artificial/pseudo ε-cleavage sites, Alcα CTF-ε1p (ε1 pseudo) and Alcα CTF-ε2p (ε2 pseudo), were expressed in cells along with Alcα CTF. The p3-Alcα in media was analyzed with MALDI-TOF/MS to determine the minor/major (p3-Alcα38/p3-Alcα35 or γ2/γ1) ratio. Production levels of p3-Alcα did not largely change between Alcα CTF-ε and Alcα CTF-εp. Surprisingly, Alcα CTF-ε1p significantly increased the minor/major ratio of γ-cleaved products, and Alcα CTF-ε2p remarkably decreased the minor/major ratio (**Fig. 6A**), suggesting that alternation of γ-cleavage position may be affected by movement of the ε-cleavage position by one amino acid.

We confirmed this phenomenon with Alcβ CTF and Alcγ CTF truncated with pseudo-ε-sites (**Fig. 6B and C**). Production levels of p3-Alcβ derived from Alcβ CTF-ε positions were ~60% of those of Alcβ CTF (note "*Intens*" shown in **Fig. 6B**), but no significant differences in minor/major (γ2/γ1) ratio were observed between the Alcβ CTF-ε and Alcβ CTF-εp. Only Alcβ CTF-ε2p demonstrated a small but significant increase in the minor/major ratio of γ-cleavage. Production levels of p3-Alcγ derived from Alcγ CTF-ε positions were unchanged across all Alcγ CTFs with physiological and pseudo-ε-sites. All three pseudo-ε-sites of Alcγ

CTF significantly increased the minor/major (γ2/γ1) ratio of γ-cleavage products. In Alcβ CTF, several MS signals were detected along with the γ1 and γ2 signals (**Fig. 6B**). These signals do not reflect products derived from Alcβ CTFs because cells without expression of Alcβ CTFs also generated these signals (**Fig. S8**).

Overall, in Alcs or at least in Alcα and Alcγ, movement of the physiological ε-cleavage position by one amino acid may be a possible mechanism to induce the alteration of γ-cleavage site dominance, although the changes to cellular condition that are able to induce the movement from the physiological sites of ε-cleavage remain unknown. Importantly, no additional γ-cleavage sites were generated from the artificial/non-physiological positions of ε-cleavage used for all Alcs. Thus, again, the initial ε-cleavage site is not necessarily prerequisite to determine the position of γ-cleavage.

Discussion

In a previous study, the magnitude of Alcs γ-cleavage alteration in cells expressing FAD-linked PS1 mutants varied and differed from APP [17], suggesting that the determination of initial intramembrane ε-cleavage may not be necessarily prerequisite to cleave at a specific γ-cleavage site in the case of Alcs. We determined two to three ε-cleavage sites each in Alcα, Alcβ and Alcγ, as observed in APP and Notch [20–24]. One represented the major ε-site at which Alc CTF was predominantly cleaved. We found that some FAD-linked PS1 mutations affected the ratio of minor to major ε-cleavage, but this alteration to the ratio of minor to major γ-cleavage was not always apparent. Similarly, some FAD-linked PS1 mutations did not remarkably influence in the selection of ε-site but significantly affected the γ-cleavage site. These properties of Alcs intramembrane cleavage by γ-secretase differ from those of APP, in which changes to the ratio of minor to major γ-cleavage were consistent with changes to the ratio of minor to major ε-cleavage [10,12,25]. Therefore, we propose that the mechanism of intramembrane cleavage by γ-secretase is not regulated identically between Alcs and APP. Such relationship between the γ-site position and the ε-cleavage site in Alcs was demonstrated with several different types of experiment, and our current findings suggest that the endophenotype of γ-secretase malfunction appears to affect either γ-cleavage or ε-cleavage position in Alcs.

Notably, Alcs CTF possessing artificial/non-physiological pseudo-ε-sites at their C-termini did not generate novel γ-sites, while the ratio of minor to major γ-cleavage products changed significantly in Alcα and Alcγ, but less so in Alcβ. This observation suggests that the selection of ε-site may be relatively flexible, but the position of the γ-site is rigid, and that the minor to major ratio of the γ-cleavage site rather than the positional alteration of ε-cleavage or the minor to major ratio of ε-cleavage site is more likely to reflect the endophenotype of γ-secretase malfunction. This may be consistent with a recent report that γ-secretase malfunction tends to generate Aβ42 instead of Aβ38 which should be generated finally from APP by an entire γ-secretase [14], indicating a selection of different γ-site positions as an endophenotype of γ-secretase malfunction. Furthermore, our study indicates that the shift of physiological ε-cleavage to an unusual ε-site can alter the minor/major ratio of γ-site cleavage. It remains for future studies to determine whether such non-physiological ε-cleavage of Alcs occurs in cells.

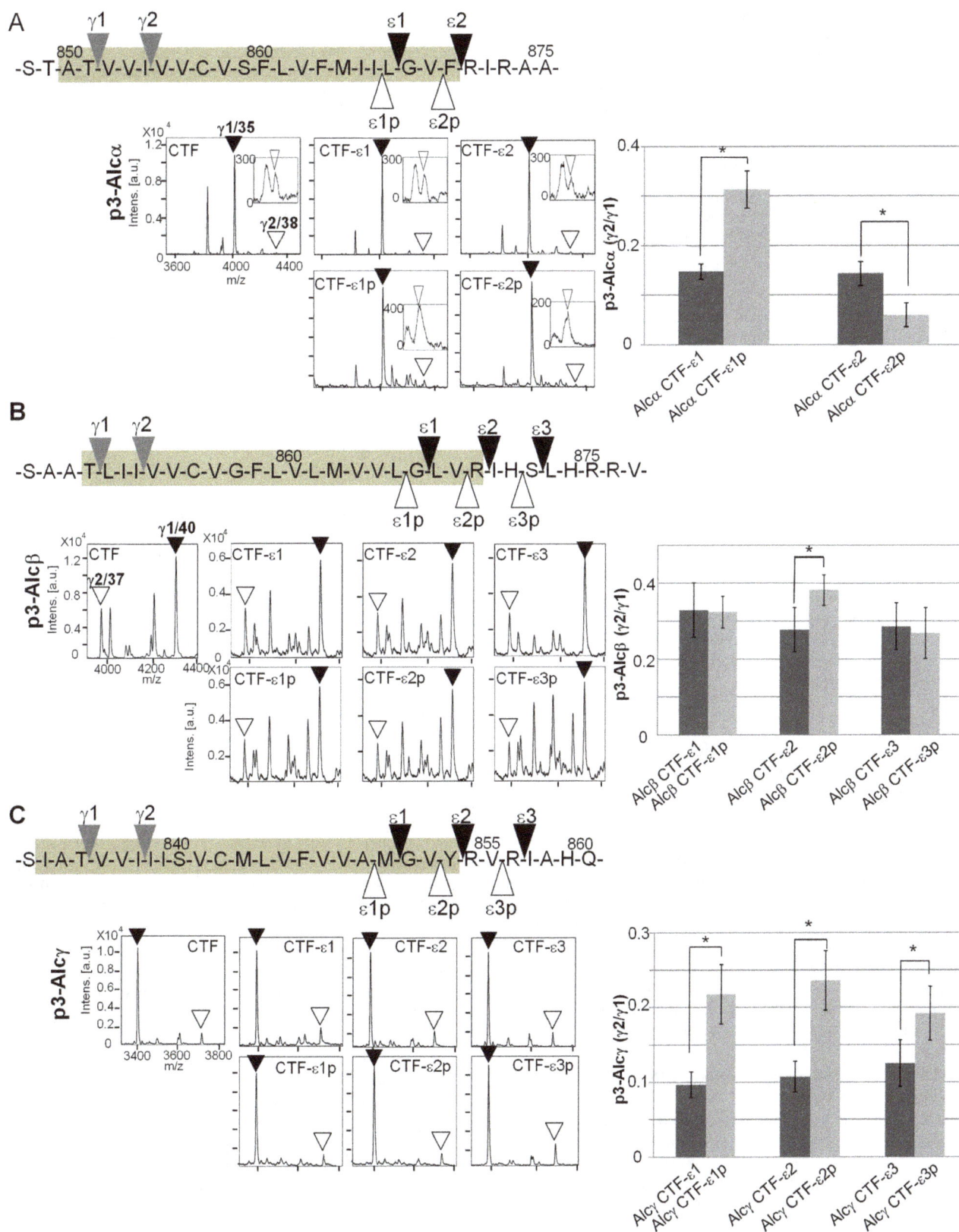

Figure 6. Alteration of Alcs γ-cleavage when physiological ε-cleavage sites are replaced with non-physiological/pseudo-ε-cleavage sites. A. Positions of the physiological major and minor (ε1 and ε2) and pseudo- (ε1p and ε2p) ε-cleavage sites (upper left) are shown along with the physiological major and minor γ-cleavage sites (γ1 and γ2). The shaded amino acid sequence indicates a putative membrane-embedded region. Non-physiological/pseudo-ε-cleavage sites were designed by shifting one residue toward the N terminal of the physiological ε-cleavage sites. Representative MS spectra of p3-Alcα secreted by HEK293 cells expressing Alcα CTF, Alcα CTF-ε1, Alcα CTF-ε2, Alcα CTF-ε1p, or Alcα CTF-ε2p are shown (lower left panels). The major species p3-Alcα2N+35 with γ1 site (γ1/35, closed arrowheads) and minor species p3-Alcα2N+38 with γ2 site (γ2/

38, open arrowheads) are indicated. The peak area of p3-Alcα2N+38 was compared with that of p3-Alcα2N+35, and the ratios (p3-Alcα2N+38/p3-Alcα2N+35) are indicated as γ2/γ1 (right panel). The spectra of minor species p3-Alcα38 are enlarged in windows in which intensities of 200, 300, and 400 on the y-axis correspond to 0.02, 0.03 and 0.04 in the original panels. **B.** Positions of the physiological major and minor (ε1, ε2, and ε3) and pseudo- (ε1p, ε2p, and ε3p) ε-cleavage sites (upper left) are shown along with the physiological major and minor γ-cleavage sites (γ1 and γ2). Representative MS spectra of p3-Alcβ secreted by HEK293 cells expressing Alcβ CTF, Alcβ CTF-ε1, Alcβ CTF-ε2, Alcβ CTF-ε3, Alcβ CTF-ε1p, Alcβ CTF-ε2p, or Alcβ CTF-ε3p are shown (lower left). The major species p3-Alcβ40 withγ1 site (γ1/40, closed arrowheads) and minor species p3-Alcβ37 with γ2 site (γ2/37, open arrowheads) are indicated. The peak area of p3-Alcβ37 was compared with that of p3-Alcβ40, and the ratios (p3-Alcβ37/p3-Alcβ40) are indicated asγ2/γ1 (right panel). **C.** Positions of the physiological major and minor (ε1, ε2, and ε3) and pseudo- (ε1p, ε2p, and ε3p) ε-cleavage sites (upper left) are shown along with the physiological major and minor γ-cleavage sites (γ1 and γ2). Representative MS spectra of p3-Alcγ secreted by HEK293 cells expressing Alcγ CTF, Alcγ CTF-ε1, Alcγ CTF-ε2, Alcγ CTF-ε3, Alcγ CTF-ε1p, Alcγ CTF-ε2p, or Alcγ CTF-ε3p are shown (lower left). The major p3-Alcγ31 with γ1 site (γ1/31, closed arrowhead) and minor p3-Alcγ34 with γ2 site (γ2/34, open arrowhead) are indicated. The peak area of p3-Alcγ34 was compared with that of p3-Alcγ31, and the ratios (p3-Alcγ34/p3-Alcγ31) are indicated as γ2/γ1 (right panel). (**A–C**) The ratios of products from the pseudo-site were compared to those from the respective physiological sites. Statistical analysis was performed by Student's t test (mean ± S.E., n = 4, *P<0.05).

The present findings may be an important step to revealing the mechanism of γ-secretase malfunction in SAD, which differs from that in FAD. Because altered γ-secretase processing was observed as an endophenotype of p3-Alcα in CSF of some SAD subjects [18], Alcs may be more sensitive substrates to detect γ-secretase malfunction [26].

Materials and Methods

Plasmid Construction and Stable Cell Lines Expressing PS1

The human Alcadein cDNAs, hAlcα1, hAlcβ, and hAlcγ and plasmids encoding human PS1 cDNAs have been described [17]. FAD-linked mutations were introduced by PCR-based site-directed mutagenesis to generate pcDNA3.1-PS1L166P, pcDNA4-PS1R278T, and pcDNA4-PS1A434C. HEK293 cells were transfected with these plasmids, and cells stably expressing PS1 were cloned as described [17].

Antibodies

Rabbit polyclonal anti-Alcα antibody UT135 was raised against a peptide composed of Cys plus the sequence between positions 839 and 851 (NPHPFAVVPSTAT+C) of human Alcα. Rabbit polyclonal anti-Alcβ antibody UT143 was raised against a GST-fusion protein containing the sequence between positions 819 and 847 (FLHRGHQPPPEMAGHSLASSHRNSMIPSA) of human Alcβ. Rabbit polyclonal anti-Alcγ antibody UT166 was raised against a peptide composed of Cys plus the sequence between positions 823 and 834 (C+IQHSSVVPSIAT) of human Alcγ. These Alc-specific antibodies were specific to their respective p3-Alc targets with the exception of UT166, which exhibited cross-reactivity to p3-Alcα (data not shown). These antibodies were used to isolate and detect p3-Alc [17]. The monoclonal anti-FLAG antibody (M2) was purchased from Sigma-Aldrich.

MALDI-TOF/MS and -MS/MS Analysis of p3-Alc Secreted into the Culture Medium

HEK293 cells (8–9×10^6) were transfected with plasmids (6 μg) in Lipofectamine 2000 according to the manufacturer's protocol (Invitrogen) for 24 h. The p3-Alcα, p3-Alcβ, and p3-Alcγ that were secreted into the medium (10 ml) were recovered by immunoprecipitation in the presence of protease inhibitor cocktail (5 μg/ml chymostatin, 5 μg/ml leupeptin, and 5 μg/ml pepstatin) as described [17] using the polyclonal anti-p3-Alcα UT135 (4 μg of affinity purified IgG), polyclonal anti-p3-Alcβ UT143 (100 μl of serum), and polyclonal anti-p3-Alcγ UT166 (100 μl of serum) antibodies, respectively, and Protein G-Sepharose beads. The beads were sequentially washed with Wash buffer I (10 mM Tris-HCl (pH 8.0), 140 mM NaCl, 0.1% (w/v) n-octyl-D-glucoside,

0.025% (w/v) sodium azide) and Wash buffer II (10 mM Tris-HCl (pH 8.0), 0.025% (w/v) sodium azide), and samples were then eluted with a solution of trifluoroacetic acid/acetonitrile/water (1:20:20) saturated with sinapinic acid. The samples were dried on a target plate, and MALDI-TOF/MS analysis was performed using an UltraflexII TOF/TOF (Bruker Daltonics, Bremen, Germany). Molecular masses were calibrated using the peptide calibration standard (Bruker Daltonics) [17,18]. The quantitative accuracy of mass spectrometric analysis with immunoprecipitation was confirmed previously [17], and molecular masses of p3-Alc species measured with MALDI-TOF/MS were compared with theoretical values to confirm the accuracy of mass spectrometric analysis (**Table S1**). Furthermore, we confirmed that the quantity of peptides is not affected by coexistence with the increased amount of other peptides, suggesting specific ion suppression of peptide does not occur in this assay (**Figs. S9 and S10**).

Membrane Incubation for Substrate Cleavage by γ-secretase (in vitro γ-secretase Assay)

To detect Alc ICD, HEK293 cells stably expressing wild-type PS1 or PS1 with a FAD-linked mutation were transfected with plasmid (6 μg) in Lipofectamine 2000 according to the manufacturer's protocol (Invitrogen). After 20-h culture of cells, the γ-secretase inhibitor DAPT (10 μM, 3,5-(Difuorophenyl)acetyl-L-alanyl-L-2-phenylglycine t-butyl ester) was added to the medium, and cells were cultured for an additional 4 h. The cells were then harvested and lysed in 500 μl of homogenizing buffer (20 mM HEPES, 150 mM NaCl, 10% glycerol, 5 mM EDTA, 5 mM EGTA) by passing through a 27-gauge needle 30 times on ice. After the removal of unbroken organelles and nuclei by centrifugation at 3,000 rpm for 10 min at 4°C, the membranes were precipitated by centrifugation at 100,000× g for 60 min at 4°C. The crude membrane fraction was washed once with homogenizing buffer and re-suspended in an assay buffer (20 mM HEPES, 150 mM NaCl, 10% glycerol, 5 mM EDTA, 5 mM EGTA, 10 μM amastatin, 0.1 μM arphamenine A). After incubation for 2 h at 37°C for substrate cleavage by γ-secretase, the membrane suspension was subjected to centrifugation at 100,000× g for 30 min at 4°C. The supernatant including the ε-site-cleaved product with FLAG-tag was subjected to immunoprecipitation with anti-FLAG antibody and Protein G-Sepharose beads. Immunoreactive proteins were analyzed by MALDI-TOF/MS. Molecular masses of p3-Alc and Alc ICD species generated by in vitro γ-secretase assay and measured with MALDI-TOF/MS were compared with theoretical values to confirm the accuracy of mass spectrometric analysis (**Table S1**).

Quantitative Aβ Assay

Aβ40 and Aβ42 secretion into the medium were quantified by sELISA as described [27], and their net values are shown in **Table S2**.

Supporting Information

Figure S1   Schematic structure of Alc-ΔC-FLAG fusion proteins used for the *in vitro* γ-secretase assay to determine ε-cleavage sites. Cytoplasmic regions of Alcα, Alcβ and Alcγ were truncated at the indicated positions and fused to FLAG-tag sequence. Amino acid numbering corresponds to human Alcadein α1 (971 amino acids), Alcadein β (956 mino acids), and Alcadein γ (956 amino acids) [16]. Primary α-cleavage sites are indicated with open arrowheads. CTF (*light gray shading on Alc proteins*), C-terminal region of Alc cleaved by α-secretase.

Figure S2   Representative MS spectra of Alc ICD-ΔC-FLAG generated by *in vitro* γ-secretase assay with membranes derived from cells expressing wild-type PS1 or the dominant-negative PS1 mutant D385A. Membranes from cells expressing Alcα-ΔC-FLAG (left), Alcβ-ΔC-FLAG (middle), and Alcγ-ΔC-FLAG (right) in the presence of wild-type PS1 (upper) or PS1 D385A mutant (lower) were subjected to *in vitro* γ-secretase assay to generate Alc ICD-ΔC-FLAG, which was recovered by immunoprecipitation with anti-FLAG antibody and analyzed with MALDI-TOF/MS. The major product cleaved at the ε1 site (closed arrowhead) and minor products cleaved at ε2 and ε3 sites (open arrowheads) are indicated.

Figure S3   Identification of major and minor ε-cleavage sites of Alcadeins. Amino acid sequences of Alcα ICD-ΔC-FLAG generated from Alcα-ΔC-FLAG (**A**), Alcβ ICD-ΔC-FLAG generated from Alcβ-ΔC-FLAG (**B**), and Alcγ ICD-ΔC-FLAG generated from Alcγ-ΔC-FLAG (**C**) were determined by MALDI-MS/MS analysis. Left panels show the major Alc ICD-ΔC-FLAG product with N-terminal ε1 site, right (A) and middle panels (B and C) show minor products with N-terminal ε2 sites, and right panels (B and C) show additional minor products with N-terminal ε3 sites. Representative MS spectra of Alc ICD-ΔC-FLAG are shown in Fig. 1A, and the amino acid sequences determined by this study are indicated in Fig. 1B. Amino acid sequence "DYKDDDDK" indicates FLAG sequence.

Figure S4   Comparison of p3-Alc species generated by membrane incubation (*in vitro* γ-secretase assay) with those secreted by cells. Comparison of representative MS spectra of p3-Alc species secreted into culture medium by cells expressing Alc CTF (Medium) with those generated by *in vitro* γ-secretase assay with membrane fractions prepared from cells expressing Alc CTF (Membrane incubation). The peak area of minor p3-Alc (γ2, open arrowheads) was compared with that of major p3-Alc (γ1, closed arrowheads), and the minor/major (γ2/γ1) ratios are indicated (right). To examine the background signals, MS spectra of *in vitro* γ-secretase assay without incubation are shown (Membrane incubation (-)). **A.** Spectra of p3-Alcα (left panels), and the minor/major ratio (p3-Alcα2N+38/p3-Alcα2N+35) in medium and the ratio (p3-Alcα38/p3-Alcα35) generated by *in vitro* γ-secretase assay (right graph) are shown. In the *in vitro* γ-secretase assay, p3-Alcα species were predominantly generated, while p3-Alcα2N+ species were predominantly secreted into the culture medium by cells. Thus, we compared the γ2/γ1

ratios between the p3-Alcα2N+38/p3-Alcα2N+35 ratio in media and the p3-Alcα38/p3-Alcα35 ratio in membrane incubation. **B.** Spectra of p3-Alcβ (left panels), and the minor/major ratios (p3-Alcβ37/p3-Alcβ40) secreted into medium and generated by *in vitro* γ-secretase assay (right graph) are shown. **C.** Spectra of p3-Alcγ (left panels), and the minor/major ratios (p3-Alcγ34/p3-Alcγ31) secreted into medium and generated by *in vitro* γ-secretase assay (right graph) are shown. (**A–C**) Statistical analysis was performed using Student's t test (mean ± S.E., n = 4). No significant difference between medium and *in vitro* γ-secretase assay was observed.

Figure S5   Displacement of the intramembrane γ- and ε-sites of Alcα, Alcβ Alcγ and APP in cells expressing FAD-linked mutations of PS1. (**A–C**) Representative MS spectra of p3-Alc (upper) secreted by cells expressing wild-type PS1 and FAD-linked mutants of PS1, and Alc ICD (lower) generated by *in vitro* γ-secretase assay with membranes from the same cells. **A.** The p3-Alcα species secreted by HEK293 cells expressing Alcα-ΔC-FLAG were immunoprecipitated and subjected to MALDI-TOF/MS analysis. The Alcα ICD-ΔC-FLAG species generated by *in vitro* γ-secretase assay were immunoprecipitated and analyzed by MALD-TOF/MS analysis. Spectra of the minor product p3-Alcα38 (γ2) are enlarged in windows, in which intensity of 300 on the y-axis corresponds to 0.03 in the original panels. **B.** The p3-Alcβ species secreted by HEK293 cells expressing Alcβ-ΔC-FLAG were immunoprecipitated and subjected to MALDI-TOF/MS analysis. The Alcβ ICD-ΔC-FLAG species generated by *in vitro* γ-secretase assay were immunoprecipitated and analyzed by MALD-TOF/MS analysis. Spectra of minor sites (ε2 and ε3) are enlarged in windows in which intensities of 1200 and 800 on the y-axis correspond to 0.12 and 0.08, respectively, in the original panels. **C.** The p3-Alcγ species secreted by HEK293 cells expressing Alcγ-ΔC-FLAG were immunoprecipitated and subjected to MALDI-TOF/MS analysis. The Alcγ ICD-ΔC-FLAG species generated by *in vitro* γ-secretase assay were immunoprecipitated and analyzed by MALD-TOF/MS analysis. (**A–C**) Closed arrowheads indicate major γ- or ε-site cleaved products: p3-Alcα2N+35 (panel A upper) and Alcα ICDε1 (panel A lower), p3-Alcβ40 (panel B upper) and Alcβ ICD-ε1 (panel B lower), and p3-Alcγ31 (panel C upper) and Alcγ ICD-ε1 (panel C lower). Open arrowheads indicate minor γ- or ε-cleaved products: p3-Alcα2N+38 (panel A upper) and Alcα ICD-ε2 (panel A lower); p3-Alcβ37 (panel B upper), Alcβ ICD-ε2, and Alcβ ICD-ε3 (panel B lower); and p3-Alcγ34 (panel C upper), Alcγ ICD-ε2, and Alcγ ICD-ε3 (panel C lower). **D.** Representative MS spectra of AICD. The AICD-FLAG generated by *in vitro* γ-secretase assay with membranes from HEK293 cells expressing CTFβ/C99-FLAG were immunoprecipitated with anti-FLAG antibody and analyzed by MALD-TOF/MS analysis. Closed arrowheads indicate major ε1-cleaved products, and open arrowheads indicate minor ε2-cleaved products.

Figure S6   Correlation between minor/major ratios of γ-cleavage products and ε-cleavage products. Covariant analysis of γ2/γ1 ratio with the ratios of certain minor ε2 or ε3 products to major ε1 products was performed. Graphs showing the relationships between the ratio of p3-Alcα γ2/γ1 to Alcα ICD ε2/ε1 ratio (**A**), the ratio of p3-Alcβ γ2/γ1 to Alcβ ICD ε2/ε1 and ε3/ε1 ratios (**B**), the ratio of p3-Alcγ γ2/γ1 to Alcγ ICD ε2/ε1 and ε3/ε1 ratios (**C**), and the ratio Aβ42/Aβ40 to AICD ε2/ε1 ratio (**D**). wt, wild-type PS1; A434C, L166P, and R278T are FAD-linked PS1 mutants (see Figs. 3–5). R^2, correlation coefficient.

Figure S7 Schematic structure of Alc-ΔC proteins with physiological ε-cleavage sites. A. The cytoplasmic region of Alcα was truncated at the indicated major ε1 and minor ε2 sites and fused to a signal peptide (SP) sequence at the N terminal through a Met+Ala sequence composed of "2N+" species. Amino acid numbering corresponds to human Alcadeinα1 (971 amino acids). The primary α-cleavage site is indicated with a gray arrowhead, and the cleavage indicated with a broken-line arrowhead generates Alcα CTF. Positions of γ-cleavage sites are indicated with open arrowheads, and ε-sites are indicated with closed arrowheads (the larger arrowhead indicates the major ε-site, and the smaller indicates the minor ε-site). **B.** The cytoplasmic region of Alcβ was truncated at the indicated major ε1 and minor ε2 and ε3 sites and fused to a signal peptide (SP) sequence at the N terminal. Amino acid numbering corresponds to human Alcadein β (956 amino acids). The primary α-cleavage site is indicated with a gray arrowhead, and the cleavage indicated with a broken-line arrowhead generates Alcβ CTF. Positions of γ-cleavage sites are indicated with open arrowheads, and ε-sites are indicated with closed arrowheads (the larger arrowhead indicates the major ε-site, and the smaller two indicate minor ε-sites). **C.** The cytoplasmic region of Alcγ was truncated at the indicated major ε1 and minor ε2 and ε3 sites and fused to a signal peptide (SP) sequence at the N terminal. Amino acid numbering corresponds to human Alcadein γ (955 amino acids). The primary α-cleavage site is indicated with a gray arrowhead, and the cleavage indicated with a broken-line arrowhead generates Alc γ CTF. Positions of γ-cleavage sites are indicated with open arrowheads, and ε-sites are indicated with closed arrowheads (the larger arrowhead indicates the major ε-site, and the smaller two indicate minor ε-sites).

Figure S8 Identification of p3-Alcβ species secreted by cells. In Fig. 6B, the immunoprecipitation-TOF-MS study using the media of cells expressing Alcβ CTF with C-terminal truncated ε-site presented complex spectra. To identify p3-Alcβ species secreted by cells, mock media derived from cells without expression of Alcβ CTF was also analyzed, and the spectra were compared to those of cells expressing Alcβ CTF. The major p3-Alcβ40 (γ1) and minor p3-Alcβ37 (γ2) products are indicated with arrowheads. Other MS signals are not products derived from Alcβ CTF.

Figure S9 Quantitative accuracy of immunoprecipitation-mass spectrometric analysis in the presence of another peptide (I). Endogenously generated p3-Alcα 2N+35 in the presence of increased amount of synthetic p3-Alcα35 peptide was subjected to immunoprecipitation with UT135 and analyzed with MALDI-TOF/MS. **A.** Amino acid sequence of p3-Alcα2N+35 and p3-Alcα35. **B.** Representative immunoprecipitation-mass spectra of endogenous and synthetic p3-Alcα peptides. To fixed volume (2 mL) of cultured medium of HEK293 cells expressing Alcα, indicated amount (0, 0.5, 1.0, 2.0 and 4.0 ng) of synthetic p3-Alcα35 peptide was added, and subjected to immunoprecipitation. The cells secrete p3-Alcα 2N+35 (closed arrowhead) largely with small amount of p3-Alcα35 (open arrowhead) (left panel, 0 ng of synthetic peptide). **C.** Quantitative accuracy of the ratio of p3-Alcα35/p3-Alcα 2N+35. The relationship of area ratios of p3-Alcα35/p3-Alcα 2N+35 in the presence of various amounts of synthetic p3-Alcα35 peptide were analyzed. The endogenous p3-Alcα 2N+35 levels are not affected in the presence of increased amount of synthetic p3-Alcα35 peptide (B),

and the p3-Alcα35/p3-Alcα 2N+35 ratio increased proportionally with the increased amount of synthetic p3-Alcα35 peptide ($R^2 = 0.99938$ in C), indicating the quantification of a specific peptide is not affected in the presence of increased amounts of another peptide in this immunoprecipitation-mass spectrometric analysis.

Figure S10 Quantitative accuracy of immunoprecipitation-mass spectrometric analysis in the presence of another peptide (II). Synthetic p3-Alcβ37 and p3-Alcβ40 were subjected to immunoprecipitation with UT143 and analyzed with MALDI-TOF/MS. **A.** Amino acid sequence of p3-Alcβ37 and p3-Alcβ40. **B–C.** Representative immunoprecipitation-mass spectra of synthetic p3-Alcβ peptides in PBS (2 mL) containing 0.1% (W/V) bovine serum albumin (B) or cultured medium (2 mL) of mock HEK293 cells (C). To the fixed amount (1 ng) of synthetic p3-Alcβ37 peptide (open arrowhead) indicated amount (0, 0.5, 1.0, 2.0 and 4.0 ng) of synthetic p3-Alcβ40 peptide (closed arrowhead) was added, and subjected to immunoprecipitation. The cells don't secrete p3-Alcβ species but show non-specific products as signals, which are detectable in C, but not in B, along with synthetic p3-Alcβ37 (open arrowhead) and p3-Alcβ40 (closed arrowhead). **D–E.** Quantitative accuracy of the ratio of p3-Alcβ40/p3-Alcβ37. The relationship of area ratios of p3-Alcβ40/p3-Alcβ37 with various amounts of synthetic p3-Alcβ40 peptide were analyzed (panel D indicates the result of B, and panel E indicates the result of C). The synthetic p3-Alcβ37 levels are not affected in the presence of increased amount of synthetic p3-Alcβ40 peptide (B) and unknown immunoprecipitates (C), and the p3-Alcβ40/p3-Alcβ37 ratio increased proportionally with the increased amount of synthetic p3-Alcβ40 peptide ($R^2 = 0.99866$ in D and $R^2 = 0.99861$ in E), indicating the quantification of a specific peptide is not affected in the presence of increased amounts of another peptide in this immunoprecipitation-mass spectrometric analysis.

Table S1 Molecular masses observed by TOF/MS analysis and expected. Molecular masses (Da) of Alc ICD-ΔC-FLAG peptides (**upper**) and p3-Alcs (**lower**) generated by *in vitro* γ-secretase assay with cell membranes. The p3-Alcα peptide products γ1/35 and γ2/38 indicate p3-Alcα35 and p3-Alcα38, respectively, but not p3-Alcα2N+35 and p3-Alcα2N+38, which are secreted by cultured cells [17], because the *in vitro* γ-secretase assay with cultured cell membranes generates dominantly p3-Alcα species but not p3-Alcα2N+ species (see **Fig. S4A**).

Table S2 Net Aβ values quantified with sELISA. Medium Aβ40 and Aβ42 values of the studies indicated in Fig. 2D were quantified with sELISA [27], and the average values are summarized with standard deviation (n = 4).

Acknowledgments

Authors thank Dr. Maho Morishima (Hokkaido University) for critical discussions and helpful comments.

Author Contributions

Conceived and designed the experiments: YP YS HT SH TS. Performed the experiments: YP SU AK SH. Analyzed the data: YP YS HT TY SH TS. Contributed reagents/materials/analysis tools: YP SU AK SH. Wrote the paper: HT SH TS.

References

1. De Strooper B (2003) Aph-1, Pen-2, and nicastrin with presenilin generate an active γ-secretase complex. Neuron 38: 9–12.
2. Wolfe MS, Kopan R. (2004) Intramembrane proteolysis: theme and variations. Science 305: 1119–1123.
3. Lichtenthaler SF, Steiner H. (2007) Sheddases and intramembrane-cleaving proteases: RIPpers of the membrane. Symposium on regulated intramembrane proteolysis. EMBO Rep 8: 537–541.
4. Selkoe DJ (1999) Translating cell biology into therapeutic advances in Alzheimer's disease. Nature 399: A23–A31.
5. De Strooper B, Annaert W, Cupers P, Saftig P, Craessaerts K, et al. (1999) A presenilin-1-dependent γ-secretase-like protease mediates release of Notch intracellular domain. Nature 398: 518–522.
6. Zhao G, Cui MZ, Mao G, Dong Y, Tan J, et al. (2005) γ-Cleavage is dependent on ζ-cleavage during the proteolytic processing of amyloid precursor protein within its transmembrane domain. J Biol Chem 280: 37689–37697.
7. Zhao G, Tan J, Mao G, Cui MZ, Xu X. (2007) The same γ-secretase accounts for the multiple intramembrane cleavages of APP. J Neurochem 100: 1234–1246.
8. Funamoto S, Morishima-Kawashima M, Tanimura Y, Hirotani N, Saido T C, et al. (2004) Truncated carboxyl-terminal fragments of β-amyloid precursor protein are processed to amyloid β-proteins 40 and 42. Biochemistry 43: 13532–13540.
9. Qi-Takahara Y, Morishima-Kawashima M, Tanimura Y, Dolios G, Hirotani N, et al. (2005) Longer forms of amyloid β protein: implications for the mechanism of intramembrane cleavage by γ-secretase. J Neurosci 25: 436–445.
10. Takami M, Nagashima Y, Sano Y, Ichihara S, Morishima-Kawashima M, et al. (2009) γ-Secretase: Successive tripeptide and tetrapeptide release from the transmembrane domain of β-carboxyl terminal fragment. J Neurosci 29, 13042–13052.
11. Sisodia SS, St George-Hyslop PH (2002) γ-Secretase, Notch, Abeta and Alzheimer's disease: where do the presenilins fit in? Nat Rev Neurosci 4: 281–290.
12. Sato T, Dohmae N, Qi Y, Kakuda N, Misonou H, et al. (2003) Potential link between amyloid β-protein 42 and C-terminal fragment γ 49–99 of β-amyloid precursor protein. J Biol Chem 278: 24294–24301.
13. Bergmans BA, De Strooper B (2010) γ-secretases: from cell biology to therapeutic strategies. Lancet Neurol 9: 215–226.
14. Okochi M, Tagami S, Yanagida K, Takami M, Kodama TS, et al. (2012) γ-Secretase modulates and presenilin 1 mutants act differentially on presenilin/γ-secretase function to cleave Aβ42 and Aβ43. Cell Reports 3: 1–10.
15. Kakuda N, Shoji M, Arai H, Furukawa K, Ikeuchi T, et al. (2012) Altered γ-secretase activity in mild cognitive impairment and Alzheimer's disease. EMBO Mol Med 4: 344–352.
16. Araki Y, Tomita S, Yamaguchi H, Miyagi N, Sumioka A, et al. (2003) Novel cadherin-related membrane proteins, Alcadeins, enhance the X11-like protein-mediated stabilization of amyloid β-protein precursor metabolism. J Biol Chem 278: 49448–49458.
17. Hata S, Fujishige S, Araki Y, Kato N, Araseki M, et al. (2009) Alcadein cleavages by amyloid β-precursor protein (APP) α- and γ-secretases generate small peptides, p3-Alcs, indicating Alzheimer disease-related γ-secretase dysfunction. J Biol Chem 284: 36024–36033.
18. Hata S, Fujishige S, Araki Y, Taniguchi M, Urakami K, et al. (2011) Alternative processing of γ-secretase substrates in commom forms of mild cognitive impairment and Alzheimer's disease: evidence for γ-secretase dysfunction. Ann Neurol 69: 1026–1031.
19. Araki Y, Miyagi N, Kato N, Yoshida T, Wada S, et al. (2004) Coordinated metabolism of Alcadein and amyloid β precursor protein regulates FE65-dependent gene transactivation. J Biol Chem 279: 24343–24354.
20. Selkoe DJ, Wolfe MS. (2007) Presenilin: running with scissors in the membrane. Cell 131: 215–221.
21. Okochi M, Fukumori A, Jiang J, Itoh N, Kimura R, et al. (2006) Secretion of the Notch-1 Aβ-like peptide during Notch signaling. J Biol Chem 281: 7890–7898.
22. Gu Y, Misonou H, Sato T., Dohmae N. Takio K, et al. (2001) Distinct intramembrane cleavage of the β-amyloid precursor protein family resembling γ-secretase-like cleavage of Notch. J Biol Chem 276: 35235–35238.
23. Yu C, Kim SH. Ikeuchi T, Xu H, Gasparini L, et al. (2001) Characterization of a presenilin-mediated amyloid precursor protein carboxyl-terminal fragment γ. Evidence for distinct mechanisms involved in γ-secretase processing of the APP and Notch1 transmembrane domains. J Biol Chem 276: 43756–43760.
24. Weidemann A, Eggert S, Reinhard FB, Vogel M, Paliga K, et al. (2002) A novel ε-cleavage within the transmembrane domain of the Alzheimer amyloid precursor protein demonstrates homology with Notch processing. Biochemistry 41: 2825–2835.
25. Chavez-Gutierrez L, Bammens L, Benilova I, Vandersteen A, Benurwar M, et al. (2012) The mechanism of γ-secretase dysfunction in familial Alzheimer disease. EMBO. J. 31: 2261–2274.
26. Hata S, Taniguchi M, Piao Y, Ikeuchi T, Fagan AM, et al. (2012) Multiple γ-secretase product peptides are coordinately increased in concentration in the cerebrospinal fluid of a subpopulation of sporadic Alzheimer's disease subjects. Mol Neurodegener 7: 16.
27. Mizumaru C, Saito Y, Ishikawa T, Yoshida T, Yamamoto T, et al. (2009) Suppression of APP-containing vesicle trafficking and production of β-amyloid by AID/DHHC-12 protein. J Neurochem 111: 1213–1224.

5

Identification of Naturally Processed Hepatitis C Virus-Derived Major Histocompatibility Complex Class I Ligands

Benno Wölk[1,2,9], Claudia Trautwein[3,9], Benjamin Büchele[1], Nadine Kersting[1], Hubert E. Blum[1], Hans-Georg Rammensee[3], Andreas Cerny[4], Stefan Stevanovic[3], Darius Moradpour[1,5,¶], Volker Brass[1,¶]*

1 Department of Medicine II, University of Freiburg, Freiburg, Germany, 2 Institute of Virology, Hannover Medical School, Hannover, Germany, 3 Department of Immunology, University of Tübingen, Tübingen, Germany, 4 Clinical Pharmacology and Clinical Immunology/Allergology, Inselspital, University of Bern, Bern, Switzerland, 5 Division of Gastroenterology and Hepatology, Centre Hospitalier Universitaire Vaudois, University of Lausanne, Lausanne, Switzerland

Abstract

Fine mapping of human cytotoxic T lymphocyte (CTL) responses against hepatitis C virus (HCV) is based on external loading of target cells with synthetic peptides which are either derived from prediction algorithms or from overlapping peptide libraries. These strategies do not address putative host and viral mechanisms which may alter processing as well as presentation of CTL epitopes. Therefore, the aim of this proof-of-concept study was to identify naturally processed HCV-derived major histocompatibility complex (MHC) class I ligands. To this end, continuous human cell lines were engineered to inducibly express HCV proteins and to constitutively express high levels of functional HLA-A2. These cell lines were recognized in an HLA-A2-restricted manner by HCV-specific CTLs. Ligands eluted from HLA-A2 molecules isolated from large-scale cultures of these cell lines were separated by high performance liquid chromatography and further analyzed by electrospray ionization quadrupole time of flight mass spectrometry (MS)/tandem MS. These analyses allowed the identification of two HLA-A2-restricted epitopes derived from HCV nonstructural proteins (NS) 3 and 5B (NS3$_{1406-1415}$ and NS5B$_{2594-2602}$). In conclusion, we describe a general strategy that may be useful to investigate HCV pathogenesis and may contribute to the development of preventive and therapeutic vaccines in the future.

Editor: Antonio Bertoletti, Singapore Institute for Clinical Sciences, Singapore

Funding: German Bundesministerium für Bildung und Forschung (01 KI 9951) European Commission (QLK2-CT1999-00356 and QLK2-CT2002-01329) Swiss National Science Fondation (SNF 3200B0-103874) European Union INTERREG-IV-2009-FEDER-Hepato-Regio-Net Deutsche Forschungsgemeinschaft (Forschergruppe FOR 1202) Swiss National Science Foundation (SNF 31003A-138484) Deutsche Forschungsgemeinschaft (BR 3440/2-1) Deutsche Forschungsgemeinschaft (SFB 900) The funders had no role in study design, data collection and analysis, decision to publish, or preparation of the manuscript.

Competing Interests: The authors have declared that no competing interests exist.

* E-mail: volker.brass@uniklinik-freiburg.de

9 These authors contributed equally to this work.

¶ These authors also contributed equally to this work.

Introduction

With an estimated 120–180 million chronically infected individuals, HCV is a leading cause of chronic hepatitis, liver cirrhosis and hepatocellular carcinoma worldwide [1]. Antiviral therapy has improved considerably with the introduction of pegylated interferon-α and ribavirin as well as, more recently, the first generation of directly acting antivirals. However, many patients still do not respond to or cannot tolerate antiviral therapy. In addition, HCV continues to be transmitted in certain areas of the world [2]. Therefore, the development of preventive and therapeutic vaccines against hepatitis C is of major public health importance [3].

Innate and adaptive immune responses to HCV have been studied in great detail [4]. Resolution of acute hepatitis C correlates with the induction of strong and broad CD4+ and CD8+ T cell responses [5]. However, the majority of patients fail to eliminate HCV and develop chronic infection (reviewed in [5,6]). The high genetic variability of HCV significantly contributes to the escape from the immune system and complicates the development of an efficient vaccine [7]. Nevertheless, more recent data indicate that there is protective immunity against HCV [4].

A critical step for the understanding of the immunopathogenesis of HCV infection and HCV clearance is the presentation of viral epitopes on MHC class I molecules from infected cells. Most of the currently available experimental systems are limited, since an *a priori* defined set of synthetic peptides is used to either externally load target cells or to expand epitope-specific CD8+ T cells which are then used in downstream readout applications.

Therefore, the aim of this study was to identify specific MHC class I ligands which are naturally processed and presented by cells expressing HCV proteins. To this end, we engineered continuous human cell lines to inducibly express HCV proteins and to constitutively express high levels of functional HLA-A2. MHC class I molecules were isolated from large-scale cultures of these cell lines, followed by elution and identification of naturally processed CTL epitopes. This proof-of-concept study allowed the identification of two naturally processed HCV-derived MHC class I ligands. Although both epitopes have been described previously

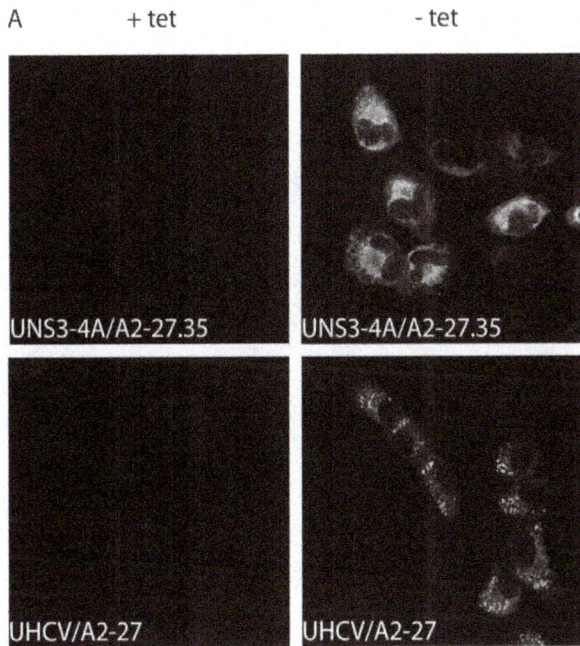

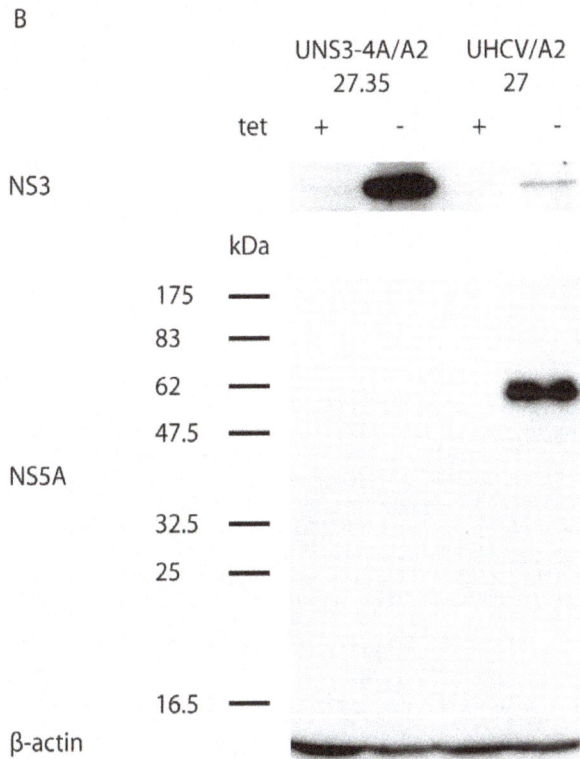

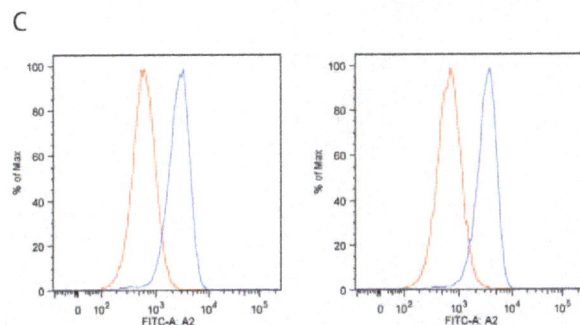

Figure 1. Cell lines UNS3-4A/A2-27.35 and UHCV/A2-27. (A) Indirect immunofluorescence microscopy of UNS3-4A/A2-27.35 and UHCV/A2-27 cells cultured for 48 h in the presence (+ tet) or absence (− tet) of tetracycline. Monoclonal antibody 1B6 against HCV NS3 was used as primary antibody. (B) Immunoblot analysis of UNS3-4A/A2-27.35 and UHCV/A2-27 cells were cultured for 48 h in the presence (+ tet) or absence (− tet) of tetracycline. Monoclonal antibodies 1B6 against HCV NS3 or 11H against NS5A were used as primary antibodies. (C) Left panel: HLA-A2 surface expression of UNS3-4A/A2-27.35 (blue histogram) compared to the founder cell line UNS3-4A-24 (red histogram). Right panel: HLA-A2 surface expression of UHCV/A2-27 (blue histogram) compared to the founder cell line UHCVcon-57.3 (red histogram).

by conventional T-cell dependent methods, this novel approach has the potential to identify novel and unconventional epitopes.

Results

Generation of stable cell lines inducibly expressing HCV proteins and constitutively expressing functional HLA-A2

To identify naturally processed HCV MHC class I ligands by mass spectrometry (MS), as described in this manuscript, U-2 OS human osteosarcoma-derived cell lines UNS3-4A-24 and UHCV-con-57.3 [8,9] were used. These cells allow for tight, tetracycline-regulated expression of the HCV nonstructural protein 3-4A (NS3-4A) complex or of the entire viral polyprotein, respectively. However, U-2 OS cells express low levels of endogenous HLA-A2 [10]. To enhance MHC class I presentation of HCV antigens, UNS3-4A-24 and UHCVcon-57.3 cells were engineered to overexpress HLA-A2. Clones UNS3-4A/A2-27.35 and UHCV/A2-27 were used in this study. As shown in Figure 1 (panels A and B), these cells allow for tightly regulated expression of HCV proteins, as documented by immunoblot and indirect immuno-fluorescence microscopy. NS3 was identified in both cell lines at the expected molecular mass of 70 kDa. Detection of NS5A at the expected molecular mass of 56 kDa demonstrates correct viral polyprotein processing in UHCV/A2-27 cells. Furthermore, cells were analyzed for HLA-A2 surface expression by flow cytometry (Fig. 1C). Compared to the founder cells, both cell lines showed an enhanced level of HLA-A2 surface expression. Furthermore, strong HLA-A2 expression could also be detected after withdrawal of tetracycline from the culture medium and induction of HCV protein expression (data not shown). This is a prerequisite for the elution of HCV proteins from MHC class I molecules isolated from these cells and demonstrates that HCV protein expression does not interfere with HLA-A2 surface expression.

Targeting of UNS3-4A/A2-27.35 and UHCV/A2-27 cells by HCV-specific CD8+ CTL

To explore whether UNS3-4A/A2-27.35 and UHCV/A2-27 cells can serve as targets for HCV-specific CTL, cell lines were cocultured with CD8+ T cell lines that have been raised to recognize epitopes localized in NS3 (B7, CINGVCWTV) or NS5B (B22, ALYDVVTKL), respectively. T cell activation was assessed by intracellular IFN-γ staining and flow cytometry. As shown in Figure 2A, CTL lines were activated specifically after external loading with the corresponding synthetic peptide. More important, coculture of UNS3-4A/A2-27.35 and UHCV/A2-27 cells with CD8+ T cell lines demonstrated their potential to endogenously express and functionally present HCV epitopes (Fig. 2B). Coculture of UNS3-4A/A2-27.35 and UHCV/A2-27 cells leads to efficient activation of the CD8+ T cell line which recognizes the B7 epitope located in NS3. By contrast, and as expected, the

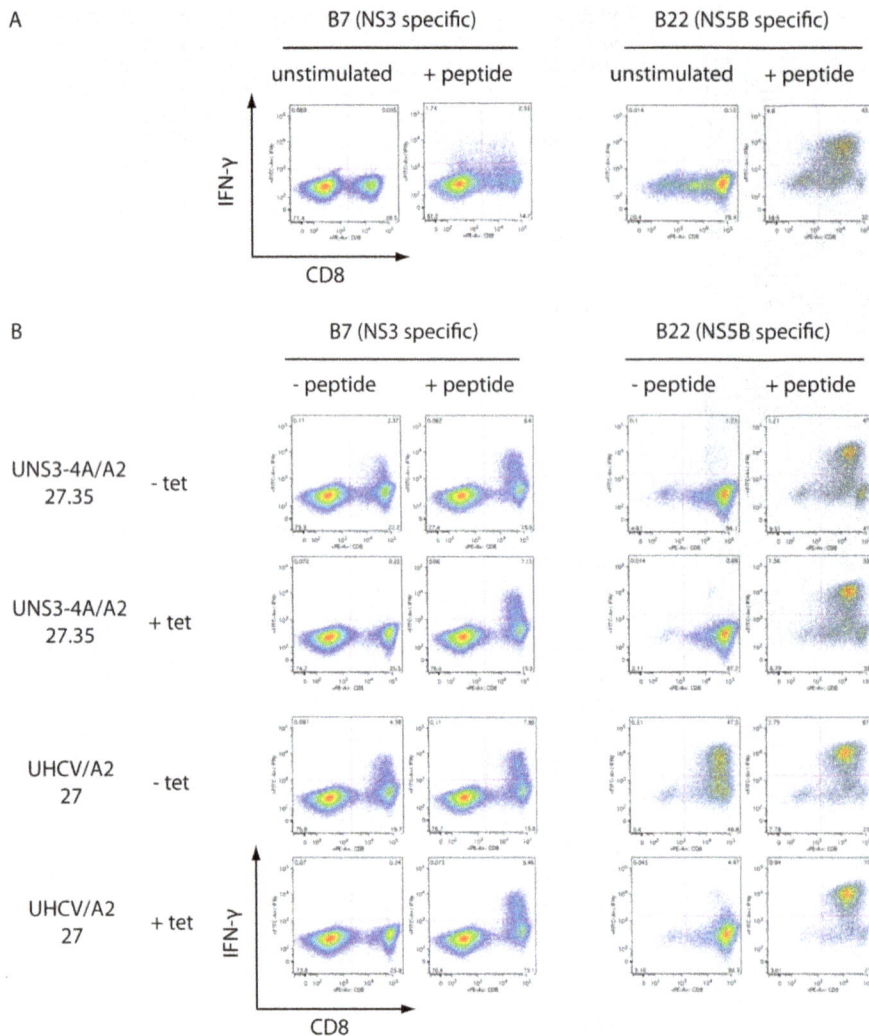

Figure 2. Targeting of UNS3-4A/A2-27.35 and UHCV/A2-27 cells by HCV-specific CD8+ CTL. CD8+ T cell lines B7 and B22, targeting epitopes CINGVCWTV derived from NS3 and ALYDVVTKL localized in NS5B, respectively, were used for coculture experiments. (**A**) Stimulation of T cell lines with the corresponding synthetic peptide leads to their activation, as indicated by the expression of IFN-γ. For all dot blots in this figure, CD8 is displayed on the abscissa and IFN-γ expression on the ordinate. (**B**) UNS3-4A/A2-27.35 and UHCV/A2-27 cells cultured for 48 h in the presence (+ tet) or absence (− tet) of tetracycline were cocultured with CD8+ T cell lines. In the presence of tetracycline, no substantial activation of the T cell lines could be observed. However, strong activation was observed after loading of UNS3-4A/A2-27.35 and UHCV/A2-27 cells with the corresponding synthetic peptide. By contrast, the NS3-specific B7 cell line was activated efficiently by UNS3-4A/A2-27.35 and UHCV/A2-27 cells cultured in the absence of tetracycline while the NS5B-specific B22 cell line could be stimulated only by UHCV/A2-27 cells, which upon tetracycline withdrawal express the entire HCV polyprotein. When target cells were cultured in the absence of tetracycline, T cell activation was enhanced only slightly by external peptide loading.

CD8+ T cell line specific for the B22 epitope in NS5B recognizes only UHCV/A2-27 cells. No CTL reactivity was observed when target cells were cultured in the presence of tetracycline, i.e. when they did not express HCV proteins. These results validate both cell lines as functional targets for HCV-specific CD8+ T cells and demonstrate that naturally processed HCV epitopes are efficiently presented on HLA-A2 molecules in this model system.

Large-scale expansion of UNS3-4A/A2-27.35 and UHCV/A2-27 cells, isolation of MHC class I molecules and peptide elution

To obtain the required amounts of HLA-A2 with naturally processed ligands, UNS3-4A/A2-27.35 and UHCV/A2-27 cells were expanded to large scale, yielding cell pellets of 60 ml each. From these, 1.5 and 7.0 nmol MHC class I molecules were

isolated, respectively. The yield of HLA molecules (as determined by Edman degradation) corresponds to 25 and 117 pmol per gram of cells, respectively. According to our experience, such values are typical for tumor cells that do not have abundant HLA expression [11]. The higher yield from UHCV/A2-27 cells presumably correlates with higher HLA-A2 expression. Subsequently, ligands were eluted from MHC class I molecules, as described in the Materials and Methods section and as illustrated schematically in Figure 3.

Detection of HCV-derived peptides in MHC class I ligands isolated from UNS3-4A/A2-27.35 and UHCV/A2-27 cells

Naturally processed MHC class I ligands of UNS3-4A/A2-27.35 and UHCV/A2-27 cells were separated by high performance liquid chromatography (HPLC) and analyzed by electro-

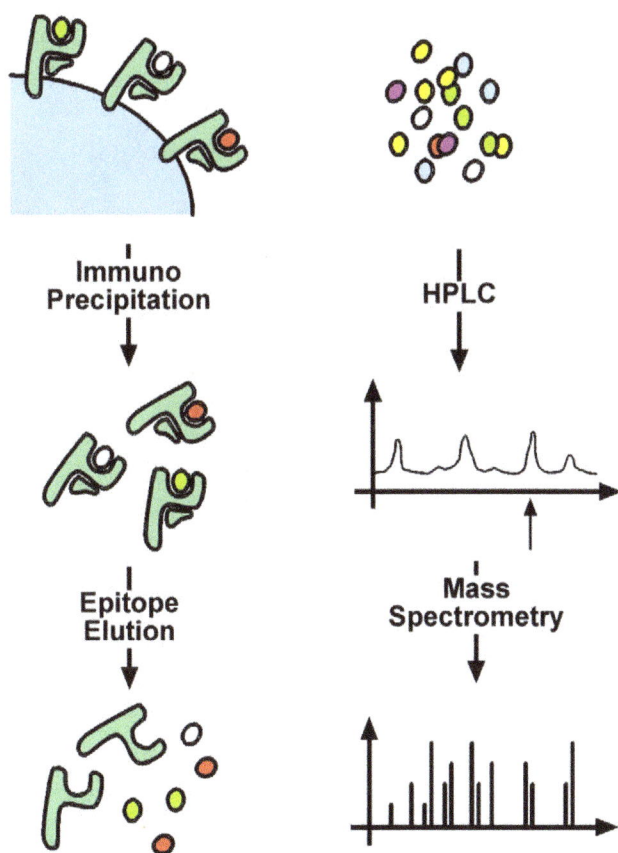

Figure 3. Identification of naturally processed HLA-A2 ligands. Following large-scale expansion, cell pellets were lysed and MHC class I molecules with bound ligands were purified by immunoprecipitation. Subsequently, ligands were eluted and separated by high performance liquid chromatography (HPLC). Fractions of interest were sequenced by mass spectrometry.

spray ionization (ESI) quadrupole time of flight (Q-TOF) mass spectrometry (MS)/tandem MS, as described in the Materials and Methods section. HPLC and MS of predicted synthetic peptides allowed to screen for epitopes of interest in ligand preparations and subsequent sequence verification by online liquid chromatography (LC) MS analysis. With this approach, two naturally processed MHC class I ligands could be identified. The peptide isolated from UNS3-4A/A2-27.35 cells is derived from NS3 (HCV polyprotein position 1406–1415; $NS3-4A_{1406–1415}$) with the amino acid sequence KLVALGINAV (Fig. 4). The epitope isolated from UHCV/A2-27 cells is derived from NS5B (HCV polyprotein position 2594–2602; $NS5B_{2594–2602}$) with the amino acid sequence ALYDVVSKL (Figures S1 and S2) that is cross-recognized by epitope ALYDVVTKL-specific T cells (Fig. 2) in spite of the threonine instead of serine in position 2600.

Discussion

In this proof-of-concept study, we identified two naturally processed HCV-derived MHC class I ligands. These were isolated from cell lines UNS3-4A/A2-27.35 and UHCV/A2-27, allowing the tightly regulated expression of HCV NS3-4A and of the entire viral polyprotein, respectively, as well as the constitutive expression of functional HLA-A2. These cell lines were specifically recognized by HCV-specific CTL, indicating efficient presentation of

naturally processed MHC class I ligands. MHC class I molecules were purified by immunoprecipitation. Ligands were eluted, separated by HPLC and sequenced by ESI Q-TOF-MS/MS. Among other MHC class I ligands, two naturally processed HCV epitopes, $NS3_{1406–1415}$ and $NS5B_{2594–2602}$ were identified. Interestingly, both had been described previously, based on prediction algorithms [12,13].

HCV replication occurs within the cytoplasm of infected cells. Translation and processing of the viral polyprotein take place in close association with membranes of the endoplasmic reticulum. HCV proteins are then subjected to cellular antigen-processing pathways, resulting in the presentation of HCV epitopes on the cell surface. It has been estimated that approx. 2,000–10,000 molecules of a protein are required to allow the presentation of one antigen [14]. In fact, antigen processing involves a highly complex interplay of multiple steps and factors. Degradation by the proteasome, interaction with chaperones such as calnexin, incorporation into the peptide loading complex, involving other chaperones such as tapasin, peptide trimming by aminopeptidases, loading onto empty MHC I molecules, and, finally, transport across the secretory pathway to the cell surface are pivotal steps that are tightly coordinated during this process. Several approaches typically used to identify MHC I epitopes bypass this complex process, which could be one reason why they eventually fail to identify relevant epitopes. These approaches include (i) the induction and identification of T cells through the stimulation with defined synthetic peptides and (ii) the generation of epitope-specific T cell lines which are subsequently tested for their reactivity with cells expressing a given antigen. There are several potential limitations of these systems as they are restricted by the use of synthetic peptides or depend on the presence of functional HCV-specific T cells.

Against this background, a direct T cell-independent identification of naturally processed MHC class I epitopes would be desirable. In this context, we have demonstrated that peptide elution from HCV protein-expressing cell lines and subsequent direct sequencing by ESI Q-TOF MS/MS is a feasible and powerful approach. Using this experimental setting, UNS3-4A/A2-27.35 and UHCV/A2-27 cells turned out to be valuable tools which could be further employed to study HCV antigen processing and presentation in the future. The identification of two known HCV HLA-A2 ligands that are localized in NS3 and NS5B demonstrates their authentic processing and presentation in vivo. Furthermore, a more elaborate approach, using tracer substances such as isotope-labeled peptides, might hold promise for the quantification of an epitope relative to the complete repertoire of presented ligands. Few reports on Epstein-Barr virus-encoded proteins investigated how the amount of presented MHC I antigen complex could influence the efficiency of recognition by CD8+ T cells [15,16]. Therefore, in addition to the proof-of-principle of this particular experimental setting, the identification of the two epitopes by this novel approach underlines their importance as natural targets for HCV-specific T cells. CTL responses against these epitopes could be of particular importance to control viral infection and may be included as targets in future vaccination strategies.

In principle, the direct sequencing of MHC I ligands should allow to identify epitopes after posttranslational protein modifications such as glycosylation, phosphorylation and proteolytic processing as well as unconventional epitopes derived from alternative reading frames or RNA splicing that are not detected by the current conventional methods. Future studies aimed at identifying naturally processed HCV-derived MHC class I ligands may provide novel insights into epitope processing and presenta-

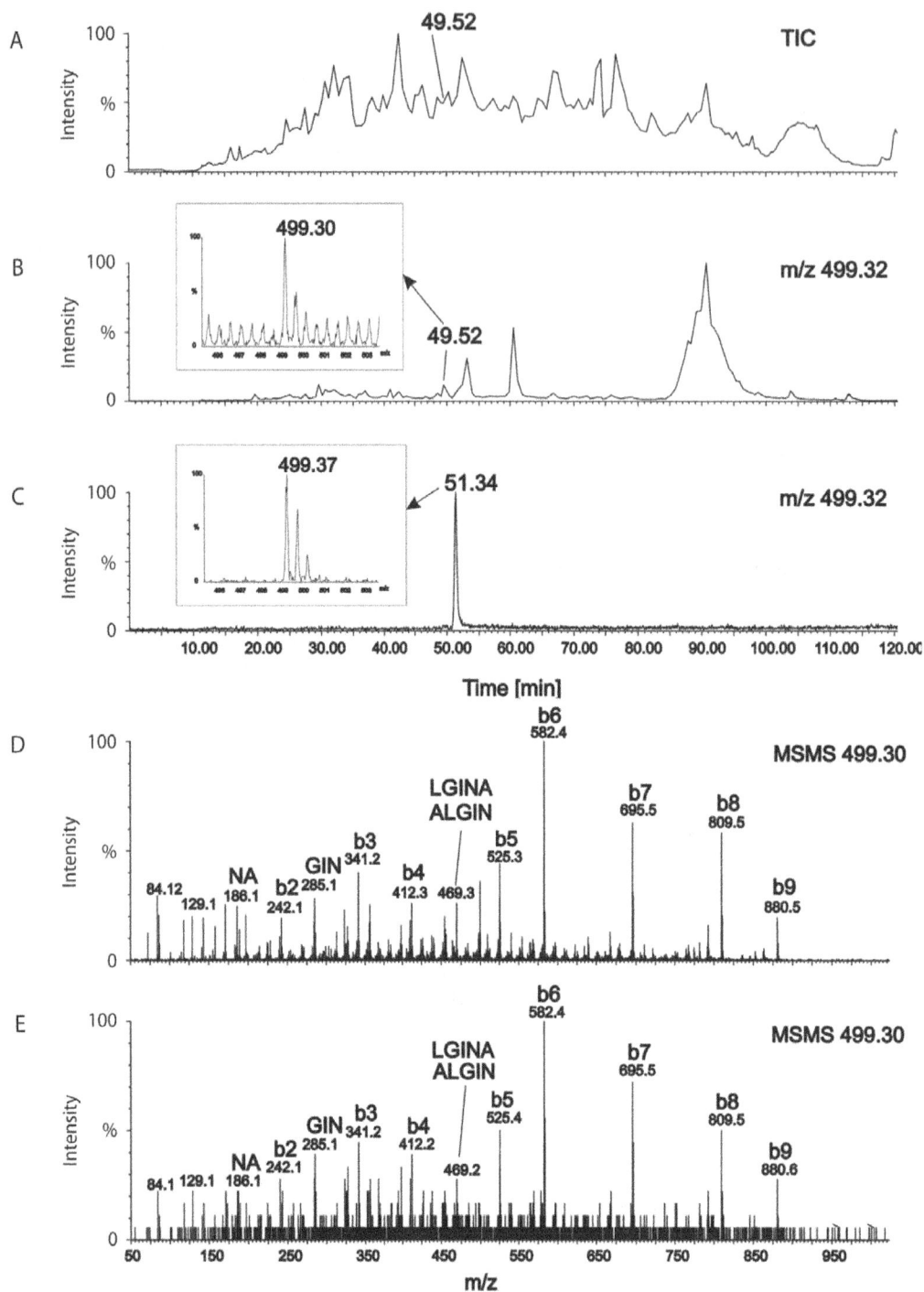

Figure 4. Detection and characterization of peptide NS3$_{1406-1415}$ in MHC class I ligands isolated from UNS3-4A/A2-27.35 cells. Naturally processed MHC class I ligands of UNS3-4A/A2-27.35 cells were separated by high performance liquid chromatography (HPLC) and analyzed by mass spectrometry. (A) A chromatogram of the total ion current (TIC) is shown. (B) The mass chromatogram of ionized peptides with m/z = 499.3 reveals a peak at retention time t = 49.5 min that correlates with (C) a signal of the synthetic peptide NS3$_{1406-1415}$ (KLVALGINAV) which elutes after 51.3 min. In all graphs, HPLC retention times are shown on the abscissa and relative signal intensity on the ordinate. Small insets in panels B and C show mass spectra of peptides eluted at the indicated time points. (D) MSMS spectrum of the synthetic peptide NS3$_{1406-1415}$ (KLVALGINAV). (E) MSMS spectrum of the natural peptide isolated from UNS3-4A/A2-27.35 cells, revealing amino acid sequence KLVALGINAV identical to the synthetic peptide shown in panel A. Identified amino acid sequences of peptide fragments are indicated on the top of the peaks. Due to the protonated N-terminal lysine residue, the b series of peptide fragments is dominating the MSMS spectra.

tion as well as recognition, thereby contributing to the understanding of HCV pathogenesis.

Materials and Methods

Establishment and characterization of inducible cell lines

U-2 OS human osteosarcoma-derived cell lines inducibly expressing the HCV NS3-4A complex (UNS3-4A-24) or the entire HCV polyprotein (UHCVcon-57.3) derived from the HCV H77 isolate (genotype 1a) have been described previously [8,9]. These cells were stably transfected with a cytomegalovirus (CMV) promoter-driven genomic HLA-A2 construct to augment MHC class I expression, as described previously [10]. In brief, cells were cotransfected with pCMV/HLA-A2 and pTK-Hyg (Clontech, Palo Alto, CA) followed by selection with 500 μg/ml G418 (for the selection of the tetracycline-regulated transactivator, tTA), 1 μg/ml puromycin (for the selection of the HCV expression construct driven by a tTA-dependent promoter), and 100 μg/ml hygromycin (for the selection of the HLA-A2 expression construct). Stable transfectants were cloned and screened for high-level HLA-A2 expression by flow cytometry, as specified below. Tightly regulated expression of HCV proteins was confirmed by immunoblot analyses which were performed as described previously [17]. Clones UNS3-4A/A2-27.35 and UHCV/A2-27 were selected for further experiments.

Immunofluorescence microscopy

Indirect Immunofluorescence was performed as described previously [17]. The specific antibodies against NS3 (1B6) and NS5A (11H) have been described elsewhere [8,18].

Peptides and antibodies for flow cytometry

Peptides with the amino acid sequence CINGVCWTV and ALYDVVTKL were synthesized with free amino and carboxyl termini by standard Fmoc chemistry (Genaxxon BioScience, Ulm, Germany). The peptides were dissolved and diluted as described previously [19]. Fluorescein isothiocyanate- or phycoerythrin-labeled anti-HLA-A2, anti-CD8 and anti-IFN-γ antibodies as well as isotype-matched control antibodies were from Becton Dickinson (Heidelberg, Germany) and were used following the manufacturer's recommendations.

CD8+ T cell lines

Blood samples were obtained from two HLA-A2-positive patients with chronic HCV infection after written informed consent and in accordance with the 1975 Declaration of Helsinki, federal guidelines as well as the local ethics committee as described previously [20]. T cell lines raised against the epitopes CINGVCWTV and ALYDVVTKL, localized within NS3 and NS5B, respectively, were referred to as B7 and B22.

Intracellular IFN-γ staining

Intracellular IFN-γ staining was performed essentially as described [19]. In brief, T cells (2×10^5 per 96-well) were stimulated either directly with peptides (10 μg/ml) or by UNS3-4A/A2-27.35 and UHCV/A2-27 cell lines in the presence of 50 U/ml human recombinant interleukin 2 (Hoffmann La Roche, Basel, Switzerland) and 1 μl/ml brefeldin A (Becton Dickinson, Heidelberg, Germany). As a positive control, UNS3-4A/A2-27.35 and UHCV/A2-27 were externally loaded with the corresponding synthetic peptide before the coculture. Cocultivation experiments of T cell lines with UNS3-4A/A2-27.35 and UHCV/A2-27 cells were performed at an effector-to-target ratio of 1:1. After 5 h of incubation (37°C, 5% CO_2), cells from each well were blocked

with IgG_1 antibodies and stained with antibodies against CD8. After permeabilization with Cytofix/Cytoperm (Becton Dickinson, Heidelberg, Germany), cells were stained with antibodies against IFN-γ and fixed in 2% paraformaldehyde. Samples were acquired on a FACS Canto II flow cytometer (Becton Dickinson, Heidelberg, Germany) and analyzed with FlowJo v8.8.6 software (TreeStar, Ashland, OR).

Cell culture and large-scale expansion

UNS3-4A/A2-27.35 and UHCV/A2-27 cells were grown in Dulbecco's modified Eagle medium (DMEM), 10% fetal bovine serum, 500 μg/ml G418, 1 μg/ml puromycin, 100 μg/ml hygromycin and 1 μg/ml tetracycline. For large-scale expansion, cells were initially expanded in 15-cm culture dishes and then seeded into a forty-tray array (Nunc Cell Factories, Nalge Nunc International, Naperville, IL). Tetracycline was withdrawn to induce expression of the HCV proteins, followed one week later by detachment with Accutase (PAA, Pasching, Austria), harvesting and washing of the cell pellets.

Isolation of MHC class I molecules, peptide elution and sequencing

HLA-presented peptides were obtained by immunoprecipitation of HLA molecules from cell lines UNS3-4A/A2-27.35 and UHCV/A2-27. One volume of 2× lysis buffer containing PBS, 0.6% CHAPS, and complete protease inhibitor (Roche Diagnostics, Basel, Switzerland) was added to shock-frozen cell pellets. Subsequently, cells were homogenized by stirring for 1 h at 4°C. After four times 30 s sonication, the sample was stirred again for 1 h and subsequently centrifuged at $3,000 \times g$, 4°C for 20 min, and afterwards at $150,000 \times g$, 4°C for 1 h, to remove cell debris. Finally, the supernatant was passed through a 0.2-μm filter (Sartorius, Göttingen, Germany).

For immunoprecipitation, lysates were applied for at least 12 h to a CNBr-activated Sepharose 4B column (40 mg Sepharose per mg tissue; GE Healthcare, München, Germany) to which the HLA-A-, HLA-B-, and HLA-C-specific antibody W6/32 had been coupled (1 mg antibody per mg tissue), as described by the manufacturer. After binding, the column was rinsed with 250 ml PBS and subsequently with 500 ml double-distilled water. For elution and dissociation of bound HLA-peptide complexes, the column was shaken at least four times in one bed volume of 0.1% trifluoroacetic acid (TFA) for 20 min at room temperature. The four TFA eluates were subsequently collected and combined. Finally, the HLA-presented peptides were isolated by ultrafiltration through a Centricon 10-kDa cutoff membrane (Millipore, Billerica, MA). For LC-MS, samples were freeze-dried and resuspended in 0.1% formic acid.

HLA-extracted peptide pools were separated by reverse-phase HPLC (Ultimate Dionex, Amsterdam, Netherlands) and analyzed by nano-ESI MS on a Q-TOF MS/tandem MS (Micromass, Manchester, UK), as described [21]. Fragment spectra were analyzed manually and database searches (National Center for Biotechnology Information) were carried out using Multiple Alignment System for Protein Sequences Based on Three-way Dynamic Programming (MASCOT, http://www.matrixscience.com). Mass chromatograms of ionized synthetic peptides $NS3_{1406-1415}$ (KLVALGINAV) and $NS5B_{2594-2602}$ (ALYDVVSKL) were used as references to identify peaks of interest.

Supporting Information

Figure S1 Detection of peptide $NS5B_{2594-2602}$ in MHC class I ligands isolated from UHCV/A2-27 cells. Naturally processed MHC class I ligands of UHCV/A2-27 cells were separated by high performance liquid chromatography (HPLC) and analyzed by mass spectrometry. (A) Mass chromatogram of the total ion current (TIC). (B) In the mass chromatogram for m/z = 504.3 a signal peak is detected at retention time t = 63.5 min which was further analyzed by tandem mass spectrometry (MS/MS) (cf. Fig. S2).

Figure S2   Identification of the NS5B$_{2594-2602}$ epitope in UHCV/A2-27 cell-derived MHC class I ligands. (A) UHCV/A2-27 cell-derived peptide (cf. Fig. S1) and (B) the synthetic peptide NS5B$_{2594-2602}$ (ALYDVVSKL) were both detected with m/z = 504.3 at a retention time of t = 63.2 min and subsequently analyzed by tandem mass spectrometry (MS/MS). Fragmentation spectra of the synthetic peptide and the natural peptide are shown in panels A and B, respectively. Analysis of the natural peptide reveals amino acid sequence ALYDVVSKL and is identical to that of the synthetic peptide. Identified amino acid sequences of peptide fragments are indicated on top of the peaks. Due to the protonated lysine residue close to the C-terminal residue, the y series of peptide fragments is dominating the MSMS spectra.

Acknowledgments

We gratefully acknowledge Natalia Ivashkina, Christel Gremion and Benno Grabscheid for their contributions in the early phase of this project, Anja Wahl for excellent technical assistance, as well as Christoph Neumann-Haefelin and Robert Thimme for critical discussions and support.

Author Contributions

Conceived and designed the experiments: BW CT NK HEB HR AC SS DM VB. Performed the experiments: BW CT BB NK SS VB. Analyzed the data: BW CT BB NK HEB HR AC SS DM VB. Contributed reagents/materials/analysis tools: BW CT NK HEB HR AC SS DM VB. Wrote the paper: BW CT BB NK HEB AC SS DM VB.

References

1. Nature Outlook (2011) Hepatitis C. Nature 474: S1–S21.
2. Miller FD, Abu-Raddad LJ (2010) Evidence of intense ongoing endemic transmission of hepatitis C virus in Egypt. Proc Natl Acad Sci U S A 107: 14757–14762.
3. Thimme R, Neumann-Haefelin C, Böttler T, Blum HE (2008) Adaptive immune responses to hepatitis C virus: from viral immunobiology to a vaccine. Biol Chem 389: 457–467.
4. Rehermann B (2009) Hepatitis C virus versus innate and adaptive immune responses: a tale of coevolution and coexistence. J Clin Invest 119: 1745–1754.
5. Bowen DG, Walker CM (2005) Adaptive immune responses in acute and chronic hepatitis C virus infection. Nature 436: 946–952.
6. Seeff LB (2000) Why is there such difficulty in defining the natural history of hepatitis C? Transfusion 40: 1161–1164.
7. Walker CM (2010) Adaptive immunity to the hepatitis C virus. Adv Virus Res 78: 43–86.
8. Wölk B, Sansonno D, Kräusslich HG, Dammacco F, Rice CM, et al. (2000) Subcellular localization, stability, and trans-cleavage competence of the hepatitis C virus NS3-NS4A complex expressed in tetracycline-regulated cell lines. J Virol 74: 2293–2304.
9. Schmidt-Mende J, Bieck E, Hügle T, Penin F, Rice CM, et al. (2001) Determinants for membrane association of the hepatitis C virus RNA-dependent RNA polymerase. J Biol Chem 276: 44052–44063.
10. Gremion C, Grabscheid B, Wölk B, Moradpour D, Reichen J, et al. (2004) Cytotoxic T lymphocytes derived from patients with chronic hepatitis C virus infection kill bystander cells via Fas-FasL interaction. J Virol 78: 2152–2157.
11. Stickel JS, Stickel N, Hennenlotter J, Klingel K, Stenzl A, et al. (2011) Quantification of HLA class I molecules on renal cell carcinoma using Edman degradation. BMC Urol 11: 1.
12. Cerny A, McHutchison JG, Pasquinelli C, Brown ME, Brothers MA, et al. (1995) Cytotoxic T lymphocyte response to hepatitis C virus-derived peptides containing the HLA A2.1 binding motif. J Clin Invest 95: 521–530.
13. Lechner F, Wong DK, Dunbar PR, Chapman R, Chung RT, et al. (2000) Analysis of successful immune responses in persons infected with hepatitis C virus. J Exp Med 191: 1499–1512.
14. Hillen N, Stevanovic S (2006) Contribution of mass spectrometry-based proteomics to immunology. Expert Rev Proteomics 3: 653–664.
15. Crotzer VL, Christian RE, Brooks JM, Shabanowitz J, Settlage RE, et al. (2000) Immunodominance among EBV-derived epitopes restricted by HLA-B27 does not correlate with epitope abundance in EBV-transformed B-lymphoblastoid cell lines. J Immunol 164: 6120–6129.
16. Tellam J, Fogg MH, Rist M, Connolly G, Tscharke D, et al. (2007) Influence of translation efficiency of homologous viral proteins on the endogenous presentation of CD8+ T cell epitopes. J Exp Med 204: 525–532.
17. Moradpour D, Kary P, Rice CM, Blum HE (1998) Continuous human cell lines inducibly expressing hepatitis C virus structural and nonstructural proteins. Hepatology 28: 192–201.
18. Brass V, Bieck E, Montserret R, Wölk B, Hellings JA, et al. (2002) An amino-terminal amphipathic alpha-helix mediates membrane association of the hepatitis C virus nonstructural protein 5A. J Biol Chem 277: 8130–8139.
19. Thimme R, Oldach D, Chang KM, Steiger C, Ray SC, et al. (2001) Determinants of viral clearance and persistence during acute hepatitis C virus infection. J Exp Med 194: 1395–1406.
20. Jo J, Aichele U, Kersting N, Klein R, Aichele P, et al. (2009) Analysis of CD8+ T-cell-mediated inhibition of hepatitis C virus replication using a novel immunological model. Gastroenterology 136: 1391–1401.
21. Lemmel C, Weik S, Eberle U, Dengjel J, Kratt T, et al. (2004) Differential quantitative analysis of MHC ligands by mass spectrometry using stable isotope labeling. Nat Biotechnol 22: 450–454.

6

An Enzyme-Generated Fragment of Tau Measured in Serum Shows an Inverse Correlation to Cognitive Function

Kim Henriksen[1]*, Yaguo Wang[2], Mette G. Sørensen[1], Natasha Barascuk[1], Joyce Suhy[3], Jan T. Pedersen[4], Kevin L. Duffin[5], Robert A. Dean[5], Monika Pajak[1], Claus Christiansen[1], Qinlong Zheng[2], Morten A. Karsdal[1]

1 Nordic Bioscience Biomarkers and Research, Herlev, Denmark, 2 Nordic Bioscience, Beijing, China, 3 CCBR-Synarc, San Francisco, California, United States of America, 4 Neurodegeneration, H. Lundbeck A/S, Copenhagen, Denmark, 5 Translational Sciences Department, Eli Lilly and Company, Indianapolis, Indiana, United States of America

Abstract

Objective: Alzheimer's disease (AD) is a devastating neurological disease characterized by pathological proteolytic cleavage of tau protein, which appears to initiate death of the neurons. The objective of this study was to investigate whether a proteolytic fragment of the tau protein could serve as blood-based biomarker of cognitive function in AD.

Methods: We developed a highly sensitive ELISA assay specifically detecting an A Disintegrin and Metalloproteinase 10 (ADAM10)-generated fragment of tau (Tau-A). We characterized the assay in detail with to respect specificity and reactivity in healthy human serum. We used samples from the Tg4510 tau transgenic mice, which over-express the tau mutant P301L and exhibit a tauopathy with similarities to that observed in AD. We used serum samples from 21 well-characterized Alzheimer's patients, and we correlated the Tau-A levels to cognitive function.

Results: The Tau-A ELISA specifically detected the cleavage sequence at the N-terminus of a fragment of tau generated by ADAM10 with no cross-reactivity to intact tau or brain extracts. In brain extracts from Tg4510 mice compared to wt controls we found 10-fold higher levels of Tau-A ($p<0.001$), which indicates a pathological relevance of this marker. In serum from healthy individuals we found robust and reproducible levels of Tau-A, indicating that the analyte is present in serum. In serum from AD patients an inverse correlation ($R^2 = 0.46$, $p<0.001$) between the cognitive assessment score (Mattis Dementia Rating Scale (MDRS)) and Tau-A levels was observed.

Conclusion: Based on the hypothesis that tau is cleaved proteolytically and then released into the blood, we here provide evidence for the presence of an ADAM10-generated tau fragment (Tau-A) in serum. In addition, the levels of Tau-A showed an inverse correlation to cognitive function, which could indicate that this marker is a serum marker with pathological relevance for AD.

Editor: John C S Breitner, McGill University/Douglas Mental Health Univ. Institute, Canada

Funding: Natasha Barascuk received funding from the Ministry of Science, Technology and Education, and Mette Sørensen received funding from the Danish Research Foundation (Den Danske Forskningsfond). The funders had no role in study design, data collection and analysis, decision to publish, or preparation of the manuscript.

Competing Interests: MAK and CC own stock in Nordic Bioscience A/S. KH, YW, MGS, NB, MP, CC and MAK are employees of Nordic Bioscience A/S. KLD and RAD are employees of Eli Lilly and Company. JTP is an employee of H. Lundbeck A/S and JS is an employee of CCBR-Synarc.

* E-mail: kh@nordicbioscience.com

Introduction

Alzheimer's disease (AD) is a devastating neurological disease, which with the ever-increasing age of the population is expected to explode in numbers. AD is characterized by global cognitive decline including language breakdown. At present, treatment is limited to alleviation of the symptoms and disease modifying approaches have so far failed [1–3]. A contributing factor to the lack of success within drug development is the absence of blood-based biomarkers, which indicate disease progression and thereby can help the selection of patients for clinical trials [4].

Hence, methods allowing monitoring of neurodegeneration in AD, i.e. before onset of cognitive loss, are intensely sought, as these are essential to design clinical trials assessing the potential of drugs to prevent progression of AD [4].

Cerebrospinal fluid (CSF) biomarkers have provided diagnostic value for AD; however, their application is limited owing to the invasiveness of lumbar puncture [4].

Potential candidate biomarkers are protein fragments which reflect specific cleavage sites in proteins, and, due to their smaller size, may pass the Blood-Brain-Barrier (BBB) and thereby be detected in serum [5]. Importantly, these smaller protein fragments may yield more information than their intact counterparts because they have been degraded by specific enzymes, which may be an important feature of AD [6,7].

In AD, the pathological processing of the protein Tau by proteases is of great interest [8], as this appears to be a key correlate of neuronal cell death [6]. Proteolytic cleavage of tau is mediated by several different proteases, such as caspases and calpains [8]. However, several other proteases also appear to play a role in neuronal degeneration even though they mainly have been associated with secretase functions [8].

We hypothesized that a link between plaques and NFTs involves a process in which the intracellular tau protein is exposed to extracellular or even circulating secretases, such as ADAM10, i.e. during neuronal apoptosis [6]. Secondly, we hypothesized that this secretase-mediated cleavage of tau would lead to the generation of fragments which could be used as biomarkers of AD.

We thus aimed to develop a useful serum assay monitoring a tau degradation fragment generated by ADAM10, a putative α-secretase [8] and assessed the pathological relevance of this assay by its ability to detect tau degradation fragments in rodent samples, as well as human serum samples collected from both healthy individuals and from Alzheimer's patients.

Materials and Methods

In Vitro cleavage for mass spectrometry

Protease cleavage was performed by mixing 100 µg tau and 1 µg of enzyme (ADAM10) MMP buffer (100 mM Tris-HCl, 100 mM NaCl, 10 mM $CaCl_2$, 2 mM Zn acetate, pH 8.0) and incubating for 7 days. Finally the cleavage was verified by visualization using the SilverXpress® Silver Staining Kit (cat. no. LC6100, Invitrogen, Carlsbad, Ca, USA) according to the manufacturer's instructions.

Peptide identification

Peptide fragments in the *in vitro* cleaved samples were identified using matrix-assisted laser desorption time of flight mass spectrometry (MALDI-TOF MS) and liquid chromatography coupled to electro spray ionization (ESI) tandem mass spectrometry (LC-MS/MS). MALDI-TOF samples were purified using C18 zip-tips (cat.no.ZTC18SO24, Millipore, Billerica, MA, USA) according to specifications and 0.1 µg of material was eluted onto a MTP 384 ground steel target plate (Bruker-Daltonics, Bremen, Germany). MALDI tandem mass spectra were recorded on a Bruker ultraflex MALDI-TOF/TOF mass spectrometer (Bruker-Daltonics, Bremen, Germany) in positive ion reflector mode. Mass spectra were externally calibrated in the m/z range of $800-4000$ using peptides generated by tryptic digestion of bovine β-lactoglobulin. The m/z software "Flexanalysis" (Bruker-Daltonics, Bremen, Germany) was used to analyze spectra. LCMS samples were ultra-filtrated to remove proteins above 10 kDa, the pH was adjusted to 2.0 using formic acid, and a 4 µL sample was analyzed by LC-MS/MS. LC was performed on a nanoACQUITY UPLC BEH C18 column (Waters, Milford, MA, USA) using a formic acid/acetonitrile gradient. MS and MS/MS were performed on a Synapt High Definition Mass Spectrometry quadruple time of flight MS (QUAD-TOF; Waters, Milford, MA, USA), with acquisition range of 350-1600 m/z in MS and 50-2000 m/z, in MS/MS. The software "ProteinLynx Global SERVER (PLGS)" (Waters, Milford, MA, USA) was used to analyze spectra and generate peak lists. To identify peptides, MS and MS/MS data were searched against a tau (FASTA) protein database using the Mascot 2.2 (Matrix Science, Boston, MA, USA) software with either the MALDI-TOF/TOF or ESI-QUAD-TOF settings.

Selection of peptide for immunizations

The first six amino acids of each free end of the sequences identified by MS were regarded as neo-epitopes generated by the protease in question. All obtained protease-generated sequences were analyzed for homology and distance to other cleavage sites and then blasted for homology using the NPS@:network protein sequence analysis, and from these analyses three unique cleavage fragments were selected.

Reagents and peptides

All reagents were standard high-quality chemicals from companies such as Merck and Sigma Aldrich. The synthetic peptides used for monoclonal antibody production and validation were: Immunogenic peptide:

TPRGAAPPGQ-GGC-KLH (Keyhole-Limpet-Hemocyanin)
Selection/screening peptide
TPRGAAPPGQ
De-selection peptide:
ATPRGAAPPGQ

Deselection peptides elongated with one amino acid in the N-terminus were purchased from the Chinese Peptide Company, Beijing, China. Peptide conjugation reagents were produced by Pierce (Thermofisher, Denmark).

Development of the antibody and ELISA

The methods used for monoclonal antibody development were as previously described [9,10].

ELISA methodology

In preliminary experiments, we optimized the reagents, their concentrations and the incubation periods by performing several checkerboard analyses (data not shown).

The final Tau-A ELISA was developed as follows: A 96-well ELISA plate pre-coated with streptavidin was further coated with 6 ng/ml of the synthetic peptide TPRGAAPPGQ-Biotin dissolved in Tris-BT buffer at 20°C for 30 min by constant shaking at 300 rpm. The plate was washed five times in washing buffer and 20 µl of sample was added, followed by 100 µl of peroxidase conjugated anti-human mAb-Tau-A solution (50 ng/ml). The plate was incubated for 1 h at 20°C in assay buffer (150 mM Trizma, 1% BSA, 0.05% Tween-20, 0.36% Bronidox L5) during which time it was shaken at 300 rpm.

The plate was again washed five times followed by the addition of 100 µl tetramethylbenzinidine (TMB) (Kem-En-Tec cat.438OH). The plate was incubated for 15 min in darkness and shaken at 300 rpm. In order to cease the reaction, 100 µl of stopping solution (95–97% H_2SO_4, Merck Cat. No.: 1.00731) was added and the plate was analysed in the ELISA reader at 450 nm with 650 nm as the reference.

Buffers used for the ELISAs

Buffer used for dissolving the coating peptide was composed of the following: (50 mM PBS, 137 mM, 1% BSA, 0.05% Tween-20, 0.9% EDTA, 0.36% Bronidox L5, 10% sorbitol, pH = 7), and reaction stopping buffer composed of 0.1% H_2SO_4

ELISA-plates used for the assay development were Streptavidin-coated from Roche cat.: 11940279. All ELISA plates were analysed with the ELISA reader from Molecular Devices, SpectraMax M, (CA, USA).

Standards

A standard curve was performed by serial dilution of the synthetic selection peptide. Synthetic standard concentrations for

Tau-A were 0, 0.59, 1.17, 2.34, 4.69, 9.38, 18.75, 37.5, 75, 150, and 300 ng/ml.

Samples for testing native reactivity of the antibodies

For assay development and validation, serum and plasma from 15 healthy adult volunteers aged 23–45 years of both genders were used. We also tested serum samples from mice and rats to determine the level of interspecies cross reactivity.

Animal samples

Tissues including brain, liver, muscle, colon, kidney, lung, skin and pancreas isolated from 5 six-month-old Sprague Dawley rats and 5 brains from each of either the wildtype or Tg4510 mice were flash-frozen in liquid nitrogen and pulverized using a Bessman pulverizer. The "powder" was transferred to a vial and weighed. Extraction buffer (50 mM Tris-HCl, 50 mM HEPES, 1 mM EDTA, 0.5% sodium deoxycholate, 15% glycerol, protease inhibitor cocktail (Roche cat#05056489001), pH 8.3) was added at 1 mL buffer/250 mg tissue. The lysate was cleared by sonication. After sonication the debris was spun at 4°C/5 min/ 10000 rpm and the supernatants were collected and stored at −80°C until further use.

Protein concentrations were determined using the DC Protein Assay (Biorad).

In Vitro cleavage of tissues

Protease cleavage was performed by mixing 100 µg of tissue extract and 1 µg of ADAM10 in MMP buffer (100 mM Tris-HCl, 100 mM NaCl, 10 mM $CaCl_2$, 2 mM Zn acetate, pH 8,0) and incubating for 7 days. Finally, the cleavage was verified by western blotting and ELISA analysis.

Western Blotting

20 µg of each rat tissue extract and 100 µg of each mouse tissue extract was loaded onto an SDS-PAGE gel, and the gel was run, followed by transfer of the proteins to nitrocellulose membranes. Ponceau Red staining was then used to verify equal protein loading on the membranes. The levels of Tau-A fragments and total tau protein were detected by incubation with the primary antibodies diluted to 100 ng/mL in TBS-T containing skim milk powder [11]. A secondary antibody recognizing mouse IgG conjugated to horse-radish peroxidase was then added, and finally the blot was visualized using enhanced chemiluminescence as previously described [11].

Human samples

Aliquots of serum collected at baseline were obtained from Alzheimer's disease (AD) patients (n = 21) that participated in a clinical trial entitled "*A Prospective, Randomized Start, Multicenter, Double-Blinded, Concurrently Placebo-Controlled Study to Evaluate the Effect of Flow-Regulated Ventriculoperitoneal Shunting on Progression of Alzheimer's disease: An Investigation of the Safety and Effectiveness of the COGNIshunt^{TM} CNS Fluid Shunt System*" sponsored by Eunoe, Inc. (Pleasanton, CA). The trial is registered on clinicaltrials.gov with the identified: NCT00056628. Characteristics: Age at onset 70(SD+/−7), Females/Males (16/6), baseline MDRS score 112(SD+/−12), Intact tau measured in baseline CSF samples (collected at shunt implantation) was 849 pg/mL (SD+/−1009 pg/mL). Further details of the study were previously reported by Silverberg et al [12]. Subject inclusion required a clinical diagnosis of probable AD based on National Institute of Neurological Disorders and Stroke–Alzheimer's Disease and Related Disorders Association (NINCDS-ADRDA) criteria [13] and a Mini-Mental State Examination (MMSE) score between 15 and 24, inclusive. Blood collected at baseline by venipuncture was allowed to clot and serum was prepared by centrifugation. Serum was stored frozen at −70°C until analyzed. Results derived from analysis of serum was compared to baseline Mattis Dementia Rating Scale (MDRS) scores obtain before initiation of study intervention [14,15].

Ethics

Animal samples were from a previously published study [16].

Study subjects and either a family member or a Durable Power of Attorney (if one existed) gave written informed consent including a special addendum permitting the use of stored specimens for AD research [12]. According to Danish legislation no ethical approval is needed to perform the analysis of serum samples collected in a prior study, and a waiver describing this is on file at Nordic Bioscience.

Statistical analysis

For assay validation, optical density was fitted against analyte concentration applying a four-parameter logistic regression to the calibration curve. Average, standard deviations, percentage coefficient of variation (%CV), and differences from theoretical values were calculated for all standards and samples. Quantitative data were analysed using GraphPad Prism 5 (GraphPad Software, San Diego, CA, USA). Significant differences between means were determined using Student's two-tailed unpaired t-test. Normality of the data was ensured using a Shapiro-Wilkes Test. Data were expressed as mean (or geometric means) ± standard error of the mean and differences were considered significant at a p level of 0.05 or lower.

Results

Characterization of the Tau-A ELISA assay

An antibody recognizing the ADAM10 generated cleavage sequence of tau (TPRGAAPPGQ) was raised, and used for development of an ELISA (Tau-A). As seen in figure 1, the assay is specific for the cleavage site, as an extension of the sequence by one amino acid led to loss of reactivity (Figure 1A). Further validation of specificity using ADAM10 degraded recombinant tau or brain extracts confirmed the specificity towards cleaved tau (Figure 1B–C). In addition, tissue profiling confirmed that tau primarily originates from the brain. The lower limit of detection (LLOD) was determined to be 2.9 ng/mL and the upper limit of detection (ULOD) was 226.3 ng/mL. The assay was technically robust and was able to detect Tau-A levels in human serum and plasma, as well as mouse and rat serum within dilution ranges of 1+2 to 1+6 (Table 1). In addition, a linear spiking recovery was obtained within the above described dilution ranges (data not shown). The intra-assay coefficient of variation was 5.8%, while the inter-assay CV% was 12.6%. No loss of reactivity was observed following 5 consecutive freeze-thaw cycles.

Biological validation of the Tau-A ELISA

Analysis using the ELISA revealed that brains from the animal model of tauopathy, the Tg4510 mice, had 10-fold higher levels of Tau-A than their corresponding wild type (wt) controls (Figure 1E). Western blotting also showed that Tg4510 mice had very high levels of Tau-A, while the control mice had very little (Figure 1F), even if equal amounts of protein were loaded (data not shown). Re-probing the blot with an antibody against total tau showed almost equal levels of total tau, although the intensity of the high molecular weight bands was greater in the Tg4510, as expected from the model [16].

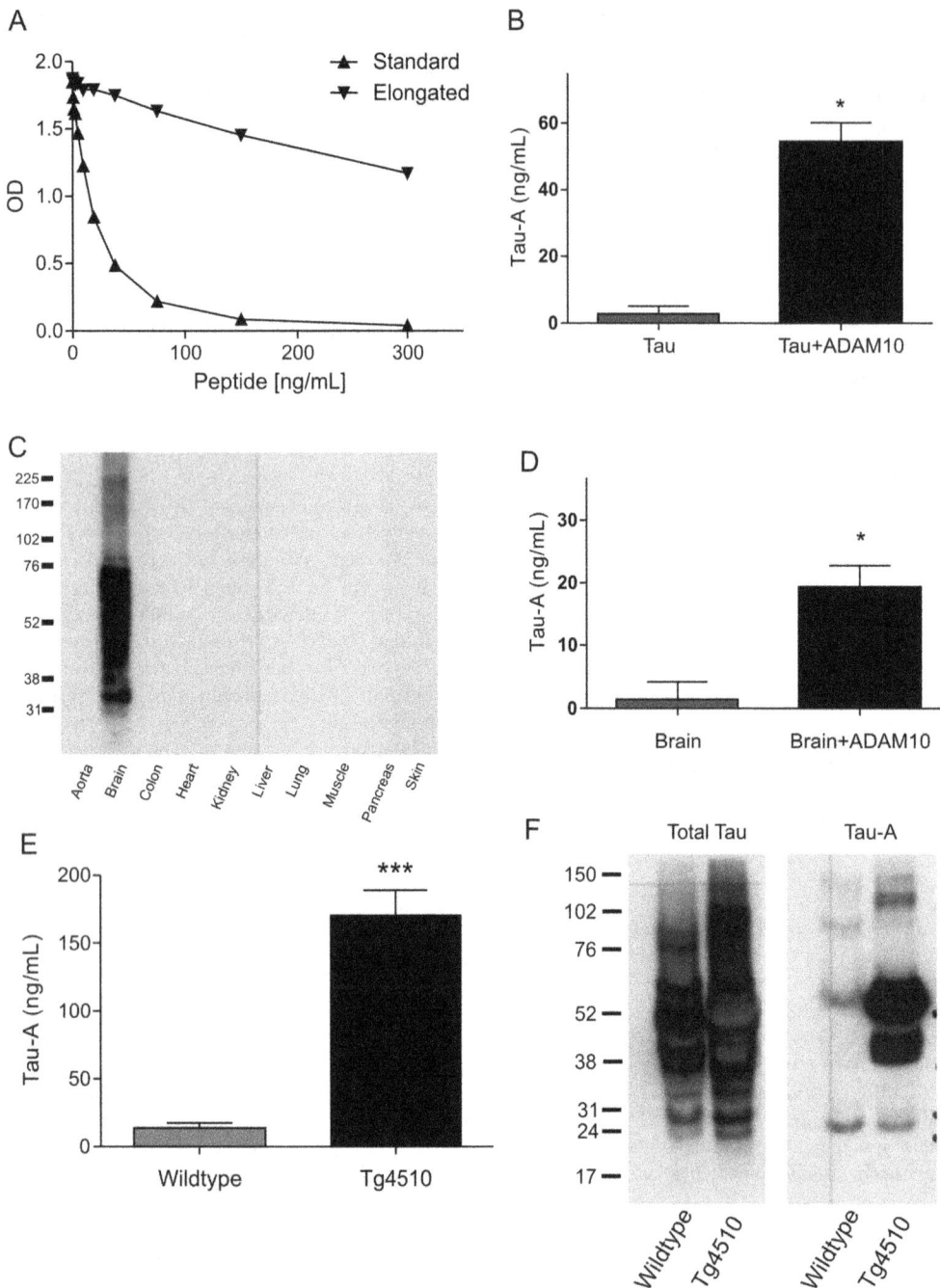

Figure 1. Characteristics and biological validation of the Tau-A assay. A) Standard curves with either the selection peptide or the elongated peptide at concentrations of 0, 0.59, 1.17, 2.34, 4.69, 9.38, 18.75, 37.5, 75, 150, and 300 ng/ml. B) Measurement of Tau-A fragments in *in vitro* digests of tau. C) Western blots of extracted tissues, D) ELISA measurement of brain extracts in the presence or absence of ADAM10. E) Tau-A levels in extracted brains from either control or Tg4510 mice (a model of Alzheimer's disease) measured using the ELISA. F) Western blots comparing brain extracts from wild type and Tg4510 mice. Left: a western blot conducted with an antibody recognizing intact tau (MAB3420 Chemicon). Right: Western blot conducted with our in-house antibody (NB191).

Tau-A levels correlate with MDRS score

To investigate whether a relationship between the marker and AD disease stage, we correlated the tau levels in the AD patients to scores obtained using the Mattis Dementia Rating Scale [17], and highly interestingly we found a significant (p = 0.003) and inverse relationship between MDRS and Tau-A (Figure 2). No correlations to other parameters, such as intact tau in CSF or age could be observed in this small cohort (data not shown).

Discussion

Potential serum and/or plasma-based markers for AD have been investigated extensively, yet a single biomarker with a correlation to cognitive function remains to be identified [4].

In this study, we developed an ELISA assay detecting fragments of tau generated by ADAM10 with the exclusive aim of analyzing pathological tau processing in serum. Firstly, we confirmed that

Table 1. Dilution recovery measured in human samples and calculated in % of lowest dilution possible.

Tau-A ng/mL	HS1 45.2	HS2 33.7	HS3 15.8	Average	HP1 48.0	HP2 30.3	HP3 36.9	Average
Diluted 1+2	100	100	100		-	-	-	
Dilution 1+3	90	94	118		100	100	100	
Dilution 1+4	90	96	114		103	112	98	
Dilution 1+5	80	83	108		107	116	111	
Dilution 1+6	81	78	107		111	102	109	
Dilution 1+7	-	-	-		104	95	97	
Mean	85	88	112	95	107	106	104	105

Tau-A ng/mL	RS1 135.8	RS2 48.6	RS3 83.3	Average	MS1 24.4	MS2 14.1	MS3 13.6	Average
Dilution 1+2	100	100	100		100	100	100	
Dilution 1+3	106	113	98		105	84	111	
Dilution 1+4	80	107	94		93	87	111	
Dilution 1+5	83	105	94		108	103	137	
Dilution 1+6	80	98	101		-	-	-	
Mean	87	106	97	97	102	91	111	101

Values in bold represent data points outside the accepted variation and are excluded from the calculations as they are measured too close to the lower limit of detection (LLOD) value (2.9 ng/mL). HS = Human Serum, HP = Human Plasma. RS = Rat Serum, MS = Mouse Serum.

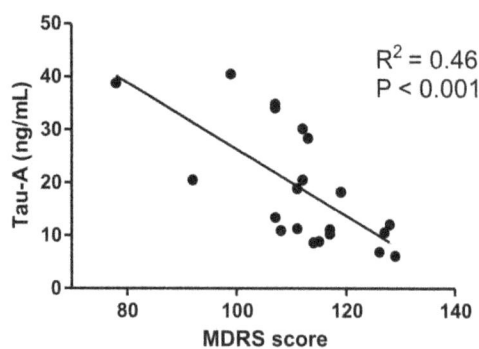

Figure 2. Pathology related changes in Tau-A levels. Inverse correlation between Tau-A and the Mattis Dementia Rating Scale (MDRS).

the antibody was highly specific towards an ADAM10 generated fragment of tau, and that this fragment was detectable in serum. In terms of absolute amount of the analyte in blood, this is unknown, although it likely is low; however, a determination of these will require further study, as the assay employs a synthetic standard.

Importantly, this marker showed an inverse correlation with MDRS, indicating that it is related to loss of cognitive function. A biochemical marker monitoring proteolytic processing of tau in serum, is of interest as this is a key process in the development of cognitive deficits [6,8]. It is also the first serum tau biomarker which correlates to cognitive function, thereby showing promise for the application of this marker in future studies. While other studies have indicated that Aβ levels in plasma are correlated to cognitive function, albeit in elderly without dementia [18], this is still the first study measuring tau processing

While neo-epitopes are nothing new in AD, as measurements of Aβ42 and phosphorylated tau have been reported as neo-epitopes formed as a consequence of disease [4], selective screening of serum for *in vitro* generated tau fragments had not previously been published.

The combination of ADAM10 and tau selected for this work was based on the novel hypothesis that during progression of AD, tau will be exposed to secretase-mediated cleavage either directly in the brain or as fragments generated by other brain proteases, which then become secondarily processed as they enter the circulation; however, this requires further studies. We were able to detect a specific ADAM10-generated tau peptide fragment in serum, as well as highly elevated levels in the brains of the Tg4510 mice, a model of tauopathies known to demonstrate tau

aggregation [16]. These observations suggest ADAM10 processing of tau is a relevant process in neuronal death during tauopathies, such as AD, although the exact mechanism of action still remains to be identified.

Using mass spectrometry we identified the sequence [470]TPRGAAPPGQ which is derived from the brain isoform of tau and highly specific for tau. Interestingly, this neo-epitope resides in a part of the tau sequence which is speculated to be removed by processing during the progression of AD, and thus may be of pathological relevance [19].

To ensure that the identified tau fragment was specific to neurons, and not found in other tissues, we performed a series of extractions, and we found that total tau is primarily expressed in brain, and that the addition of ADAM10 led to significant cleavage of intact tau. Tau-A was not detected in any other intact or digested tissues, supporting the tissue- specificity of the fragment. In brain extracts of the tauopathy model, the Tg4510 mice, we found 10-fold elevated levels of Tau-A compared with healthy controls, indicating pathological relevance.

Importantly, we found that Tau-A levels correlated inversely with MDRS, a finding which supports that Tau-A to loss of cognitive ability [17,20], albeit in a fairly small group of patients. Whether this correlation directly reflects loss of neurons requires further studies using more direct measurements of neuronal loss, which could be a key aspect in determining whether this marker can be applied to identify earlier progressors, such a those with prodromal AD. On the other hand we found no correlation to total tau in CSF or to age (data not shown), but these findings may be explained by the small sample size. While tau pathology is clearly related to AD, it is well-established that pathological tau processing occurs in several different forms of dementia, and hence further studies in other dementias as well as cognitively normal elderly are needed to further characterize the potential of this marker.

In summary, we have developed the first serum-based assay detecting pathological fragments of tau, and importantly this fragment was directly and inversely related to cognitive function. We speculate that the assay could be a useful and practical tool for the diagnosis of neuronal loss, and could monitor the efficacy of treatment and progression of AD.

Author Contributions

Conceived and designed the experiments: KH MAK JS QZ CC. Performed the experiments: NB. Analyzed the data: KH MGS. Contributed reagents/materials/analysis tools: JTP KLD RAD. Wrote the paper: KH MAK.

References

1. Corbett A, Smith J, Ballard C (2012) New and emerging treatments for Alzheimer's disease. Expert Rev Neurother. 12: 535–543.
2. Hampel H, Lista S, Khachaturian ZS (2012) Development of biomarkers to chart all Alzheimer's disease stages: the royal road to cutting the therapeutic Gordian Knot. Alzheimers Dement. 8: 312–336.
3. Blennow K (2010) Biomarkers in Alzheimer's disease drug development. Nat Med. 16: 1218–1222.
4. Cummings JL. (2011) Biomarkers in Alzheimer's disease drug development. Alzheimers Dement.7: e13–e44.
5. Wang Y, Sorensen MG, Zheng Q et al. (2012) Will Post-translational Modifications of Brain Proteins Provide Novel Serological Markers for Dementias? Int J Alz Disease 2012:209409.
6. Reifert J, Hartung-Cranston D, Feinstein SC (2011) Amyloid {beta}-Mediated Cell Death of Cultured Hippocampal Neurons Reveals Extensive Tau Fragmentation without Increased Full-length Tau Phosphorylation. J Biol Chem. 286: 20797–20811.
7. Karsdal MA, Henriksen K, Leeming DJ et al (2010) Novel combinations of Post-Translational Modification (PTM) neo-epitopes provide tissue-specific biochem-

ical markers-are they the cause or the consequence of the disease? Clin Biochem. 43: 793–804.
8. De SB. (2010) Proteases and proteolysis in Alzheimer disease: a multifactorial view on the disease process. Physiol Rev. 90: 465–494.
9. Barascuk N, Veidal SS, Larsen L et al. (2010) A novel assay for extracellular matrix remodeling associated with liver fibrosis: An enzyme-linked immunosorbent assay (ELISA) for a MMP-9 proteolytically revealed neo-epitope of type III collagen. Clin Biochem. 43: 899–904.
10. Gefter ML, Margulies DH, Scharff MD. (1977) A simple method for polyethylene glycol-promoted hybridization of mouse myeloma cells. Somatic Cell Genet. 3: 231–236.
11. Henriksen K, Gram J, Schaller S et al. (2004) Characterization of osteoclasts from patients harboring a G215R mutation in ClC-7 causing autosomal dominant osteopetrosis type II. Am J Pathol. 164: 1537–1545.
12. Silverberg GD, Mayo M, Saul T et al. (2008) Continuous CSF drainage in AD: results of a double-blind, randomized, placebo-controlled study. Neurology. 71: 202–209.
13. McKhann G, Drachman D, Folstein M et al. (1984) Clinical diagnosis of Alzheimer's disease: report of the NINCDS-ADRDA Work Group under the

auspices of Department of Health and Human Services Task Force on Alzheimer's Disease. Neurology. 34: 939–944.

14. Fama R, Sullivan EV, Shear PK et al. (1997) Selective cortical and hippocampal volume correlates of Mattis Dementia Rating Scale in Alzheimer disease. Arch Neurol. 54: 719–728.

15. Smith GE, Ivnik RJ, Malec JF et al. (1994) Psychometric Properties of the Mattis Dementia Rating Scale. Assessment. 1: 123–132.

16. Sahara N, DeTure M, Ren Y et al. (2013) Characteristics of TBS-Extractable Hyperphosphorylated Tau Species: Aggregation Intermediates in rTg4510 Mouse Brain. J Alzheimers Dis. 33: 249–63.

17. Schmidt R, Freidl W, Fazekas F et al. (1994) The Mattis Dementia Rating Scale: normative data from 1,001 healthy volunteers. Neurology. 44: 964–966.

18. Yaffe K, Weston A, Graff-Radford NR et al. (2011) Association of plasma beta-amyloid level and cognitive reserve with subsequent cognitive decline. JAMA. 305: 261–266.

19. Barten DM, Cadelina GW, Hoque N et al. (2011) Tau transgenic mice as models for cerebrospinal fluid tau biomarkers. J Alzheimers Dis. 24:127–141.

20. Rascovsky K, Salmon DP, Hansen LA, Galasko D (2008) Distinct cognitive profiles and rates of decline on the Mattis Dementia Rating Scale in autopsy-confirmed frontotemporal dementia and Alzheimer's disease. J Int Neuropsychol Soc. 14: 373–383.

7

Characterization of Antibacterial and Hemolytic Activity of Synthetic Pandinin 2 Variants and Their Inhibition against *Mycobacterium tuberculosis*

Alexis Rodríguez[1,2], Elba Villegas[2], Alejandra Montoya-Rosales[3], Bruno Rivas-Santiago[3], Gerardo Corzo[1]*

1 Departamento de Medicina Molecular y Bioprocesos, Instituto de Biotecnología, Universidad Nacional Autónoma de México, Cuernavaca Morelos, México, 2 Centro de Investigación en Biotecnología, Universidad Autónoma del Estado de Morelos, Cuernavaca, Morelos, México, 3 Medical Research Unit-Zacatecas, Mexican Institute of Social Security, UIMZ-IMSS, Zacatecas, Mexico

Abstract

The contention and treatment of *Mycobacterium tuberculosis* and other bacteria that cause infectious diseases require the use of new type of antibiotics. Pandinin 2 (Pin2) is a scorpion venom antimicrobial peptide highly hemolytic that has a central proline residue. This residue forms a structural *"kink"* linked to its pore-forming activity towards human erythrocytes. In this work, the residue Pro14 of Pin2 was both substituted and flanked using glycine residues (P14G and P14GPG) based on the low hemolytic activities of antimicrobial peptides with structural motifs Gly and GlyProGly such as magainin 2 and ponericin G1, respectively. The two Pin2 variants showed antimicrobial activity against *E. coli*, *S. aureus*, and *M. tuberculosis*. However, Pin2 [GPG] was less hemolytic (30%) than that of Pin2 [G] variant. In addition, based on the primary structure of Pin2 [G] and Pin2 [GPG], two short peptide variants were designed and chemically synthesized keeping attention to their physicochemical properties such as hydrophobicity and propensity to adopt alpha-helical conformations. The aim to design these two short antimicrobial peptides was to avoid the drawback cost associated to the synthesis of peptides with large sequences. The short Pin2 variants named Pin2 [14] and Pin2 [17] showed antibiotic activity against *E. coli* and *M. tuberculosis*. Besides, Pin2 [14] presented only 25% of hemolysis toward human erythrocytes at concentrations as high as 100 μM, while the peptide Pin2 [17] did not show any hemolytic effect at the same concentration. Furthermore, these short antimicrobial peptides had better activity at molar concentrations against multidrug resistance *M. tuberculosis* than that of the conventional antibiotics ethambutol, isoniazid and rifampicin. Therefore, Pin2 [14] and Pin2 [17] have the potential to be used as an alternative antibiotics and anti-tuberculosis agents with reduced hemolytic effects.

Editor: Jürgen Harder, University Hospital Schleswig-Holstein, Campus Kiel, Germany

Funding: This work was also supported by CONACyT project Grants 153606 to GC, and 83962/2007 and 106949/2008 to Elba Villegas. The funders had no role in study design, data collection and analysis, decision to publish, or preparation of the manuscript.

Competing Interests: The authors have declared that no competing interests exist.

* Email: corzo@ibt.unam.mx

Introduction

Cationic Antimicrobial Peptides (CAMPs) are components of the biological defense system of microorganisms, plants, animals and humans [1–3]. A large number of CAMPs have been characterized in invertebrates, especially from the phylum Arthropoda [4–6], and in vertebrates from the class Amphibia [7,8]. CAMPs with alpha-helical conformation share some common characteristics such as antimicrobial activities at low micromolar concentrations and alpha-helix conformation in hydrophobic environments [9–11]. They have potent antibacterial activities that made them promissory candidates to develop novel antibiotics because of their broad-spectrum activity towards multi-resistant pathogenic Gram-positive and Gram-negative bacteria, as well as towards clinically important yeasts such as *Candida albicans* [12–14].

Pin2 (FWGALAKGALKLIPSLFSSFSKKD, see table 1) is a 24-residue alpha-helical antimicrobial peptide characterized from the venom of the African scorpion *Pandinus imperator*, this peptide has antimicrobial activities towards Gram-positive and Gram-negative bacteria in the micromolar range; however, it shows hemolytic activity at similar concentrations [15]. High-resolution structure of Pin2, determined by NMR, showed that the peptide is essentially alpha-helical with a proline, which induces a structural *"kink"* in the central part of its structure [15]. The proline *"kink"* is a structural characteristic of some CAMPs that confer them high pore-forming abilities. Clear examples of such peptides are melittin [16], alamethicin [17] and pardaxin [18]. In those peptides the presence of a proline *"kink"* was correlated to a high antimicrobial activity but unfortunately also showed high hemolytic activities [17,19–21]. The elimination of the proline or its substitution in their primary structure of these peptides had different effects on their secondary structures and biological activities. For example, the substitution of the P14 for alanine in the antimicrobial peptides melittin and pardaxin increase their hemolytic and antimicrobial activities, respectively, as result of an increment in the helicity grade of such antimicrobial analogs [19]. On the other hand, CAMPs such as Magainin 2 from the African frog *Xenopus laevis*, Oxypinins from the wolf spider *Oxyopes takiobus* and Ponericins G1 from the ponerin ant, *Pachycondyla goeldii*, showed antimicrobial activities with low cytotoxic effects towards erythrocytes [22–24]. These peptides have different amino acid motifs in the central region of their primary structures. For example, magainin 2 has a single glycine in the middle of its structure, Oxypinin 2b (Oxki2b) has a GlyValGly motif, and Ponericin G has glycine residues flanking the central proline, a GlyProGly motif. Furthermore, cecropin and melittin synthetic hybrids in which the proline of melittin was changed by Cecropin residues showed no hemolytic activities in contrast to the parental peptide [25]. Likewise, the substitution of the proline residue (P14) in Pin2 for the residues Val, GlyVal, ValGly or GlyValGly reduced the hemolytic activity of Pin2 without any significant changes in its antimicrobial activity [26].

In this work, based on the low hemolytic activities shown by the antimicrobial peptides magainin 2 from *X. laevis* and Ponericin G1 from *P. goeldii*, two synthetic variants of Pin2, Pin2 [G] and Pin2 [GPG] were chemically synthesized with the aim to reduce the hemolytic activity and preserve the antibiotic activities of Pin2. In addition, two short variants of Pin2, with 14 and 17 residues, respectively, were designed and chemically synthesized with the aim to continue reducing their hemolytic but keeping their antimicrobial activities as well as to reduce the number of residues to have a low cost CAMP. Here these antimicrobial peptide variants are proposed as potential antibiotics for the clinical treatment of pathogenic bacteria including *Mycobacterium tuberculosis*.

Materials and Methods

Ethics statement

Approvals to conduct experimental protocols to study hemolysis on human red cells were approved by the Bioethics Committee of the Biotechnology Institute, where this work was done. Human red cells were from volunteer Gerardo Corzo, who signed the informed consent for this study and is also author of this report.

Microorganisms

The bacterial strains used were *Escherichia coli* (ATCC 25922) and *Staphylococcus aureus* (ATCC 25923). They were purchased directly from the American Type Culture Collection (ATCC) through The Global Bioresource Center. *Mycobacterium tuberculosis* H37Rv (ATCC 27294) [27] and *M. tuberculosis* muti-drug resistant strain (MDR) [28] were from the Medical Research Unit-Zacatecas belonging to the Mexican Institute of Social Security.

Solid phase peptide synthesis

Pin2 and Pin2 variants were chemically synthesized by a solid-phase method using the Fmoc methodology on an Applied Biosystems 433A peptide synthesizer. Fmoc-Asp(otBu) or Fmoc-Leu(otBu)-Wang resins were used to provide a free carboxyl at the C-terminus of the Pin2 and its variants. Cleavage and deprotection of peptides, from resins and from protecting side chain groups, were performed using a chemical mixture composed of 1 g crystalline phenol, 0.2 g imidazole, 1 mL thioanisol, 0.5 mL 1,2-ethanedithiol in 20 mL trifluoroacetic acid (TFA). The resin was removed by filtration, and the deprotected peptides in solution were precipitated using cold ethyl ether. The precipitated peptides were washed twice with cold diethyl ether to remove remaining scavengers and protecting groups. Each crude synthetic peptide was then dissolved in a 10% aqueous acetonitrile solution and separated by reverse phase HPLC (RP-HPLC) on a semipreparative C_{18} column (10×250 mm, Vydac, USA). Cationic exchange

Table 1. Physicochemical properties of Pin2 and Pin2 variants.

Peptide	Sequence	GRAVY	AMF	μHrel	Q	RT (min)	Molecular Weight (Da)	
							Theoretical	Experimental[§]
Pin2	FWGALAKGALKLIPSLFSSFSKKD	0.329	0.4283	0.48	+3	30.6	2612.1	2612.0
Pin2 [G]	FWGALAKGALKLIGSLFSSFSKKD	0.379	0.4296	0.49	+3	40.1	2572.0	2572.0
Pin2 [GPG]	FWGALAKGALKLIGPGSLFSSFSKKD	0.273	0.4369	0.28	+3	27.6	2726.2	2727.0
Pin2 [14]	FWGLKGLKFSKKL	−0.357	0.4271	0.51	+5	18.5	1680.1	1680.0
Pin2 [17]	FWGLKGLKGPGKFSKKL	−0.435	0.4453	0.35	+5	17.2	1891.3	1891.3

GRAVY, Sequence Grand average of hydropathicity, calculated using the Expasy ProtParam tool (http://web.expasy.org/protparam), according to Kyte and Doolittle, 1982 [44].
AMF, Average Molecular Flexibility values were calculated according to Liu et al., 2008 [45].
μHrel, Relative Hydrophobic Moment, a value of 0.5 thus indicates that the peptide has about 50% of the maximum possible amphipathicity. Calculated using HydroCalc (http://www.bbcm.univ.trieste.it/~tossi/HydroCalc/HydroMCalc.html).
Q, Net charge.
RT, Retention Time in minutes.
[§]Mass Spectrometry, ESI-MS (Finnigan LCQ[DUO] ion trap mass spectrometer, San José, CA, USA).

chromatography and C_{18} analytical RP-HPLC were further used to purify the synthetic peptides, after all purification steps, the final purity of the peptides was higher than 95%. The molecular masses of the synthetic peptides were obtained by mass spectrometry using a LCQ DUO ion trap mass spectrometer (Finnigan, San Jose, USA) with and ESI source from 2.1 to 3.1 kV.

Antimicrobial assays

Minimal inhibitory concentrations (MIC) and growth inhibition curves were obtained using pure peptides in the presence of bacteria using two different methods, agar diffusion susceptibility assays and broth microdilution assays in accordance to the procedures from the Clinical and Laboratory Standards Institute (CLSI, http://www.clsi.org).

The agar diffusion susceptibility assay was performed using 10 mL of Mueller-Hinton agar (MHA) underlay on a Petri dish plate, while at the same time, 0.1 mL aliquot of a mid-logarithmic-phase (1×10^8 CFU/mL in MHB with $A_{625nm} = 0.5$) culture, was inoculated in a sterile tube containing 9.9 mL of non solidified MHA and mixed. The content of the tube was overlaid in the previously poured MHA Petri dish. Then 5 µL aliquots of a diluted antimicrobial peptide at 300, 100, 80, 50, 37.5, 25, 18.8, 12.5, 6.25, 3.1 and 1.6 µM were subsequently loaded into the overlay gel. Samples were incubated overnight at 37°C. The antimicrobial activity was determined by measuring clear zone diameters or halos observed around each peptide concentration in cultured MH Petri dishes. Peptide MICs were defined as the lowest peptide concentration with a clear zone halo.

Broth microdilution assays were performed using stock solutions of Pin2 [G], Pin2 [GPG], Pin2 [14] or Pin2 [17] antimicrobial peptide diluted serially from 25 to 0.4 µM to a final volume of 200 µL, placed in polypropylene microtubes and vacuum dried. Next a volume of 200 µL aliquots of the bacterial suspension (1×10^5 CFU/mL) in MHB was dispensed into each of the polypropylene microtubes and mixed with the diluted antimicrobial peptide. Then each was transferred into a well of a 96-well microtiter plate and bacterial growth was evaluated by measuring absorbance every 2 h until 10 h of incubation time at 37°C. The optical density (OD) of each well was measured at 625 nm in an ELISA reader (BioRad, model 450, Hercules, CA, USA). The positive control contained only the bacterial suspension, and the negative control contained only sterile culture medium. The resulting MICs were defined as the lowest peptide concentration that showed zero visible growth or absence of growth, that is growth inhibition (100%). The minimal inhibitory concentrations (MIC) values were the mean result of three independent experiments.

Mycobacterium tuberculosis assays

The Resazurin microtitre assay plate (REMA) method was conducted to determine the M. tuberculosis susceptibility to the action of Pin2 and the different Pin2 variants studied in this work. Resazurin is an oxidation–reduction indicator and has been used to assess viability, bacterial contamination and to test for antimicrobial activity [29]. The M. tuberculosis H37Rv (ATCC 27294) and a clinically isolated multidrug resistant (MDR) strain were used in these experiments. The REMA plate method was performed in 7H9 medium containing 10% of OADC (oleic acid, albumin, dextrose and catalase) (Becton–Dickinson, Sparks, MD, USA). Resazurin sodium salt powder (Sigma) was prepared at 10% (wt/vol) in distilled water and filter-sterilized (0.22 µm). Two-fold serial dilutions of each synthetic peptide dissolved in 100 µL of 7H9-OADC culture medium were added directly in 96-well plate at concentrations ranging from 96.2 to 0.3 µg/mL.

Ethambutol and rifampicin were used as positive controls. The M. tuberculosis inoculum was prepared from a 14-day logarithmic phase culture and adjusted to 1 of the Mcfarland scale (0.76 OD, 600 nm) with 7H9-OADC culture medium, and then diluted to 1:20. Then, 100 µL of these inoculums were added to each well on the plate (final volume 200 µL). The plates were covered with proper lids and incubated at 37°C in a normal atmosphere. After 5 days of incubation, 20 µL of the Resazurin solution were added to each control well only, and incubated for 24 h at 37°C for color development. If color was developed as expected (pink for microorganism's control growth and blue for sterility control), 20 µL of the Resazurin solution were added into each well containing the peptides at different concentrations. After 24 h, a change from blue color to pink color indicated the reduction of Resazurin and therefore bacterial growth. The minimal inhibitory concentration (MIC) was defined as the lowest peptide concentration that prevented the reduction of Resazurin and therefore a color change from blue to pink. Previous studies suggest that some host defense peptides may induce dormancy or a bacteriostatic state in M. tuberculosis [30]. To examine this, 10 µL from the lowest concentration that did not reduce Resazurin were serially diluted and seeded onto 7H10 agar plates supplemented with Middlebrook OADC enrichment media and incubated for at least 21 days at 37°C, to observe if M. tuberculosis growth reestablishment occurred. All antimicrobial tests were conducted in triplicate.

Hemolytic assays

Hemolytic activity was determined by incubating suspensions of human red blood cells with serial dilutions of each selected peptides. Red blood cells were rinsed several times in PBS by centrifugation for 3 min at 3,000 g until the OD of the supernatant reached the OD of the control (PBS only). Red blood cells were counted by a hemocytometer and adjusted to $7.7 \times 10^6 \pm 0.3 \times 10^6$ cells/mL. Red blood cells were then incubated at room temperature for 1 h in 10% Triton X-100 (positive control), in PBS (blank), or with amphipathic peptides at concentrations of 0.4, 0.8, 1.6, 3.1, 6.2, 12.5 and 25 µM, only for Pin2 [14] and Pin2 [17] the 50 and 100 µM concentrations were evaluated. The samples were then centrifuged at 10,000 g for 5 min, the supernatant was separated from the pellet, and its absorbance measured at 570 nm. The relative optical density compared to that of the suspension treated with 10% Triton X-100 was defined as the percentage of hemolysis.

Circular Dichroism (CD) measurements

The CD experiments were recorded on a Jasco model J-720 spectropolarimeter (Tokyo, Japan). The different spectra were measured from 260 to 190 nm on samples in water, 20, 40 and 60% trifluoroethanol (TFE), at room temperature, with a 1-mm-pathlength cell. Data were collected at 1 nm with a scan rate of 100 nm/min and a time constant of 0.5 s. The concentration of each peptide was 150 µg/mL Data were the average of five separate recordings and were analyzed on line by the software K2d (http://www.embl.de/~andrade/k2d.html) [31,32].

Statistical analysis

The experimental values represent means ± standard deviations. The hemolytic constants (IC_{50}) were obtained using a non-linear regression where the data was fit to a Boltzmann sigmoid equation. To determine statistically significant differences, an analysis of variance (ANOVA) followed by post hoc testing using the Tukey's method was performed using the software package Prism 4 (GraphPad, Inc., USA). Statistical significances were accepted at $p < 0.05$.

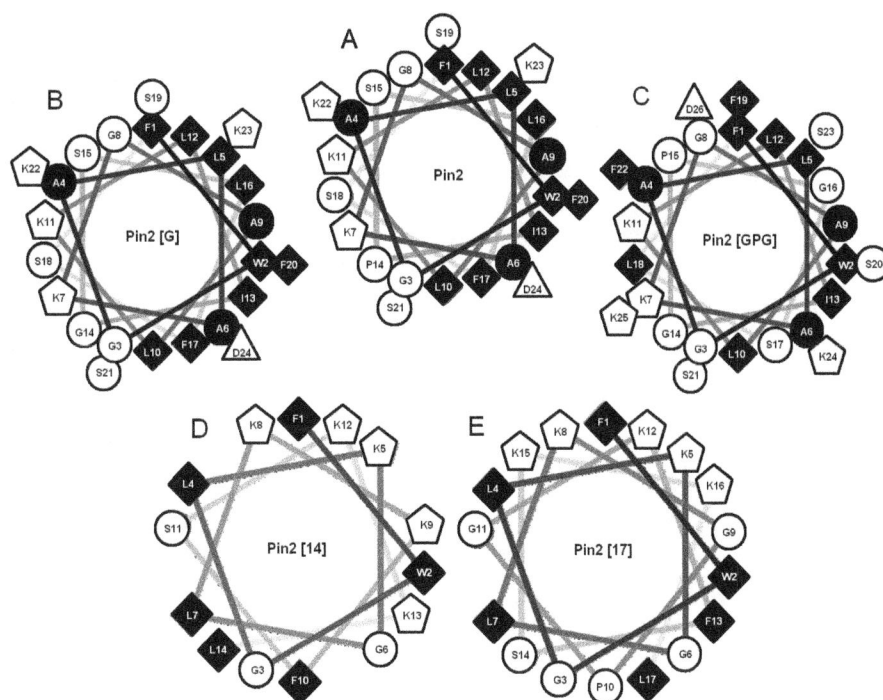

Figure 1. Helical wheel diagrams of Pin2 and Pin2 variants. Helical wheels were prepared by the software Helical Wheel Projections [56]. The hydrophobic residues are colored in black, hydrophilic and neutral residues are colored in white. A. Pin2, B. Pin2 [G], C. Pin2 [GPG], D. Pin2 [14] and E. Pin2 [17].

Results

Design of Pin2 [G], Pin2 [GPG] and their short variants

The physicochemical parameters of CAMPs such as the net charge, the amphipathic character and the hydrophobicity could be modified to improve their antimicrobial properties, inclusive cell membrane selectivity [33–35]. However, still there are several drawbacks for the development of CAMPs as therapeutic agents, among them the synthesis cost for their production is one of the most important [36–38]. Therefore in this work, we attempt to design first low hemolytic CAMPs and subsequently reduce the number or residues. Hence, the variants of Pin2 [G] and Pin2 [GPG] were designed to reduce their hemolytic activities but to preserve their antibiotic activities. Here, the proline residue (P14) was the target as previously reported [26]. Furthermore, the short variants were designed based on Pin2 [G] and Pin2 [GPG]. First, the Aspartic residue (D24) was replaced by a Lys in order to increase the net charge from +3 to +5. Second, the small hydrophobic (Ala) and the uncharged polar (Ser) residues were eliminated of the primary sequence to reduce the peptide length. To keep low hemolytic properties in the short CAMPs, the central Pro was replaced for Gly or for the GlyProGly triplet. Finally and prior to the chemical synthesis the amphipathic character of the resulted 14 and 17 amino acid sequences was optimized to values near to 50 and 30%, respectively. This was achieved by means of helical wheel projections (**Figure 1**), and by sequence search using 2,259 antimicrobial peptides with proved activity deposited in the APD2 database [39]. Also the amino acid sequence search had the goal to eliminate the casual probability to find characterized CAMPs similar to the designed short Pin2 variants. The physicochemical properties of Pin2 [G], Pin2 [GPG] and their short variants are shown in table 1. The amino acid sequences of the reported CAMPs with higher identity to Pin2 [14] and Pin2

[17] were aligned. Table 2 shows that the two short variants are only 40–54% similar to the parental peptide Pin2 and less than 50% similar to other natural and synthetic antimicrobial peptides.

Peptide chemical synthesis and purification

The chemical synthesis of Pin2 and its variants was performed using a 0.1 mmol Fmoc chemistry. The crude synthetic peptide Pin2 and its variants were uniformly purified by RP-HPLC. As a result of the amino acid substitution and deletions in the primary structure of Pin2, the elution times of the Pin2 variants under RP-HPLC were shorter compared to the parental peptide except for Pin2 [G] (**Figure 2**). All purified Pin2 variants were proved to have the expected molecular masses. For instance, the experimental *versus* the theoretical masses expected were 2,612.0/ 2,612.1 for the native Pin2, 2,572.0/2,572.0 for Pin2 [G], 2,727.0/2,726.2 for Pin2 [GPG], 1,680.0/1,680.1 for Pin2 [14], and 1,891.3/1,891.3 for Pin2 [17]. As a consequence of the substitution and deletions in the primary structure of Pin2, the retention times (RT) under RP-HPLC changed, for instance the Pin2 [G] variant showed a RT of 40.1 min indicating a more hydrophobic molecule respect to Pin2 (RT of 30.6 min). Also, the less hydrophobic character of Pin2 [GPG] variant showed a RT of 27.6 min, while the RT of the short variants Pin2 [14] and Pin2 [17] were 18.5 and 17.2 min, respectively. The hydrophobic nature of the Pin2 variants observed by the experimental RT under RP-HPLC agreed to the theoretical hydrophobic character calculated using the GRAVY values. The amino acid sequences and some physicochemical properties of these peptides are listed in table 1. Once confirmed the identity of all synthesized peptides, their secondary structure was analyzed as well as their antimicrobial and hemolytic properties were observed.

Table 2. Alignment of the sequences of Pin2 [14] and Pin2 [17] with the sequence of other antimicrobial peptides.

Peptide	UniProtKB ID	Origin	Sequence‡	AA	Identity (%)	Q	Reference
Pin2 [14]	-	Synthetic	-FWG-LKGLKKFS---KKL--------	14	-	+5	This work
Pin2	P83240	Pandinus imperator	-FWGALAKGALKLI---PSLFSSFSKKD	24	50.0	+3	[15]
CPF-SE3	P84386	Silurana epitropicalis	GFLGSLLKTGLKVG---SNLL-------	18	50.0	+3	[46]
CE-MA§	-	Synthetic	-KWK-LFKKIKFLHSAKKF--------	17	47.4	+7	[47]
Macropin 1		Macropis fulvipes	---G--FGMALKLL---KKVL-------	13	47.1	+3	[48]
Mastoparam	P69034	Mischocyttarus phthisicus	INW---LKLGKKMM---SAL-------	14	47.1	+3	[49]
Japonicin 1	B3VZU2	Rana chensinensis	-FFP-LALLCKVF---KKC--------	14	46.7	+3	[50]
Pin2 [17]	-	Synthetic	-FWG-LKGLKGP-----GKFSKKL	17	-	+5	This work
Pin2	P83240	Pandinus imperator	-FWGALAKGALKLIPSLFSSFSKKD	24	54.2	+3	[15]
Ponericin W5	P82427	Pachycondyla goeldii	-FWGALIKGAAKLIPSVVGLFKKKQ	24	50.0	+5	[23]
CPF-SE3	P84386	Silurana epitropicalis	GFLGSLLKTGLK-----VGSNLL--	18	47.6	+3	[46]
NRC-15	-	Glyptocephalus cynoglossus L	GFWGKLFKLGLHG----IGLLHLHL	21	47.6	+2	[51]
Brevinin 1RTb	D1MIZ6	Amolops ricketti	-FLGSLLGLVGKVVPTLFCKISKKC	24	45.8	+4	[52]
Brevinin 1AUa	-	Rana aurora aurora	-FLPILAGLAAKLVPKVFCSITKKC	24	44.0	+4	[53]
Maximin H39	Q58T55	Bombina maxima	-ILGPVLGLVGNAL---GGLIKKL	20	42.9	+2	[54]

‡, The sequence alignment was obtained using the on line program ClustalW2 (http://www.ebi.ac.uk/Tools/msa/clustalw2/).
AA, Number of amino acids.
Identity (%), sequence identity in percentage.
Q, Net Charge.
§, Cecropin A (residues 1–8)/Magainin 2b (residues 4–12) hybrid.

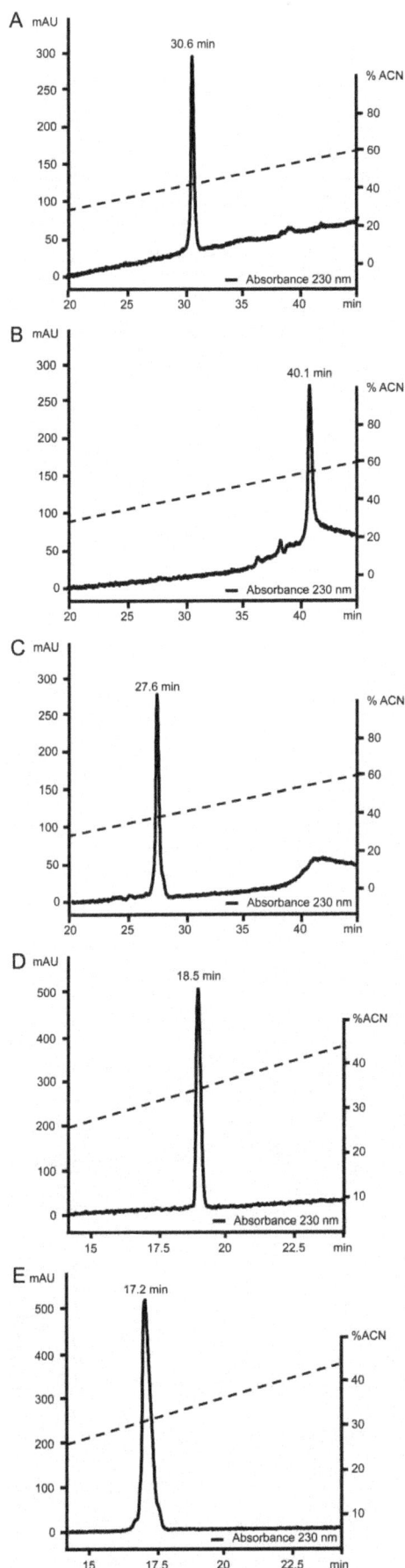

Figure 2. RP-HPLC purification of Pin2 and the variants characterized in this report. A. Pin2, B. Pin2 [G], C. Pin2 [GPG], D. Pin2 [14] and E. Pin2 [17].

Circular dichroism secondary structure analysis

The CD spectrum of Pin2 was compared to the CD spectra of the Pin2 [G] and Pin2 [GPG] variants. The secondary structure analysis was performed using CD spectral data from 190 to 260 nm, in the presence of 60% aqueous TFE, the CD spectra of Pin2 [G] and Pin2 [GPG] showed a clear ordered structure with two minimum ellipticity values at 208 and 222 nm (**Figure 3A**), indicating an alpha-helix conformation. In order to obtain more information concerning the propensity of these peptides to adopt alpha-helical structures, their CD spectra at different TFE proportions (0, 20, 40 and 60%) was acquired. The parental peptide Pin2 (**Figure 3B**) and the Pin2 [GPG] (**Figure 3D**) variant showed a clear "*random coil*" CD profile in the absence of TFE, but interestingly, the Pin2 [G] variant showed a 20% of alpha-helical structure in the absence of TFE (**Figure 3C**), suggesting a more structured Pin2 [G]. Pin2 [G] and Pin2 [GPG] adopted a maximum of 60% of alpha-helical structure at 40% TFE. Similarly, the CD spectra of the sort variants Pin2 [14] and Pin2 [17] were conducted at 0, 20, 40 and 60% TFE (**Figures 3E and 3F**). In the absence of TFE both Pin2 [14] and Pin2 [17] had clear positive ellipticities around 220 nm, and the increment in the TFE proportion induced a reduction in their ellipticity values. Furthermore, in the absence of TFE Pin2 [14] showed a strong negative ellipticity at 205 nm, and the increase of TFE enhanced its CD ellipticity value. Similar behavior was observed for Pin2 [17]. The CD deconvolution for Pin2 [14] and Pin2 [17] resulted in unstructured peptides even at the highest TFE concentration of 60%.

Antimicrobial activities in Mueller-Hinton Agar and Mueller-Hinton Broth

The Gram-positive bacterium *S. aureus* is one of the main concerns in several topical infections while the Gram-negative bacterium *E. coli* is present in many gastrointestinal infections and other infection diseases. In agar culture media, the antimicrobial activity of Pin2 and Pin2 variants were tested against *S. aureus* and *E. coli*. Pin2 [G] variant displayed the highest antimicrobial activity with a MIC value of 12.5 μM, towards both *S. aureus* and *E. coli* (**Table 3**). Both parental Pin2 and Pin2 [GPG] showed MIC values of 37.5 and 25 μM for *S. aureus*, and 18.8 and 25 μM for *E. coli*, respectively. However the short variants Pin2 [14] and Pin2 [17] had MIC values >300 and 80 μM towards *S. aureus*, respectively, but they both Pin2 [14] and Pin2 [17] were more active towards *E. coli* with MIC values of 25 μM.

In broth culture media, the bacterial growth of both *S. aureus* and *E. coli* was observed in the presence of serial concentrations of all Pin2 variants from 25 to 0.4 μM (**Figure 4**). Most of the synthetic peptides showed bactericidal or bacteriostatic antimicrobial activity against the two strains with MIC values of 12.5 and 25 μM. Pin2 [G] and Pin2 [GPG] showed MIC values of 12.5 μM for both *E. coli* and *S. aureus* (**Figures 4A, 4B, 4E and 4F**). The short variant Pin2 [14] had a MIC value of 25 μM against *E. coli* (**Figure 4C**) but a bacteriostatic effect at 25 μM against *S. aureus* (**Figure 4G**). Finally, the short variant Pin2 [17] only showed bacteriostatic effect on *E. coli* at 25 μM (**Figure 4D**), and no antimicrobial activity towards *S. aureus* was observed (**Figure 4H**). **Table 3** summarizes the MIC values observed for Pin2 and its synthetic variants.

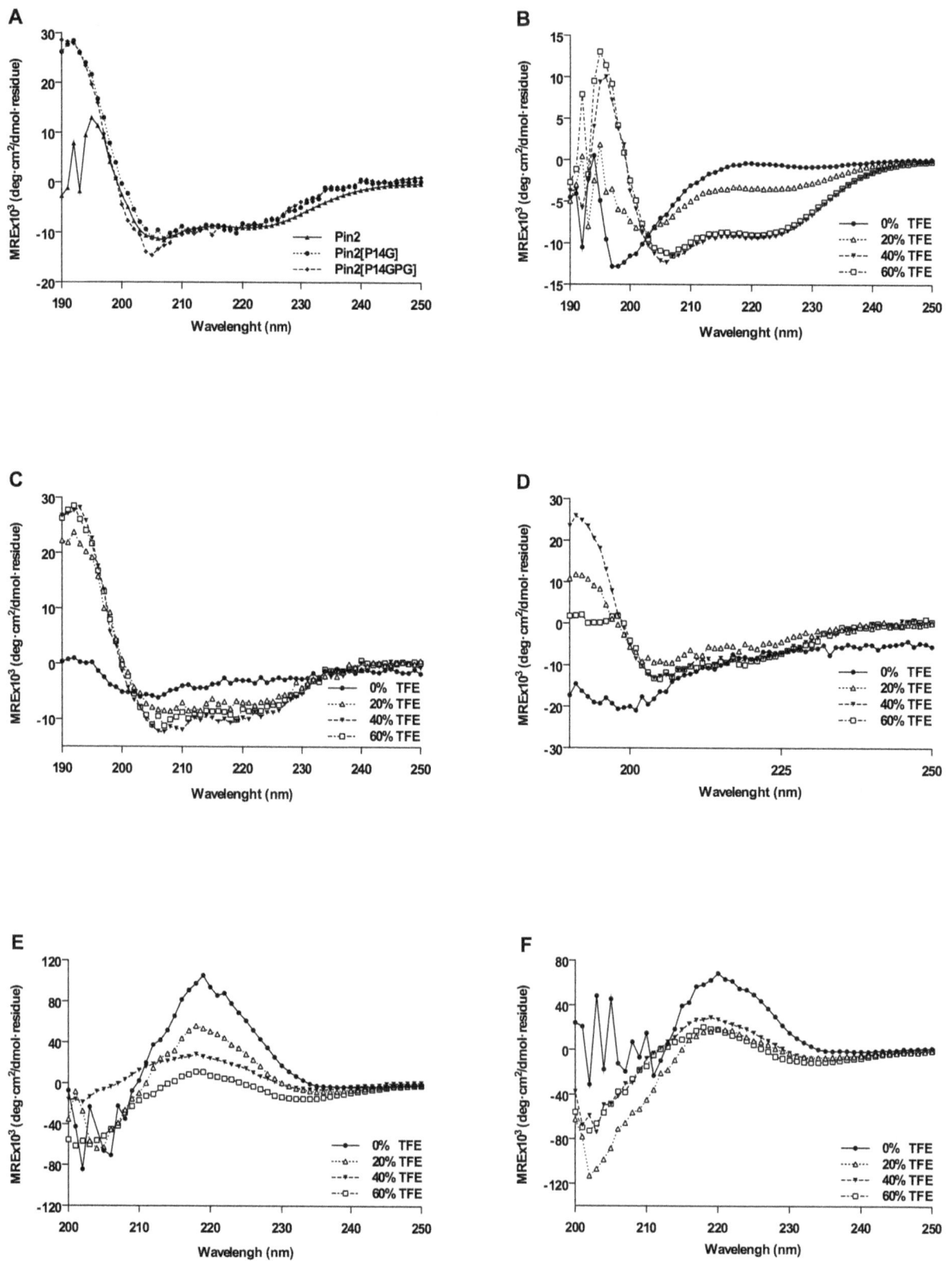

Figure 3. Circular dichroism of Pin2 and its long and short variants at different concentrations of TFE. A. CD spectra of Pin2, Pin2 [G] and Pin2 [GPG] at 60% TFE, B. Pin2, C. Pin2 [G], D. Pin2 [GPG], E. Pin2 [14], F. Pin2 [17].

Table 3. Antimicrobial and hemolytic activities of Pin2 and the Pin2 variants.

| Peptide | Assay | MIC (μM) | | Hemolysis (IC$_{50}$)‡ |
		E. coli	S. aureus	
Pin2	MHA	18.8	37.5	3.3 [1.9–5.7]
	MHB	12.5	12.5	
Pin2 [G]	MHA	12.5	12.5	1.4 [0.4–4.3]
	MHB	12.5	12.5	
Pin2 [GPG]	MHA	25	25	46.6 [34–64]*
	MHB	12.5	12.5	
Pin2 [14]	MHA	25	>300	418.4 [291–602]*
	MHB	25	>25†	
Pin2 [17]	MHA	25	80	NO
	MHB	>25†	>25	

MHA, Mueller-Hinton Agar.
MHB: Mueller Hinton Broth.
‡Mean and 95% confidence intervals, values expressed in μM.
*The numeric IC$_{50}$ value was obtained from the Boltzmann sigmoid equation fit.
†Bacteriostatic effect.
NO means no observed hemolytic activity at 100 μM.

Activity of Pin2 variants against M. tuberculosis

The antibiotic capacities of Pin2 variants were tested in two strains of *M. tuberculosis* using the REMA methodology. *M. tuberculosis* H37Rv is an ATCC strain which has been widely studied and its genome have been completely sequenced, on the other hand, *M. tuberculosis* MDR is a clinical isolated strain which possesses antimicrobial resistance against rifampicin and isoniazid, which are antibiotics used commercially for the treatment of tuberculosis. All synthetic peptide variants showed antimicrobial activities over both strains of *M. tuberculosis* (**Table 4**). The MIC values observed were from 11 to 30 μM for the *M. tuberculosis* H37Rv and from 6 to 33 μM for *M. tuberculosis* MDR. It was interesting to observe that the two short variants Pin2 [14] and Pin2 [17] had MIC values of 11.9 and 11.6 μM towards *M. tuberculosis* H37Rv and MIC values of 6 and 14.8 μM against *M. tuberculosis* MDR, respectively. These results show the feasibility for using short sequence CAMPs against pathogenic *M. tuberculosis* strains that have acquired resistance to commercial antibiotics.

Hemolytic activity of Pin2 variants

The hemolytic assays of Pin2 and the Pin2 variants on human erythrocytes (**Figure 5**) showed that the parental peptides Pin2 as the Pin2 [G] variant at 25 μM had the highest hemolytic activities of all five peptides evaluated, with a 91% and a 100% of hemolysis, respectively. However at the same concentration the Pin2 [GPG] variant showed only a 30% of hemolysis, indicating that the insertion of the GPG motif is relevant for hemolytic activity reduction. The two short variants had not hemolytic effects at 25 μM. Furthermore, they were assayed up to 100 μM observing that the peptide Pin2 [14] showed 25% of hemolysis, while Pin2 [17] did not show any hemolytic effect at such concentration. The statistical analysis is shown in table S1 in file S1.

Discussion

Proline residues are commonly founded in the central regions of alpha-helical segments of transmembrane proteins [34,35]. Sim-

ilarly to these proteins, proline plays an important role in the biological activity of CAMPs, where it imparts significant structural and functional properties [36]. For example, melittin, gaegurins, pardaxin, pipinins, brevinins and ponericins contain a proline residue in the central region of their sequences (**Table S2 in file S1**). Several reports had mentioned that this residue gives a higher capacity to develop pores in bacterial cell membranes, such characteristic is important for their antimicrobial activities, but also CAMPs containing Pro show high hemolytic activities at low concentrations, indicating a low selectivity between bacterial and mammalian cell membranes [17,19–21]. As reported here, the proline residue provides hemolytic properties to Pin2; therefore, to reduce its hemolytic properties, it was substituted by glycine, but unexpectedly Pin2 [G] kept its hemolytic properties; so, also it was flanked by two glycine residues in order to reduce their lytic activities on mammalian cells but to maintain its antimicrobial properties. Additionally, two short peptides were designed to reduce the length of Pin2. As a consequence of the substitution and deletions in the primary structure of Pin2, the physicochemical properties of the Pin2 variants changed (**Table 1**).

The CD spectrum of Pin2, Pin2 [G] and Pin2 [GPG] (**Figure 3B–D**) in absent of TFE was typical of unstructured proteins or random coil proteins. However, all three peptides Pin2, Pin2 [G] and Pin2 [GPG] showed classic α-helical structures at 60% TFE (**Figure 3B–D**) showing two negative bands of similar magnitude at 222 and 208 nm, and a positive band in the range from 190 to 200 nm typical of structured α-helical proteins [37,38]. These results suggest a high propensity to form α-helical structures in hydrogen promoting solvents, which mimics cell membrane environments. On the other hand, the CD profiles of the short peptides were unexpected even at 60% of TFE; for example, Pin2 [14] in the absence of TFE had a strong positive band displayed around 220 nm and a negative band around 205 nm (**Figure 3E**). The positive band could be related to the presence of the aromatic side chains of phenylalanine and tryptophan residues (exciton coupling) in the amine termini region of this peptide, and the negative band near to 205 nm indicates a random coil conformation [39,40]. The increment of the TFE

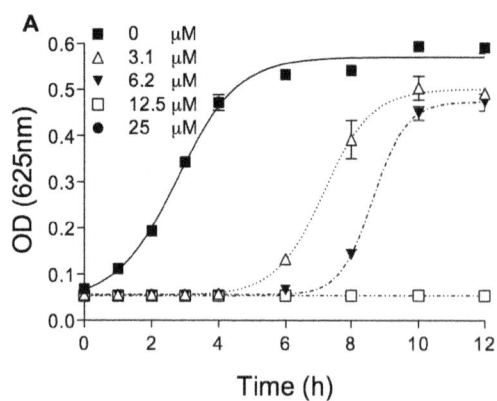

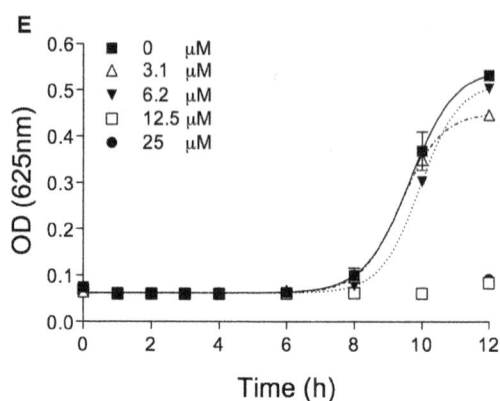

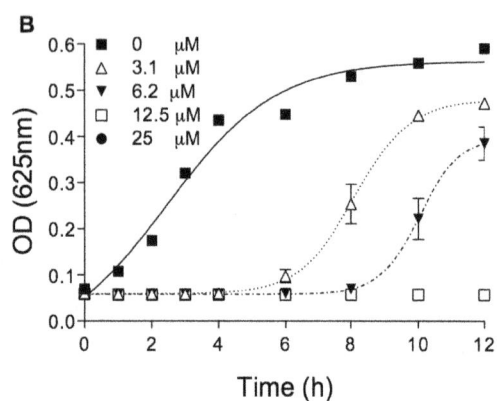

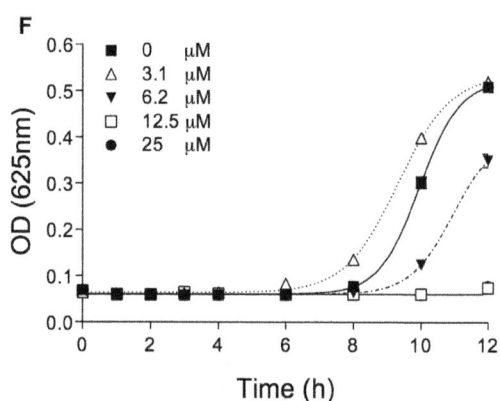

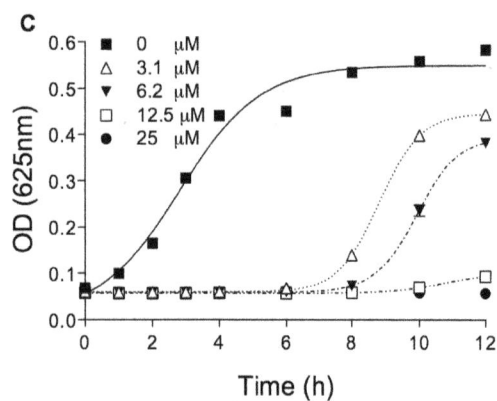

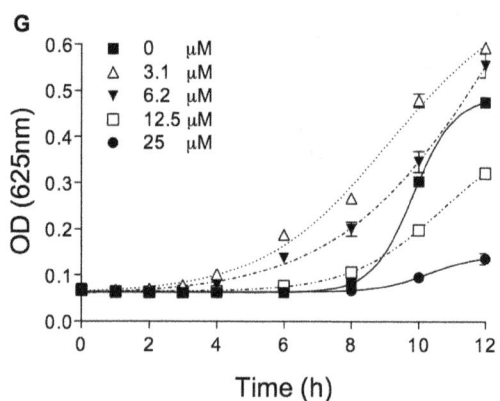

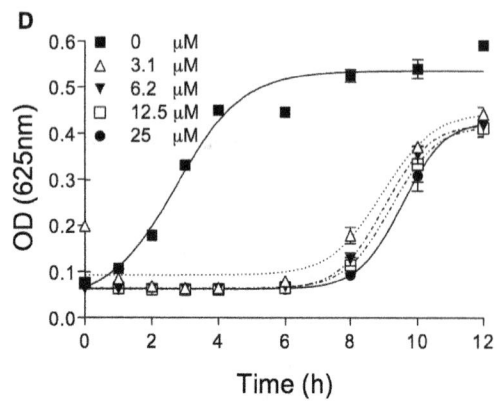

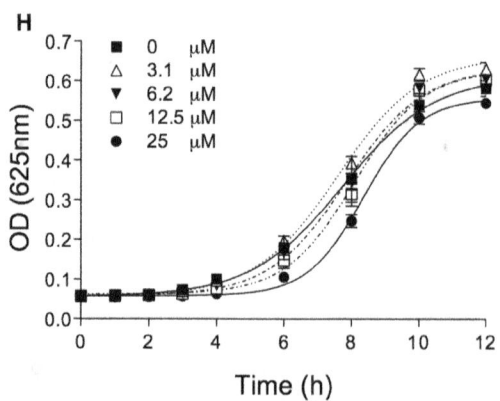

Figure 4. Antimicrobial activity of the Pin2 variants against *E. coli* **ATCC 25922 and** *S. aureus* **ATCC 25923.** *E. coli* antimicrobial activity, A. Pin2 [G], B. Pin2 [GPG], C. Pin2 [14], and D. Pin2 [17]. *S. aureus* antimicrobial activity, E. Pin2 [G], F. Pin2 [GPG], G. Pin2 [14], and H. Pin2 [17]. The concentration of peptides used was from 0.4 to 25 μM (n = 3).

proportion resulted in reduction of the positive band but the negative band, thus indicating the secondary structure induction, similar as was observed by Gopal *et al.* (2012) with the (KW)$_4$ peptide [41]. Nevertheless, since the proposed antimicrobial peptide mechanism of action occurs because structural peptide arrangements with the bacterial membrane, the CD structural modifications of Pin2 [14] may explain its antimicrobial activity. The presence of positive bands observed around at 220 nm for Pin2 [17] (**Figure 3F**) could be explained also for the same aromatic chromophore effects in Pin2 [14]. However the increment in the negative band around 203 nm, resulted for the TFE increments, could be related to a beta-hairpin structural conformation, similar to the indolicidin, a poliproline antimicrobial peptide having tryptophan residues in its primary structure [42] and also because of the presence of structure disruptor residues, such as, the presence of the GlyProGly tripled in the middle of its structure and the three additional glycine residues, Gly6, Gly9 and Gly11. Similar CD profiles compared with those of the Pin2 [17] were observed by Riemen and Waters (2010) with WSWS peptide series. These 12 mer peptides have an exciton coupling of the aromatic residues Trp2 and Trp9. For example, the WSWS peptide RWVSVNGKWISQ has a CD spectrum containing a minimum at 215 nm and a maximum at 229 nm [43]. In this report, Pin2 [17] contains the aromatic residues phenylalanine and tryptophan in close distance that could result in a minimum at 204 nm and a maximum at 220 nm; so the observed CD spectrum could be also the consequence of a random coil conformation with an aromatic exciton coupling similar to that observed for the WSWS peptides.

The antimicrobial activities of Pin2 [G] and Pin2 [GPG] were from 12.5 to 25 μM, in both MHA and MHB bacterial cultures, towards *E. coli* and *S. aureus*. Although the antimicrobial activities of Pin2 [G] and Pin2 [GPG] were quite similar to the parental peptide, the antimicrobial activity of Pin2 [G] was better than that of Pin2 [GPG] in both MHA and MHB bacterial cultures. Concerning the hemolytic activities, Pin2 [G] was more hemolytic that the parental peptide. This increase in function in Pin2 [G] might be related to its more hydrophobic character and to its more

ordered alpha-helical structure. With respect to the antimicrobial activity of the short variants, Pin2 [14] and Pin2 [17], its antimicrobial activities were acceptable towards *E. coli* in both MHA and MHB (25 μM), but very poor towards *S. aureus*. However, they were less lytic to erythrocytes.

In order to correlate the differences observed in the hemolytic activity of the different variants with their amphipathicity, we look at the helical wheel projections of Pin2 and Pin2 variants (**Figure 1**). First, both helical wheels of Pin2 and Pin2 [G] look similar with the exception that substitution of the proline residue by glycine. This substitution did not introduce a modification in the amphipathycity of the parental peptide Pin2 (**Figure 1A–B**); however, for Pin2 [GPG] it is observed that the hydrophilic part of this peptide is crowded with hydrophobic residues (**Figure 1C**, left side). Here, it is also observed that the hydrophobic face (**Figure 1C**, right side) has been disrupted because of the inclusion of the hydrophilic residues Gly15 and Ser17. Nevertheless, the residues Ser20 and Ser23 in the hydrophobic face may balance the amphipathic properties of Pin2 [GPG]. Such hydrophobic balance between the left and right side observed in the helical wheel seems to cause a reduction in the hemolytic activity of this peptide. Concerning the helical wheel projection of the short variants, these representations show the amphipathic distribution as they were previously designed, so the diminish in the hemolytic profile of the short variants might be related to the reduction of hydrophobic residues. These results could be related to the hemolytic activity assays, for instance, the peptide with more alpha-helix propensity and more hydrophobic, Pin2 [G], showed the highest antimicrobial and hemolytic activity, while, the peptide with the less pronounced alpha-helix propensity and less hydrophobic Pin2 [GPG], at the same time showed a reduction in its antimicrobial and hemolytic activity. Respect to the short Pin2 variants, the CD data analysis indicates that Pin2 [14] and Pin2 [17] are beta-structured, contrary to the predicted structure following the Schiffer-Edmundson wheel projections. The non-helical secondary structure observed in both Pin2 [14] and Pin2 [17] was not an impediment to exert antimicrobial activities. Although they were lower to their helical derived peptides for the Gram-positive and

Table 4. Antimicrobial activity of Pin2 and the Pin2 variants on two strains of *M. tuberculosis*.

Peptide	MW (Da)	*Mycobacterium tuberculosis* H37Rv		*Mycobacterium tuberculosis* MDR*	
		MIC (μg/mL)	MIC (μM)	MIC (μg/mL)	MIC (μM)
Pin2	2,612.1	57.7±22.3	22.1±8.6	86.5	33.1
Pin2 [G]	2,572.0	48.1	18.7	48.1	18.7
Pin2 [GPG]	2,762.2	80.1±24.8	29±9	48.1	17.4
Pin2 [14]	1,680.1	20±6.2	11.92±3.7	10±3.1	6±1.8
Pin2 [17]	1,891.3	22±4.9	11.65±2.59	28.0±9.8	14.8±5.2
Ethambutol§	204.3	0.5	2.5	20	97
Isoniazid	137.1	24±8.8	175.1±63.9	6±2.2	43.8±16
Rifampicin§	823.0	0.4	0.5	32	38.9

The MIC values were calculated using the Resazurin dye reduction method, 500,000 bacteria per well were evaluated.
§, MIC values reported by Rastogi, et al., 1996 [55].
*, Clinically isolated strain characterized with resistance to rifampicin and isoniazid in UIMZ-IMSS, Zacatecas, Mexico.

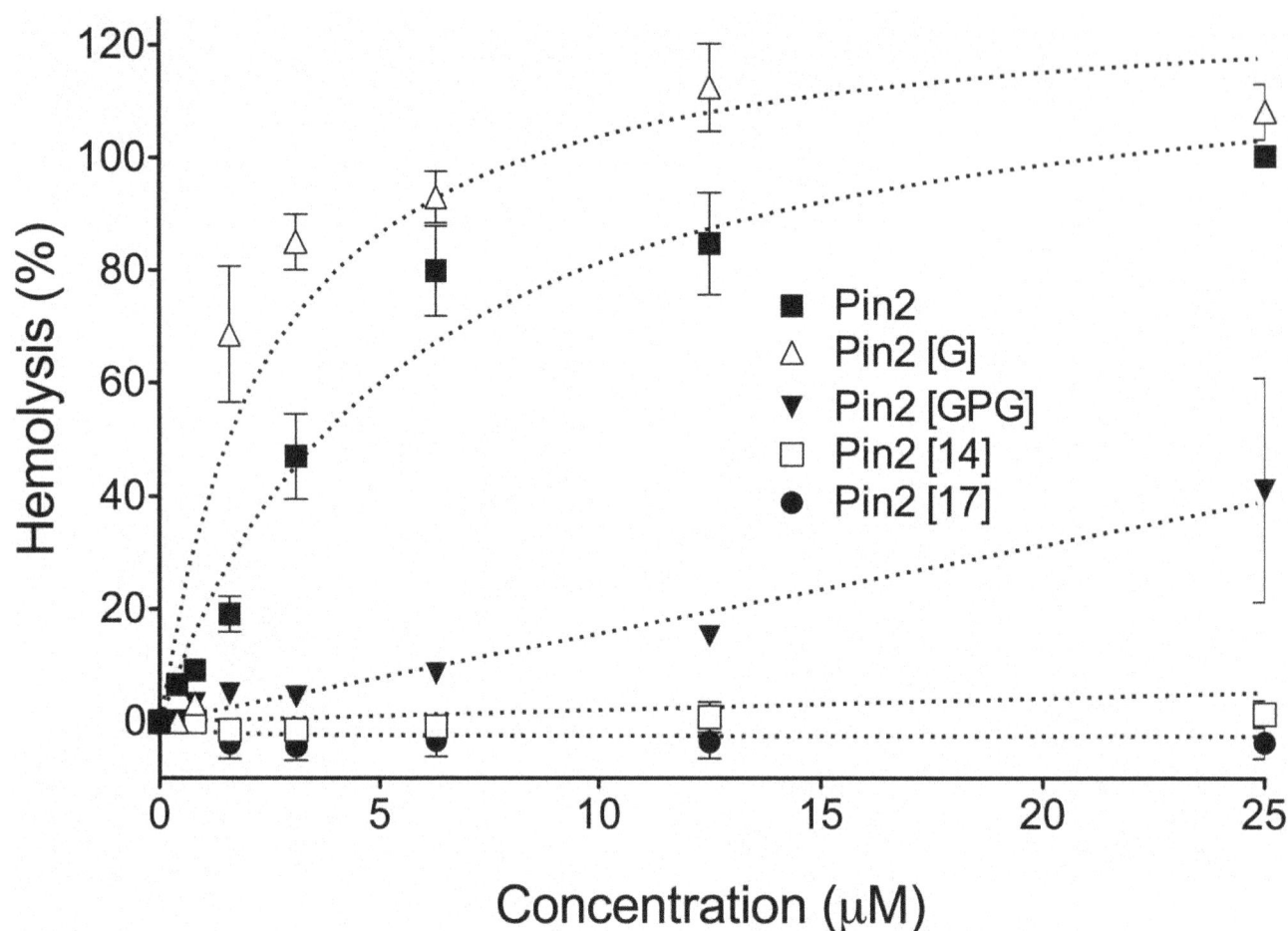

Figure 5. Hemolytic activity in human red blood cells. Data are the average of at least four independent experiments. Error bars represent the standard deviations.

Gram-negative bacteria, Pin2 [14] and Pin2 [17] were more efficient to inhibit the growth of *Mycobacterium tuberculosis*.

Concerning the activity against *M. tuberculosis*, it is notable that the larger peptides Pin2, Pin2 [G] and Pin2 [GPG] had lower inhibitory capacity than the shorter peptides Pin2 [14] and Pin2 [17]. Also, it is noteworthy that Pin2 and its variants are comparable to the antibiotics used for the treatment of tuberculosis (i.e. ethambutol, rifampicin, isoniazid) at the molar doses. This comparison is more evident in the inhibition of *M. tuberculosis* MDR, and the interest to design short antimicrobial peptides.

Here we based the design of low hemolytic and short antimicrobial peptides on observed patterns in nature and on theoretical calculations. We found a strong correlation in hydrophobicity and alpha-structured molecules with high hemolytic activity; however, the antimicrobial capacity could be sustained with low eukaryotic lytic activities in short hydrophilic antimicrobial peptides such that Pin2 [14] and Pin2 [17] with a plus of maintaining a wider antimicrobial spectrum; that is affecting Gram-positive and Gram-negative bacteria as well as multi-drug resistant *Mycobacterium tuberculosis*.

Supporting Information

File S1 Supporting tables. Table S1, Statistical analysis of variance with ANOVA of the hemolysis data, followed by post hoc testing using the Tukey's method. **Table S2**, Proline in the middle regions of hemolytic antimicrobial peptides.

Acknowledgments

We acknowledge Dr. Fernando Zamudio from Instituto de Biotecnología-UNAM for mass spectrometric determinations and QBP Ma. Rocío Patiño Maya from Instituto de Química-UNAM and Dr. Gloria Saab from Instituto de Biotecnología for CD spectra data acquisition. The patent filling advice from M.Sc. Martin Patiño and MBA Mario Trejo are also greatly acknowledged. Alexis Rodríguez was recipient of a PhD scholarship 24492 from CONACyT (172816), and now he is supported by a postdoctoral fellowship from PROMEP.

Author Contributions

Conceived and designed the experiments: GC. Performed the experiments: AR AMR. Analyzed the data: EV GC BRS. Contributed reagents/materials/analysis tools: EV BRS GC. Wrote the paper: AR GC.

References

1. Bulet P, Stöcklin R, Menin L (2004) Anti-microbial peptides: from invertebrates to vertebrates. Immunological Reviews 198: 169–184.

2. Andreu D, Rivas L (1998) Animal antimicrobial peptides: an overview. Peptide Science 47: 415–433.

3. Kamysz W (2005) Are antimicrobial peptides an alternative for conventional antibiotics? Nuclear Medicine Review Central and Eastern Europe 8: 78–86.

4. Zeng XC, Corzo G, Hahin R (2005) Scorpion venom peptides without disulfide bridges. IUBMB Life 57: 13–21.

5. Bulet P, Hetru C, Dimarcq JL, Hoffmann D (1999) Antimicrobial peptides in insects; structure and function. Developmental & Comparative Immunology 23: 329–344.

6. Moerman L, Bosteels S, Noppe W, Willems J, Clynen E, et al. (2002) Antibacterial and antifungal properties of α-helical, cationic peptides in the venom of scorpions from southern Africa. European Journal of Biochemistry 269: 4799–4810.

7. Rollins-Smith LA, Reinert LK, O'Leary CJ, Houston LE, Woodhams DC (2005) Antimicrobial peptide defenses in amphibian skin. Integrative and Comparative Biology 45: 137–142.

8. Rinaldi AC (2002) Antimicrobial peptides from amphibian skin: an expanding scenario. Current Opinion in Chemical Biology 6: 799–804.

9. Huang Y, Huang J, Chen Y (2010) Alpha-helical cationic antimicrobial peptides: relationships of structure and function. Protein and cell 1: 143–152.

10. Tossi A, Sandri L, Giangaspero A (2000) Amphipathic, α-helical antimicrobial peptides. Peptide Science 55: 4–30.

11. Nguyen LT, Haney EF, Vogel HJ (2011) The expanding scope of antimicrobial peptide structures and their modes of action. Trends in Biotechnology 29: 464–472.

12. Yanmei L, Hongyan Q, Qi X, Zhijian S, Yadong H, et al. (2012) Overview on the recent study of antimicrobial peptides: origins, functions, relative mechanisms and application Peptides 37: 2207–2215.

13. Park SC, Park Y, Hahm KS (2011) The role of antimicrobial peptides in preventing multidrug-resistant bacterial infections and biofilm formation. International Journal of Molecular Sciences 12: 5971–5992.

14. Yeung AY, Gellatly S, Hancock RW (2011) Multifunctional cationic host defence peptides and their clinical applications. Cellular and Molecular Life Sciences 68: 2161–2176.

15. Corzo G, Escoubas P, Villegas E, Barnham KJ, He W, et al. (2001) Characterization of unique amphipathic antimicrobial peptides from venom of the scorpion Pandinus imperator. Biochem J 359: 35–45.

16. Kreil G (1973) Structure of melittin isolated from two species of honey bees. FEBS Letters 33: 241–244.

17. Dathe M, Kaduk C, Tachikawa E, Melzig MF, Wenschuh H, et al. (1998) Proline at position 14 of alamethicin is essential for hemolytic activity, catecholamine secretion from chromaffin cells and enhanced metabolic activity in endothelial cells. Biochimica et Biophysica Acta-Biomembranes 1370: 175–183.

18. Oren Z, Shai Y (1996) A class of highly potent antibacterial peptides derived from pardaxin, a pore-forming peptide isolated from moses sole fish Pardachirus marmoratus. European Journal of Biochemistry 237: 303–310.

19. Dempsey CE, Bazzo R, Harvey TS, Syperek I, Boheim G, et al. (1991) Contribution of proline-14 to the structure and actions of melittin. FEBS Letters 281: 240–244.

20. Thennarasu S, Nagaraj R (1996) Specific antimicrobial and hemolytic activities of 18-residue peptides derived from the amino terminal region of the toxin pardaxin. Protein Engineering 9: 1219–1224.

21. Oren Z, Shai Y (1997) Selective lysis of bacteria but not mammalian cells by diastereomers of melittin: structure–function study. Biochemistry 36: 1826–1835.

22. Corzo G, Villegas E, Gómez-Lagunas F, Possani LD, Belokoneva OS, et al. (2002) Oxyopinins, large amphipathic peptides isolated from the venom of the wolf spider Oxyopes kitabensis with cytolytic properties and positive insecticidal cooperativity with spider neurotoxins. Journal of Biological Chemistry 277: 23627–23637.

23. Orivel J, Redeker V, Le Caer JP, Krier F, Revol-Junelles AM, et al. (2001) Ponericins, new antibacterial and insecticidal peptides from the venom of the ant Pachycondyla goeldii. Journal of Biological Chemistry 276: 17823–17829.

24. Zasloff M (1987) Magainins, a class of antimicrobial peptides from Xenopus skin: isolation, characterization of two active forms, and partial cDNA sequence of a precursor. Proc Natl Acad Sci U S A 84: 5449–5453.

25. Shin SY, Kang JH, Lee DG, Jang SY, Seo MY, et al. (1999) Influences of hinge region of a synthetic antimicrobial peptide, cecropin A(1–13)-melittin(1–13) hybrid on antibiotic activity. Bulleting of the Korean Chemistry Society 20: 1078–1084.

26. Rodriguez A, Villegas E, Satake H, Possani LD, Corzo G (2011) Amino acid substitutions in an alpha-helical antimicrobial arachnid peptide affect its chemical properties and biological activity towards pathogenic bacteria but improves its therapeutic index. Amino Acids 40: 61–68.

27. Kubica GP, Kim TH, Dunbar FP (1972) Designation of strain H37Rv as the neotype of Mycobacterium tuberculosis. International Journal of Systematic Bacteriology 22: 99–106.

28. Rivas-Santiago B, Rivas-Santiago C, Castañeda-Delgado JE, León–Contreras JC, Hancock REW, et al. (2013) Activity of LL-37, CRAMP and antimicrobial peptide-derived compounds E2, E6 and CP26 against Mycobacterium tuberculosis. International Journal of Antimicrobial Agents 41: 143–148.

29. Martin A, Camacho M, Portaels F, Palomino JC (2003) Resazurin microtiter assay plate testing of Mycobacterium tuberculosis susceptibilities to second-line drugs:

30. Rivas-Santiago B, Contreras JCL, Sada E, Hernández-Pando R (2008) The potential role of lung epithelial cells and β-defensins in experimental latent tuberculosis. Scandinavian Journal of Immunology 67: 448–452.

31. Andrade MA, Chacón P, Merelo JJ, Morán F (1993) Evaluation of secondary structure of proteins from UV circular dichroism using an unsupervised learning neural network. Protein Engineering 6: 383–390.

32. Merelo JJ, Andrade MA, Prieto A, Morán F (1994) Proteinotopic feature maps. Neurocomputing 6 443–454.

33. Wang G, Li X, Wang Z (2009) APD2: the updated antimicrobial peptide database and its application in peptide design. Nucleic acids research 37: D933–D937.

34. Sansom MS (1992) Proline residues in transmembrane helices of channel and transport proteins: a molecular modelling study. Protein engineering 5: 53–60.

35. Bywater RP, Thomas D, Vriend G (2001) A sequence and structural study of transmembrane helices. Journal of Computer-Aided Molecular Design 15: 533–552.

36. Cordes FS, Bright JN, Sansom MS (2002) Proline-induced distortions of transmembrane helices. Journal of molecular biology 323: 951–960.

37. Greenfield NJ (2007) Using circular dichroism spectra to estimate protein secondary structure. Nature protocols 1: 2876–2890.

38. Kelly SM, Price NC (2000) The use of circular dichroism in the investigation of protein structure and function. Current protein and peptide science 1: 349–384.

39. Freskgaard PO, Maartensson LG, Jonasson P, Jonsson BH, Carlsson U (1994) Assignment of the contribution of the tryptophan residues to the circular dichroism spectrum of human carbonic anhydrase II. Biochemistry 33: 14281–14288.

40. Woody RW (1994) Contributions of tryptophan side chains to the far-ultraviolet circular dichroism of proteins. European biophysics journal 23: 253–262.

41. Gopal R, Park JS, Seo CH, Park Y (2012) Applications of circular dichroism for structural analysis of gelatin and antimicrobial peptides. International journal of molecular sciences 13: 3229–3244.

42. Ladokhin AS, Selsted ME, White SH (1999) CD spectra of indolicidin antimicrobial peptides suggest turns, not polyproline helix. Biochemistry 38: 12313–12319.

43. Riemen AJ, Waters ML (2010) Positional effects of phosphoserine on β-hairpin stability. Organic & Biomolecular Chemistry 8: 5411–5417.

44. Kyte J, Doolittle RF (1982) A simple method for displaying the hydropathic character of a protein. Journal of molecular biology 157: 105–132.

45. Liu L, Fang Y, Huang Q, Pan Q, Wu J (2008) A new structure-activity relationship of linear cationic α-helical antimicrobial peptides. In: Peng Y, Weng X, editors. 7th Asian-Pacific Conference on Medical and Biological Engineering: Springer Berlin Heidelberg. pp. 167–170.

46. Conlon JM, Mechkarska M, Prajeep M, Sonnevend A, Coquet L, et al. (2012) Host-defense peptides in skin secretions of the tetraploid frog Silurana epitropicalis with potent activity against methicillin-resistant Staphylococcus aureus (MRSA). Peptides 37: 113–119.

47. Oh H, Hedberg M, Wade D, Edlund C (2000) Activities of synthetic hybrid peptides against anaerobic bacteria: aspects of methodology and stability. Antimicrobial Agents and Chemotherapy 44: 68–72.

48. Slaninová J, Mlsová V, Kroupová H, Alán L, Tůmová T, et al. (2012) Toxicity study of antimicrobial peptides from wild bee venom and their analogs toward mammalian normal and cancer cells. Peptides 33: 18–26.

49. Čeřovský V, Slaninová J, Fučík V, Hulačová H, Borovičková L, et al. (2008) New potent antimicrobial peptides from the venom of Polistinae wasps and their analogs. Peptides 29: 992–1003.

50. Jin LL, Song SS, Li Q, Chen YH, Wang QY, et al. (2009) Identification and characterisation of a novel antimicrobial polypeptide from the skin secretion of a Chinese frog (Rana chensinensis). International Journal of Antimicrobial Agents 33: 538–542.

51. Patrzykat A, Gallant JW, Seo JK, Pytyck J, Douglas SE (2003) Novel antimicrobial peptides derived from flatfish genes. Antimicrobial Agents and Chemotherapy 47: 2464–2470.

52. Wang H, Ran R, Yu H, Yu Z, Hu Y, et al. (2012) Identification and characterization of antimicrobial peptides from skin of Amolops ricketti (Anura: Ranidae). Peptides 33: 27–34.

53. Conlon JM, Sonnevend A, Davidson C, Demandt A, Jouenne T (2005) Host-defense peptides isolated from the skin secretions of the northern red-legged frog Rana aurora aurora. Developmental & Comparative Immunology 29: 83–90.

54. Liu R, Liu H, Ma Y, Wu J, Yang H, et al. (2011) There are abundant antimicrobial peptides in brains of two kinds of Bombina toads. Journal of Proteome Research 10: 1806–1815.

55. Rastogi N, Labrousse V, Goh KS (1996) In vitro activities of fourteen antimicrobial agents against drug susceptible and resistant clinical isolates of Mycobacterium tuberculosis and comparative intracellular activities against the virulent H37Rv strain in human macrophages. Current Microbiology 33: 167–175.

56. Zidovetzki R, Rost B, Armstrong DL, Pecht I (2003) Transmembrane domains in the functions of Fc receptors. Biophysical Chemistry 100: 555–575.

rapid, simple, and inexpensive method. Antimicrobial Agents and Chemotherapy 47: 3616–3619.

8

Defining the Erythrocyte Binding Domains of *Plasmodium vivax* Tryptophan Rich Antigen 33.5

Hema Bora, Rupesh Kumar Tyagi, Yagya Dutta Sharma*

Department of Biotechnology, All India Institute of Medical Sciences, New Delhi, India

Abstract

Tryptophan-rich antigens play important role in host-parasite interaction. One of the *Plasmodium vivax* tryptophan-rich antigens called PvTRAg33.5 had earlier been shown to be predominantly of alpha helical in nature with multidomain structure, induced immune responses in humans, binds to host erythrocytes, and its sequence is highly conserved in the parasite population. In the present study, we divided this protein into three different parts i.e. N-terminal (amino acid position 24–106), middle (amino acid position 107–192), and C-terminal region (amino acid position 185–275) and determined the erythrocyte binding activity of these fragments. This binding activity was retained by the middle and C-terminal fragments covering 107 to 275 amino acid region of the PvTRAg33.5 protein. Eight non-overlapping peptides covering this 107 to 275 amino acid region were then synthesized and tested for their erythrocyte binding activity to further define the binding domains. Only two peptides, peptide P4 (at 171–191 amino acid position) and peptide P8 (at 255–275 amino acid position), were found to contain the erythrocyte binding activity. Competition assay revealed that each peptide recognizes its own erythrocyte receptor. These two peptides were found to be located on two parallel helices at one end of the protein in the modelled structure and could be exposed on its surface to form a suitable site for protein-protein interaction. Natural antibodies present in the sera of the *P. vivax* exposed individuals or the polyclonal rabbit antibodies against this protein were able to inhibit the erythrocyte binding activity of PvTRAg33.5, its fragments, and these two synthetic peptides P4 and P8. Further studies on receptor-ligand interaction might lead to the development of the therapeutic reagent.

Editor: Asad U. Khan, Aligarh Muslim University, India

Funding: This work was supported by the Department of Biotechnology [BT/PR9800/MED/29/44/2007 to YDS and Senior Research Fellowships to HB and RKT] of the Government of India. The funders had no role in study design, data collection and analysis, decision to publish or preparation of the manuscript.

Competing Interests: The authors have declared that no competing interests exist.

* E-mail: ydsharma@hotmail.com

Introduction

Plasmodium vivax is the commonest human malaria parasite. It has the widest distribution throughout the tropics, subtropics, and temperate zones [1]. Due to lack of continuous in vitro culture, characterization of *P. vivax* molecules has been very slow. As a result, only fewer *P. vivax* vaccine candidate antigens are under the clinical trials as compared to *P. falciparum*. Hence, there is a need to identify the newer *P. vivax* molecules which play important role in survival of the parasite inside its host, have very limited genetic diversity, and generate protective immune responses. Parasite molecules involved in host-parasite interaction play major role in the parasite's life cycle. Molecules participating in this step may be exploited to design the therapeutic reagents which can inhibit the interaction of the parasite with its human host.

Tryptophan-rich antigens from *Plasmodium* species have been proposed as potential vaccine candidates. For the first time, tryptophan-rich antigens were characterized from *P. yoelii* where, pypAg1 and pypAg3 showed protective immune responses against infection in mice [2]. Immunization with recombinant pypAg1 reduced four to seven fold parasitemia against *P. yoelii* infection [3]. Subsequently, two such proteins termed as Tryptophan and Threonine-rich Antigen (TryThrA) and Merozoite associated Tryptophan-rich Antigen (MaTrA) were characterized from *P.* *falciparum*. These proteins had characteristics which were similar to *P. yoelii* antigens [4,5]. Further studies on TryThrA led to the identification of peptides which could bind to normal human erythrocytes and also inhibited the in vitro merozoite invasion [6]. Later, Tryptophan-rich Antigen 3 (TrpA-3,) and Lysine-Tryptophan-rich Antigen (LysTrpA) of *P.falciparum* were characterized [7].

The genome sequencing of human malaria parasite *P.vivax* and its closest animal model represented by monkey malaria parasite *P.cynomolgi* revealed that each parasite has larger number of tryptophan-rich antigens [8,9]. Characterization of these tryptophan-rich antigens is needed to develop the newer drug and vaccine targets. Earlier, we had identified the first *P. vivax* tryptophan-rich antigen and named it as PvTRAg [10]. It was followed by the characterization of many such proteins of this parasite which generated immune responses in *P. vivax* patients and did not show much genetic polymorphism in parasite population [10–17]. Six of 15 PvTRAgs, including PvTRAg33.5, have shown the erythrocyte binding activity [18]. Recently, we have reported the physico-chemical characterization and molecular modeling of PvTRAg33.5 [13]. This protein has also shown humoral and cellular immune responses in humans, and no genetic diversity in parasite population [19]. In the present study, we have defined the erythrocyte binding domains of PvTRAg33.5 and this binding activity was inhibited by the *P. vivax* patients' sera.

Materials and Methods

Materials

For antibody inhibition assay, the heparinized blood ($\sim$200 μl) was collected from the microscopically confirmed *P. vivax* malaria patients. Heparinized blood (2 ml) was also collected from the healthy lab individuals with B positive blood group for the erythrocyte binding assays. All individuals were informed about the study and their written consent was obtained for blood collection. Institutional ethical guidelines were followed during blood collection. Ethics committee of All India Institute of Medical Sciences, New Delhi, approved the study via approval number IEC/NP-342/2012 & RP-11/2012.

Cloning, Expression and Purification of Three Fragments Derived from PvTRAg33.5

The cloning, expression, and purification of recombinant PvTRAg33.5 derived from exon-2 of *pvtrag33.5* gene have been described earlier [13]. Exon-2 of PvTRAg33.5 was further divided into three different parts i.e. N-terminal region (N-PvTRAg33.5) covering 70–318 bp (24–106 amino acid residues), middle region (M-PvTRAg33.5) covering 319–577 bp (107–192 amino acid residues) and C-terminal region (C-PvTRAg33.5) covering 555–828 bp (185–275 amino acid residues) by amplifying each region separately using *Pfx* polymerase and PvTRAg33.5-pGEM®T Easy recombinant clone as template. The N-terminal fragment was PCR amplified by using primers NTF 5′-TGTAGTCGACT-CAAAGCGCAGTAG-3′ and NTR 5′- TCAAATATCTA-GAAAAATTATTCC-3′ (Restriction sites engineered in primers are underlined). After Initial denaturation of template DNA at 94°C for 10 minutes, a total of 35 cycles were carried out under following conditions; denaturation at 94°C for 30 seconds, annealing at 50°C for 30 seconds, and extension at 68°C for 30

seconds. Final extension was at 68°C for 15 minutes. The other two fragments (middle and C-terminal) were amplified under the same conditions using primers MTF 5′-TCTGGATCCTT-GAGTGATGGATAC-3′ and MTR 5′-TGAAGCCCAGTTTC-TAGATTCCTG-3′ for M-PvTRAg33.5 and CTF 5′-GTTGGATCCATGTATTGGGAT-3′ and CTR 5′-TTGTTCCTAATTGAGTCTAGAATTCC-3′ (Restriction sites engineered in primers are underlined) for C-PvTRAg33.5. The PCR products were cloned in to the pPROEX™HT expression vectors and Histidine-tagged protein was purified using immobilized metal affinity chromatography on Ni^{2+} NTA agarose column according to manufacturer's instructions (Qiagen, GmbH, Hilden, Germany). Since C-PvTRAg33.5 could not be purified, it was recloned in pGEX4T-2 vector, which adds GST-tag at the N-terminal region of the protein, and purified using Glutathione Sepharose™ 4B resin as per manufacturer's instructions (GE Healthcare Bio-Sciences AB, Uppsala, Sweden). The homogeneity of purified recombinant proteins was confirmed by SDS-PAGE as described earlier [12].

Erythrocyte Binding Assay by Cell – ELISA

Erythrocyte binding activity of tagged PvTRAg33.5 and its fragments was analyzed as described elsewhere [20]. Briefly, each well of a 96-well microtiter plate was coated with one million human erythrocytes of B positive blood group individuals and incubated overnight at 4°C. After washing with PBS and blocking with 5% BSA for 2 h at 37°C, different concentrations (0.015, 0.03, 0.062, 0.125, 0.25, 0.5, 1, and 2 μM) of purified Histidine-tagged PvTRAg33.5, N-PvTRAg33.5, M-PvTRAg33.5, or GST tagged C-PvTRAg33.5 were added. The plates were incubated for 4 h at room temperature. After washing, the plates were developed with anti-His_6 or anti-GST (for C-PvTRAg33.5) monoclonal antibodies (Sigma Aldrich, St. Louis, USA) followed

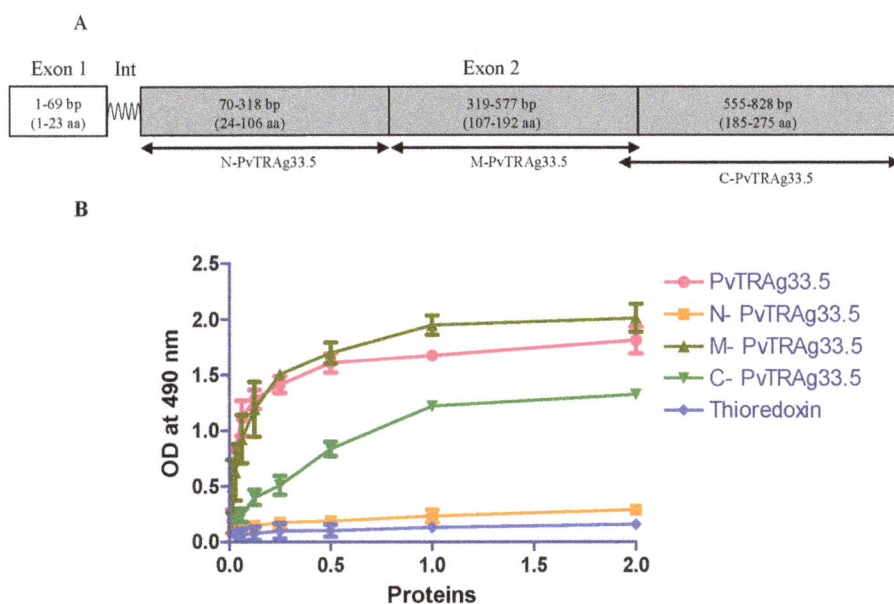

Figure 1. Determination of erythrocyte binding regions of PvTRAg33.5. (A) Schematic representation of PvTRAg33.5. Exon 1 encodes for a 23 amino acid signal peptide. Wavy lines indicate the intron. Exon 2 (shaded grey) encodes the mature protein which was fragmented in to three parts i.e. N-PvTRAg33.5, M-PvTRAg33.5 and C-PvTRAg33.5. **(B)** Cell-ELISA showing erythrocyte binding of Histidine-tagged PvTRAg33.5 and its three fragments with human erythrocytes (C-PvTRAg33.5 was GST-tagged). Increasing concentrations of the purified recombinant proteins were allowed to bind with $\sim$1 million erythrocytes in a microtiter plate and reacted with primary anti-His_6 or anti GST antibody and HRP conjugated secondary antibody. Recombinant Histidine-tagged thioredoxin from *D. desulfuricans* was used as negative control. Error bar indicates the standard deviation of mean from three experiments. Int, intron.

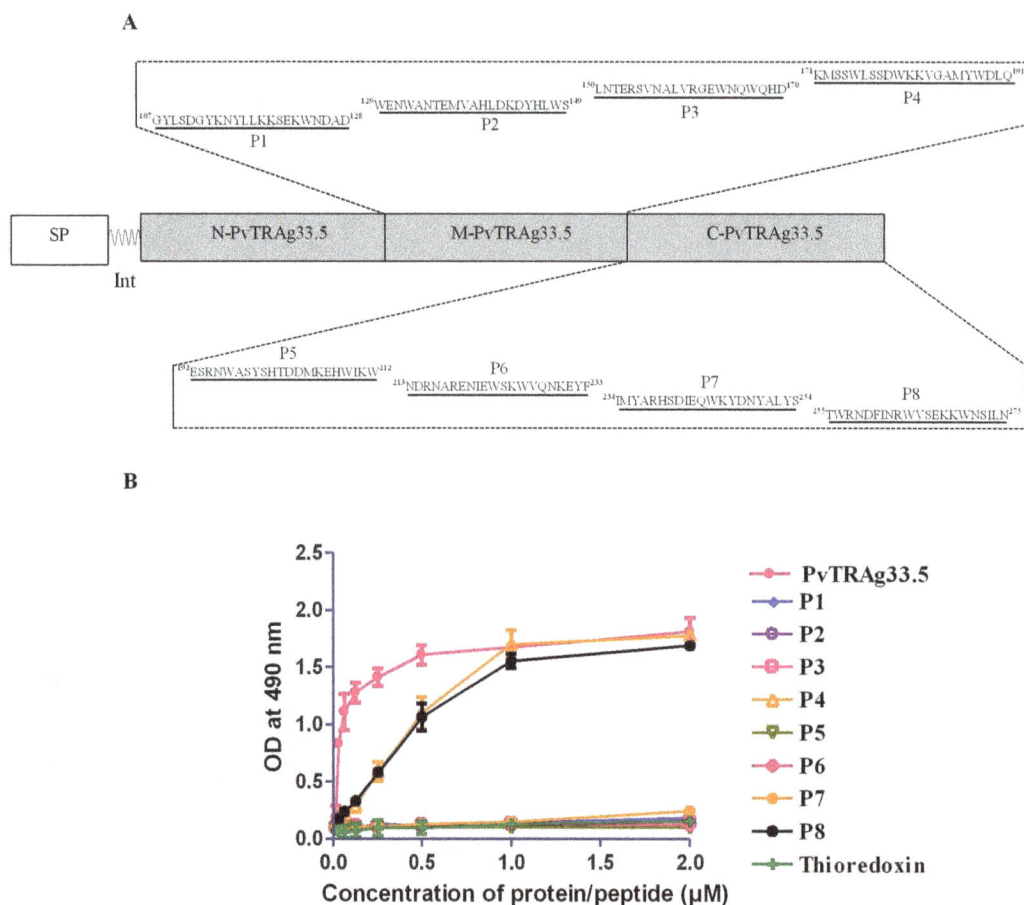

Figure 2. Determination of erythrocyte peptide binding domains of PvTRAg33.5. (A) Schematic representation of eight non-overlapping peptides designed from M-PvTRAg33.5 and C-PvTRAg33.5 fragments. Amino acid sequence and name of each peptide along with residue numbers is shown. **(B)** Cell-ELISA showing erythrocyte binding affinity of these synthetic non overlapping peptides. Increasing concentrations of these peptides were allowed to bind with erythrocytes. Reaction with primary and secondary antibodies was carried out as described in Fig. 1B. Recombinant thioredoxin from *D. desulfuricans* was used as negative controls. Mean± SD value of absorbance from three experiments is plotted. SP, signal peptide; Int, intron.

by horseradish peroxidase (HRP) conjugated anti-mouse IgG secondary antibody (Pierce Biotechnology Inc., Rockford, IL, USA) and o-phenyldiamine (OPD) substrate (Sigma-Aldrich, St. Louis, MO, USA). Optical density was recorded at 490 nm. A 23 kDa Histidine-tagged thioredoxin from *Desulfovibrio desulfuricans* [21] and recombinant GST protein were was used as negative controls. The absorbance of GST alone was subtracted from GST-tagged C-PvTRAg33.5 in binding experiments.

For antibody inhibition assay, the tagged PvTRAg33.5, its fragments or peptides (250 nM) were incubated overnight at 4°C with various dilutions (1:10, 1:50, 1:100, and 1:1000) of rabbit anti-PvTRAg33.5 antibodies and then added to erythrocytes in a micro titer plate. Plates were then developed with anti-His₆ or anti-GST monoclonal antibody as described above. Same protocol was followed with pooled *P. vivax* infected patient sera at dilutions 1:10, 1:30, 1:50, 1:75 and 1:100. The anti-PvTRAg33.5 antibodies were also affinity purified from the pooled *P. vivax* infected patient sera using CarboxyLink™ Immobilization Kit (Thermo scientific, Rockford, USA) according to manufacturer's protocol. Briefly, 2 mg of PvTRAg33.5 was coupled to the resin. After washing with PBS, bound antibody was eluted by 0.2 M Glycine-HCl buffer. The purified antibody was dialyzed in 50 mM Tris-HCl, diluted to

the same level as of the initial serum, and used in the erythrocyte binding inhibition assay.

Peptide Designing

In order to further define the erythrocyte binding region of PvTRAg33.5, non-overlapping peptides were commercially synthesized (Thermo fisher Scientific, GmbH, Germany). A stretch of six histidines was added to the C-terminal end of each peptide so that the peptide bound to erythrocytes could be detected by anti-His₆ monoclonal antibody. Erythrocyte binding activity of these peptides was assessed by Cell-ELISA, as above. The inhibition of erythrocyte binding activity of peptides by antibodies was also checked by Cell-ELISA. Experiments were performed in duplicate wells and repeated at least three times.

Competition Assay

As mentioned above, approximately one million erythrocytes coated in to each well of the 96 well microtiter plate and blocked with 5% BSA was incubated with 5 μM of peptide P4 or P8, or P4 plus P8, or 2 μM of Histidine tagged complete PvTRAg33.5 for 3 h at room temperature. After washing, 1 μM of GST-tagged complete PvTRAg33.5 was added to each well and plate was incubated for 3 h at room temperature. After washing, the plates

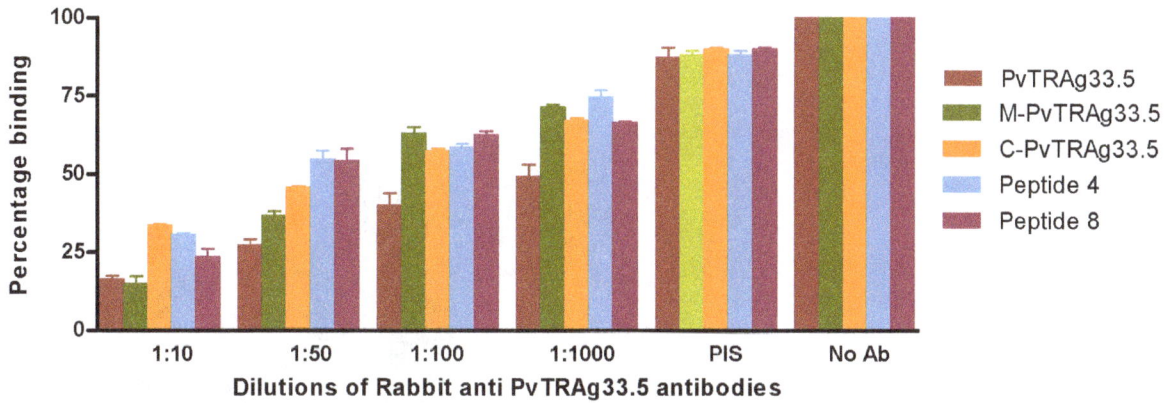

Figure 3. Inhibition of erythrocyte binding of PvTRAg33.5 derived fragments and peptides by rabbit anti-PvTRAg33.5 antibody. The tagged recombinant PvTRAg33.5, its fragments, or synthetic peptides were mixed with different dilutions of polyclonal antisera raised in rabbit against PvTRAg33.5 before adding to the microtiter plate coated with erythrocytes. Further steps of color development were same as in Fig. 1 and 2B. Binding in the absence of antibody was taken as percentage control. Error bar indicates the standard deviation of mean from three experiments. No Ab, no antibody; PIS, pre-immune rabbit sera.

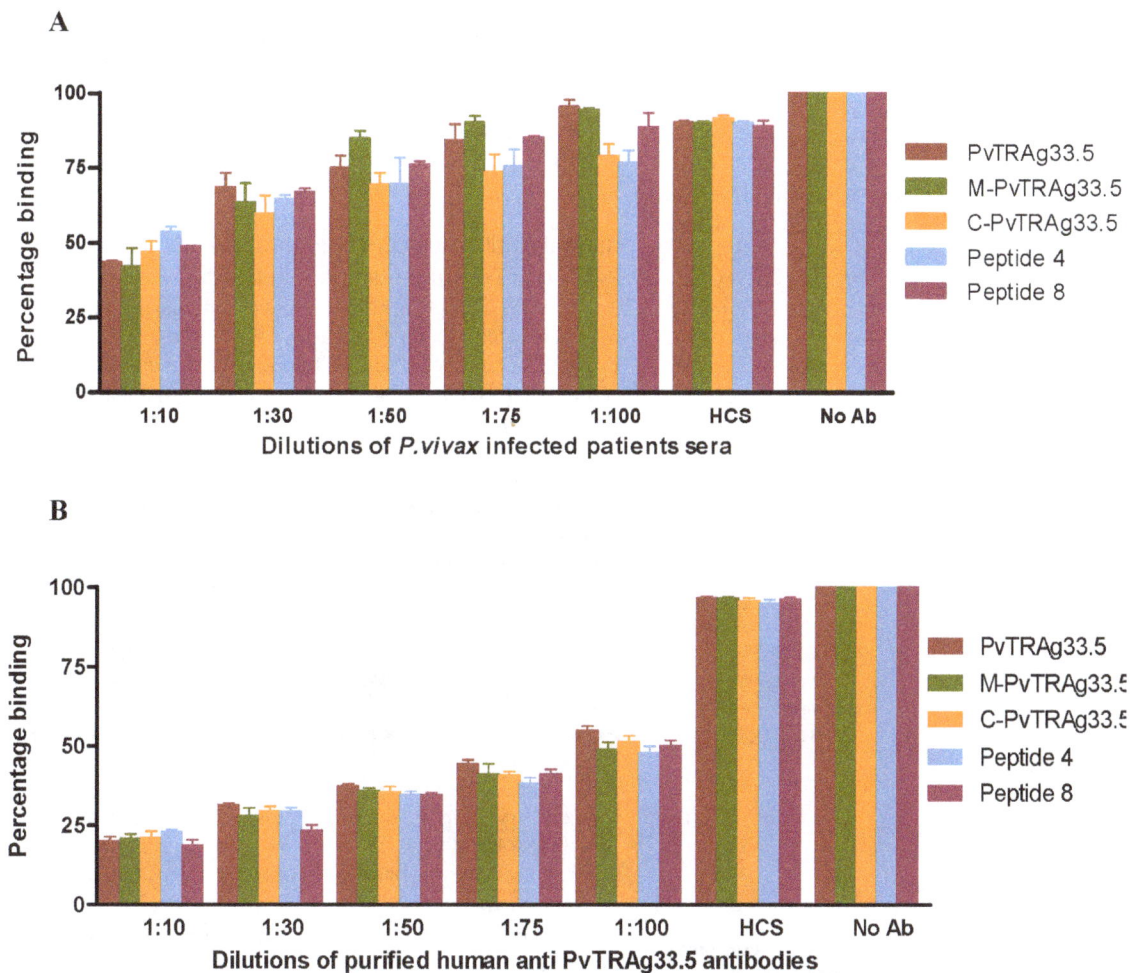

A

B

Figure 4. Inhibition of erythrocyte binding of PvTRAg33.5 derived fragments and peptides by *P. vivax* patients' sera. Pooled *P. vivax* infected patients' sera (**A**) and by purified anti-PvTRAg33.5 antibodies from these pooled patients sera (**B**) were used for these antibody mediated inhibition assays. The tagged recombinant PvTRAg33.5, its fragments, or synthetic peptides were mixed with different dilutions of *P.vivax* patients' sera or purified antibodies before adding to the microtiter plate coated with one million erythrocytes. Further steps of color development were same as in Fig. 1 and 2B. Binding in the absence of antibody is taken as percentage control. Error bar indicates the standard deviation of mean from three experiments. No Ab, no antibody; HCS, healthy human control sera.

were incubated with monoclonal antibodies against GST and processed further as described above. For (no competition) control, erythrocytes were directly incubated with GST- tagged complete PvTRAg33.5 or GST alone and plate was developed as above. The absorbance obtained for GST alone was subtracted from readings obtained for the GST- tagged protein.

Results

Defining the Erythrocyte Binding Regions of PvTRAg33.5

As shown in figure (**Fig. 1A**), the PvTRAg33.5 was divided into three parts, named as N-terminal (N-PvTRAg33.5), middle (M-PvTRAg33.5) and C-terminal (C-PvTRAg33.5) fragments. These fragments were expressed in *E.coli*, and purified recombinant peptides containing either the Histidine-tag or GST-tag were used for erythrocyte binding assay. There was an overlap of 8 amino acids between M-PvTRAg33.5 and C-PvTRAg33.5 fragments. Results showed that the middle (M-PvTRAg33.5) and C- terminal (C-PvTRAg33.5) fragments bind to human erythrocytes in a concentration dependent manner (**Fig. 1B**). The N-terminal (N-PvTRAg33.5) fragment comprising of initial 106 amino acids did not show binding to the human RBC.

Further Defining of the Erythrocyte Binding Domains of PvTRAg33.5

Since middle and C- terminal fragments of PvTRAg33.5 bind to human erythrocytes, a total of eight different non-overlapping peptides (each peptide of 21 amino acid length except peptide P1 which was 22 amino acid long) were synthesized from this region (107 to 275 amino acid position) of the protein (**Fig. 2A**). These peptides were then tested for their binding activity to the human erythrocytes. The peptide P4 (KMSSWLSSDWKKVGA-MYWDLQ) located at 171–191 amino acid position (represented in M-PvTRAg33.5 fragment) and peptide P8 ('TWRNDFINRWVSEKKWNSILN') at 255–275 amino acid position (represented in C-PvTRAg33.5 fragment), showed binding activity towards the uninfected human erythrocytes in a concentration dependent manner (**Fig. 2B**). Rest of the peptides (P1, P2, P3, P5, P6, and P7) did not bind to these erythrocytes. Complete recombinant PvTRAg33.5 protein and bacterial thioredoxin were used as positive and negative controls, respectively.

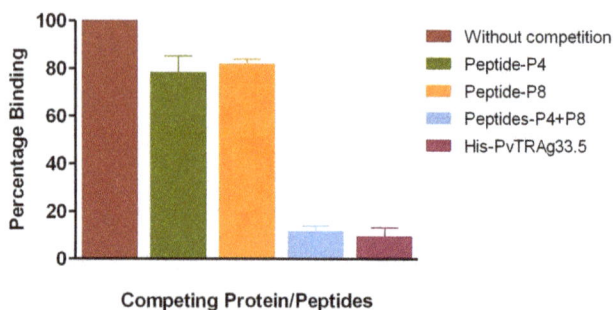

Figure 5. Competition of peptide P4 and P8 with complete PvTRAg33.5 in erythrocyte binding assay. Approximately one million erythrocytes in each well of a 96 well microtiter plate were incubated with 5 μM of peptides or 2 μM of Histidine tagged complete PvTRAg33.5 protein. In a control well, no protein or peptide was added to the erythrocytes. The plate was then incubated with 1 μM of GST-tagged complete PvTRAg33.5, and developed with anti GST monoclonal antibodies, as described in Fig. 1B. Error bar indicates the standard deviation of mean from three different experiments.

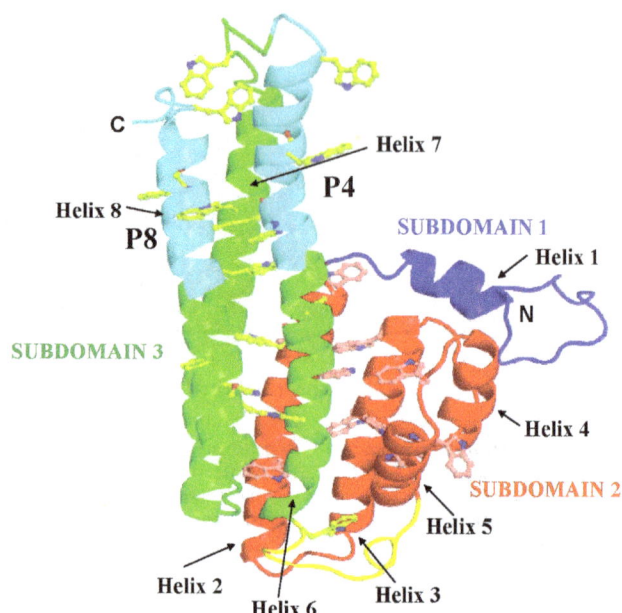

Figure 6. Location of erythrocyte binding peptide domains on modeled PvTRAg33.5. PvTRAg33.5 has three subdomains. Subdomain 1 (blue), Subdomain 2 (red), and subdomain 3 (green). The helices of different subdomains are shown in respective colors and coiled structure as loops. Erythrocyte binding peptides P4 (in helix 6) and P8 (in helix 8) are shown in cyan color.

Natural Antibodies Inhibit Erythrocyte Binding Activity of PvTRAg33.5 Fragments and its Peptides

The M-PvTRAg33.5, C-PvTRAg33.5, and peptides P4 and P8 were incubated separately with various dilutions of Rabbit anti-PvTRAg33.5 antibody before allowing their binding to uninfected erythrocytes. Results showed that this antibody inhibited binding of these recombinant or synthetic peptides to human erythrocytes in a dilution dependent manner (**Fig. 3**). Similarly, natural antibodies present in the *P. vivax* patients' sera also inhibited their binding to these erythrocytes (**Fig. 4**). Pre-immune rabbit sera or sera from healthy uninfected individuals did not affect the binding. In both the experiments, shown in figures 3 and 4, it was observed that the rabbit sera inhibited the erythrocyte binding at a higher dilution as compared with the patients' sera. This could arise if the rabbit sera had higher titers of antibodies than the patients' sera or patient sera had some non-specific antibodies. For this purpose, the anti-PvTRAg33.5 antibodies were affinity purified from the pooled *P.vivax* patients' sera and used for inhibition assay. It was observed that purified anti-PvTRAg33.5 antibodies from patients' sera inhibited the erythrocyte binding activity of these recombinant and synthetic peptides more efficiently than the unpurified sera (**Fig. 4B**). This difference in binding inhibition by the purified antibody was statistically significant (P<0.05).

Peptide P4 and P8 Recognize their own Erythrocyte Receptor

Erythrocyte binding of peptide P4 and P8 has been established above. Each of these two peptides was then allowed to compete with complete recombinant (GST-tagged) PvTRAg33.5 protein in erythrocyte binding assay by Cell-ELISA. Results showed a partial inhibition in binding activity of complete PvTRAg33.5 to RBC, if each peptide was allowed to compete with it separately. But the combined mixture of P4 and P8 peptides was able to inhibit the binding of the complete PvTRAg33.5 to RBC almost completely (**Fig. 5**). This complete inhibition of erythrocyte binding by the mixture of these two peptides was similar to that of the control where competition was allowed between recombinant Histidine-tagged PvTRAg33.5 and the GST- tagged PvTRAg33.5 complete proteins (**Fig. 5**). These results therefore suggest that each peptide recognizes its own receptor on human erythrocyte.

Discussion

Tryptophan-rich proteins of malarial parasites play important role in parasite survival, as reported in case of murine malaria caused by *P. yoelii* [2,3,22]. *Plasmodium vivax* contains the largest number of such proteins and earlier we have been able to characterize some of them, including PvTRAg33.5 [10–19]. Here, we have characterized this protein further and shown that it binds to the uninfected host erythrocytes through two different peptide regions which come closer on the surface of the protein in its 3D structure, and this binding is inhibited by the patients' sera.

Previous studies have shown that tryptophan-rich antigens of *Plasmodium* species interact with uninfected host erythrocytes [6,12,22]. Since we have already shown that PvTRAg33.5 binds to the host erythrocytes [18], we wanted to define the binding domains of this protein further. Hence, PvTRAg33.5 was divided in to three different parts i.e. N-terminal region (amino acid position 24–106), middle (amino acid position 107–192), and C-terminal region (amino acid position 185–275) and tested them for their erythrocyte binding activity. Middle and C-terminal fragments showed the erythrocyte binding activity (**Fig. 1**). Further exclusion of non-binding region came from the synthetic peptide binding studies where peptide P4 (amino acid position 171–191) and peptide P8 (amino acid position 255–275) showed erythrocyte binding. Thus we were able to define two peptide domains in PvTRAg33.5 which were 63 amino acids apart and showed specific binding to host RBCs (**Fig. 2B**). These two peptides P4 and P8 were localized in the modeled tertiary structure of PvTRAg33.5 [13] where peptide P4 was found to be located towards the end of the helix 6 as well as the beginning of the coil joining helix 6 and 7 whereas peptide P8 is located towards the posterior region of helix 8 as well as the C terminal loop (**Fig. 6**). Both these peptides are located in subdomain 3 of the protein on two parallel helices. These peptide sequences may be exposed to the surface of the protein to form a suitable binding site for interaction.

Enzymatic treatments to remove specific surface receptors have helped in the identification of molecules which interact with the parasite proteins and mediate the crucial step of host cell invasion by the parasite [23–28]. During our earlier studies, we had ruled out the involvement of glycophorins, sialglycoproteins and band-3 protein of erythrocytes as PvTRAg33.5 receptor(s). This is because the erythrocytes treated with trypsin, chymotrypsin and neuraminidase did not show any effect on their binding affinity towards PvTRAg33.5 [18]. This is similar to PfRH5 (Pf reticulocyte homolog 5) where binding occurs with a molecule which is neither sialoglycoprotein nor band-3 [29,30]. Further studies are required

to identify RBC receptors for this protein. However, cross-competition of PvTRAg33.5 with other RBC binding PvTRAgs revealed that it has two erythrocyte receptors. One of its RBC receptor is shared by PvTRAg38 and other is shared by PvATRAg74. On the other hand, both of its erythrocyte receptors were shared by PvTRAg35.2, PvTRAg69.4, and PvTRAg [18]. Since PvTRAg33.5 has two erythrocyte binding domains (P4 and P8), it is possible that each peptide domain recognizes its own RBC receptor. This was proven by the competition assay where each peptide was able to partially compete with PvTRAg33.5 during erythrocyte binding. However, in combination (peptides P4 and P8 together) they completely inhibited the PvTRAg33.5 binding to erythrocytes (**Fig. 5**). This suggests that each peptide recognizes its own RBC receptor but both domains are required for efficient binding. This receptor-ligand interaction between PvTRAg33.5 (or its peptides) and erythrocyte protein is blocked by the PvTRAg33.5 antibodies whether raised in rabbit or purified from the *P. vivax* patients' sera (**Fig. 3, 4**).

Earlier, we have shown that the PvTRAg33.5 sequence is highly conserved in the *P. vivax* parasite population [19]. Therefore, the erythrocyte binding domains (P4 and P8 sequences) will also remain conserved among field isolates. As mentioned above, the natural antibodies could block this peptide binding to its host erythrocyte. This information on host-parasite interaction then could be utilized to develop the immunotherapeutic reagents. These peptides thus can form the basis of a minimal subunit based, multiepitopic, multistage anti malarial vaccine. An alternative approach which utilizes the chemical synthesis of peptides that resemble key regions of proteins on the pathogen's surface or have role in host-parasite interaction can also be deployed to develop the therapeutic agents [31]. This approach has been used for the peptides derived from various *P. falciparum* proteins showing binding affinity to host erythrocytes and their role in RBC invasion [6,32–37].

In conclusion, two domains at amino acid position 171–191 (P4) and 255–275 (P8) of PvTRAg33.5 bind to host erythrocyte where each domain recognizes its own erythrocyte receptor. This binding is inhibited by polyclonal sera raised against PvTRAg33.5 as well as by the natural antibodies produced during *P. vivax* infection. Further studies are required to identify its RBC receptor which can lead to the development of a therapeutic reagent.

Acknowledgments

We are grateful to Dr. Punit Kaur for her help in analyzing the 3D structure, Shalini Narang for preparing the manuscript. We acknowledge the facility of Biotechnology Information System (BTIS) of the department.

Author Contributions

Obtained ethical clearance for use of human blood: YDS. Obtained project grant: YDS. Conceived and designed the experiments: YDS. Performed the experiments: HB RKT. Analyzed the data: HB RKT YDS. Contributed reagents/materials/analysis tools: HB. Wrote the paper: HB YDS.

References

1. Garcia LS (2010) Malaria. Clin Lab Med 30: 93–129.
2. Burns JM Jr, Dunn PD, Russo DM (1997) Protective immunity against *Plasmodium yoelii* malaria induced by immunization with particulate blood-stage antigens. Infect Immun 65: 3138–3145.
3. Burns JM Jr, Adeeku EK, Dunn PD (1999) Protective immunization with a novel membrane protein of *Plasmodium yoelii*-infected erythrocytes. Infect Immun 67: 675–680.

4. Uhlemann AC, Oguariri RM, McColl DJ, Coppel RL, Kremsner PG, et al. (2001) Properties of the *Plasmodium falciparum* homologue of a protective vaccine candidate of *Plasmodium yoelii*. Mol Biochem Parasitol 118: 41–48.
5. Ntumngia FB, Bouyou-Akotet MK, Uhlemann AC, Mordmuller B, Kremsner PG, et al. (2004) Characterisation of a tryptophan-rich *Plasmodium falciparum* antigen associated with merozoites. Mol Biochem Parasitol 137: 349–353.
6. Curtidor H, Ocampo M, Rodriguez LE, Lopez R, Garcia JE, et al. (2006) *Plasmodium falciparum* TryThrA antigen synthetic peptides block in vitro

merozoite invasion to erythrocytes. Biochem Biophys Res Commun 339: 888–896.

7. Ntumngia FB, Bahamontes-Rosa N, Kun JF (2005) Genes coding for tryptophan-rich proteins are transcribed throughout the asexual cycle of *Plasmodium falciparum*. Parasitol Res 96: 347–353.

8. Carlton JM, Adams JH, Silva JC, Bidwell SL, Lorenzi H, et al. (2008) Comparative genomics of the neglected human malaria parasite *Plasmodium vivax*. Nature 455: 757–763.

9. Tachibana S, Sullivan SA, Kawai S, Nakamura S, Kim HR, et al. (2012) *Plasmodium cynomolgi* genome sequences provide insight into *Plasmodium vivax* and the monkey malaria clade. Nat Genet 44: 1051–1055.

10. Jalah R, Sarin R, Sud N, Alam MT, Parikh N, et al. (2005) Identification, expression, localization and serological characterization of a tryptophan-rich antigen from the human malaria parasite *Plasmodium vivax*. Mol Biochem Parasitol 142: 158–169.

11. Alam MT, Bora H, Mittra P, Singh N, Sharma YD (2008) Cellular immune responses to recombinant *Plasmodium vivax* tryptophan-rich antigen (PvTRAg) among individuals exposed to vivax malaria. Parasite Immunol 30: 379–383.

12. Alam MT, Bora H, Singh N, Sharma YD (2008) High immunogenecity and erythrocyte-binding activity in the tryptophan-rich domain (TRD) of the 74-kDa *Plasmodium vivax* alanine-tryptophan-rich antigen (PvATRAg74). Vaccine 26: 3787–3794.

13. Bora H, Garg S, Sen P, Kumar D, Kaur P, et al. (2011) *Plasmodium vivax* tryptophan-rich antigen PvTRAg33.5 contains alpha helical structure and multidomain architecture. PLoS One 6: e16294.

14. Garg S, Chauhan SS, Singh N, Sharma YD (2008) Immunological responses to a 39.8 kDa *Plasmodium vivax* tryptophan-rich antigen (PvTRAg39.8) among humans. Microbes Infect 10: 1097–1105.

15. Mittra P, Singh N, Sharma YD (2010) *Plasmodium vivax*: immunological properties of tryptophan-rich antigens PvTRAg 35.2 and PvTRAg 80.6. Microbes Infect 12: 1019–1026.

16. Siddiqui AA, Bora H, Singh N, Dash AP, Sharma YD (2008) Expression, purification, and characterization of the immunological response to a 40-kilodalton *Plasmodium vivax* tryptophan-rich antigen. Infect Immun 76: 2576–2586.

17. Siddiqui AA, Singh N, Sharma YD (2007) Expression and purification of a *Plasmodium vivax* antigen - PvTARAg55 tryptophan- and alanine-rich antigen and its immunological responses in human subjects. Vaccine 26: 96–107.

18. Tyagi RK, Sharma YD (2012) Erythrocyte Binding Activity Displayed by a Selective Group of *Plasmodium vivax* Tryptophan Rich Antigens Is Inhibited by Patients' Antibodies. PLoS One 7: e50754.

19. Zeeshan M, Bora H, Sharma YD (2012) Presence of memory T cells and naturally acquired antibodies in *Plasmodium vivax* malaria-exposed individuals against a group of tryptophan-rich antigens with conserved sequences. J Infect Dis 207: 175–185.

20. Espinosa AM, Sierra AY, Barrero CA, Cepeda LA, Cantor EM, et al. (2003) Expression, polymorphism analysis, reticulocyte binding and serological reactivity of two *Plasmodium vivax* MSP-1 protein recombinant fragments. Vaccine 21: 1033–1043.

21. Sarin R, Sharma YD (2006) Thioredoxin system in obligate anaerobe *Desulfovibrio desulfuricans*: Identification and characterization of a novel thioredoxin 2. Gene 376: 107–115.

22. Burns JM, Adeeku EK, Belk CC, Dunn PD (2000) An unusual tryptophan-rich domain characterizes two secreted antigens of *Plasmodium yoelii*-infected erythrocytes. Mol Biochem Parasitol 110: 11–21.

23. Haynes JD, Dalton JP, Klotz FW, McGinniss MH, Hadley TJ, et al. (1988) Receptor-like specificity of a *Plasmodium knowlesi* malarial protein that binds to Duffy antigen ligands on erythrocytes. J Exp Med 167: 1873–1881.

24. Duraisingh MT, Maier AG, Triglia T, Cowman AF (2003) Erythrocyte-binding antigen 175 mediates invasion in *Plasmodium falciparum* utilizing sialic acid-dependent and -independent pathways. Proc Natl Acad Sci U S A 100: 4796–4801.

25. Reed MB, Caruana SR, Batchelor AH, Thompson JK, Crabb BS, et al. (2000) Targeted disruption of an erythrocyte binding antigen in *Plasmodium falciparum* is associated with a switch toward a sialic acid-independent pathway of invasion. Proc Natl Acad Sci U S A 97: 7509–7514.

26. Sahar T, Reddy KS, Bharadwaj M, Pandey AK, Singh S, et al. (2011) *Plasmodium falciparum* reticulocyte binding-like homologue protein 2 (PfRH2) is a key adhesive molecule involved in erythrocyte invasion. PLoS One 6: e17102.

27. Gilberger TW, Thompson JK, Triglia T, Good RT, Duraisingh MT, et al. (2003) A novel erythrocyte binding antigen-175 paralogue from *Plasmodium falciparum* defines a new trypsin-resistant receptor on human erythrocytes. J Biol Chem 278: 14480–14486.

28. Lobo CA, Rodriguez M, Reid M, Lustigman S (2003) Glycophorin C is the receptor for the *Plasmodium falciparum* erythrocyte binding ligand PfEBP-2 (baebl). Blood 101: 4628–4631.

29. Crosnier C, Bustamante LY, Bartholdson SJ, Bei AK, Theron M, et al. (2011) Basigin is a receptor essential for erythrocyte invasion by *Plasmodium falciparum*. Nature: 534–537.

30. Rodriguez M, Lustigman S, Montero E, Oksov Y, Lobo CA (2008) PfRH5: a novel reticulocyte-binding family homolog of *plasmodium falciparum* that binds to the erythrocyte, and an investigation of its receptor. PLoS One 3: e3300.

31. Corradin G, Kajava AV, Verdini A (2010) Long synthetic peptides for the production of vaccines and drugs: a technological platform coming of age. Sci Transl Med 2: 50rv53.

32. Curtidor H, Urquiza M, Suarez JE, Rodriguez LE, Ocampo M, et al. (2001) *Plasmodium falciparum* acid basic repeat antigen (ABRA) peptides: erythrocyte binding and biological activity. Vaccine 19: 4496–4504.

33. Obando-Martinez AZ, Curtidor H, Arevalo-Pinzon G, Vanegas M, Vizcaino C, et al. (2010) Conserved high activity binding peptides are involved in adhesion of two detergent-resistant membrane-associated merozoite proteins to red blood cells during invasion. J Med Chem 53: 3907–3918.

34. Obando-Martinez AZ, Curtidor H, Vanegas M, Arevalo-Pinzon G, Patarroyo MA, et al. (2010) Conserved regions from *Plasmodium falciparum* MSP11 specifically interact with host cells and have a potential role during merozoite invasion of red blood cells. J Cell Biochem 110: 882–892.

35. Pinzon CG, Curtidor H, Bermudez A, Forero M, Vanegas M, et al. (2008) Studies of *Plasmodium falciparum* rhoptry-associated membrane antigen (RAMA) protein peptides specifically binding to human RBC. Vaccine 26: 853–862.

36. Puentes A, Garcia J, Vera R, Lopez R, Suarez J, et al. (2004) Sporozoite and liver stage antigen *Plasmodium falciparum* peptides bind specifically to human hepatocytes. Vaccine 22: 1150–1156.

37. Urquiza M, Rodriguez LE, Suarez JE, Guzman F, Ocampo M, et al. (1996) Identification of *Plasmodium falciparum* MSP-1 peptides able to bind to human red blood cells. Parasite Immunol 18: 515–526.

9

Novel Protocol for the Chemical Synthesis of Crustacean Hyperglycemic Hormone Analogues — An Efficient Experimental Tool for Studying Their Functions

Alessandro Mosco[1]*, Vientsislav Zlatev[2], Corrado Guarnaccia[2], Sándor Pongor[2], Antonella Campanella[1], Sotir Zahariev[2], Piero G. Giulianini[1]

1 Department of Life Sciences, University of Trieste, Trieste, Italy, **2** International Centre for Genetic Engineering and Biotechnology, AREA Science Park, Trieste, Italy

Abstract

The crustacean Hyperglycemic Hormone (cHH) is present in many decapods in different isoforms, whose specific biological functions are still poorly understood. Here we report on the first chemical synthesis of three distinct isoforms of the cHH of *Astacus leptodactylus* carried out by solid phase peptide synthesis coupled to native chemical ligation. The synthetic 72 amino acid long peptide amides, containing L- or D-Phe^3 and (Glp^1, D-Phe^3) were tested for their biological activity by means of homologous *in vivo* bioassays. The hyperglycemic activity of the D-isoforms was significantly higher than that of the L-isoform, while the presence of the N-terminal Glp residue had no influence on the peptide activity. The results show that the presence of D-Phe^3 modifies the cHH functionality, contributing to the diversification of the hormone pool.

Editor: Emanuele Buratti, International Centre for Genetic Engineering and Biotechnology, Italy

Funding: Financial support was provided by a grant assigned to E.A. Ferrero from the Ministry of Education, University and Research (MIUR, http://prin.miur.it/) (project PRIN 2008 – 2008BZFFFY "Modulazione ormonale e riconoscimento individuale nel comportamento agonistico dei Crostacei Decapodi"). The funders had no role in study design, data collection and analysis, decision to publish, or preparation of the manuscript.

Competing Interests: The authors have declared that no competing interests exist.

* E-mail: alessandro.mosco@phd.units.it

Introduction

The crustacean Hyperglycemic Hormone is a neuropeptide synthesized and secreted by the X organ-sinus gland complex (XO-SG) located in the eyestalks of decapod crustaceans. Its main role concerns the regulation of the hemolymphatic glucose level, but additional functions include growth control, reproduction, ion balance, stress responses [1]. cHH belongs to a family of crustacean peptides that includes the Molt Inhibiting Hormone (MIH), the Vitellogenin Inhibiting Hormone (VIH) and the Mandibular Organ Inhibiting Hormone (MOIH). All these peptides are structurally related, having six conserved cysteines that form three intramolecular disulfide bridges [2]. The family can be further classified into two subfamilies on the basis of the primary structure and the preprohormone peptide organization. The cHH subfamily or type I possesses a cryptic sequence of unknown function, named cHH precursor-related peptide, in the unprocessed precursor [3], while the MIH/VIH/MOIH subfamily or type II peptides that are longer and more variable in length, have a Gly at position 12 and lack the cHH precursor-related peptide [4] [5] [6]. Mature cHHs have a length of 72 to 73 amino acid residues, and an homology among the different species ranging from 40 to 99% [7]. They have an amidated C-terminus and, in many cases, pyroglutamate as N-terminal blocking group [1] [2]. Moreover, the phenylalanyl residue at position 3 of the sequence can be found in either L- or D-configuration [8].

C-terminal amidation is the most significant post-translational processing, because of its influence on cHH bioactivity, and the lower functionality showed by non-amidated peptides can be explained by a lower binding affinity to the cHH receptor due to their negative C-terminus charge [9] [10].

Another post-translational modification concerns the cyclization of the Glu residue at the N-terminal end, which is found in cHHs of brachyuran crab species and crayfish, but not in shrimps [11]. The amino acid sequence of both N-terminus blocked and free cHHs is almost identical, whereas the N-terminal residue can be glutamine or pyroglutamate. In *C. maenas* both isoforms proved to possess a similar biological activity and the same hemolymphatic clearance rates. Thus seems that N-terminal cyclization has no obvious biological function [12] [13], but it may well be that the presence of a blocked N-terminus protects the cHH against peptidases, extending its half-life like in other peptides [14].

D-amino acids have been found only in the peptides of some invertebrates and amphibians. These include a toxin from a spider venom, a snail neuropeptide, some crustacean hormones of the cHH family, and opiate and antimicrobial peptides from amphibian skin [15]. To date L and D stereoisomers have been reported only for the cHH of Astacoidea (*Astacus leptodactylus*, *Cherax destructor*, *H. americanus*, *Orconectes limosus*, *Procambarus bouvieri*, *Procambarus clarkii*), the stereoinversion concerning always the Phe^3 [16] [8] [17–20]. Little is known about the role of isomerization and if the two stereoisomers have the same functions or exhibit distinctive bioactivities. The presence of D-Phe^3 seems to confer a higher hyperglycemic activity to the cHH. In an *in vivo* heterologous bioassay, *O. limosus* was injected with L- or D-cHHA extracted from SG of *H. americanus*, and purified by reverse-phase high performance liquid chromatography (RP-HPLC). The peak value of glycemia was in the same range for both isoforms,

30.5±6.7 mg/dL and 34.3±5.6 for the L and D form respectively, but the time course was different, with the D-cHHA inducing a later response, having the maximal hyperglycemic peak at 3–4 h instead of 2 h as for the L-cHHA, and a slower return to the baseline, being the hyperglycemic effect still detectable at 8 h [8]. A similar time course was found in *A. leptodactylus* after injection of homologous D-cHH, purified by RP-HPLC from a sinus gland extract, which induced a higher hyperglycemia and had an extended response, the hemolymph glucose level being significantly higher both at 3 h and 8 h, when compared to the concentration elicited by the L-cHH. The same study proved that cHH is involved in the control of osmoregulation, and that the different stereoisomers may have different target tissues and/or receptors. Indeed, all the purified cHHs were able to increase hemolymph Na⁺ concentration, but only D-cHH induced a significant raise in hemolymph osmolality [19]. The higher efficacy of the D isomer in exerting a biological activity was proved also for the native cHH of *P. clarkii*, purified from the XO-SG complex. In *in vitro* experiments it was shown that the cHH is able to inhibit ecdysteroid biosynthesis in cultured Y-organs, suggesting that this neuropeptide may play a role in the crustacean molting process. Of the two isoforms tested, the D-cHH showed a ten fold higher inhibitory effect on ecdysteroidogenesis as well as being more potent in inducing hyperglygemia [16]. On the contrary, in the closely related Mexican crayfish *P. bouvieri*, whose cHH share an identical amino acid sequence with those of *P. clarkii*, the two isoforms showed the same ability to induce hyperglycemia in destalked animals [17].

The present study describes the hyperglycemic activity of 3 analogues of cHH of the crayfish *A. leptodactylus*, prepared for the first time by chemical synthesis, demonstrating the attractive possibility to synthetically produce milligram amounts of cHH in a specific desired form for future studies deciphering its functions.

Results

Peptide synthesis

The sequence of cHH1-72 and of its isomers (Fig. 1) contain several potentially problematic sites for the SPPS with Fmoc/tBu strategy. It contains six cysteine residues (prone to racemization), nine aspartic acid residues (possible aspartimide formation, followed by isomerisation, racemization and piperidide formation) and a very hydrophobic region in the C-terminus part since 10 residues on the last 17 aa (56–72) are Leu, Ile, or Val (possible poor couplings and Fmoc-deprotection, due to secondary structure formation on the resin).

In our approach the sequence was divided in two parts, cHH1-38 and cHH39-72 in order to use a native chemical ligation reaction at Cys³⁹ to obtain the full length 72mer peptide (Fig. 1) First, side chain and N-terminus Boc-protected cHH1-38 peptides (Fig. 1, A) were synthesized by SPPS using Fmoc-methodology on a very acid labile chlorotrityl resin. Second, protected peptide carboxylic acids were cleaved from the resin with hexafluoroisopropanol (HFIP) in dichloromethane (DCM) (Fig. 1, B). At the end (Fig. 1, C), the crude protected peptides were esterified to corresponding 4-acetamidophenyl-thioesters, according to published procedure [21]. After acid deprotection (Fig. 1, D), with thiol free cleavage mixture, all 3 peptides thioaryl esters were purified by RP-HPLC. C-terminal cHH39-72 peptide amide was synthesized on Tentagel Sieber amide resin (Merck), a resin for Fmoc SPPS derivatized by xanhydrylamine linker which gives peptide amide fragments upon cleavage with trifluoroacetic acid (TFA) [22]. After cleavage and deprotectection [TFA, triisopropylsilane (TIPS), H₂O, 3,6-dioxa-1,8-octanedithiol (DODT),

phenol, (81%/3%/3%/8%/5%) for 4 h] the C-terminal peptide amide was purified by semipreparative RP-HPLC. The purity/identity of all peptides was verified by liquid chromatography electrospay ionization mass spectrometry (ESI-MS) (Table 1, Fig. 2). Even though the synthesis of cHH39-72 went smoothly (the purity of crude peptide was >55%), the deprotected and purified peptide possesses a very low solubility in aqueous buffers (due to its high hydrophobic AAs content), which complicates both purification/recovery and followed step of NCL reactions (Fig. 1, E). Despite the several solubilization conditions tried [guanidine hydrochloride (GuHCl), urea, sodium dodecyl sulfate, 2,2,2-trifluoroethanol, HFIP, and its mixtures] a substantial cHH39-72 precipitation occurred in all NCL conditions tested, negatively affecting the final yields of purified and folded (see experimental part) cHH1-72 isomers. Different approaches such as introducing on C-terminus solubilizing cleavable tails are actually under evaluation. Purified linear 72 amino acids long peptides were folded (Fig. 1, F) in basic condition in the presence of oxidized/reduced glutathione or Cys/Cys2 and purified to homogeneity by RP-HPLC with yields between 1 and 3%.

Bioassays

The biological assays on *A. leptodactylus* were performed as time-course experiments to assess potential biological activity differences among the three chiral isoforms of the synthetic peptide (Fig. 1, 1, 2 and 3) and to compare them to the native hormone. The two chiral synthetic isoforms of cHH showed significant different hyperglycemic responses in term of both time course/kinetic pattern and absolute values of circulating glucose (Fig. 3).

Injection of 1.7 pmol/g live weight of each of the three analogue peptides revealed the following results: synthetic D-cHH induced a quick hyperglycemic response already detectable after 1 h and a strong hyperglycemic response with the maximum peak of 95.4±8.1 mg/dL glucose after 4 h, high values lasting at 8 h (76±12.1 mg/dL) and glycemia returning to its nearby basal level, 8.9±1 mg/dL, after 24 h (Fig. 3, a). Glp-D-cHH mirrored the time course of D-cHH, with the strongest hyperglycemic response of 96.1±5.6 mg/dL glucose after 4 h, high values lasting at 8 h (62.9±11.8 mg/dL) and the glycemia returning to its basal level, 2.9±0.5 mg/dL, after 24 h (Fig. 3, c). L-cHH induced a lower hyperglycemic response with the maximum peak of 52±6.5 mg/dL glucose after 2 h, and the glycemia returning to its nearby basal level, 5.5±0.4 mg/dL, after 8 h (Fig. 3, b). Injection of the purified SG-cHH was comparable to that of the synthetic L-cHH, the maximum (44±4.5 mg/dL) being reached after 2 h, and the glycemia returning to its basal level, 5±1.1 mg/dL after 24 h (Fig. 3, d). The comparison of the glycemia elicited by each of the analogues during the first 8 hours post-injection revealed significant differences between time points (Kruskal-Wallis $p<0.01$) and the Bonferroni adjusted p values of post-hoc pairwise comparisons at different times compared to initial values were significant. At 24 h, only the hyperglycemia induced by the D-cHH was still significantly different from that of time 0 h. The comparison among the glycemic values elicited by the different peptides 4 or 8 hours post-injection revealed significant differences (4 h – Kruskal-Wallis chi-squared = 29.5491, df = 3, $p = 1.72 \cdot 10^{-6}$ ***; 8 h – Kruskal-Wallis chi-squared = 21.7912, df = 3, $p = 7.209 \cdot 10^{-5}$ ***). The Bonferroni adjusted p values of post hoc pairwise comparisons revealed two peptide groups, D-cHH and Glp-D-cHH on one hand eliciting significantly higher glycemic response than L-cHH and SG-cHH on the other hand. To evaluate the overall hyperglycemic potentiality of the three synthetic peptides and the native one, the maximum values recorded in the experiments for each specimen from 1 h to 8 h were plotted in Fig. 4. The D-cHH and the Glp-D-cHH (means

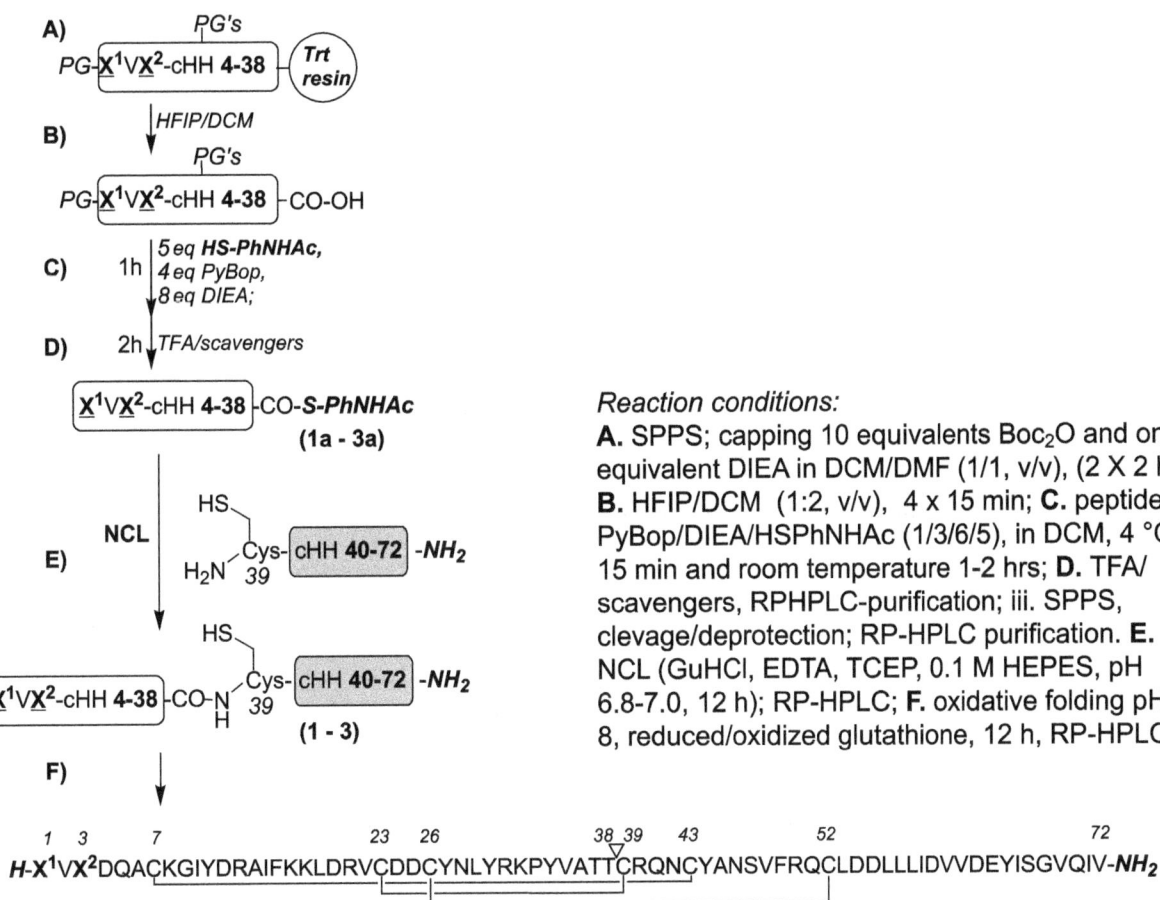

Reaction conditions:
A. SPPS; capping 10 equivalents Boc$_2$O and one equivalent DIEA in DCM/DMF (1/1, v/v), (2 X 2 h); **B.** HFIP/DCM (1:2, v/v), 4 x 15 min; **C.** peptide/ PyBop/DIEA/HSPhNHAc (1/3/6/5), in DCM, 4 °C, 15 min and room temperature 1-2 hrs; **D.** TFA/ scavengers, RPHPLC-purification; iii. SPPS, clevage/deprotection; RP-HPLC purification. **E.** NCL (GuHCl, EDTA, TCEP, 0.1 M HEPES, pH 6.8-7.0, 12 h); RP-HPLC; **F.** oxidative folding pH 8, reduced/oxidized glutathione, 12 h, RP-HPLC.

**Native cHH:** **X^1**=Gln/pGlu; **X^2** =L-Phe/D-Phe

**Synthetic cHH (this study):**

1. QVF-cHH: **X^1**=Gln; **X^2** =L-Phe; **2.** QVf-cHH: **X^1**=Gln; **X^2** =D-Phe and **3.** pEVf-cHH: **X^1**=pGlu; **X^2** =D-Phe

The triangle between residues 38 (Thr) and 39 (Cys) represent the place for native chemical ligation (NCL)

f = D-Phe; pE = Glp (residue of pyroglutamic acid); PG = protecting group

Figure 1. Scheme of the SPPS coupled to NCL to obtain the full length peptides. The leading segments containing a C-terminal thioester, QVF-cHH4-38, QVdF-cHH4-38, pEVdF-cHH4-38, and the cHH39-72 following segment with a free N-terminal cysteine were synthesized by SPPS. The NCL reaction between the leading and following segments returned the full length 72mer cHH isomers, which were further subjected to oxidative folding to obtain the folded peptides.

of max glycemia of, respectively, 100.5±9.3 and 98.7±6 mg/dL) showed again their higher capacity to mobilize the glucose compared to both L-cHH and SG-cHH (means of max glycemia of, respectively, 52.4±4 and 52.7±9.4 mg/dL).

Discussion

The cHH plays a major role in controlling the hemolymphatic glucose level, but it also has significance in regulating molt, reproduction and homeostasis, nevertheless the putative distinct functions exerted by the different circulating cHH forms are still poorly understood. To study the biological role of the cHH family peptides, large amounts of them are required, and in the past years different approaches were tried in order to obtain sufficient peptide quantities. To achieve this purpose different technologies were used: extraction from natural sources [23] [24] [25],

production by recombinant DNA technology [26] [27] [10], and by chemical synthesis [28].

The purification of native neuropeptides by RP-HPLC has a low yield, about 2–4 μg (270–489 pmol) per sinus gland in _C. maenas_ [29], but the invaluable advantage is to obtain naturally folded peptides, complete of all the post-translational modifications. DNA recombinant technology proved to be useful to obtain crustacean peptides in greater quantities to enable further investigation of their physiological functions. Recombinant cHHs from _Metapenaeus ensis_ [30], _Nephrops norvegicus_ [31], _Penaeus japonicus_ [9], _Macrobrachium rosenbergii_ [32], and _A. leptodactylus_ [10], were produced using bacterial expression systems. All these works reported the insertion of fusion tags for purification of target proteins by affinity chromatography. These additional sequences may interfere with the receptor binding which results in a lowered biological activity [9]. Moreover, recombinant peptides are never

Table 1. ESI-MS data of cHH analogues (**1–3**), thioesters (**1a–3a**) and cHH (39–72)-amide.

	cHH (1–72)		cHH aryl thioester (1–38)	cHH (39–72)-amide
	calc.(av.)/found (Da)		calc.(av.)/found (Da)	calc.(av.)/found (Da)
1	L-Phe^3-cHH:	1a	QVF-cHH-4-38:	
	8407.7/8406.0		4670.4/4669.2	
2	D-Phe^3-cHH:	2a	QVf-cHH-4-38:	3904.5/3903.0
	8407.7/8406.2		4670.4/4669.9	
3	Glp-D-Phe^3-cHH:	3a	pEVf-cHH-4-38:	
	8390.7/8385.0		4653.4/4651.2	

obtained with an amidated C-terminus, thus having a lower functionality. Therefore, this important post-translational modification must be added later through a specific enzymatic reaction [9] [10], while another post-translational modification, the pyroglutamilation of the N-terminus, is prevented by the presence of N-terminal tags. To overcome the problems posed by peptide expression in *Escherichia coli*, an eukaryotic expression system was developed. The cHH from *Penaeus monodon* was successfully expressed in the yeast *Pichia pastoris*, from which it was secreted to the culture medium already correctly refolded, and from it directly purified by RP-HPLC. The recombinant Pem-cHH was capable of inducing hyperglycemia in eyestalk-ablated prawns, although with a lower potency compared with the native peptide, due probably to the lack of C-terminal amidation [33].

Chemical synthesis enables a level of control on protein composition that greatly exceeds that attainable with ribosome-mediated biosynthesis, allowing post-translational modifications like C- and N-terminus blockage and insertion of D-amino acids, while native chemical ligation allows to overcome the limit of the peptide length posed by SPPS [34]. This approach was applied to synthesize the MIH from *P. clarkii*. The synthetic peptide was comparable to the sinus gland extracted Prc-MIH in the inhibition of ecdysteroid secretion by *in vitro* cultured Y-organs [28].

Our study was aimed at the production of several isoforms of the cHH of *A. leptodactylus*, by means of SPPS coupled with the native chemical ligation, in order to verify the effect that N-terminus pyroglutamylation and isomerization of the Phe^3 have on peptide functionality through *in vivo* homologous bioassays.

Changes in amino acid chirality add more variety to the pool of synthesized peptides. The presence of a D-amino acid may lead to structural modifications as the peculiar type II′β-turn structure found in the snail Gly-D-Phe-Ala sequence of achatin I [35] or the β-turn of amphibian opioid peptides that is involved in the receptor recognition [36]. The L-isomers of these peptides do not bind to the corresponding receptors, thus having no biological function. In other cases D-peptides were found to be far more potent in eliciting a biological response. Frog dermorphin, when injected into the rat brain, is about one thousand more potent then morphine in inducing long-lasting analgesia [37]. This behaviour may be also the result of a resistance to peptidases, as L-dermorphin is rapidly degraded, while the D-peptide is not hydrolyzed [15].

Our study shows that D-cHH was more potent than L-cHH in inducing glycemia, with a quick response detectable after 1 h and the maximum peak of 95.4±8.1 mg/dL glucose after 4 h from injection. On the contrary, L-cHH had its maximum peak at 2 h after injection, with hyperglycemic peak of 52±6.5 mg/dL. The superiority of D-cHH over the L-counterpart is probably due to a

change in the secondary structure of the peptide that may increase the cHH affinity for receptors located on its target organs. This behavior is in agreement with previous studies. Indeed, D-cHH purified from SG of *A. leptodactylus* exhibited a higher hyperglycemic response compared to the L-isomer [19]. The hyperglycemic effect triggered by D-cHH was extended in time: the glycemia recorded at 24 h was still significantly elevated ($p = 1.1 \cdot 10^{-3}$) respect to basal levels (time 0). This result is surprising, if we consider that cHH is cleared from the hemolymph quite dynamically, being of 10 min the half life reported in *C. maenas* [38]. The prolonged effect could be due to a slower dissociation constant from the receptor, as consequence of the conformational variation in 3D structure of D-isomers, or due to a higher resistance to peptidases. A low, but significant hyperglycemia at 24 h after injection with native D-cHH was already found in *A. leptodactylus* [19], an effect corroborated by our findings. The different hyperglycemic time-course of the L and D isomers was also shown in a heterologous bioassay, where the injection of D-cHH from *H. americanus* into *O. limosus* produced a longer hyperglycemic response than that caused by equal doses of L-cHH, even if the maximal hyperglycemic responses were comparable [8]. The Glp-D-cHH was more potent than the N-terminus blocked native hormone of *A. leptodactylus*, inducing a higher hyperglycemic outcome. Native cHH circulates in the hemolymph as a mixture of L- and D-isoforms, of which the D-cHH is the less abundant, in *C. destructor* amounting to 30–40% of the L-isoform [20]. The hemolymph content of cHH reflects the ratio of cells synthesizing the two isomers. In the X-organ of the crayfish *O. limosus*, a cluster of approximately 30 cells producing cHH was identified, of which only about 8 cells synthesize the D-cHH [18]. Therefore, it is conceivable that the minor activity of the native peptide extract was due to the higher proportion of the L-isoform which, as we have seen, has a lower activity. Our results show that the isomerization of the Phe^3 affects the cHH functionality modifying its capability to mobilize the glucose reserves, and being cHH a pleiotropic hormone involved, besides glucose regulation, in secretion of digestive enzymes [39], lipid metabolism [40], osmoregulation [19] [41], growth [29] [42], and reproduction [43] [44], it can be also assumed that the two isoforms may play different roles in various physiological contexts. In the crayfish *P. clarkii* both stereoisomers exhibit hyperglycemic activity, but the D-isomer showed a more potent inhibition of ecdysteroid synthesis in *in vitro* cultured Y-organ [16]. The involvement of cHH in the control of osmoregulation was proved in the crab *P. marmoratus* where cHH extracted from SG increased the Na$^+$ influx in perfused posterior gills from crabs acclimated to diluted seawater [45]. The ability of this neuropeptide to exert an effect on osmolality and Na$^+$ concentration of the hemolymph was

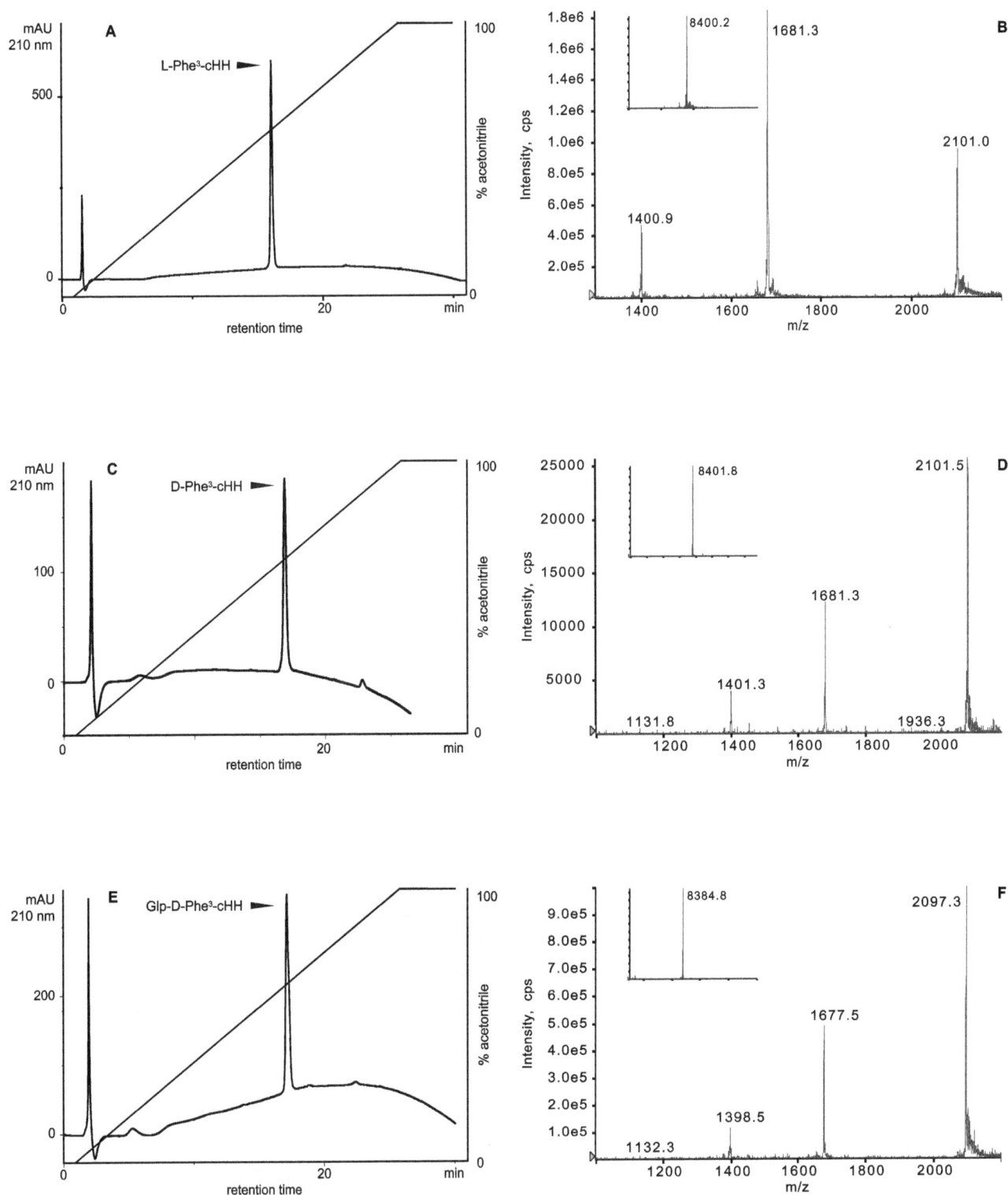

Figure 2. RP-HPLC profiles and ESI-MS spectra of the synthesized cHH isomers. Reverse Phase HPLC profiles of the purified refolded synthetic polypeptides and their corresponding ESI-MS multiply charged ion and deconvoluted spectra (insets): L-Phe^3-cHH (A and B), D-Phe^3-cHH (C and D), Glp-D-Phe^3-cHH (E and F). Mobile Phase A: 0.1% TFA in water. Mobile Phase B: 0.1% TFA in MeCN. Gradient: 0–100% B over 25 min. Column: Gemini 5 C18 2.0×150 mm.

proved in *in vivo* bioassays on *A. leptodactylus*. The injection of D-cHH raised both hemolymph osmolality and Na$^+$ concentration after 24 h, while L-cHH had no significant effect on osmolality, but increased Na$^+$ concentration, however with a lower rise [19].

Besides isomerization, we studied the effect of another post-translational modification, pyroglutamylation from the N-terminal glutamine residue of cHH. It has been already proved in crabs that pyroglutamyl-blocked and unblocked cHHs exhibited identical

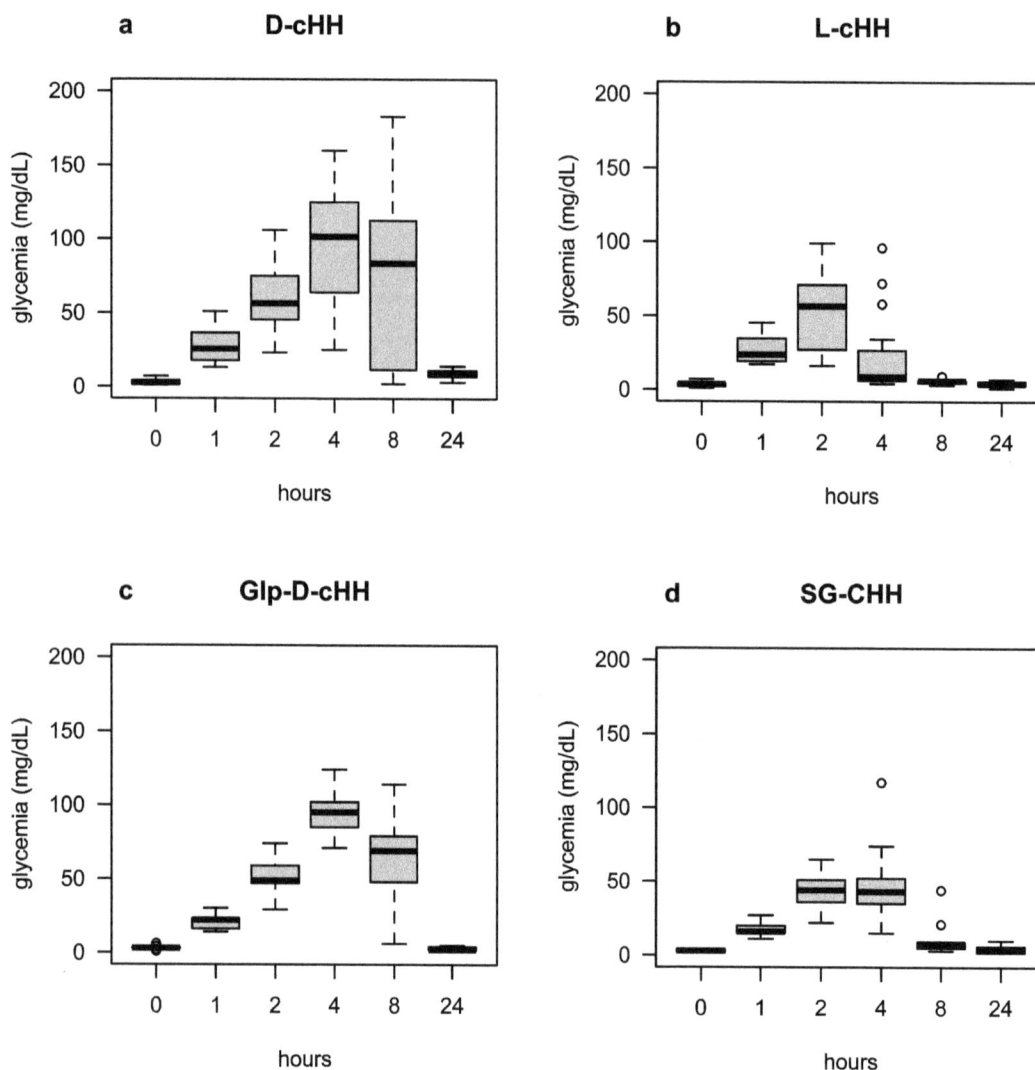

Figure 3. Time course of the induced glycemia after injection of the synthetic and wild type cHHs. a) Time course of hemolymph glycemia after injection of 1.7 pmol/g live weight of D-cHH. The maximum peak of 95.4±8.1 mg/dL glucose is reached after 4 h, high values lasting at 8 h (76±12.1 mg/dL), when the maximum values were also recorded, and the glycemia returning to its nearby basal level, 8.9±1 mg/dL, after 24 h. N = 20. b) Time course of hemolymph glycemia after injection of 1.7 pmol/g live weight of L-cHH. L-cHH induced a slower hyperglycemic response with the maximum peak of 52±6.5 mg/dL glucose after 2 h, and the glycemia returning to its nearby basal level, 5.5±0.4 mg/dL, after 8 h. N = 16. c) Time course of hemolymph glycemia after injection of 1.7 pmol/g live weight of Glp-D-cHH. The time course of Glp-D-cHH shows the stronger hyperglycemic response of 96.1±5.6 mg/dL glucose after 4 h, high values lasting at 8 h (62.9±11.8 mg/dL) and the glycemia returning to its basal level, 2.9±0.5 mg/dL, after 24 h. N = 9. d) Time course of hemolymph glycemia after injection of 1.7 pmol/g live weight of SG-cHH. The hyperglycemic time course after the injection of 1.7 pmol/g live weight of purified SG-cHH was comparable to that of synthetic D-cHH and Glp-D-cHH peptides, but the maximum (44±4.5 mg/dL) was reached after 2 h and the glycemia returning to its basal level, 5±1.1 mg/dL after 24 h. N = 9.

ability to induce hyperglycemia *in vivo* and to inhibit ecdysteroidogenesis *in vitro*, and showed similar clearance/degradation rates in the hemolymph. Both cHH variants are present in the sinus glands at a constant ratio, the blocked cHH being the major form, and the same proportion is maintained in the hemolymph [12]. In a study of the biosynthesis of cHH performed as time-course experiments with pulse-chase incubations, on *in vitro* cultured X-organ SG complexes from *O. limosus*, cyclization resulted incomplete as glutaminyl L-cHH represented about 30% of the blocked peptide. Because cyclization of 70–80% of the cHH occurs within 8 h, then remaining constant, it was supposed that this modification is catalyzed by a putative glutaminyl cyclase during the axonal transport [46]. Our results confirm the bioactivity data already found in crabs, because the hyperglycemic

activity showed by blocked cHH, Glp-D-cHH, was not significantly different ($p > 0.9$) at any time of the time course from that of D-cHH, whose N-terminus is free. The similar behavior of these two forms suggests that the pyroglutamate block at the N-terminus has no biological function, at least with respect to hyperglycemic activity or to peptide half-life, though it is widely accepted that N-terminal blocked peptides are more resistant to aminopeptidases [14].

Among the protein post-translational modifications, which enrich the diversity of the peptides encoded by the same gene, the fascinating modification of chirality is the less understood. Our data indicate that SPPS coupled to native chemical ligation is a suitable strategy for the synthesis of crustacean peptides of the cHH family, being the only one that allows the introduction of all

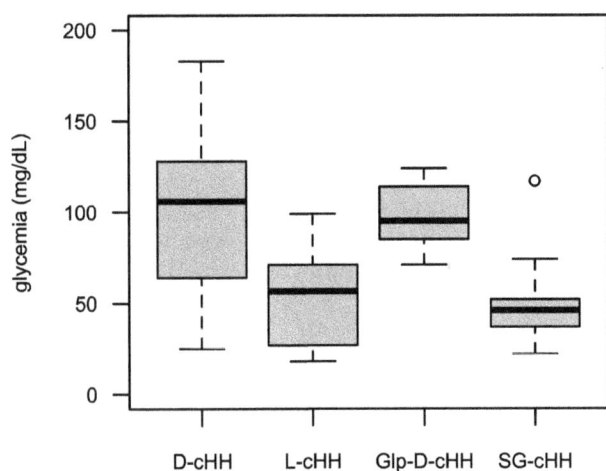

Figure 4. Boxplots of the maximum glucose values recorded for the synthetic and native peptides. Both D-cHH and Glp-D-cHH induced a strong hyperglycemia with a mean peak glucose concentration of, respectively, 100.5±9.3 and 98.7±6 mg/dL, while the hyperglycemic effect of GS-cHH and of L-cHH was lower, with a mean of max glycemia of 52.7±9.4 mg/dL and of 52.4±6.4 mg/dL respectively.

the post-translational modifications necessary to obtain a peptide identical to the native hormone, and therefore it is an appropriate alternative to DNA recombinant technology. The optimization of the present synthesis protocol can make available cHH peptides for large scale *in vivo* and *in vitro* experiments that could lead to a better understanding of cHH role and structure-activity relationships.

Materials and Methods

Reagents

9-Fluorenylmethoxycarbonyl (Fmoc) protected amino acid building blocks were ordered from Novabiochem, Bachem, and Fluka. Resins for SPPS were from Novabiochem and Fluka. Benzotriazol-1yl-oxy-tris-(pyrrolidino)-phosphonium-hexafluorophosphate (PyBOP) and 2-(6-Chloro-1H-benzotriazole-1-yl)-1,1,3,3-tetramethylaminium hexafluorophosphate (HCTU) activators were from NovaBiochem. Anhydrous N,N'-dimethylformamide (DMF) and DCM were from Romil; phenol was from AnalR; piperidine (PIP) and HFIP were from IRIS Biotech; N-diisopropylethylamine (DIPEA) and TFA were from Biosolve; TIPS from Alfa Aesar; DODT, diethylether, petroleum ether and dichloroethane (DCE) were from Sigma; 4-acetamidothiophenol (HSPhNHAc) was from Merck.

Peptide synthesis

Peptides were synthesized at 0.1 mmol scale by solid phase method using Fmoc/tBu chemistry. Both N- and side chains protected peptide carboxylic acids from N-terminal sequence of cHHs (1–38) were assembled on Fmoc-Thr(But)-TGT-resin (Fluka, substitution ~0.2 mmoles/g). cHH39-72 amide was synthesized at 0.3 mmol scale on TentaGel Sieber amide resin (Novabiochem, 0.2 mmol/g).

Side chain protecting groups were: Trityl (Trt for Asn, Cys, Gln), t-Butyl (tBu for Ser, Thr), t-Butyloxycarbonyl (Boc for Lys, Tyr, Trp), tert-Butylester (OtBu for Asp, Glu), 2,2,4,6,7-pentamethyldihydro-benzofuran-5-sulfonyl (Pbf for Arg). The Fmoc amino acid (four molar excess with respect to the initial resin substitution), the coupling reagent (HCTU), and the base

(DIPEA), in a ratio of 1:1:2, were dissolved in DMF to give a final amino acid concentration of 0.3 M. Systematic double coupling was adopted to add residues using PyBOP as coupling reagent instead of HCTU. Fmoc protecting groups were removed by a 20% (v/v) solution of PIP in DMF containing 0.1 M 1-hydroxybenzotriazole to decrease aspartimide formation. In order to minimize Cys racemization, 4 equivalents for 2 h of Fmoc-Cys(Trt)-OPfp in DMF was employed in the first coupling. The second coupling was performed for 1 h with four molar excess of Fmoc-Cys(Trt)-OH, HCTU, 2,6-dimethylpyridine, in 1:1:2 ratio as a 0.3 M solution in DCE/DMF 1:1 (v/v), as described [47].

Cleavage and side chain deprotection of cHH39-72 peptide amide resin was performed for 4 h with 20 mL/g resin of TIPS/ H_2O/DODT/phenol/TFA (3%/3%/8%/5%/81%), followed by precipitation and washing with diethyl ether and drying. Crude cHH39-72 peptide amide were reduced by 10 molar excess tris(2-carboxyethyl)phosphine (TCEP) in 0.1 M Tris pH 8 with 6 M GuHCl and purified by semipraparative RP-HPLC (Jupiter C4 10×250 mm, Phenomenex) using 0.5% gradient slope from 0.1% TFA in water (mobile phase A) to 0.1% TFA in 98% acetonitrile (MeCN) (mobile phase B).

Cleavage of all three protected cHH1-38 peptide isoforms was performed separately by 33% HFIP in DCM (4 times with 2 resin volume for 15 min). The solvents were evaporated, the rest was washed with Et2O/hexane (1/3) and dried. Thioester formation of cHH1-38 isoforms was performed according published procedure [21]. Briefly, 200 mg (~0.027 mmol) each of protected cHH1-38 was suspended in 1 ml DCM containing 27.9 mg (0.216 mmol, 8 eq.) DIPEA, mixed for 2 min at room temperature under nitrogen, then cooled to 4°C; dry PyBop 56.2 mg (0.108 mmol, 4 eq.) was added, mixed vigorously for 0.5 min and then 22.7 mg (0.135 mmol, 5 eq.) dry HSPhNHAc (purity >99%, freshly prepared by TCEP reduction) was added under nitrogen. The reaction mixture was vigorously mixed for additional 4 min and then occasionally in total for 2 h. The solvent was evaporated and the rest (oil) was mixed with 3 ml Et2O and 6 ml petroleum ether, the solid was centrifuged, washed again with 2 ml petroleum ether and dried *in vacuo*. The crude peptide thioesters were fully deprotected for 3 h with H_2O/TIPS/Phenol/TFA (2.5%/2.5%/5%/90%, 20 mL/g resin) and then precipitated/washed with diethyl ether. The crude peptide thioesters were dissolved and purified by semipreparative RP HPLC as described above. Pure (according ESI-MS, API 150EX, ABSciex) fractions were pooled and feeze-dried.

Native chemical ligation

The native chemical ligations of the thioester peptides cHH1-38 with cHH39-72 were performed dissolving the two peptide components at concentration ~5–10 mg/mL in ligation buffer [(0.5 M 4-(2-hydroxyethyl)-1-piperazineethanesulfonic acid (HEPES), 6 M GuHCl, 50 mM TCEP, 20 mM ethylenediaminetetraacetic acid (EDTA)] at pH 6.8–7.0. The N-terminal peptide was used in slight molar excess (~10%) with respect to the counterpart. The reactions went overnight at RT under gentle stirring.

Purification and oxidative folding

The NCL reaction mixtures were analyzed by LC-MS and purified by RP-HPLC in two different mobile phase systems. Linear reduced L-cHH and D-cHH were purified from the NCL mix by RP-HPLC using the classical acidic (0.1% TFA) mobile phase system on a semipreparative column Jupiter C4 10×250 mm (Phenomenex) using a gradient of 0.5% B/min and the product was eluted at ~40% MeCN. The pure fractions (according ESI-MS) of L-cHH and D-cHH were pooled and freeze dried. The peptides were re-dissolved at 0.1 mg/ml and subjected to oxidative folding in 0.25 M Tris-HCl, 5 mM reduced glutathione, 1 mM oxidized glutathione, 1 mM EDTA buffer adjusted at pH 8 under argon. The refolded L-cHH and D-cHH were purified by RP-HPLC on a semipreparative column Jupiter C4 10×250 mm using a gradient of 0.5% B/min. The product was eluted at ~38% MeCN. The Glp-D-cHH NCL reaction mixture was instead purified using the same column but in neutral pH conditions (Mobile Phase A was 20 mM triethylammonium acetate in water, Mobile Phase B was 20 mM triethylammonium acetate, 80% acetonitrile) and a gradient of 0.75% B/min from 15% to 60% B. Pure (according to ESI-MS) fractions were pooled and freeze-dried. Purified (according to ESI-MS) fraction of Glp-D-cHH at pH 7 was diluted to 0.1 mg/mL, directly folded by air oxidation overnight under gentle stirring and then freeze dried. The loss of 6 Da in the molecular masses indicating the completeness of the three disulfide bonds formation was verified by ESI-MS in all cases.

Extraction and purification of native Asl-cHH

Sinus glands from 40 eyestalks of *A. leptodactylus* were collected and 100 μL of extraction solution (90% MetOH, 9% acetic acid, 1% H_2O) was added. After sonication, the sample was centrifuged at 12 000× g for 10 min at 4°C, and the supernatant collected. The pellet was resuspended in 100 μL of the extraction solution, centrifuged again, and the two supernatants were mixed together. The extract was purified on a HPLC system (Gilson) equipped with a Zorbax SB-C18 4.6×150 mm column from Agilent Technologies Inc. (DE, USA) thermostated at 25°C. Mobile Phase A was 0.1% TFA in water, Mobile Phase B was 0.1% TFA in acetonitrile. The separation was done using a gradient of 0–100% B in 60 min at 1 mL/min. Collected fractions were analysed on a API150EX single quadrupole mass spectrometer (ABSciex), and those containing the expected molecular mass of 8383.5 Da, corresponding to the peptide with N-terminus pyroglutamate, were pooled and liophylised.

Biological assays with cHH isoforms

Adult *A. leptodactylus*, with an average weight of 41±5,9 g, were obtained from a local dealer. The animals were kept in 120 L tanks with closed circuit filtered and thoroughly aerated tapwater, at room temperature. Animals were fed with fish pellets three times per week. Crayfish were anesthetized in ice for 15 min and then bilaterally eyestalk ablated 48 h before the start of the experiment, in order to avoid any possible interference due to the endogenous cHH. To assay the hyperglycemic activity, lyophilized RP-HPLC fractions of synthetic or native peptides were resuspended in 50 μL sterile phosphate buffered saline (PBS). Peptide concentration was determined by UV absorbance at 280 nm using calculated ε values of 14815 M^{-1} cm^{-1} for the peptide oxidized form. The extinction coefficient was computed using the ProtParam programme on the ExPASy server [48]. Each animal was treated with a peptide amount equal to 1.7 pmol/g live weight. Before injection, peptide samples were diluted with sterile PBS and 100 μL of this solution was injected. Animals were bled at 0, 1, 2, 4, 8 and 24 h after injection. Hemolymph glucose level was quantified using the glucose oxidase method (One Touch glucose test kit – Lifescan).

Statistics

For glycemia elicited at the various times, the normality of data was checked with a Shapiro-Wilk test and homogeneity of variance for each time across groups and for each peptide across times were checked with a Bartlett test. Analyses were conducted using non-parametric statistics, i.e. Kruskal-Wallis rank sum test with post-hoc Wilcoxon rank sum test pairwise comparisons with Bonferroni correction, since the null hypothesis of the Shapiro Wilk and/or the Bartlett tests could not be rejected. The box and whiskers plots were drawn with the boxplot command. In the text, values are expressed as mean ± standard error. Statistical analyses were performed using R version 2.9.2 software [49].

Author Contributions

Conceived and designed the experiments: AM CG SP SZ PG. Performed the experiments: AM VZ CG AC PG. Analyzed the data: AM CG PG. Contributed reagents/materials/analysis tools: SP PG. Wrote the paper: AM CG SP SZ PG.

References

1. Fanjul-Moles ML (2006) Biochemical and functional aspects of crustacean hyperglycemic hormone in decapod crustaceans: review and update. Comp Biochem Physiol C Toxicol Pharmacol 142: 390–400.
2. Giulianini PG, Edomi P (2006) Neuropeptides controlling reproduction and growth in Crustacea: a molecular approach. In: Satake H, ed. Invertebrate neuropeptides and hormones: basic knowledge and recent advances. Trivandrum Kerala: Transworld Research Network. pp 225–252.
3. Weidemann W, Gromoll J, Keller R (1989) Cloning and sequence analysis of cDNA for precursor of a crustacean hyperglycemic hormone. FEBS Lett 257: 31–34.
4. de Kleijn DP, Sleutels FJ, Martens GJ, Van Herp F (1994) Cloning and expression of mRNA encoding prepro-gonad-inhibiting hormone (GIH) in the lobster *Homarus americanus*. FEBS Lett 353: 255–258.
5. Klein JM, Mangerich S, de Kleijn DP, Keller R, Weidemann WM (1993) Molecular cloning of crustacean putative molt-inhibiting hormone (MIH) precursor. FEBS Lett 334: 139–142.
6. Chen SH, Lin CY, Kuo CM (2005) In silico analysis of crustacean hyperglycemic hormone family. Mar Biotechnol (NY) 7: 193–206.
7. Marco HG (2004) Unity and diversity in chemical signals of arthropods: the role of neuropeptides in crustaceans and insects. International Congress Series 1275: 126–133.
8. Soyez D, Van Herp F, Rossier J, Le Caer JP, Tensen CP, et al. (1994) Evidence for a conformational polymorphism of invertebrate neurohormones. D-amino acid residue in crustacean hyperglycemic peptides. J Biol Chem 269: 18295–18298.
9. Katayama H, Ohira T, Aida K, Nagasawa H (2002) Significance of a carboxyl-terminal amide moiety in the folding and biological activity of crustacean hyperglycemic hormone. Peptides 23: 1537–1546.
10. Mosco A, Edomi P, Guarnaccia C, Lorenzon S, Pongor S, et al. (2008) Functional aspects of cHH C-terminal amidation in crayfish species. Regul Pept 147: 88–95.
11. Huberman A, Aguilar MB, Navarro-Quiroga I, Ramos L, Fernández I, et al. (2000) A hyperglycemic peptide hormone from the Caribbean shrimp *Penaeus (litopenaeus) schmitti*. Peptides 21: 331–338.
12. Chung JS, Webster SG (1996) Does the N-terminal pyroglutamate residue have any physiological significance for crab hyperglycemic neuropeptides? Eur J Biochem 240: 358–364.
13. Chung JS, Wilkinson MC, Webster SG (1998) Amino acid sequences of both isoforms of crustacean hyperglycemic hormone (CHH) and corresponding precursor-related peptide in *Cancer pagurus*. Regul Pept 77: 17–24.
14. Rink R, Arkema-Meter A, Baudoin I, Post E, Kuipers A, et al. (2010) To protect peptide pharmaceuticals against peptidases. J Pharmacol Toxicol Methods 61: 210–218.
15. Kreil G (1997) D-amino acids in animal peptides. Annu Rev Biochem 66: 337–345.
16. Yasuda A, Yasuda Y, Fujita T, Naya Y (1994) Characterization of crustacean hyperglycemic hormone from the crayfish (*Procambarus clarkii*): multiplicity of molecular forms by stereoinversion and diverse functions. Gen Comp Endocrinol 95: 387–398.
17. Aguilar MB, Soyez D, Falchetto R, Arnott D, Shabanowitz J, et al. (1995) Amino acid sequence of the minor isomorph of the crustacean hyperglycemic hormone (CHH-II) of the Mexican crayfish *Procambarus bouvieri* (Ortmann): presence of a D-amino acid. Peptides 16: 1375–1383.
18. Soyez D, Laverdure A-M, Kallen J, Van Herp F (1998) Demonstration of a cell-specific isomerization of invertebrate neuropeptides. Neuroscience 82: 935–942.

19. Serrano L, Blanvillain G, Soyez D, Charmantier G, Grousset E, et al. (2003) Putative involvement of crustacean hyperglycemic hormone isoforms in the neuroendocrine mediation of osmoregulation in the crayfish *Astacus leptodactylus*. J Exp Biol 206: 979–988.

20. Bulau P, Meisen I, Reichwein-Roderburg B, Peter-Katalinic J, Keller R (2003) Two genetic variants of the crustacean hyperglycemic hormone (CHH) from the Australian crayfish, *Cherax destructor*: detection of chiral isoforms due to posttranslational modification. Peptides 24: 1871–1879.

21. von Eggelkraut-Gottanka R, Klose A, Beck-Sickinger AG, Beyermann M (2003) Peptide [alpha]thioester formation using standard Fmoc-chemistry. Tetrahedron Lett 44: 3551–3554.

22. Sieber P (1987) A new acid-labile anchor group for the solid-phase synthesis of C-terminal peptide amides by the Fmoc method. Tetrahedron Lett 28: 2107–2110.

23. Webster SG, Keller R (1986) Purification, characterisation and amino acid composition of the putative moult-inhibiting hormone (MIH) of *Carcinus maenas* (Crustacea, Decapoda). J Comp Physiol B 156: 617–624.

24. Soyez D, Van Deijnen JE, Martin M (1987) Isolation and characterization of a vitellogenesis-inhibiting factor from sinus glands of the lobster, *Homarus americanus*. J Exp Biol 244: 479–484.

25. Kegel G, Reichwein B, Weese S, Gaus G, Peter-Katalinic J, et al. (1989) Amino acid sequence of the crustacean hyperglycemic hormone (CHH) from the shore crab, *Carcinus maenas*. FEBS Lett 255: 10–14.

26. Ohira T, Nishimura T, Sonobe H, Okuno A, Watanabe T, et al. (1999) Expression of a recombinant Molt-Inhibiting Hormone of the Kuruma prawn *Penaeus japonicus* in *Escherichia coli*. Biosci Biotechnol Biochem 63: 1576–1581.

27. Edomi P, Azzoni E, Mettulio R, Pandolfelli N, Ferrero EA, et al. (2002) Gonad-inhibiting hormone of the Norway lobster (*Nephrops norvegicus*): cDNA cloning, expression, recombinant protein production, and immunolocalization. Gene 284: 93–102.

28. Sonobe H, Nishimura T, Sonobe M, Nakatsuji T, Yanagihara R, et al. (2001) The molt-inhibiting hormone in the American crayfish *Procambarus clarkii*: its chemical synthesis and biological activity. Gen Comp Endocrinol 121: 196–204.

29. Chung JS, Webster SG (2003) Moult cycle-related changes in biological activity of moult-inhibiting hormone (MIH) and crustacean hyperglycaemic hormone (CHH) in the crab, *Carcinus maenas*. From target to transcript. Eur J Biochem 270: 3280–3288.

30. Gu PL, Yu KL, Chan SM (2000) Molecular characterization of an additional shrimp hyperglycemic hormone: cDNA cloning, gene organization, expression and biological assay of recombinant proteins. FEBS Lett 472: 122–128.

31. Mettulio R, Edomi P, Ferrero EA, Lorenzon S, Giulianini PG (2004) The crustacean hyperglycemic hormone precursors a and b of the Norway lobster differ in the preprohormone but not in the mature peptide. Peptides 25: 1899–1907.

32. Ohira T, Tsutsui N, Nagasawa H, Wilder MN (2006) Preparation of two recombinant crustacean hyperglycemic hormones from the giant freshwater prawn, *Macrobrachium rosenbergii*, and their hyperglycemic activities. Zoolog Sci 23: 383–391.

33. Treerattrakool S, Udomkit A, Eurwilaichitr L, Sonthayanon B, Panyim S (2003) Expression of biologically active crustacean hyperglycemic hormone (CHH) of *Penaeus monodon* in *Pichia pastoris*. Mar Biotechnol (NY) 5: 373–379.

34. Nilsson BL, Soellner MB, Raines RT (2005) Chemical synthesis of proteins. Annu Rev Biophys Biomol Struct 34: 91–118.

35. Kamatani Y, Minakata H, Iwashita T, Nomoto K, In Y, et al. (1990) Molecular conformation of achatin-I, an endogenous neuropeptide containing D-amino acid residue: X-Ray crystal structure of its neutral form. FEBS Lett 276: 95–97.

36. Tancredi T, Temussi PA, Picone D, Amodeo P, Tomatis R, et al. (1991) New insights on μ/δ selectivity of opioid peptides: Conformational analysis of deltorphin analogues. Biopolymers 31: 751–760.

37. Broccardo M, Erspamer V, Falconieri Erspamer G, Improta G, Linari G, et al. (1981) Pharmacological data on dermorphins, a new class of potent opioid peptides from amphibian skin. Br J Pharmacol 73: 625–631.

38. Chung JS, Webster SG (2005) Dynamics of in vivo release of molt-inhibiting hormone and crustacean hyperglycemic hormone in the shore crab, *Carcinus maenas*. Endocrinology 146: 5545–5551.

39. Sedlmeier D (1988) The crustacean hyperglycemic hormone (CHH) releases amylase from the crayfish midgut gland. Regul Pept 20: 91–98.

40. Santos EA, Nery LE, Keller R, Gonçalves AA (1997) Evidence for the involvement of the crustacean hyperglycemic hormone in the regulation of lipid metabolism. Physiol Zool 70: 415–420.

41. Chung JS, Webster SG (2006) Binding sites of crustacean hyperglycemic hormone and its second messengers on gills and hindgut of the green shore crab, *Carcinus maenas*: a possible osmoregulatory role. Gen Comp Endocrinol 147: 206–213.

42. Chung JS (2010) Hemolymph ecdysteroids during the last three molt cycles of the blue crab, *Callinectes sapidus*: quantitative and qualitative analyses and regulation. Arch Insect Biochem Physiol 73: 1–13.

43. de Kleijn DP, Janssen KP, Waddy SL, Hegeman R, Lai WY, et al. (1998) Expression of the crustacean hyperglycemic hormones and the gonad-inhibiting hormone during the reproductive cycle of the female American lobster *Homarus americanus*. J Endocrinol 156: 291–298.

44. Tsutsui N, Katayama H, Ohira T, Nagasawa H, Wilder MN, et al. (2005) The effects of crustacean hyperglycemic hormone-family peptides on vitellogenin gene expression in the kuruma prawn, *Marsupenaeus japonicus*. Gen Comp Endocrinol 144: 232–239.

45. Spanings-Pierrot C, Soyez D, Van Herp F, Gompel M, Skaret G, et al. (2000) Involvement of crustacean hyperglycemic hormone in the control of gill ion transport in the crab *Pachygrapsus marmoratus*. Gen Comp Endocrinol 119: 340–350.

46. Ollivaux C, Soyez D (2000) Dynamics of biosynthesis and release of crustacean hyperglycemic hormone isoforms in the X-organ-sinus gland complex of the crayfish *Orconectes limosus*. Eur J Biochem 267: 5106–5114.

47. Han Y, Albericio F, Barany G (1997) Occurrence and minimization of cysteine racemization during stepwise solid-phase peptide synthesis. J Org Chem 62: 4307–4312.

48. Gasteiger E, Hoogland C, Gattiker A, Duvaud S, Wilkins MR, et al. (2005) Protein identification and analysis tools on the ExPASy server. In: Walker JM, ed. The proteomics protocols handbook. Totowa NJ: Humana Press. pp 571–607.

49. R Development Core Team (2010) R: a language and environment for statistical computing. Vienna: R Foundation for Statistical Computing.

10

Rgg-Associated SHP Signaling Peptides Mediate Cross-Talk in Streptococci

Betty Fleuchot[1,2¤], Alain Guillot[1,2], Christine Mézange[1,2], Colette Besset[1,2], Emilie Chambellon[1,2], Véronique Monnet[1,2], Rozenn Gardan[1,2]*

1 INRA, UMR1319 MICALIS, Jouy en Josas, France, **2** AgroParistech, UMR MICALIS, Jouy en Josas, France

Abstract

We described a quorum-sensing mechanism in the streptococci genus involving a short hydrophobic peptide (SHP), which acts as a pheromone, and a transcriptional regulator belonging to the Rgg family. The *shp/rgg* genes, found in nearly all streptococcal genomes and in several copies in some, have been classified into three groups. We used a genetic approach to evaluate the functionality of the SHP/Rgg quorum-sensing mechanism, encoded by three selected *shp/rgg* loci, in pathogenic and non-pathogenic streptococci. We characterized the mature form of each SHP pheromone by mass-spectrometry. We produced synthetic peptides corresponding to these mature forms, and used them to study functional complementation and cross-talk between these different SHP/Rgg systems. We demonstrate that a SHP pheromone of one system can influence the activity of a different system. Interestingly, this does not seem to be dependent on the SHP/Rgg group and cross-talk between pathogenic and non-pathogenic streptococci is observed.

Editor: Michael M. Meijler, Ben-Gurion University of the Negev, Israel

Funding: This study was supported by the Institut National de la Recherche Agronomique (INRA) and the Ministère de l'Education Nationale de la Recherche et de la Technologie (MENRT). The PAPPSO platform received the financial support from the Ile de France regional council and from CEMAGREF. The funders had no role in study design, data collection and analysis, decision to publish, or preparation of the manuscript.

Competing Interests: The authors have declared that no competing interests exist.

* E-mail: rozenn.gardan@jouy.inra.fr

¤ Current address: Laboratoire de Chimie Bactérienne, Aix Marseille University-CNRS UMR7283, Marseille, France

Introduction

Quorum-sensing (QS) is a cell-cell communication mechanism that allows bacteria to control gene expression in a co-ordinated manner at a population scale. It involves the detection of an autoinducer signal that is synthesized, and actively or passively secreted; it is detected, or triggers a response, when its extracellular concentration reaches a threshold or quorum. This sensing leads cells to modulate expression of the gene targets of the mechanism. QS controls various important functions including for example virulence in *Pseudomonas aeruginosa* [1] and *Staphylococcus aureus* [2], biofilm development in *Pseudomonas putida* [3] and *S. aureus* [2] and sporulation in bacilli [4]. The autoinducers of Gram-negative bacteria mainly belong to the family of acyl-homoserine lactones [5] whereas in those of Gram-positive bacteria are peptides [6,7].

Peptide autoinducers are either detected indirectly from the extracellular medium or directly in the intracellular medium. In the first case, the peptides are sensed by the histidine kinase of a two component system. This leads to the modification of the phosphorylation status of the histidine kinase and then of its cognate response regulator which modulates the expression of its target genes. This mechanism controls many functions, including triggering competence for natural transformation in *Streptococcus pneumoniae* and *Bacillus subtilis* [8]. In the second case, the peptides are imported into the cell by an oligopeptide transporter and once inside, interact with a transcriptional regulator or a Rap protein; conjugation in *Enterococcus faecalis* and virulence in *Bacillus cereus* [9] are controlled by this type of mechanism.

Many of these QS mechanisms have been deciphered in detail by more than 40 years of studies, such that we now understand quite well how a cell communicates with its siblings. A QS issue that has emerged more recently is the communication between different strains of the same species and even between species. Work on this subject has led to the definition of pherotypes or specificity groups, and amino acid sequence polymorphism has been documented for the signaling peptides and their receptors. All bacteria that belong to one pherotype can sense the peptides synthesized by members of the same pherotype but not the peptides synthesized by members of the others. Pherotypes have been defined for different mechanisms: Agr in *S. aureus* [10,11], ComCDE in *S. pneumoniae* [12], ComQXPA in *B. subtilis* [13,14] and PapR/PlcR in *B. cereus* [15,16]. These studies are of significance for at least two reasons: i) a better knowledge of the interaction between the signaling peptides and their receptors may allow intervention, based on synthetic peptides, and this would be of particular value for the regulation of virulence factors as demonstrated in *S. aureus* [17]; ii) deciphering this diversity and evolution may help understand ecological adaptation by bacteria [18].

We recently discovered a QS mechanism that relies on a transcriptional regulator of the Rgg family and a small hydrophobic peptide (SHP) detected in the intracellular medium [19,20]. We studied the *shp/ster_1358* (*rgg1358*) locus of *Streptococcus thermophilus* LMD-9, where the two genes are transcribed divergently, and showed that SHP1358 is secreted, matured and imported back into the cell by the oligopeptide transporter Ami.

Table 1. Bacterial strains used in this study.

Bacterial strain and genotype		Resistance[a]	Description[b]	Source or reference
Streptococcus thermophilus **LMD-9 and derivates**				
	LMD-9	Wild-type		[38]
	TIL1038	*blp*::P$_{shp1299}$-*luxAB aphA3*	Km	pGICB004a::P$_{shp1299}$ → LMD-9	This study
	TIL1042	*amiCDE::erm blp*::P$_{shp1299}$-*luxAB aphA3*	Km/Erm	TIL1389 DNA → TIL1038	This study
	TIL1047	*shp1299::erm* [c]	Erm	PCR fragment *shp1299::erm* → LMD-9	This study
	TIL1048	Δ*ster_1299 blp*::P$_{shp1299}$-*luxAB aphA3*	Km	pGICB004a::P$_{shp1299}$ → TIL1160	This study
	TIL1052	*shp1299::erm blp*::P$_{shp1299}$-*luxAB aphA3*	Km/Erm	pGICB004a::P$_{shp1299}$ → TIL1047	This study
	TIL1160	Δ*ster_1299*		pG⁺host9::updown.*ster_1299* → LMD-9	This study
	TIL1165	*blp*::P$_{shp1358}$-*luxAB*			[21]
	TIL1200	Δ*shp1358 blp*::P$_{shp1358}$-*luxAB*			[21]
	TIL1345	*blp*::*gbs1555 shp1555-luxAB aphA3*	Km	pGICB004a::*gbs1555 shp1555* → LMD-9	This study
	TIL1380	*blp*::*shp1555-luxAB aphA3*	Km	pGICB004a::*shp1555* → LMD-9	This study
	TIL1381	*amiCDE::erm blp*::*gbs1555 shp1555-luxAB aphA3*	Km/Erm	TIL1389 DNA → TIL1345	This study
	TIL1382	*blp*::*gbs1555* P$_{shp1555}$-*luxAB aphA3*	Km	pGICB004a::*gbs1555* P$_{shp1555}$ → LMD-9	This study
	TIL1383	*blp*::SMU.1509 *shp1509-luxAB aphA3*	Km	pGICB004a::SMU.1509 *shp1509* → LMD-9	This study
	TIL1384	*blp*::SMU.1509 P$_{shp1509}$-*luxAB aphA3*	Km	pGICB004a::SMU.1509 P$_{shp1509}$ → LMD-9	This study
	TIL1385	*amiCDE::erm blp*::SMU.1509 *shp1509-luxAB aphA3*	Km/Erm	TIL1389 DNA → TIL1383	This study
	TIL1386	*blp*::*shp1509-luxAB aphA3*	Km	pGICB004a::*shp1509* → LMD-9	This study
	TIL1389	*amiCDE::erm*	Erm	PCR fragment *amiCDE::erm* → LMD-9	This study
Streptococcus mutans				
	UA159	Wild-type ATCC 700610			[39]
Streptococcus agalactiae				
	NEM316	Wild-type			[40]
Escherichia coli				
	TG1 *repA⁺*	TG1 derivative with *repA* gene integrated into the chromosome			[19]

[a]Km and Erm indicate resistance to kanamycin and erythromycin, respectively.
[b]Arrows indicate construction by transformation with chromosomal DNA or plasmid.
[c]*shp1299* is annotated *ster_1298* in Genbank.

Then, the mature form of SHP1358 interacts with Rgg1358 enabling Rgg1358 to control, positively, the expression of two targets, *shp1358* and *ster_1357* [21]. The *ster_1357* gene encodes a secreted cyclic peptide of unknown function. A similar mechanism has been suggested for the *shp/stu0182* (*rgg0182*) locus of *S. thermophilus* strain LMG18311 [22] and has been confirmed for two SHP/Rgg systems in *Streptococcus pyogenes*: one is an activator and the other a repressor involved in biofilm development [23,24]. The phylogenetic tree of Rgg-like proteins indicates that Rgg are widespread in Gram-positive bacteria but that *shp*-associated *rgg* genes are only found in the streptococci genus. This genus contains various species, including commensal bacteria of the human microbiome, the GRAS (generally recognized as safe) bacterium, *S. thermophilus*, used for the manufacture of dairy products, but also human pathogenic bacteria such as *S. pneumoniae*, *S. agalactiae*, *S. pyogenes* and *S. mutans* [25]. We found 68 *shp/rgg* copies, 28 of which encode a unique amino acid sequence, although the sequences of all these SHP pheromones are generally similar. Nearly all streptococci genomes contain one copy, but some streptococci have multiple copies, for example *S. thermophilus* strain LMD-9 has six. This phylogenetic study of Rgg amino acid sequences led to their classification into three groups. In groups I and II, the SHPs have a conserved glutamate and

aspartate, respectively, and the *shp* and *rgg* genes are transcribed divergently. In group III, the *shp* genes are located downstream from the *rgg* genes, in a convergent orientation and the SHPs have a glutamate or an aspartate residue [21].

Different streptococci species can meet in raw milk [26], the human oral microbiome [27,28] and gastrointestinal tract [29]. We therefore investigated whether there is interspecies cross-talk via SHP peptides. We first studied the functionality of the SHP/Rgg cell-cell communication mechanism associated with three different *shp/rgg* loci, one from each of the three groups, in three distinct streptococci species. The mature form of each SHP was identified in the extracellular medium. We then used synthetic peptides to study the specificity of the interaction between the SHPs and the Rgg regulators of different groups and species. We demonstrate cross-talk between SHP/Rgg systems belonging to distinct groups and different species.

Materials and Methods

Bacterial Strains and Growth Conditions

The bacterial strains used in this study are listed in Table 1. *S. thermophilus*, *S. agalactiae* and *S. mutans* strains were grown at 30, 37 or 42°C in M17 medium (Difco) supplemented with 10 g l⁻¹

Table 2. Plasmids used in this study.

Plasmid	Description[a,b]	Source or reference
pG⁺host9	Erm, Ts plasmid	[41]
pG⁺host9::updown.ster_1299	Erm, pG⁺host9 derivative, for ster_1299 gene remplacement by double cross-over integration	This study
pGICB004	Erm, Ts plasmid allowing the integration of transcriptional fusions to the luxAB reporter genes at the blp locus in S. thermophilus	[21]
pGICB004a	Erm, pGICB004 derivative containing the aphA3 gene, conferring kanamycin resistance, upstream from the luxAB genes	This study
pGICB004a::P$_{shp1299}$	Erm, Km, pGICB004a derivative used to introduce a P$_{shp1299}$-luxAB transcriptional fusion into S. thermophilus LMD-9	This study
pGICB004a::gbs1555 shp1555	Erm, Km, pGICB004a derivative used to introduce a gbs1555 shp1555-luxAB transcriptional fusion into S. thermophilus LMD-9	This study
pGICB004a::gbs1555 P$_{shp1555}$	Erm, Km, pGICB004a derivative used to introduce a gbs1555 P$_{shp1555}$-luxAB transcriptional fusion into S. thermophilus LMD-9	This study
pGICB004a::shp1555	Erm, Km, pGICB004a derivative used to introduce a shp1555-luxAB transcriptional fusion into S. thermophilus LMD-9	This study
pGICB004a::shp1509	Erm, Km, pGICB004a derivative used to introduce a shp1509-luxAB transcriptional fusion into S. thermophilus LMD-9	This study
pGICB004a::SMU.1509 shp1509	Erm, Km, pGICB004a derivative used to introduce a SMU.1509 shp1509-luxAB transcriptional fusion into S. thermophilus LMD-9	This study
pGICB004a::SMU.1509 P$_{shp1509}$	Erm, Km, pGICB004a derivative used to introduce a SMU.1509 P$_{shp1509}$-luxAB transcriptional fusion into S. thermophilus LMD-9	This study

[a]Ts indicates that the plasmid encodes a thermosensitive RepA protein.
[b]Km and Erm indicate resistance to kanamycin and erythromycin, respectively.

lactose (M17lac) or in a chemically defined medium (CDM) without shaking, under atmospheric air and with a ratio of air space to liquid of approximately 90% [30]. *Escherichia coli* strains were grown at 30 or 37°C in Luria-Bertani (LB) broth with shaking. Agar (1.5%) was added to the media as appropriate. When required, antibiotics were added to the media at the following final concentrations: erythromycin, 200 μg ml⁻¹ for *E. coli* and 5 μg ml⁻¹ for *S. thermophilus*; and kanamycin, 1 mg ml⁻¹ for *S. thermophilus*. The optical density at 600 nm of the cultures was measured with an Uvikon 931 spectrophotometer (Kontron).

DNA Manipulation and Sequencing

Restriction enzymes, T4 DNA ligase (New England Biolabs), and Phusion DNA polymerase (Finnzymes) were used according to the manufacturers' instructions. Standard methods were used for DNA purification, restriction digestion, PCR, ligation and sequencing. The oligonucleotides, purchased from Eurogentec, are listed in Table S1. PCR amplifications were carried out in a GeneAmp PCR System 2720 (Applied Biosystems) and all amplified fragments were purified with a Wizard purification kit (Promega). Plasmids were extracted with QIAprep spin miniprep kits (Qiagen). The *E. coli* strain TG1 *repA⁺* was used as the host for cloning experiments. *S. thermophilus* was transformed using natural competent [31] or electrocompetent [19] cells. The plasmids used are listed in Table 2.

Construction of Mutant Strains

The overlapping PCR method was used to delete the *shp1299* and the *amiCDE* genes as follows. Briefly, the erythromycin (erm) cassette was amplified by PCR with oligonucleotides Erm-F and Erm-R and pG⁺host9 as the template and fused by PCR to fragments located upstream and downstream from the *shp1299* gene and to fragments located upstream from the *amiC* gene and downstream from the *amiE* gene. Upstream and downstream

fragments of both *shp1299* and *ami* genes were amplified with oligonucleotides shp1299_up-F/shp1299_up-R and shp1299_down-F/shp1299_down-R and amiCDE_up-F/amiC-DE_up-R, amiCDE_down-F/amiCDE_down-R. The resulting fused PCR fragments were used to transform natural competent cells of strain LMD-9 leading to the construction of strains TIL1047 (*shp1299::erm*) and TIL1389 (*amiCDE::erm*). Strain TIL1160 (Δster_1299) was constructed by deleting an internal fragment of the gene by a double crossover event using pG⁺host9. Briefly, oligonucleotides ster_1299-SpeI with ster_1299-EcoRIA and ster_1299-EcoRIB with ster_1299-HindIII were used to amplify upstream and downstream fragments from the *ster_1299* gene; the resulting two fragments were double digested with the restriction enzymes *Spe*I+*Eco*RI and *Eco*RI+*Hind*III, respectively, and ligated between the *Spe*I and *Hind*III restriction sites of pG⁺host9. The resulting plasmid, pG⁺host9::updown.*ster_1299*, was used to transform electrocompetent cells of strain LMD-9. Integration and excision of the plasmid led to the deletion of the *ster_1299* gene.

Constructions of Strains Containing *luxAB* Reporters

First, plasmid pGICB004a, a derivative of pGICB004, was constructed to facilitate integration of transcriptional fusions to the *luxAB* reporter genes into the *blp* locus in *S. thermophilus* LMD-9 by natural transformation and double crossover events. For this purpose, the *aphA3* cassette was amplified with oligonucleotides AphA3-F and AphA3-R using plasmid pKa as the template [32]. The resulting fragment was inserted at the *Sma*I restriction site, downstream from the *luxAB* genes in pGICB004. To study the expression of the *shp1299*, *shp1555* and *shp1509* genes in various genetic backgrounds, derivatives of pGICB004a were constructed and used to transform various strains of *S. thermophilus* as described below. The plasmid, pGICB004a::P$_{shp1299}$, was constructed as follows. The *shp1299* promoter was amplified with oligonucleotides Pshp1299-EcoRI and Pshp1299-SpeI, double digested with

Table 3. The *shp/rgg* loci used in this study.

	Species	Strain	Locus name[b]	Rgg name	SHP name[c]	SHP sequence[d]
Group I[a]	*Streptococcus agalactiae*	NEM316	*shp/gbs1555*	Rgg1555	SHP1555	MKKINKALLFTLIMDILIIVGG
Group II[a]	*Streptococcus thermophilus*	LMD-9	*shp/ster_1358*	Rgg1358	SHP1358	MKKQILLTLLLVVFEGIIVIVVG
	Streptococcus mutans	UA159	*shp/SMU.1509*	Rgg1509	SHP1509	MRNKIFMTLIVVLETIIIIGGG
Group III[a]	*Streptococcus thermophilus*	LMD-9	*shp/ster_1299*	Rgg1299	SHP1299	MKKVIAIFLFIQTVVVIDIIIFPPFG

[a]Group number of the SHP-associated Rgg according to the classification described in Fleuchot *et al.* [21].
[b]The *shp* gene is followed by the Genbank id of the *rgg* genes.
[c]The *shp* genes are not annotated in Genbank but were identified using BactgeneSHOW [20], except for the *shp* gene associated with *ster_1299*, which is annotated *ster_1298* in the genome of *S. thermophilus* strain LMD-9. Consequently, all the *shp* gene products are indicated with the term "SHP" followed by the number of the cognate *rgg* gene in Genbank. To unify the nomenclature, the *ster_1298* gene product was renamed SHP1299.
[d]The sequences of the synthetic peptides used in this study are underlined.

the restriction enzymes *Spe*I and *Eco*RI and ligated between the same restriction sites of pGICB004a. *Sca*I-linearized pGIC-B004a::P*shp1299* was used to transform competent cells of strains LMD-9, TIL1047 and TIL1160 leading to strains TIL1038 (*blp*::P*shp1299*-*luxAB aphA3*), TIL1052 (*shp1299*::*erm blp*::P*shp1299*-*luxAB aphA3*) and TIL1048 (Δ*ster_1299 blp*::P*shp1299*-*luxAB aphA3*), respectively. The plasmids pGICB004a::*gbs1555 shp1555*, pGIC-B004a::*gbs1555* P*shp1555* and pGICB004a::*shp1555* were constructed similarly and in these cases, the PCR fragments ligated into each plasmid were amplified with oligonucleotides GBS-SpeI/GBS-EcoRI, GBS-SpeI/GBSshp-EcoRI and GBSrgg-SpeI/GBS-EcoRI, respectively. Natural transformation of strain LMD-9 with each linearized plasmid lead to construction of strains TIL1345 (*blp*::*gbs1555 shp1555*-*luxAB aphA3*), TIL1382 (*blp*::*gbs1555* P*shp1555*-*luxAB aphA3*) and TIL1380 (*blp*::*shp1555*-*luxAB aphA3*), respectively. Similarly, pGICB004a::*shp1509* was constructed by ligating a PCR fragment amplified with oligonucleotides SMUrgg-SpeI/SMU-EcoRI and double digested with EcoRI and SpeI, into pGICB004a. The SMU.1509 gene contains a *Eco*RI restriction site, so pGICB004a::SMU.1509 *shp1509* and pGIC-B004a::SMU.1509 P*shp1509* were constructed in two steps. First, the downstream part of SMU.1509 was amplified with oligonucleotides SMU-SpeI/SMU-2, double digested with *Eco*RI and *Spe*I, and ligated between the same restriction sites of pGICB004a. Second, the two fragments containing the fusions to the *shp* promoter were amplified with oligonucleotides SMU-1/SMU-

EcoRI and SMU-1/SMUshp-EcoRI, digested with *Eco*RI and ligated into the same restriction site of pGICB004a already containing the downstream part of SMU.1509. Linearized pGICB004a::*shp1509*, pGICB004a::SMU.1509 *shp1509* and pGICB004a::SMU.1509 P*shp1509* were then used to transform competent cells of strain LMD-9 leading to strains TIL1386 (*blp*::*shp1509*-*luxAB aphA3*), TIL1383 (*blp*::SMU.1509 *shp1509*-*luxAB aphA3*) and TIL1384 (*blp*::SMU.1509 P*shp1509*-*luxAB aphA3*), respectively. Finally, TIL1042 (*amiCDE*::*erm blp*::P*shp1299*-*luxAB aphA3*), TIL1381 (*amiCDE*::*erm blp*::*gbs1555*::*shp1555*-*luxAB aphA3*) and TIL1385 (*amiCDE*::*erm blp*::SMU.1509::*shp1509*-*luxAB aphA3*) were constructed by transforming competent cells of strains TIL1038, TIL1345 and TIL1383, respectively, with chromosomal DNA from strain TIL1389 (*amiCDE*::*erm*).

LC-MS/MS

S. thermophilus strain LMD-9 carrying the *shp1299* gene, *S. agalactiae* strain NEM316 carrying the *shp1555* gene, *S. mutans* strain UA159 carrying the *shp1509* gene, *S. thermophilus* strains TIL1345 and TIL1383 carrying the *shp* genes of *S. agalactiae* and *S. mutans*, respectively, were grown in CDM, and the culture supernatants were recovered at the end of the exponential phase. Aliquots of 5 µl of supernatant were loaded on a Pepmap C18 column (length 150 mm, 75 µm ID, 100 Å; Dionex, Voisins-le-Bretonneux) and analyzed on-line by mass spectrometry on a LTQ-Orbitrap Discovery apparatus (Thermo Fischer, San Jose).

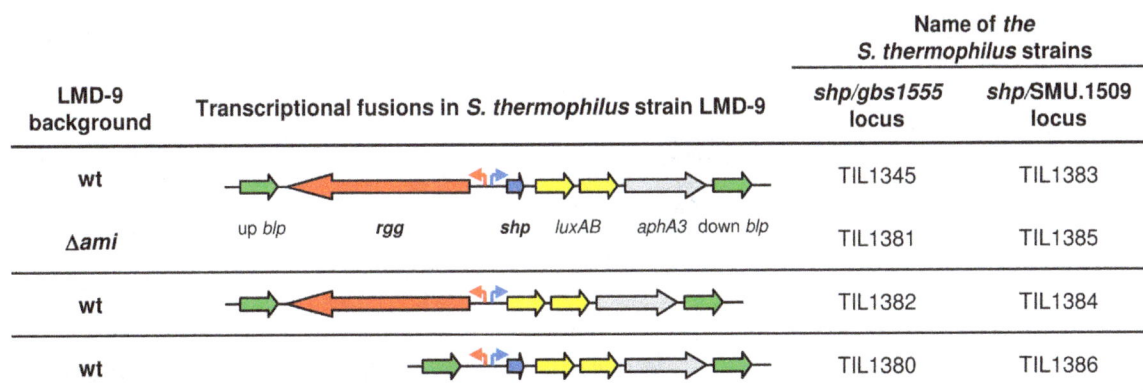

Figure 1. Description of strains containing P*shp*-*luxAB* transcriptional fusions in various genetic backgrounds. These strains were constructed in *S. thermophilus* strain LMD-9 and used to study the expression of the *shp* genes of *S. agalactiae* strain NEM316 (*shp/gbs1555* locus) and *S. mutans* strain UA159 (*shp/SMU.1509* locus) in the presence and absence of the corresponding *shp* and *rgg* genes and in the presence and absence of the *ami* genes of *S. thermophilus* strain LMD-9.

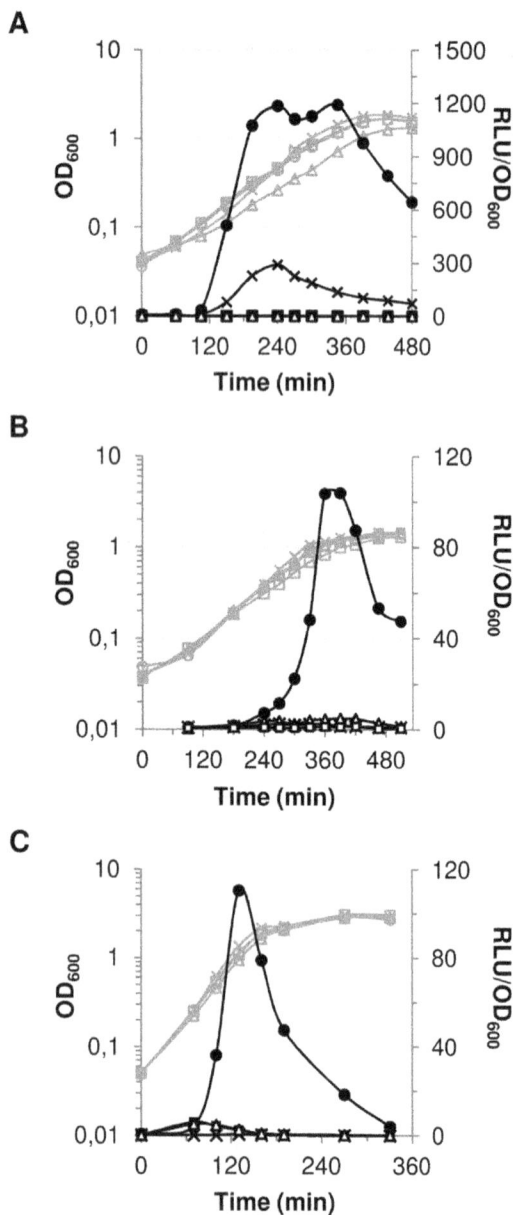

Figure 2. Growth and luciferase activities of strains containing P_{shp}-luxAB fusions in various genetic backgrounds. Growth curves (OD_{600}) are presented in gray and relative luciferase activities (RLU/OD_{600}) in black. Growth and relative luciferase activities of derivatives of *S. thermophilus* strain LMD-9 grown in CDM and containing P_{shp}-luxAB fusions of the loci *shp/gbs1555* of *S. agalactiae* (A), *shp*/SMU.1509 of *S. mutans* (B) and *shp/ster_1299* of *S. thermophilus* strain LMD-9 (C). The genetic backgrounds are indicated as follows: (●) the *shp* and *rgg* genes of the locus tested and the *ami* gene of *S. thermophilus* are present (▲) the cognate *shp* gene of the locus studied is not present, (■) the cognate *rgg* gene of the locus studied is not present and, (×) the *ami* genes of *S. thermophilus* are not present. Experiments were done at 30°C for the *shp/gbs1555* and the *shp*/SMU.1509 loci and at 42°C for the *shp/ster_1299* locus. Data shown are representative of three independent experiments.

The masses of the separated molecules were first analyzed with the high resolution accuracy (10 ppm) of the Orbitrap mass analyser. Then, selected ions were fragmented in the trap by collision

induced dissociation (CID) and the ion daughters were analyzed at low accuracy (250 ppm) in the linear ion trap (LTQ).

We manually extracted the ion current signals (XIC) of the masses of all monocharged peptides corresponding to the C-terminal fragments of the SHP precursors ranging from LIIVGG to FTLIMDILIIVGG for SHP1555, from IIIGGG to IVVLE-TIIIIGGG for SHP1509 and from FPPFG to VVIDIIIFPPFG for SHP1299. Using the sequences of the three streptococcal genomes, we checked that these peptides could not be encoded by genes other than the *shp* genes. We also checked that the XIC detected fulfilled two different criteria: (i) the retention time of the XIC detected was compatible with the hydrophobicity (GRAVY index) of the corresponding peptide and (ii), the XIC signal was absent from the supernatant of a strain that did not encode the SHP. Then, selected ions were fragmented. This approach was not successful for *S. thermophilus* expressing the *shp1509* gene of *S. mutans* (TIL1383), so a more targeted and sensitive approach was used. We searched specifically for the peptide ETIIIIGGG which has a predicted mass of 872.51 Da. First, we fragmented all the ions with a mass of 872.51+/−2 Da during the chromatographic separation and analyzed the fragments in the LTQ. Second, we extracted the MS2 XIC with a mass of 740, corresponding to the fragment b7. This transition was chosen on the basis of previous fragmentation data obtained with SHP1358 of the *S. thermophilus* strain LMD-9 [21]. The patterns of fragmentation were analyzed to identify one with all ion daughters that fitted well with the sequence of the peptide sought.

To estimate the concentration of mature SHP1358 (EGII-VIVVG) in the supernatant of *S. thermophilus* strain LMD-9, we used the corresponding heavy form of the mature peptide [NH2-EGII[V_$C^{13}N^{15}$]IVVG-OH] (Thermo, Scientific) dissolved in 5% CH_3CN and 0.1% trifluoracetic acid, as an internal standard. The heavy peptide was added to *S. thermophilus* LMD-9 supernatant at a final concentration of 10 ng ml^{-1}. The area obtained with the heavy peptide was measured and used to calculate the amount of the natural peptide.

Luciferase Assay

Cells were grown overnight at 42°C in CDM. Cultures of 50 ml of CDM were then inoculated at an OD_{600} of 0.05 and incubated at the appropriate temperature, *i.e.* 30, 37 or 42°C. Aliquots of 1 ml were sampled at regular intervals until the culture reached the stationary phase and analyzed as described previously [21]. Synthetic peptides (Table 3), stored in lyophilized form and prepared in 5% formic acid, were added as appropriate to a final concentration of 1 μM at the beginning of the cultures. Purities of crude preparations were above 90%. Results are reported in Relative Luminescence Units divided by the OD_{600} (RLU/OD_{600}). *S. thermophilus* strains TIL1345, TIL1383, TIL1038 and TIL1165 were used in cultures at 30, 37 and 42°C to assess which of these was the optimal temperature for the expression of *shp1555*, *shp1509*, *shp1299* and *shp1358*, respectively. It appeared to be 30°C for *shp1555* and *shp1509* and 42°C for *shp1299* and *shp1358* (data not shown).

Results

Selection of Relevant *shp/rgg* Loci for the Study of Cross-talk in Streptococci

To study cross-talk among streptococci via SHP signaling peptides, we chose four *shp/rgg* loci found in the three SHP-associated Rgg phylogenetic groups [21] (Table 3). For group I, we chose the locus *shp/gbs1555* (*rgg1555*) of *Streptococcus agalactiae* strain NEM316, present in all sequenced strains of *S. agalactiae*. The role

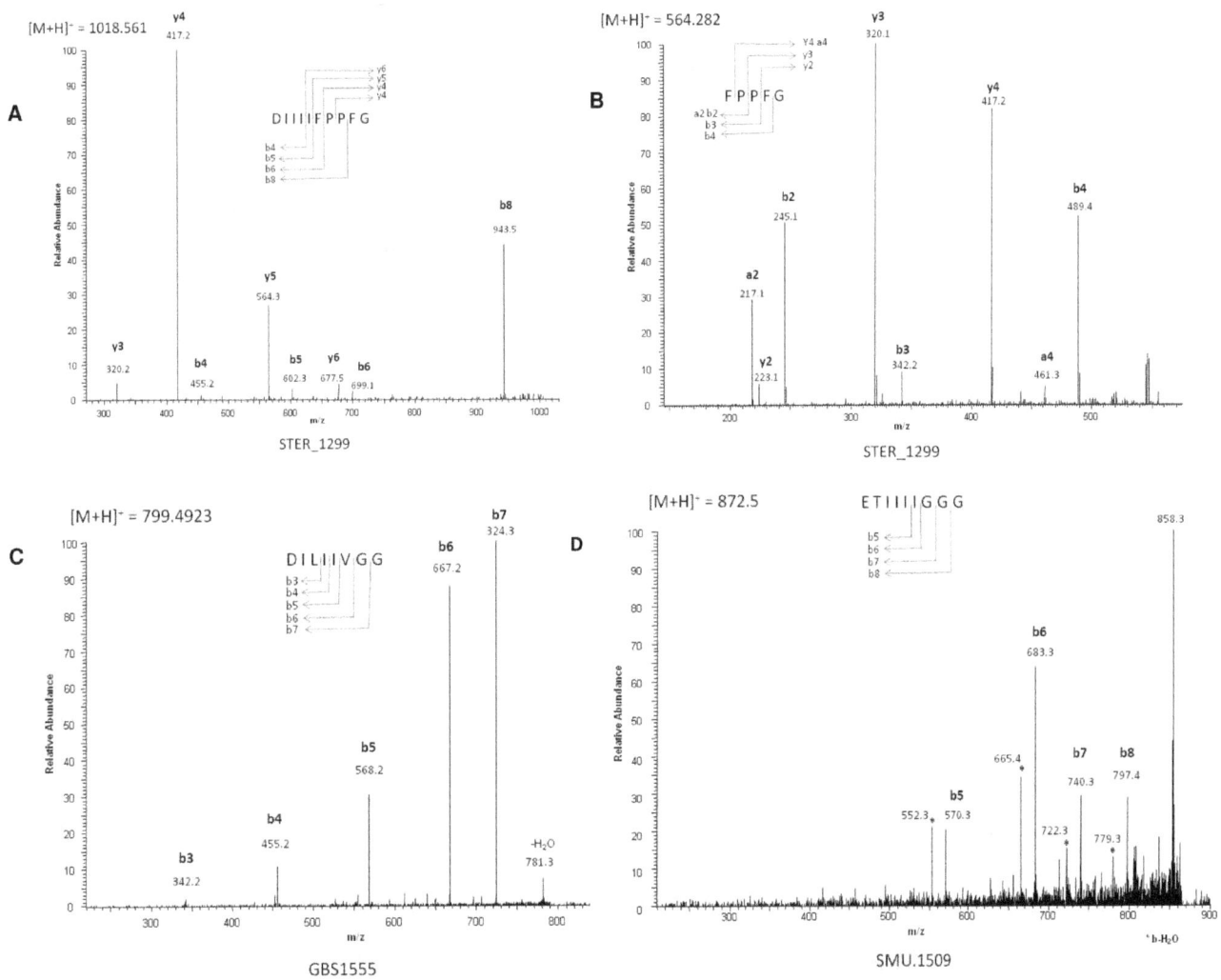

Figure 3. Fragmentation spectra of the ions of mature forms of SHP1299, SHP1555 and SHP1509. Fragmentation of the ions m/z 1018.56 (A) and m/z 564.28 (B) identified in the supernatant of cultures of *S. thermophilus* strain LMD-9. Fragmentation of the ions m/z 799.49 (C) identified in the supernatant of cultures of *S. agalactiae* strain NEM316 and m/z 872.5 (D) identified in the supernatant of cultures of *S. mutans* strain UA159. All ions were analyzed in the linear ion trap.

of Gbs1555, also called RovS, in virulence has been studied but without taking into account the existence of its cognate SHP [33]. Moreover, the amino acid sequence of the predicted mature SHP of *S. agalactiae* is identical to those of the SHPs of *Streptococcus dysgalactiae* subsp. *equisimilis* and of SHP2 of *S. pyogenes* (Table S2). For group II, we chose the locus *shp/ster_1358* (*rgg1358*) of *S. thermophilus* strain LMD-9, already studied in detail [19,21], and the locus *shp/SMU.1509* (*rgg1509*) of *Streptococcus mutans* strain UA159 present in all sequenced strains of *S. mutans*. For group III, we chose the locus *shp/ster_1299* (*rgg1299*) of *S. thermophilus* strain LMD-9 also found in strain JIM8232, in *Streptococcus oralis* strain SK60 and *Streptococcus tigurinus* strain 1368. These loci thus correspond to three SHP/Rgg mechanisms that have not previously been studied, including two in pathogenic streptococci of two different streptococci groups: mutans (*S. mutans*) and pyogenic (*S. agalactiae*). These pathogenic streptococci are found in the same niche in the oral cavity [28] and are therefore likely to encounter each other, and also, at least briefly, *S. thermophilus*, a species of the salivarius group that is one of the two starters used to produce yogurt. First, we studied the QS mechanisms of the loci that had not previously been studied.

SHP/Rgg Mechanisms in Different Species of Streptococci Function Similarly

Analysis of the *shp/ster_1358* locus of *S. thermophilus* showed that the SHP1358 peptide, the Rgg1358 transcriptional regulator and the Ami oligopeptide transporter are essential for a QS mechanism that positively controls the transcription of the *shp1358* gene, creating a positive feedback loop [21]. We tested whether the auto-induction of *shp* gene expression was conserved in the SHP/Gbs1555 system of *S. agalactiae* strain NEM316, the SHP/SMU.1509 system of *S. mutans* strain UA159 and the SHP/Ster_1299 system of *S. thermophilus* strain LMD-9 (Table 3). We evaluated the activity of the *shp* promoter of each locus in *S. thermophilus* strain LMD-9, in the presence and absence of the genes encoding the three partners, SHP, Rgg and Ami. Thus, for locus *shp/ster_1299* of *S. thermophilus*, a P*shp1299*-*luxAB* fusion was introduced into the wild-type strain LMD-9 and the Δ*rgg1299*, Δ*shp1299* and Δ*amiCDE* isogenic mutants. For *shp/gbs1555* of *S.*

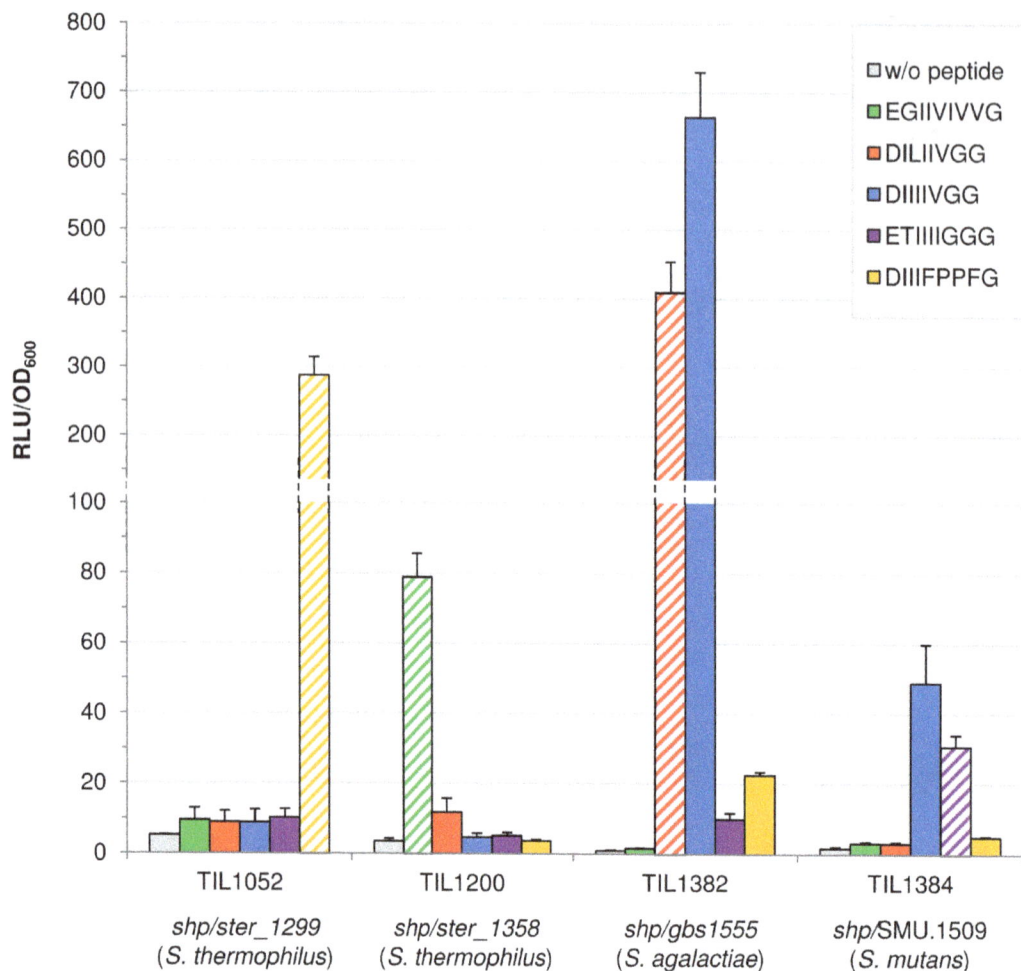

Figure 4. Cross-complementation of the *shp/rgg* loci with synthetic SHP pheromones. Maximum relative luciferase activities of the reporter strains TIL1052 (*shp1299::erm blp::P_shp1299-luxAB aphA3*), TIL1200 (*Δshp1358 blp::P_shp1358-luxAB*), TIL1382 (*blp::gbs1555::P_shp1555-luxAB aphA3*) and TIL1384 (*blp::SMU.1509::P_shp1509-luxAB aphA3*) grown in the absence (grey) or in the presence of synthetic SHP peptides added at the beginning of the culture to a concentration of 1 μM: EGIIVIVVG (green), DILIIVGG (red), DIIIIVGG (blue), ETIIIIGGG (purple), DIIIFPPFG (yellow). The legitimate SHP synthetic peptide associated to the locus studied is hatched in each case.

agalactiae and *shp*/SMU.1509 of *S. mutans*, the promoter of the *shp* gene and the *shp* genes were fused, independently, to the *luxAB* genes with or without the cognate *rgg* gene and then introduced into *S. thermophilus* strain LMD-9 or its isogenic *ΔamiCDE* mutant (Figure 1). For all three loci, luciferase activity was detected when *rgg*, *shp* and *amiCDE* genes were all present (TIL1345, TIL1383, TIL1038, Figure 2A–C). If one of the genes was absent, the expression of the three P_shp promoters was undetectable (Figure 2A–C) except for the *shp/gbs1555* locus of *S. agalactiae* studied in the *Δami* genetic background (TIL1381, Figure 2A): the relative luciferase activity in this case was one quarter of that in the wild-type genetic background (TIL1345). These experiments clearly demonstrated that the SHP pheromone, the Rgg regulatory protein of each locus and the Ami transporter of strain LMD-9 are required for strong expression of the three *shp* genes in the condition tested.

The pattern of activity of the different *shp* promoters differed: expression of luciferase activity started either in the middle (*shp/gbs1555* and *shp/ster_1299*, Figure 2A and C, respectively) or at the end (*shp*/SMU.1509 - Figure 2B) of the exponential growth phase. Once the maximal activity was reached, it was maintained until the end of the exponential growth phase for P_shp1555 (Figure 2A)

but was transient for the two other promoters studied, P_shp1509 and P_shp1299 (Figure 2B and C).

These analyses indicate that the SHP/Rgg systems of both pathogenic and non-pathogenic streptococci function in a similar way and appear to be temperature (data not shown, see Materials and Methods) and growth phase-dependent in *S. thermopilus* strain LMD-9.

The Mature Forms of SHP are Released by Cleavage of the C-terminal Part of their Precursor in Front of a Conserved Acid Residue

The mature form of the SHP1299 of *S. thermophilus* was sought directly in the supernatant of the wild-type strain LMD-9. Without purification or concentration, direct analysis of the supernatant of *S. thermophilus* LMD-9 by LC-MS/MS identified two masses corresponding to the fragments, DIIIFPPFG and FPPFG, of the SHP1299 precursor. These sequences were confirmed by fragmentation (Figure 3A–B). We used mass spectrometry to identify the sequences of the mature forms of the SHP1555 of *S. agalactiae* and SHP1509 of *S. mutans* in the supernatants of the *S. thermophilus* strains TIL1345 and TIL1383. A mass corresponding to the octapeptide DILIIVGG was identified as the mature form of the *S. agalactiae* SHP1555 produced by *S. thermophilus*. The sequence of this peptide was also confirmed by fragmentation. No mass corresponding to fragments of the precursor SHP1509 of *S. mutans* was found in supernantant of *S. thermophilus* strain TIL1383. A similar method was used to identify mature SHP peptides in the supernatants of the wild-type strains *S. agalactiae* NEM316 and *S. mutans* UA159. The production of the octapeptide DILIIVGG was confirmed for *S. agalactiae* (Figure 3C) but once again, we did not find any mass corresponding to fragments of the SHP1509 of *S. mutans*. Therefore, we predicted by analogy, that the mature SHP peptide produced by *S. mutans* was ETIIIIGGG and we used a more sensitive mass spectrometry approach (based on MS2) to detect this peptide in the supernatant of *S. mutans* UA159. This approach successfully detected one mass corresponding to this peptide (Figure 3D).

To check that the longest peptides identified by mass spectrometry for the three loci were active, synthetic peptides with these three sequences were produced. These synthetic peptides were added to cultures of reporter strains containing a P_{shp}-*luxAB* fusion of the corresponding locus but not encoding the cognate SHP. In all cases, the synthetic peptide functionally complemented the reporter strain (Figure 4 - TIL1052, TIL1382 and TIL1384 hatched bars).

We used mass spectrometry to evaluate the amount of SHP1358 naturally present in cultures of *S. thermophilus* LMD-9 at the end of the exponential phase of growth, which is when its gene is maximally expressed. The heavy form of SHP1358 [NH2-EGII[V_C^{13}N^{15}]IVVG-OH] (Thermo, Scientific) was used as an internal standard. The concentration of SHP present in the culture supernatant was estimated to be $7+/-3$ ng ml^{-1}.

The SHP Pheromones Allow Cross-talk between Streptococci

SHP/Rgg QS mechanisms are widespread among species of streptococci and the amino acid sequences of the various Rgg and SHP proteins are similar (Table S2). We therefore investigated the existence of cross-talk phenomena. We used functional complementation experiments to determine whether *shp/rgg* loci can be regulated by mature SHPs with an amino acid sequence different from that of its cognate mature SHP. The four selected loci (Table 3) were studied in four reporter strains that cannot produce the cognate SHP pheromone: TIL1052 (*shp1299::erm blp::*$P_{shp1299}$-*luxAB aphA3*), TIL1200 (Δ*shp1358 blp::*$P_{shp1358}$-*luxAB*), TIL1382 (*blp::gbs1555::*$P_{shp1555}$-*luxAB aphA3*) and TIL1384 (*blp::SMU.1509::*$P_{shp1509}$-*luxAB aphA3*). Five synthetic peptides were used: four corresponding to these *shp/rgg* loci (Table 3) and one with the sequence DIIIIVGG, which is identical to the SHP of *S. pyogenes*, called SHP3, and differs from one amino acid to that of *S. agalactiae* (Table S2). These five peptides were added independently to cultures of the four reporter strains. For the two loci of *S. thermophilus* (TIL1052 and TIL1200), no significant induction of luciferase activity was detected with any of the non-cognate synthetic peptides. However, for the two loci from the two pathogenic streptococci (TIL1382 and TIL1384), luciferase activity was induced more strongly by the illegitimate peptide DIIIIVGG than by the cognate peptide (hatched bars); the other synthetic SHPs had no detectable effects (Figure 4).

Discussion

SHP/Rgg cell-cell communication mechanism has been deciphered using the SHP/Rgg1358 locus of *S. thermophilus* from group II [21] and the SHP2/Rgg2 and SHP3/Rgg3 of *S. pyogenes* from

group I as models [23,24]. We have increased the number of examples of this mechanism by studying another locus in *S. thermophilus* (group III), one of *S. agalactiae* (group I) and one of *S. mutans* (group II). We confirmed that the *shp* genes are targets of the mechanism that relies on Rgg, SHP and Ami. These validations were performed in *S. thermophilus* i.e. in a heterologous background for the *shp/rgg* loci of *S. agalactiae* and *S. mutans*. Thus, *S. thermophilus* is able to secrete, process and import SHPs from other species efficiently. Except for the *shp1555* gene of *S. agalactiae* in the *ami*-deleted mutant, no expression of the three *shp* genes was observed in the *shp-*, *rgg-* or *ami*-deleted mutants although *S. thermophilus* strain LMD-9 contains six *shp/rgg* loci, including at least two that are active in our conditions. This implies that the interactions between SHP and Rgg and between Rgg and its DNA target are highly specific. The expression of the *shp1555* gene in the *ami*-deleted mutant was only one quarter of that in the wild-type genetic background; possibly, the precursor SHP is able to activate the Rgg regulator and bypass the *ami* deletion or the SHP precursor is processed intracellularly and able to activate the Rgg regulator. The SHP/Rgg mechanism for the *S. mutans* and *S. agalactiae* systems need to be confirmed in their homologous backgrounds. Appropriate experiments are in progress with the *shp/gbs1555* locus of *S. agalactiae* (D. Perez-Pascual, unpublished results). It is extremely likely that the proposed mechanism involving the *shp*, *rgg* and *ami* genes in these pathogenic bacteria will be confirmed, but it would be interesting to document the kinetics of expression of *shp*. In *S. thermophilus*, expression of the *shp* gene of *S. agalactiae* started early during the exponential phase of growth whereas that of the *shp* gene of *S. mutans* started at the beginning of the stationary phase. If confirmed in the homologous background, it will suggest that there are additional components contributing to the control the expression of the *shp* genes. In conclusion, this validation of the cell-cell communication mechanism for three new *shp/rgg* loci, including one from a group that had not previously been studied (group III), suggests that the main components (SHP, Rgg and Ami) are conserved in all SHP/Rgg mechanisms.

The mature forms of SHP1299, SHP1555 and SHP1509 identified directly in *S. thermophilus*, *S. agalactiae* and *S. mutans* supernatants were each the C-terminal part of the precursor and start with an acid amino acid (Asp or Glu). These amino acids are conserved in nearly all SHP identified from streptococci genome sequences; the exceptions are one in *S. thermophilus* strain LMG18311 (Rgg Stu0182-associated SHP) and another in strain CNRZ1066 (Rgg Str0182-associated SHP) that contain a Cys residue at this position [21]. Naturally secreted SHP1358 also starts with a Glu [21]. The activity of SHP2 in *S. pyogenes* is maintained if the Asp amino acid at this position is substituted with a Glu, but not with an amide-bearing residue [23]. Therefore, all mature SHPs are expected to have an Asp or Glu at their N terminus. These conserved residues seem to be required for the recognition of the precursor by the protease involved in their maturation and the activity of the mature SHP. The Eep membrane protease is involved in the production of mature SHP by *S. thermophilus* [21] and *S. pyogenes* [23]. It has not been established whether or not this role is direct. Nevertheless, Eep-encoding genes are present in all streptococci genomes, and map in a conserved environment, so it is highly probable that this role is common to all streptococci. The amino acid sequence of the SHP1555 of *S. agalactiae* produced by *S. thermophilus* and by *S. agalactiae* were identical, consistent with the conservation of the role of Eep, and the maturation more generally. In *Enterococcus faecalis*, the sex pheromones are matured by Eep [34,35,36], but there is no conservation of such acid amino acids. The maturation of the

XIP, another family of signaling peptides produced by streptococci and that are involved in the triggering of competence, seems less well conserved. Indeed, Eep is involved in the production of the XIP of *S. thermophilus* but not in that of *S. mutans*. The sequences of the XIP peptides are less conserved than those of the SHP peptide, and this may explain the involvement of different proteases in their maturation.

We detected a shorter mature form of SHP1299 in the supernatant of *S. thermophilus*. Shorter forms were not detected in other supernatants but it does not indicate that there are not present in small amounts. This probably means that these linear non-modified peptides are subject to degradation by, at least, aminopeptidases present in the extracellular medium. We have already observed such N-terminal degradation with ComS [37] indicating the existence of a significant aminopeptidase activity at the surface of streptococci.

The cross-talk experiments with four *shp/rgg* loci and five synthetic SHPs showed a generally high specificity of the SHP/Rgg interaction. Only one peptide, SHP3 from *S. pyogenes*, was able to induce the expression of the *shp* genes from the other species. This result for *S. agalactiae* was not surprising because the two peptides differ only at the third residue, and the difference is very minor (DI<u>II</u>IVGG/DIL<u>II</u>VGG). *S. pyogenes* encodes both peptides, and they can stimulate the expression of their targets to similar levels [23]. The cross-talk result with *S. mutans* was more surprising: the amino acid sequences of the SHPs are more divergent and they do not belong to the same group despite both containing a hydrophobic stretch of isoleucines (DI<u>III</u>VGG/ET<u>III</u>GGG). The SHP of *S. agalactiae* was not able to cross talk with the *S. mutans* system and *vice versa* indicating that the presence of the four isoleucine residues are critical only for the *S. mutans* system, and that the specificity of the interaction is complex. Mature SHP3 peptide can be produced by three different species of streptococci *i.e. pyogenes*, *pneumoniae* and *thermophilus*, and the mature SHP peptide of *S. agalactiae* can be produced by two other species of streptococci, *pyogenes* (SHP2) and *S. dysgalactiae* (Table S2). This suggests that if present in the same environment, these streptococci can potentially interact with each other through their SHP/Rgg systems. It would be interesting to investigate this possibility with co-cultures in an ecosystem model. Such interactions may be of great significance to the co-operation or competition between streptococci species.

Acknowledgments

We thank David Pérez-Pascual for critical reading of the manuscript.

Author Contributions

Conceived and designed the experiments: BF RG VM AG. Performed the experiments: BF AG CM CB EC. Analyzed the data: BF RG VM AG EC. Contributed reagents/materials/analysis tools: AG CM CB EM. Wrote the paper: BF RG AG VM.

References

1. Schuster M, Greenberg EP (2006) A network of networks: quorum-sensing gene regulation in *Pseudomonas aeruginosa*. Int J Med Microbiol 296: 73–81.
2. Novick RP, Geisinger E (2008) Quorum sensing in Staphylococci. Annu Rev Genet 42: 541–564.
3. Dubern JF, Lugtenberg BJ, Bloemberg GV (2006) The *ppuI-rsaL-ppuR* quorum-sensing system regulates biofilm formation of *Pseudomonas putida* PCL1445 by controlling biosynthesis of the cyclic lipopeptides putisolvins I and II. J Bacteriol 188: 2898–2906.
4. Higgins D, Dworkin J (2012) Recent progress in *Bacillus subtilis* sporulation. FEMS Microbiol Rev 36: 131–148.
5. Fuqua C, Greenberg EP (2002) Listening in on bacteria: acyl-homoserine lactone signalling. Nat Rev Mol Cell Biol 3: 685–695.
6. Altstein M (2004) Peptide pheromones: an overview. Peptides 25: 1373–1376.
7. Lazazzera BA (2001) The intracellular function of extracellular signaling peptides. Peptides 22: 1519–1527.
8. Claverys JP, Prudhomme M, Martin B (2006) Induction of competence regulons as a general response to stress in gram-positive bacteria. Annu Rev Microbiol 60: 451–475.
9. Rocha-Estrada J, Aceves-Diez AE, Guarneros G, de la Torre M (2010) The RNPP family of quorum-sensing proteins in Gram-positive bacteria. Appl Microbiol Biotechnol 87: 913–923.
10. Ji G, Beavis R, Novick RP (1997) Bacterial interference caused by autoinducing peptide variants. Science 276: 2027–2030.
11. Jarraud S, Lyon GJ, Figueiredo AM, Lina G, Vandenesch F, et al. (2000) Exfoliatin-producing strains define a fourth agr specificity group in *Staphylococcus aureus*. J Bacteriol 182: 6517–6522.
12. Pozzi G, Masala L, Iannelli F, Manganelli R, Havarstein LS, et al. (1996) Competence for genetic transformation in encapsulated strains of *Streptococcus pneumoniae*: two allelic variants of the peptide pheromone. J Bacteriol 178: 6087–6090.
13. Tran LS, Nagai T, Itoh Y (2000) Divergent structure of the ComQXPA quorum-sensing components: molecular basis of strain-specific communication mechanism in *Bacillus subtilis*. Mol Microbiol 37: 1159–1171.
14. Tortosa P, Logsdon L, Kraigher B, Itoh Y, Mandic-Mulec I, et al. (2001) Specificity and genetic polymorphism of the *Bacillus* competence quorum-sensing system. J Bacteriol 183: 451–460.
15. Bouillaut L, Perchat S, Arold S, Zorrilla S, Slamti L, et al. (2008) Molecular basis for group-specific activation of the virulence regulator PlcR by PapR heptapeptides. Nucleic Acids Res 36: 3791–3801.
16. Slamti L, Lereclus D (2005) Specificity and polymorphism of the PlcR-PapR quorum-sensing system in the *Bacillus cereus* group. J Bacteriol 187: 1182–1187.
17. Gordon CP, Williams P, Chan WC (2012) Attenuating *Staphylococcus aureus* Virulence Gene Regulation: a Medicinal Chemistry Perspective. J Med Chem [Epub ahead of print].
18. Stefanic P, Decorosi F, Viti C, Petito J, Cohan FM, et al. (2012) The quorum sensing diversity within and between ecotypes of *Bacillus subtilis*. Environ Microbiol 14: 1378–1389.
19. Ibrahim M, Guillot A, Wessner F, Algaron F, Besset C, et al. (2007) Control of the transcription of a short gene encoding a cyclic peptide in *Streptococcus thermophilus*: a new quorum-sensing system? J Bacteriol 189: 8844–8854.
20. Ibrahim M, Nicolas P, Bessières P, Bolotin A, Monnet V, et al. (2007) A genome-wide survey of short coding sequences in streptococci. Microbiology 153: 3631–3644.
21. Fleuchot B, Gitton C, Guillot A, Vidic J, Nicolas P, et al. (2011) Rgg proteins associated with internalized small hydrophobic peptides: a new quorum-sensing mechanism in streptococci. Mol Microbiol 80: 1102–1119.
22. Henry R, Bruneau E, Gardan R, Bertin S, Fleuchot B, et al. (2011) The *rgg0182* gene encodes a transcriptional regulator required for the full *Streptococcus thermophilus* LMG18311 thermal adaptation. BMC Microbiol 11: 1–13.
23. Chang JC, LaSarre B, Jimenez JC, Aggarwal C, Federle MJ (2011) Two Group A streptococcal peptide pheromones act through opposing Rgg regulators to control biofilm development. PLoS Pathog 7: 1–16.
24. Lasarre B, Aggarwal C, Federle MJ (2012) Antagonistic Rgg regulators mediate quorum sensing via competitive DNA binding in *Streptococcus pyogenes*. MBio 3: e00333–00312.
25. Kawamura Y, Hou XG, Sultana F, Miura H, Ezaki T (1995) Determination of 16S rRNA sequences of *Streptococcus mitis* and *Streptococcus gordonii* and phylogenetic relationships among members of the genus Streptococcus. Int J Syst Bacteriol 45: 406–408.
26. Zadoks RN, Gonzalez RN, Boor KJ, Schukken YH (2004) Mastitis-causing streptococci are important contributors to bacterial counts in raw bulk tank milk. J Food Prot 67: 2644–2650.
27. Dewhirst FE, Chen T, Izard J, Paster BJ, Tanner AC, et al. (2010) The human oral microbiome. J Bacteriol 192: 5002–5017.
28. Chen T, Yu WH, Izard J, Baranova OV, Lakshmanan A, et al. (2010) The Human Oral Microbiome Database: a web accessible resource for investigating oral microbe taxonomic and genomic information. Database 2010: baq013.
29. Gevers D, Knight R, Petrosino JF, Huang K, McGuire AL, et al. (2012) The Human Microbiome Project: a community resource for the healthy human microbiome. PLoS Biol 10: e1001377.

30. Letort C, Juillard V (2001) Development of a minimal chemically-defined medium for the exponential growth of *Streptococcus thermophilus*. J Appl Microbiol 91: 1023–1029.

31. Gardan R, Besset C, Guillot A, Gitton C, Monnet V (2009) The oligopeptide transport system is essential for the development of natural competence in *Streptococcus thermophilus* strain LMD-9. J Bacteriol 191: 4647–4655.

32. Débarbouillé M, Arnaud M, Fouet A, Klier A, Rapoport G (1990) The *sacT* gene regulating the *sacPA* operon in *Bacillus subtilis* shares strong homology with transcriptional antiterminators. J Bacteriol 172: 3966–3973.

33. Samen UM, Eikmanns BJ, Reinscheid DJ (2006) The transcriptional regulator RovS controls the attachment of *Streptococcus agalactiae* to human epithelial cells and the expression of virulence genes. Infect Immun 74: 5625–5635.

34. An FY, Sulavik MC, Clewell DB (1999) Identification and characterization of a determinant (*eep*) on the *Enterococcus faecalis* chromosome that is involved in production of the peptide sex pheromone cAD1. J Bacteriol 181: 5915–5921.

35. An FY, Clewell DB (2002) Identification of the cAD1 sex pheromone precursor in *Enterococcus faecalis*. J Bacteriol 184: 1880–1887.

36. Chandler JR, Dunny GM (2008) Characterization of the sequence specificity determinants required for processing and control of sex pheromone by the intramembrane protease Eep and the plasmid-encoded protein PrgY. J Bacteriol 190: 1172–1183.

37. Gardan R, Besset C, Gitton C, Guillot A, Fontaine L, et al. (2013) The extracellular life cycle of ComS, the competence stimulating peptide of *Streptococcus thermophilus*. J Bacteriol: 195: 1845–1855.

38. Makarova K, Slesarev A, Wolf Y, Sorokin A, Mirkin B, et al. (2006) Comparative genomics of the lactic acid bacteria. Proc Natl Acad Sci U S A 103: 15611–15616.

39. Ajdic D, McShan WM, McLaughlin RE, Savic G, Chang J, et al. (2002) Genome sequence of *Streptococcus mutans* UA159, a cariogenic dental pathogen. Proc Natl Acad Sci U S A 99: 14434–14439.

40. Glaser P, Rusniok C, Buchrieser C, Chevalier F, Frangeul L, et al. (2002) Genome sequence of *Streptococcus agalactiae*, a pathogen causing invasive neonatal disease. Mol Microbiol 45: 1499–1513.

41. Biswas I, Gruss A, Ehrlich SD, Maguin E (1993) High-efficiency gene inactivation and replacement system for gram-positive bacteria. J Bacteriol 175: 3628–3635.

11

Pyrokinin β-Neuropeptide Affects Necrophoretic Behavior in Fire Ants (*S. invicta*), and Expression of β-NP in a Mycoinsecticide Increases Its Virulence

Yanhua Fan[1,3], Roberto M. Pereira[2], Engin Kilic[3], George Casella[3], Nemat O. Keyhani[3]*

1 Biotechnology Research Center, Southwest University, Beibei, Chongqing, People's Republic of China, **2** Department of Entomology and Nematology, University of Florida, Gainesville, Florida, United States of America, **3** Department of Microbiology and Cell Science, University of Florida, Gainesville, Florida, United States of America

Abstract

Fire ants are one of the world's most damaging invasive pests, with few means for their effective control. Although ecologically friendly alternatives to chemical pesticides such as the insecticidal fungus *Beauveria bassiana* have been suggested for the control of fire ant populations, their use has been limited due to the low virulence of the fungus and the length of time it takes to kill its target. We present a means of increasing the virulence of the fungal agent by expressing a fire ant neuropeptide. Expression of the fire ant (*Solenopsis invicta*) pyrokinin β -neuropeptide (β-NP) by *B. bassiana* increased fungal virulence six-fold towards fire ants, decreased the LT_{50}, but did not affect virulence towards the lepidopteran, *Galleria mellonella*. Intriguingly, ants killed by the β-NP expressing fungus were disrupted in the removal of dead colony members, i.e. necrophoretic behavior. Furthermore, synthetic C-terminal amidated β-NP but not the non-amidated peptide had a dramatic effect on necrophoretic behavior. These data link chemical sensing of a specific peptide to a complex social behavior. Our results also confirm a new approach to insect control in which expression of host molecules in an insect pathogen can by exploited for target specific augmentation of virulence. The minimization of the development of potential insect resistance by our approach is discussed.

Editor: Jae-Hyuk Yu, University of Wisconsin – Madison, United States of America

Funding: This work was supported in part by a USDA grant (2010-34135-21095) and a University of Florida IFAS Innovation Award to NOK. The funders had no role in study design, data collection and analysis, decision to publish, or preparation of the manuscript.

Competing Interests: The authors have declared that no competing interests exist.

* E-mail: keyhani@ufl.edu

Introduction

The spread of fire ants is considered a classic example of world-wide biological invasions of a species into previously unoccupied habitats with the potential to result in significant ecosystem alterations. The red imported fire ant (*Solenopsis invicta*), native to South America, is considered by the World Conservation Unit as one of the top 100 worst invasive alien species, and its detrimental impact on humans, domestic and wild animals, agriculture, and ecosystems is well-documented [1,2,3]. It is a major invasive pest insect to almost the entire Southeastern United States and continues to expand it range north and westwards causing agricultural and ecosystem disruptions that extend from crop losses to declines of native species [4]. Fire ant have continued to spread despite the treatment of over 56 million hectares with Mirex bait alone and tons of other chemical insecticides [5], which themselves have significant damaging environmental consequences. Biological control of fire ants using entomopathogenic fungi, such as *Beauveria bassiana*, offers a more environmentally friendly alternative to chemical pesticides [6,7,8,9]. The use of entomopathogenic fungi, however, has met with limited success partially due to the relatively long time (3–10 days) it can take for the fungus to kill target insects. Ants have posed a particular challenge due to communal behaviors such as grooming and nest cleaning which can decrease the efficacy of microbial agents [10]. Previous work has shown that the potency of fungal insecticides can be improved

[11]. Expression of a 70 amino acid scorpion (*Androctonus australis*)-derived neurotoxin in the fungal insect pathogen, *Metarhizium anisopliae*, increased its toxicity 9-fold against *Aedes aegyptii* as compared to its wild-type parent [12]. Here, we sought to use a different approach, namely to express host molecules, e.g. hormones or neuropeptides, in the fungal pathogen. As the fungus targets the insect, it will produce the host molecule, disrupting the normal endocrine or neurological balance of the host. The desired outcome is to make the target (fire ant) more susceptible to the invading fungus, thus increasing the potency of the fungal agent. As candidates for expression in the fungus we sought to use a recently described strategy whereby peptides that participate in a critical host physiological process are used [13]. Depending upon the molecule (peptide) chosen, in theory, the increased virulence can, to a particular degree, be host specific, thus minimizing non-target effects.

The pyrokinin/pheromone biosynthesis activating neuropeptide (PBAN) family consists of insect neurohormones characterized by the presence of a C-terminal FXPRL amine sequence [14,15]. First isolated from the cockroach, *Leucophaea maderae*, as a myotropic (visceral muscle contraction stimulatory) peptide, members of this peptide family are widely distributed within the Insecta, where depending upon the species, they function in a diverse range of physiological processes that includes stimulation of pheromone biosynthesis, melanization, acceleration of pupariation, and induction and/or termination of diapause [16,17,18]. In the natural insect host, these peptides are C-terminal amidated, a modification often required for their activity. In Lepidoptera, the PBAN peptide is encoded on a translated ORF that is subsequently processed (cleaved) to yield diapause hormone (DH), and the α-, β-, and γ-neuropeptides, along with the PBAN peptide itself (which is found between the β- and γ-neuropeptides). More recently, isolation of a cDNA sequence for the fire ant, *S. invicta*, led to the identification of PBAN and related peptide homologs [19]. Analysis of the ORF revealed the presence of DH, as well β- and γ-neuropeptide homologs, but no α-neuropeptide.

Here, we assessed the impact of expressing the β-NP peptide in the fungal insect pathogen *B. bassiana*. Our data show a decrease in both the lethal dose (LD_{50}) and lethal time (LT_{50}) it takes to kill target fire ants in the β-NP expressing strain as compared to its wild-type parent. The effect was host specific, and no increase in virulence was noted when the strain was tested against the greater wax moth, *Galleria mellonella*. By using a host molecule the chances of resistance are minimized due to the simple fact that the fungal-expressed peptide represents a host molecule that is regulated in both tissue specific and developmental patterns. Any mutations that could compensate for the increased dose given by the fungus during infection would be significantly compromised, indeed, potentially dependent upon the fungus for proper development. Unexpectedly, we observed that the cadavers of ants killed by the β-NP expressing *B. bassiana* strain were treated differently, i.e. removed slower, than controls or those killed by the WT fungus. Experiments testing the effects of synthetic peptides on cadaver removal or necrophoretic behavior resulted in another serendipitous result, namely that ant cadavers treated with the β-NP-NH_2 peptide were removed much more rapidly than β-NP or control treated cadavers. The implications of these results in terms of biological control of ants and chemical sensing are discussed.

Materials and Methods

Construction of expression vector and fungal transformation

The *S. invicta* pyrokinin β-neuropeptide (β-NP, QPQFTPRL) was fused to a 28-amino acid signal peptide derived from the *B. bassiana* chitinases-1 (*chit1*) gene [20] and cloned under control of the *B. bassiana* glyceraldehyde phosphate dehydrogenase promoter (P_{gpd-Bb}). Primer pairs P1/P2 (5'-GTTGGGTATGCTCCG-GCGCG, & 5'-GGTTGTTATTGATTAAAAGG) were used to amplify $P_{gpdA-Bb}$ using *B. bassiana* genomic DNA as templates. The *B. bassiana* chitinase (*Bbchit1*) derived signal peptide (SP) was obtained with the primer pair P3/P4; (P3, 5'-CCTTTTAAT-CAATAACAACCATGGCTCCTTTTCTTCAAAC & P4, 5'-TTAGAGGCGGGGGGTAAACTGGGGCTGTCGCGGCGC-CAAGGGCGAGG) using *B. bassiana* genomic DNA as the template and with the β-NP coding sequence incorporated into primer P4. These primers were designed containing a 20 bp overlap sequences between $P_{gpdA-Bb}$ and SP: β-NP. The desired construct ($P_{gpdA-Bb}$:β-NP) was produced via primer-less assembly in a reaction mixture containing: 5 µl 5×Phusion Taq polymerase buffer, 2 µl 2.5 mM dNTP, 30 ng $P_{gpdA-Bb}$, 30 ng SP: β-NP, 0.4 U Phusion Taq DNA polymerase, total volume 25 µl. PCR reaction cycling conditions: 98°C (2 min); followed by 25 cycles of: 98°C (20 s), 56°C (30 sec), 72 (1 min); and 72°C (5 min). Primer pair P1 & P4 were used to obtain the $P_{gpdA-Bb}$:*SP*- β-NP fragment using the assembled product as template. The obtained fragments were cloned into pDrive vector (Qiagen) and verified by sequencing. $P_{gpdA-Bb}$:SP-β-NP was subcloned from pDrive vector via *EcoR*I restriction sites into pUC-Bar, yielding pUC-Bar-$P_{gpdA-Bb}$:SP-β-NP. This plasmid was linearized with *Xba*I and transformed into *B.*

bassiana competent cells as described [21]. The resultant strain was labeled Bb::spβ-NP_{gpd}.

Purification and identification of β-NP from fungi cultures using HPLC and MS/MS

In order to verify (extracellular) β-NP production in the recombinant *B. bassiana* strain, fungal cultures (Bb::spβ-NP_{gpd} and the WT parent) were grown first grown in SDBY (Sabouraud dextrose broth with 0.5% yeast extract) for 2 d, after which 1.5 g of washes cells were transferred to Czapek-dox broth (50–100 ml) for 3 days. Fungal cells were removed by centrifugation, the resultant supernatant filtered through a 0.22 µm filter, and the supernatant samples subsequently lyophilized and stored at −20°C until used. Lyophilized samples were rehydrated in 3.0 ml of water containing 0.1% TFA, and applied onto a C_{18} reverse phase SepPak column. The column was washed with 0.1% TFA and peptides were eluted with 80% acetonitrile-0.1% TFA. The eluted fraction (in acetonitrile) was dried in a SpeedVac, resuspended in water-0.1% TFA (0.5 ml) and chromatographed on a C_{18} reversed phase HPLC column with eluting factions monitored via absorbance at 214 nm. Fractions eluting at the same retention time as an initial run using synthetic β-NP used as a standard, were collected, dried with a fine stream of N_2, rehydrated to 0.2 ml with water-0.1% TFA, and rechromato-graphed as above. Fractions were collected as above, dried under N_2 and analyzed by LC-MS/MS (University of Florida, Dept. of Chemistry, analytical Services). A standard curve using synthetic β-NP was made in order to quantify the amount of peptide in the sample.

Insect Bioassays

S. invicta colonies were collected from the field, separated from the soil by drip flotation and maintained in Fluon-coated trays with a diet consisting of 10% sucrose solution, a variety of freeze-killed insects, fruits and vegetables, and chicken eggs. Fungal cultures were grown on potato dextrose agar (PDA). Plates were incubated at 26°C for 14–21 d, and aerial conidia (spores) were harvested by flooding or scraping the plates with sterile distilled H_2O containing 0.05% Tween 80. Spores concentrations were determined by direct count using a hemocytometer and adjusted to the desired concentration for use (typically between 10^6–10^8 conidia/ml). Two types of bioassays were used to assess the virulence of the fungal strains: (1) "classical bioassay" using *S. invicta* workers. Test groups of ants (25/chamber) were inoculated with fungal suspensions (concentrations ranging from 10^6–10^8 conidia/ml) using a spray tower as described [22]. The ants were housed in plastic cups (ø = 6 cm) whose sides had been coated with Fluon and topped with a perforated lid. Ants were given 10% sucrose solutions in 1.5 ml Eppendorf tubes with a cotton plug. Experiments were performed at 26°C and mortality was recorded daily. Controls were treated with Tween-80 and the mortality assays were repeated at least three times. (2) "Mock mini-mound" assays. Larger scale bioassays were performed using larger test chambers (ø = 19 cm). Test chambers contained a small Petri dish (ø = 3 cm) containing moist dental plaster, that served as the nest for the mini-mound. Ants (0.5 gm, ~2,000 individuals) including 3–4 dealate reproductive females were placed in the test chamber that included a 10% sucrose solution in an Eppendorf tube. Treatments and assay conditions were identical to the classical bioassay. Duplicate samples were performed for each experiment and the entire assay repeated three times with independent batches of fungal spores. For all experiments, a χ^2-test was first used to determine homogeneity among variance of the repeats ($p < 0.05$). Further statistical analysis of the mortality was performed using SPSS which was used to estimate the median lethal time (LT_{50}), the median lethal concentration (LC_{50}), fiducial limits and other regression parameters.

Necrophoretic behavior assays

Assay chambers and methods were based upon a previously described protocol [23]. Briefly, the conical end of a 15 ml polypropylene tube (nest) was cut off and connected via a short tubing (ø = 8 mm, 10 cm long) to a round plastic container (ø = 19 cm, foraging arena) into which a hole had been punched out in the bottom at the middle of the container. Test ants (0.1 gm, ~400 ants with at least one dealate) were placed in the assay chamber and allowed to equilibrate for 1–2 hr before the experiment was initiated. Three separate experimental protocols were employed. (1) Freeze killed; ants killed by the WT *B. bassiana* strain, and ants killed by the Bb::spβ-NP$_{gpd}$ strain were presented to untreated ants. For the freeze-killed ants, ants were placed at −80°C for 15 min, and then placed at R.T. for 24 hr before use. For fungal-killed ants, infections were performed as described above and the dead ants removed daily. Test ants were derived from those that died on day 4 post-infection. To measure necrophoretic behavior, the test items (5–10 dead ants) were placed in a ring around (1 cm from) the nest entrance. The time interval between introduction and removal of each item was recorded up to a time limit of 600–800 minutes. The number of test objects that were not moved within this interval was also recorded. (2) The effect of infection on necrophoretic behavior was probed by presenting WT- or Bb::spβ-NP$_{gpd}$-killed ants to (a) uninfected ants, (b) WT-infected ants, or (c) Bb::spβ-NP$_{gpd}$-infected ants. Ants were infected (5×10^7 conidia/ml) with the fungal strains 2 d prior to testing. Test objects (dead ants) were prepared and tested as described above. (3) The effect of synthetic peptides on ant cadaver removal was evaluated by having three peptides; (a) β-NP (QPQFTPRL, no C-terminal amidation), (b) β-NP-NH₂ (QPQFTPRL-NH₂), and (c) QAGVTGHA-NH₂ (control 8 amino acid amidated peptide) synthesized (GenScript, Piscataway, NJ). Freeze-killed ants (15 min at −80°C, allowed to thaw for 15 min R.T.) were immersed in 100 nM solutions (resuspended in sterile distilled H₂O) of the test peptide or H₂O alone for 30 s, and then allowed to air-dry for 30 min on a Kim-wipe towel. Necrophoretic behavior to the ants was measured as described above. P-values were obtained from an analysis of variance (1 or 2 way-ANOVA) for each data set, using a permutation test to guard against possible non-normality. 10,000 permutations were used for each test statistic. The unknown (i.e. never moved test objects) data had no effect on the analysis.

Results

Construction and bioassay of β-NP expressing *B. bassiana*

The fire ant β-NP, comprised of the eight-amino acid sequence, QPQFTPRL, was expressed in *B. bassiana* via transformation of an expression vector containing a constitutive *B. bassiana*-derived gpd-promoter, and the nucleotide sequence corresponding to the β-NP peptide fused to a 28-amino acid signal sequence derived from the *B. bassiana* chitinase (*chit1*) gene to produce strain Bb::spβ-NP$_{gpd}$. Heterologous expression of the peptide was confirmed by partial purification and mass spectrometry analysis of culture supernatants. These data indicated the production of a non-amidated β-NP peptide by the fungus at a concentration of ~0.2–0.4 µM.

Both classical worker group and mock mound assays were used to assess the virulence of WT and β-NP expressing *B. bassiana*

strains. Bb::spβ-NP$_{gpd}$ was much more potent (P<0.001) than WT, causing 50% mortality against fire ants after 5 days post-infection with an LD_{50} of $1.5 \pm 0.9 \times 10^7$ conidia/ml compared to an LD_{50} of $1.0 \pm 0.7 \times 10^8$ conidia/ml for the WT parent. Thus, it takes 6–7-fold fewer conidia to provide the same level of mortality. Expressing β-NP also significantly reduced survival times (Fig. 1). At a concentration of 2×10^7 conidia/ml, the mean lethal time to achieve 50% mortality (LT_{50}) was reduced from 177 ± 11 hr for the WT to 122 ± 5 hr for the β-NP expressing strain; representing ~30% reduction in the mean survival time (P<0.01). At lower spore concentrations (4×10^6 conidia/ml) the effect was even more dramatic, with the WT LT_{50} reaching 211 ± 23 hr and the β-NP expressing strain 135 ± 7 hr (P<0.001). In order to determine whether expression of the fire ant β-NP would affect virulence towards other insect, bioassays were performed with several other insect species. No significant difference was noted between the virulence of the WT and Bb::spβ-NP$_{gpd}$ strains towards the lepidopteran host, *Galleria mellonella*, in which the LT_{50} values were 158 ± 5 hr and 166 ± 8 hr, for the WT and β-NP expressing strains, respectively (P>0.05). Similarly, no difference was noted between the WT and β-NP expressing strains when tested against the tobacco hornworm, *Manduca sexta*.

Alternations in ant social behavior mediated by β-NP

In the course of performing mock fire ant mound experiments we noted that ants infected with the Bb::spβ-NP$_{gpd}$ strain appeared altered in their necrophoretic, or disposal of the dead, behavior (Fig. 2). Whereas mock-treated and WT *B. bassiana*-infected ants disposed of their dead in well defined "bone piles", Bb::spβ-NP$_{gpd}$-infected ants appeared to have randomly scattered piles of dead throughout the assay chamber, although typically at the periphery. In order to further probe this observation, we examined the responses of workers to nestmate corpses by placing corpses near the nest entrance in an experimental arena, and monitoring the time taken to remove the corpses. Workers moved ants killed by the WT *B. bassiana* strain faster than freeze-killed ants (~24 hr old, P<0.01), but not those killed by the Bb::spβ-NP$_{gpd}$ strain, which showed a wider variation, but was not significantly different from the response to the freeze-killed ants (Fig. 3). Thus, expression of

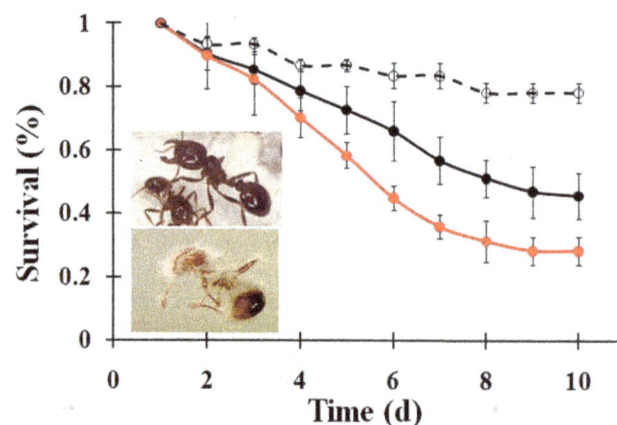

Figure 1. Fire ant bioassays infected with either the WT (●, black line) or Bb::spβ-NP$_{gpd}$ (●, red line) *B. bassiana* strains and buffer treated controls (○, dashed line). The percent survival of *S. invicta* treated with 4×10^6 conidia/ml of each strain over the indicated time course is presented. These data were used to calculate the LT_{50} values, and a concentration curve was used to determine the LD_{50} values. Inset, top panel, normal uninfected ant; bottom panel, *B. bassiana* infected ant (14 d).

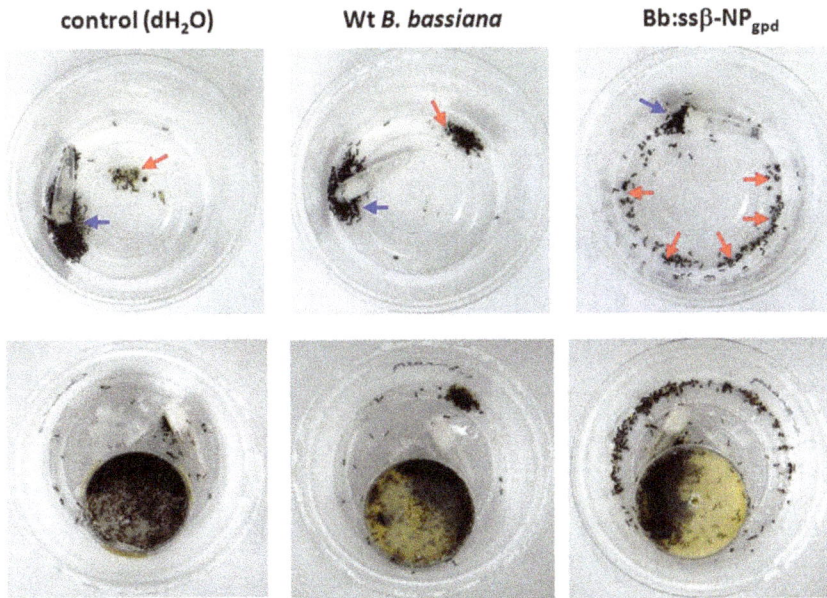

Figure 2. Distribution of dead ants after infection with WT and Bb::spβ-NPgpd B. bassiana strains in mock mound assays. Top panels, ants in arenas with 10% sucrose, but no nest area, bottom panels, test arenas containing 10% sucrose and nest area. Blue arrows in top panels indicate the location of the majority of live ants, in the bottom panels, the most of the live ants are in the nest area (small petri dish in test arena). Red arrows indicate the presence of well defined "bone piles" in the control and WT B. bassiana infected ants, and a more random distribution of dead ants in the β-NP expressing B. bassiana infected ants.

the β-NP peptide appeared to delay removal of corpses. The large variation in removal time observed with Bb::spβ-NP$_{gpd}$-infected ants may be due to differences in levels of β-NP expression in infected ants resulting from differential fungal growth within individual ant hosts. The infection state of the ants themselves did not appear to make a significant difference (P = 0.86). When WT or Bb::spβ-NP$_{gpd}$-infected ants were presented with either WT- or Bb::spβ-NP$_{gpd}$-killed ants, they moved the Bb::spβ-NP$_{gpd}$-killed ants more slowly than WT-killed ones (P<0.001, Fig. 4). These experiments confirmed that ants killed by Bb::spβ-NP$_{gpd}$ were treated differently than WT-killed ants, which were more rapidly removed regardless of the infection state of the ants themselves. This finding has potentially important application consequences since it may increase the lethality of the fungus in field applications due to reduced removal of cadavers which would increase the contact time and possible dispersal of the fungal agent within mounds.

In order to further probe the effects of β-NP, a series of synthetic peptides were examined. Since both pheromonotropic and myotropic activity of pyrokinin/PBAN peptides have been demonstrated via topical application of the peptides onto insects [24], we sought to determine the effects of β-NP-NH$_2$ (C-terminal amidated), β-NP (non-amidated peptide), and a control amidated peptide (QAGVTGHA-NH$_2$) on the necrophoretic behavior of the fire ants. Freeze-killed ants were immersed in a 100 nM solution of the tested synthetic peptides and presented to untreated ants. Surprisingly, we found that ant corpses treated with the β-NP-NH$_2$ peptide were moved significantly faster than buffer treated, β-NP-treated ants, or ants treated with a control eight-amino acid amidated peptide (P<0.001, Fig. 5). β-NP-treated ants were not moved any slower than control or buffer treated ants, although their distribution and the number of ants that were never removed within our assay conditions was larger for the β-NP treatment than for any other treatment examined.

Discussion

The reconstruction of the global invasion history of fire ants, from introduction into the United States from their native South American range to their subsequent spread to newly colonized habitats worldwide, has highlighted the unintended perils and risks associated with the interconnected nature of global trade and

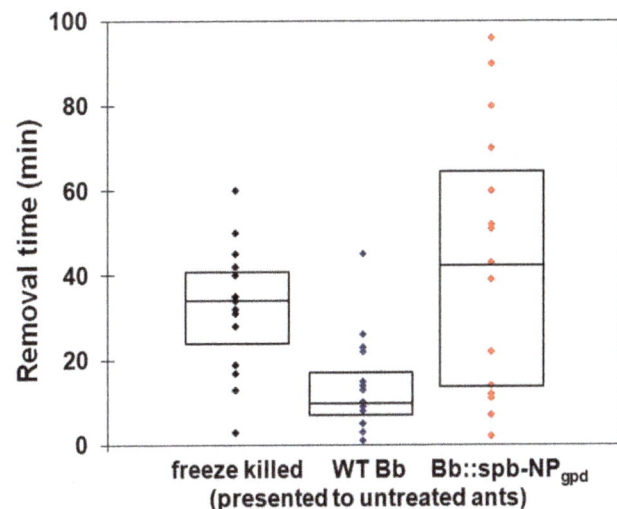

Figure 3. Responses of S. invicta to dead ants. Plots of the times between introduction and removal of items in the test arena. Boxes are bounded by the first quartile, median, and third quartile. Movement times of freeze-killed and fungal-killed, WT and Bb::spβ-NP$_{gpd}$, ants presented to untreated ants. WT-killed ants were moved significantly more quickly than either freeze-killed or Bb::spβ-NP$_{gpd}$-killed ants (P = 0.0014).

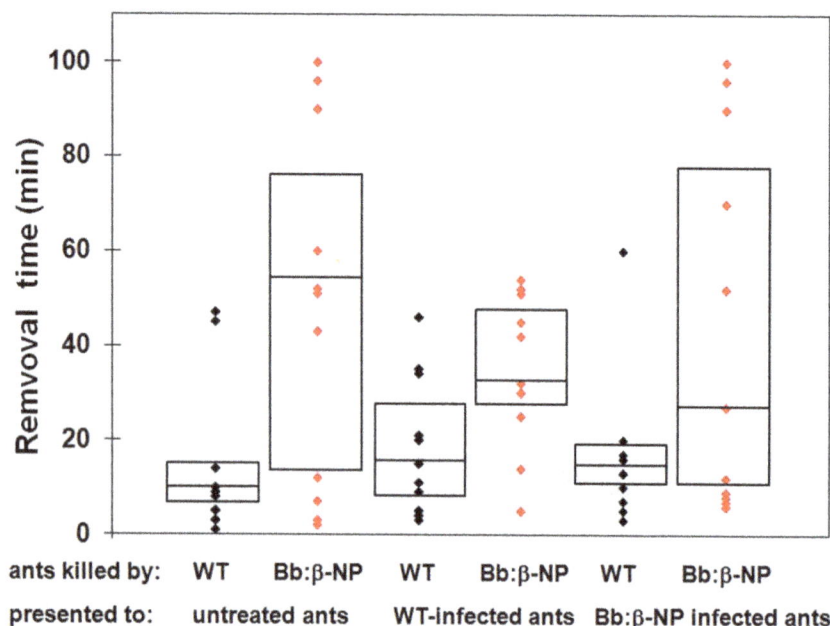

Figure 4. Movement times of WT- or Bb::spβ-NP$_{gpd}$-killed ants by untreated ants, WT-infected (2 d prior to assay) ants, or Bb::spβ-NP$_{gpd}$-infected ants. No significant differences were noted between live ants treatments (P = 0.86), with WT-killed ants moved significantly faster than Bb::spβ-NP$_{gpd}$-killed ants regardless of the tested infection state of the ants doing the moving (P<0.001). This difference was consistent over the live ant treatments (interaction P-value = 0.53).

travel [25]. The destructive nature of fire ant establishment and spread into new ecosystems has led to intense efforts at their control or eradication, in which, even fire ant detecting dogs have been employed [26]. The use of chemical pesticides has failed to stem the spread of fire ants, resulted in the emergence of pesticide resistance, and has not been without controversy [5,27]. Thus, there has been much interest in the use of biological control strategies for fire ant control ranging from release of various parasites including mites and phorid flies, to the use of viruses, microsporidia, nematodes, and fungi [6]. The use entomopathogenic fungi, such as *B. bassiana*, although promising in several

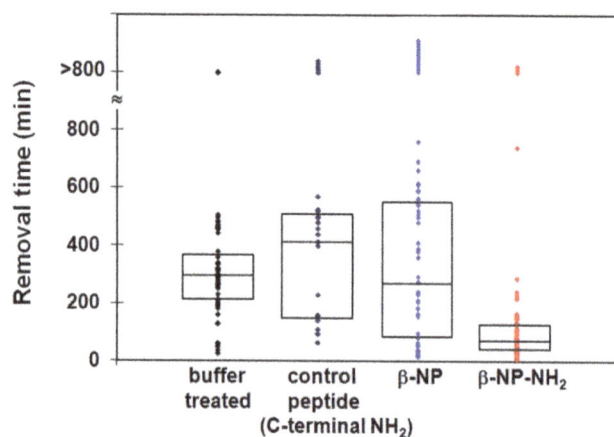

Figure 5. Movement times of dead ants treated with synthetic peptides and presented to untreated ants. Freeze-killed ants were immersed in 100 nM solution of either a control amidated peptide (QAGVTGHA-NH$_2$), β-NP (QPQFTPRL), or β-NP-NH$_2$ and presented to untreated ants. β-NP-NH$_2$-treated ants were moved significantly faster than the control and other peptide treatments (P<0.001).

studies, has thus far met with limited success [6,28,29]. Although newer formulation technologies have increased their field efficacy, the relatively slow kill rate of these fungi coupled to ant behavioral responses such as grooming and corpse removal continue to pose significant obstacles to the use of entomopathogenic fungi [10,30].

Recent efforts have demonstrated success in increasing the virulence of entomopathogenic fungi. Expression of a scorpion toxin in *M. anisopliae* increased virulence 22-fold towards the tobacco hornworm, *Manduca sexta*, and 9-fold against the mosquito *Aedes aegypti*. Expression of the same toxin in *B. bassiana* was shown to increase its virulence towards a variety of hosts including the pine caterpillar, *Dendrolimus punctatus* [31]. The heterologous expression of toxins, however, has not been without controversy, and the potential for the development of resistance to the toxin remains. We sought, therefore, to develop a different strategy for increasing virulence by using host molecules against the host from which it came [13]. There are several important features of this strategy; first a suitable insect molecule (peptide) must be identified. Numerous insect derived peptides have already been suggested or even employed for insect control [32,33,34]. Depending upon the peptide chosen (its distribution and orthology), in principle, various levels of selectivity can be obtained. Although it should be emphasized that such selectivity would need to be verified via experimental analysis, data concerning the spectrum of targets for a number of insect peptides proposed to be used for insect control already exists (see references above). Second, using a host molecule should minimize issues concerning the development of resistance primarily because any mechanism for potentially developing resistance to the host molecule is likely to severely compromise the host. In the case of fire ants, resistance development is even less likely since only queens produce progeny, thus selection occurs within a very small population.

In this report, we improved the virulence of a *B. bassiana* strain to fire ants by expressing a fire ant (neuro-) peptide in the fungal pathogen. Increased virulence in the β-NP expressing fungal strain was noted in both standard and mock mound assays. The increased virulence was specific and no effects were detected against a Lepidopteran hosts (*Galleria mellonella* and *Manduca sexta*), indicating that target-specific virulence can be achieved. This has significant potential for fungal strain improvement and regulatory agencies approval for insect control applications.

Unexpectedly, we noted an altered behavioral pattern in ants infected with the β-NP expressing strain. Rather than forming organized corpse piles as seen in uninfected and wild-type infected ant assays, the dead appeared to remain dispersed throughout the assay chambers. Removal of dead nestmates is thought to limit the potential spread of pathogens, particularly within a social society, and is a common behavior in many ants species. Our observation pointed to altered behavioral effects resulting from application of the β-NP expressing strain on fire ants. These behavioral effects were further probed using the various fungal strains as well as synthetic peptides. Experiments using synthetic peptides indicated that: (1) worker ants have chemosensory perception mechanisms that are able to discriminate between surface peptides, and (2) the β-NP-NH$_2$ peptide, but not the non-amidated form, can act as a semiochemical specifically eliciting enhanced necrophoretic behavior. However, although ants killed by the β-NP expressing fungus were moved slower than WT killed or controls, treatment of dead ants with the synthetic β-NP peptide did not show any significant differences in movement times of the dead ants as compared to the controls, i.e. the biological application displayed a phenotype not observed in the application of the synthetic peptide. There are several possible explanations for these results. First, in the biological application, β-NP would be expressed both within the ant (as the infection proceeds) as well as without. Internal expressed β-NP could then have agonistic interactions with the host's amidated peptide and/or receptor(s) or disrupt other host physiological processes that in turn affect (cadaver) recognition cues. Second, the fungal infection may elicit or suppress microbial pathogen/infection detection mechanisms which would not occur when the synthetic peptide is administered externally. Finally, the slower removal times observed using *B. bassiana* β-NP-expressing killed ants could be due to a combination of fungal factors (i.e. fungal produced enzymes, toxins, volatiles, or other compounds) that act in conjunction with the presence of β-NP to affect cadaver removal.

As a facultative parasite, our results expand the realm of examination between *B. bassiana* and their insect targets, which represents a model system in which molecular and cellular dissection of the host-pathogen interaction is beginning to emerge [35,36,37,38,39]. Our results also open up a new avenue of research with respect to the role and functions of PBAN/pyrokinin peptides in insects, linking them with a complex social behavior. A number of chemical stimuli have previously been reported to act as signals for mediating dead nestmate recognition and removal, i.e. necrophoretic behavior [23,40]. In particular, increasing concentrations of decomposition products, especially fatty acids such as myristoleic, palmitoleic, oleic, and linoleic acids, appear to be major stimuli in eliciting necrophoretic behavior. It has also been proposed that chemical stimuli that elicit removal of nestmate corpses are present on both live and dead ants, however, live ants contain additional compounds that mask these signals, which are subsequently lost or dissipated upon death. [23]. To date, there are no reports on a peptide acting as a necrophoretic modulating semiochemical. Intriguingly, the draft genome of *S. invicta* has revealed over 400 potential odorant receptor (OR) loci (of which 297 appear to be intact), one of the largest repertoire of such receptors found in insects thus far [41]. Although most ORs are thought to bind hydrophobic and/or volatile compounds and chemicals, it is interesting to speculate that within the *S. invicta* OR set there may be members that can recognize peptides (and discriminate between C-terminal amidated and non-amidated) peptides. This report links β-NP-NH$_2$ and necrophoretic behavior. Topical application of PBAN/pyrokinins are known to induce pheromotropic and myotropic activity in live insects [24], however, the full range of their physiological activities remains obscure. Members of the PBAN/pyrokinin family can apparently act as necrophorectic-eliciting cues on dead insects, expanding the potential physiological and sanitary roles of these peptides.

Acknowledgments

The authors wish to thank S. Greenhut for technical assistance, and to Drs. R. K. Van der Meer and M.-Y. Choi for helpful discussions related to the presented work.

Author Contributions

Conceived and designed the experiments: YF RMP NOK. Performed the experiments: YF RMP EK. Analyzed the data: YF RMP GC NOK. Contributed reagents/materials/analysis tools: GC. Wrote the paper: NOK. Contributed input to the writing of the manuscript: YF RMP EK GC.

References

1. Allen CR, Demarais S, Lutz RS (1994) Red imported fire ant impact on wildlife - an overview. Texas Journal of Science 46: 51–59.

2. Harris M, Knutson A, Calixto A, Dean A, Brooks L, et al. (2003) Impact of red imported fire ant on foliar herbivores and natural enemies. Southwestern Entomologist. pp 123–134.

3. Jemal A, Hughesjones M (1993) A review of the red imported fire ant (*Solenopsis invicta* Buren) and its impacts on plant, animal, and human health. Preventive Veterinary Medicine 17: 19–32.

4. Callcott AMA, Collins HL (1996) Invasion and range expansion of imported fire ants (Hymenoptera: Formicidae) in North America from 1918–1995. Florida Entomologist 79: 240–251.

5. Williams DF, Collins HL, Oi DH (2001) The red imported fire ant (Hymenoptera: Formicidae) an historical perspective of treatment programs and the development of chemical baits for control. American Entomologist 47: 146–159.

6. Williams DF, Oi DH, Porter SD, Pereira RM, Briano JA (2003) Biological control of imported fire ants. American Entomologist 49: 150–163.

7. Oi DH, Porter SD, Valles SM, Briano JA, Calcaterra LA (2009) Pseudacteon decapitating flies (Diptera: Phoridae): Are they potential vectors of the fire ant pathogens *Kneallhazia* (= *Thelohania*) *solenopsae* (Microsporidia: Thelohaniidae) and *Vairimorpha invictae* (Microsporidia: Burenellidae)? Biological Control 48: 310–315.

8. Pereira RM (2003) Areawide suppression of fire ant populations in pastures: Project update. Journal of Agricultural and Urban Entomology 20: 123–130.

9. Riggs NL, Lennon L, Barr CL, Drees BM, Cummings S, et al. (2002) Community-wide red imported fire ant management programs in Texas. Southwestern Entomologist. pp 31–41.

10. Oi DH, Pereira RM (1993) Ant behavior and microbial pathogens (Hymenoptera, Formicidae). Florida Entomologist 76: 63–74.

11. St Leger RJ, Wang CS (2010) Genetic engineering of fungal biocontrol agents to achieve greater efficacy against insect pests. Applied Microbiology and Biotechnology 85: 901–907.

12. Wang C, St. Leger R (2007) A scorpion neurotoxin increases the potency of a fungal insecticide. Nature Biotechnology 25: 1455–1456.

13. Fan Y, Borovsky D, Hawkings C, Ortiz-Urquiza A, Keyhani NO (2011) Exploiting host molecules to augment the virulence of mycoinsecticides. Nature Biotechnology: In press.

14. Nachman RJ, Roberts VA, Dyson HJ, Holman GM, Tainer JA (1991) Active conformation of an insect neuropeptide family. Proceedings of the National Academy of Sciences of the United States of America 88: 4518–4522.

15. Rafaeli A (2009) Pheromone biosynthesis activating neuropeptide (PBAN): Regulatory role and mode of action. General and Comparative Endocrinology 162: 69–78.

16. Holman GM, Cook BJ, Nachman RJ (1986) Primary structure and synthesis of a blocked myotropic neuropeptide isolated from the cockroach, *Leucophaea maderae*. Comparative Biochemistry and Physiology C-Pharmacology Toxicology & Endocrinology 85: 219–224.

17. Raina AK, Jaffe H, Kempe TG, Keim P, Blacher RW, et al. (1989) Identification of a neuropeptide hormone that regulates sex-pheromone production in female moths. Science 244: 796–798.

18. Teal PEA, Abernathy RL, Nachman RJ, Fang NB, Meredith JA, et al. (1996) Pheromone biosynthesis activating neuropeptides: Functions and chemistry. Peptides 17: 337–344.

19. Choi MY, Meer RKV (2009) Identification of a new member of the PBAN family of neuropeptides from the fire ant, Solenopsis invicta. Insect Molecular Biology 18: 161–169.

20. Fang WG, Leng B, Xiao YH, Jin K, Ma JC, et al. (2005) Cloning of Beauveria bassiana chitinase gene Bbchit1 and its application to improve fungal strain virulence. Applied and Environmental Microbiology 71: 363–370.

21. Zhang S, Fan Y, Xia YX, Keyhani NO (2010) Sulfonylurea resistance as a new selectable marker for the entomopathogenic fungus Beauveria bassiana. Appl Microbiol Biotechnol 87: 1151–1156.

22. Pereira RM, Stimac JL, Alves SB (1993) Soil antagonism affecting the dose-response of workers of the red imported fire ant, Solenopsis invicta, to Beauveria bassiana conidia. Journal of Invertebrate Pathology 61: 156–161.

23. Choe DH, Millar JG, Rust MK (2009) Chemical signals associated with life inhibit necrophoresis in Argentine ants. Proceedings of the National Academy of Sciences of the United States of America 106: 8251–8255.

24. Nachman RJ, Teal PEA, Radel PA, Holman GM, Abernathy RL (1996) Potent pheromonotropic/myotropic activity of a carboranyl pseudotetrapeptide analogue of the insect pyrokinin/PBAN neuropeptide family administered via injection or topical application. Peptides 17: 747–752.

25. Shoemaker D, Ascunce MS, Yang CC, Oakey J, Calcaterra L, et al. (2011) Global invasion history of the Fire Ant Solenopsis invicta. Science 331: 1066–1068.

26. Lien YY, Lin HM, Chi WL, Lin CC, Tseng YC, et al. (2011) Fire ant-detecting canines: a complementary method in detecting Red Imported Fire Ants. Journal of Economic Entomology 104: 225–231.

27. Buhs JB (2002) The fire ant wars - Nature and science in the pesticide controversies of the late twentieth century. Isis 93: 377–400.

28. Oi DH, Pereira RM, Stimac JL, Wood LA (1994) Field applications of Beauveria bassiana for control of the red imported fire ant (Hymenoptera, Formicidae). Journal of Economic Entomology 87: 623–630.

29. Stimac JL, Pereira RM, Alves SB, Wood LA (1993) Mortality in laboratory colonies of Solenopsis invicta (Hymenoptera, Formicidae) treated with Beauveria bassiana (Deuteromycetes). Journal of Economic Entomology 86: 1083–1087.

30. Bextine BR, Thorvilson HG (2004) Novel Beauveria bassiana delivery system for biological control of the red imported fire ant. Southwestern Entomologist 29: 47–53.

31. Wang CS, Lu DD, Pava-Ripoll M, Li ZZ (2008) Insecticidal evaluation of Beauveria bassiana engineered to express a scorpion neurotoxin and a cuticle degrading protease. Applied Microbiology and Biotechnology 81: 515–522.

32. Gade G, Hoffmann KH (2005) Neuropeptides regulating development and reproduction in insects. Physiological Entomology 30: 103–121.

33. Gade G, Goldsworthy GJ (2003) Insect peptide hormones: a selective review of their physiology and potential application for pest control. Pest Management Science 59: 1063–1075.

34. Borovsky D (2003) Trypsin-modulating oostatic factor: a potential new larvicide for mosquito control. Journal of Experimental Biology 206: 3869–3875.

35. Zhang SZ, Xia YX, Kim B, Keyhani NO (2011) Two hydrophobins are involved in fungal spore coat rodlet layer assembly and each play distinct roles in surface interactions, development and pathogenesis in the entomopathogenic fungus, Beauveria bassiana. Molecular Microbiology 80: 811–826.

36. Zhang SZ, Xia YX, Keyhani NO (2011) Contribution of the gas1 gene of the entomopathogenic fungus Beauveria bassiana, encoding a putative glycosylpho-sphatidylinositol-anchored beta-1,3-glucanosyltransferase, to conidial thermo-tolerance and virulence. Applied and Environmental Microbiology 77: 2676–2684.

37. Bidochka MJ, Clark DC, Lewis MW, Keyhani NO (2010) Could insect phagocytic avoidance by entomogenous fungi have evolved via selection against soil amoeboid predators? Microbiology-Sgm 156: 2164–2171.

38. Lewis MW, Robalino IV, Keyhani NO (2009) Uptake of the fluorescent probe FM4-64 by hyphae and haemolymph-derived in vivo hyphal bodies of the entomopathogenic fungus Beauveria bassiana. Microbiology-Sgm 155: 3110–3120.

39. Wanchoo A, Lewis MW, Keyhani NO (2009) Lectin mapping reveals stage-specific display of surface carbohydrates in in vitro and haemolymph-derived cells of the entomopathogenic fungus Beauveria bassiana. Microbiology-Sgm 155: 3121–3133.

40. Howard DF, Tschinkel WR (1976) Aspects of necrophoric behavior in red imported fire ant, Solenopsis invicta. Behaviour 56: 157–180.

41. Wurm Y, Wang J, Riba-Grognuz O, Corona M, Nygaard S, et al. (2011) The genome of the fire ant Solenopsis invicta. Proceedings of the National Academy of Sciences of the United States of America 108: 5679–5684.

12

Proteolytic Processing of Angiotensin-I in Human Blood Plasma

Diana Hildebrand, Philipp Merkel, Lars Florian Eggers, Hartmut Schlüter*

University Medical Centre Hamburg-Eppendorf, Institute of Clinical Chemistry, Mass Spectrometric Proteomics, Hamburg, Germany

Abstract

In mammalian species, except humans, N-terminal processing of the precursor peptide angiotensin I (ANG-1-10) into ANG-2-10 or ANG-3-10 was reported. Here we hypothesize that aminopeptidase-generated angiotensins bearing the same C-terminus as ANG-1-10 are also present in humans. We demonstrate the time dependent generation of ANG-2-10, ANG-3-10, ANG-4-10, ANG-5-10 and ANG-6-10 from the precursor ANG-1-10 by human plasma proteins. The endogenous presence of ANG-4-10, ANG-5-10 and ANG-6-10 in human plasma was confirmed by an immuno-fluorescence assay. Generation of ANG-2-10, ANG-3-10 and ANG-4-10 from ANG-1-10 by immobilized human plasma proteins was sensitive to the cysteine/serine protease inhibitor antipain. The metal ion chelator EDTA inhibited Ang-6-10-generation. Incubation of the substrates ANG-3-10, ANG-4-10 and ANG-5-10 with recombinant aminopeptidase N (APN) resulted in a successive N-terminal processing, finally releasing ANG-6-10 as a stable end product, demonstrating a high similarity concerning the processing pattern of the angiotensin peptides compared to the angiotensin generating activity in plasma. Recombinant ACE-1 hydrolyzed the peptides ANG-2-10, ANG-3-10, ANG-4-10 and ANG-5-10 into ANG-2-8, ANG-3-8, ANG-4-8 and ANG-5-8. Since ANG-2-10 was processed into ANG-2-8, ANG-4-8 and ANG-5-8 by plasma proteases the angiotensin peptides bearing the same C-terminus as ANG-1-10 likely have a precursor function in human plasma. Our results confirm the hypothesis of aminopeptidase mediated processing of ANG-1-10 in humans. We show the existence of an aminopeptidase mediated pathway in humans that bypasses the known ANG-1-8-carboxypeptidase pathway. This expands the knowledge about the known human renin angiotensin system, showing how efficiently the precursor ANG-1-10 is used by nature.

Editor: Michael Bader, Max-Delbrück Center for Molecular Medicine (MDC), Germany

Funding: This work was funded by the BMBF (Bundesministerium für Bildung und Forschung, Grant: 0315341B) and by LEXI (Landesexzellenzinitiative, Hamburg). The funders had no role in study design, data collection and analysis, decision to publish, or preparation of the manuscript.

Competing Interests: The authors have declared that no competing interests exist.

* E-mail: hschluet@uke.de

Introduction

The inactive prohormone decapeptide angiotensin I (Ang-1-10) is a key member of the renin angiotensin system (RAS), one of the most important blood pressure and homeostasis regulating systems [1]. ANG-1-10 is released from the circulating preprohormone angiotensinogen by the protease renin. Until today many proteases with the ability to process ANG-1-10 further into angiotensin peptides with different or even opposing physiological actions have been identified. Many of them play a crucial role in the regulation of blood pressure and homeostasis, but are also reported to be involved in other physiological processes like inflammation [2,3] cell proliferation [4] or the regulation of neuronal processes [5,6].

The peptide hormone angiotensin II (ANG-1-8) acts as a strong vasoconstrictor but also modulates many other physiological functions by binding to the AT1- or AT2-receptor [1]. ANG-1-8 can be generated by carboxyterminal proteolysis catalyzed by angiotensin converting enzyme-1 (ACE-1) or human mast cell chymase [7]. The vasodilator ANG-1-7 is known to antagonize many physiological effects of ANG-1-8 and can be generated from ANG-1-10 directly [8] as well as from ANG-1-8 and ANG-1-9 [9].

All angiotensin peptides mentioned above are the product of C-terminal cleavage of ANG-1-10. In humans, N-terminal processing of angiotensin peptides by aminopeptidases has only been reported for degradation of ANG-1-8 resulting in the angiotensin

peptides ANG-2-8 or ANG-3-8 (also known as AIII and AIV) which are released from ANG-1-8 by aminopeptidase A (APA) and aminopeptidase N (APN) respectively [1]. As regulators of blood pressure both of them play a role in the brain and the central nervous system [10,11]. The generation of these angiotensin peptides requires the initial C-terminal cleavage of ANG-1-10 by carboxypeptidases like ACE-1 or chymase to form ANG-1-8.

In rats and cats angiotensin peptides deriving from exclusive N-terminal proteolytic cleavage of ANG-1-10 by aminopeptidases were already detected. Such angiotensin peptides contain the same C-terminus as ANG-1-10. The nonapeptide ANG-2-10, octapeptide ANG-3-10 and the hexapeptide ANG-4-10 were described to be generated in the rat [12,13]. The physiological actions of ANG-3-10 were mainly investigated in the cat [14,15,16]. Takai *et al.* found that ANG-5-10 was generated by rat tissues but not by human tissues [17].

In humans little is known about the presence of these angiotensin peptides and their formation to the best of our knowledge. Recently Velez *et al.* proposed that ANG-3-10 is generated proteolytically by human podocytes and showed that ANG-2-10 is generated from ANG-1-10 by human glomerular endothelial cells [18]. The authors also postulated that ANG-3-10 was generated by APN from ANG-2-10. However, it has yet not been found if all of these peptides deriving from aminopeptidase activity are generated by human plasma proteases. Hence here we followed the question whether angiotensin peptides that contain the intact C-terminus of ANG-1-10 are generated in human plasma by aminopeptidases and if these peptides are detectable in blood plasma.

Materials and Methods

Ethical Statement

For each condition, volunteers were recruited for this study. According to the requirements of our ethics committee of the medical association Hamburg (Ethikkommission der Ärztekammer Hamburg, Germany) the participants provided signed informed consent.

All procedures concerning the experiments with murine plasma were performed in accordance with protocols approved by the Institutional Animal Care and Research Advisory Committee (Syddansk Universitet Odense, Biomedicinsk laboratorium approval number 157).

Preparation of Blood Plasma

For incubation experiments human venous citrate blood (ratio 1:9, blood to sodium citrate 3,13% (Eifelfango)) was obtained by catheterization from the cubital vein of a healthy male volunteer (Age: 50 years, blood pressure: normal, <120/80 mmHg).

For the detection of angiotensin peptides in human plasma and the incubation experiments with ANG-1-10 a volume of 50 ml venous citrate blood was drawn by catheterization from the cubital vein of a healthy female volunteer (Age: 27, blood pressure: normal, <120/80 mmHg). From this blood sample an aliquot of 10 ml was saved for the incubation experiments. The rest of the blood sample was immediately mixed with protease inhibitor cocktail (including 2 mM AEBSF, 0.3 µM aprotinin, 130 µM bestatin, 1 mM EDTA, 14 µM E-64 and 1 µM leupeptin (Sigma-Aldrich)) in a ratio of 1:50. This sample served for the detection of angiotensin peptides by an immuno-fluorescence assay.

Heparinized (10 IU/ml) mouse and rat blood (200 µl) was obtained from the caudal vein. Plasma was isolated from all blood samples by centrifugation (4–16 K, Sigma) 4000×g for 15 min.

Immobilization of Plasma Proteins

For the immobilization of plasma proteins CNBr-activated Sepharosebeads® 6MB (GE Healthcare) were used. Protein immobilization was perfomed as described in the product information sheet for CNBr-activated Sepharosebeads® (Sigma Aldrich) except for modifications described in the supporting information part "immobilization of plasma proteins" (Methods S1).

Incubation of Immobilized Plasma Proteins

Incubation of immobilized plasma proteins and the control samples (glycine derivatized Sepharosebeads® without immobilized proteins and heat inactivated immobilized plasma proteins for the incubation with ANG-1-10) was started by addition of the individual angiotensin peptides (final concentration of 10^{-5} M, solved in HPLC-grade water, Lichrosolve, Merck) to the beads. Incubation was carried out at 37°C on a rotating shaker. The final reaction volume of the samples was 30 µl. At defined incubation times aliquots (3 µl) were taken from the reaction mixtures, diluted in a ratio of 1:10 in 0.2% (v/v) formic acid/HPLC-grade Water and analyzed by LC-ESI-QQQ-MS (6430 Series, Agilent Technologies) or MALDI-MS (Reflex IV, Bruker).

Incubation of Immobilized Human Plasma Proteins with ANG-1-10 in the Presence of Protease Inhibitors

Immobilized plasma proteins were separately preincubated for 5 min with the following inhibitors: 200 µM AEBSF (Applichem), 50 µM antipain, 150 µM bestatin, 10 µM captopril, 100 µM chymostatin (Sigma-Aldrich), 100 µM EDTA (Bio-Rad). The control was incubated with immobilized plasma proteins in the absence of inhibitors. The incubation with immobilized plasma proteins was carried out as described in "Incubation of immobilized plasma proteins". The incubation was started by addition of ANG-1-10 to a final concentration of 10^{-5} M to the immobilized plasma proteins.

Incubation of Non-immobilized Plasma Proteins

For incubation of non-immobilized plasma a reaction volume of 200 µl non-immobilized undiluted and diluted plasma (1:100 in HPLC-grade water) was used per sample. Incubation was started by addition of ANG-1-10 to a final concentration of 10^{-5} M. Incubation was carried out at 37°C on a rotating shaker. As a control 200 µl of a 10^{-5} M ANG-1-10 solution without plasma was incubated under the same conditions. At defined incubation times (0 h, 0.25 h, 0.5 h, 1 h, 2 h, 4 h, 6 h, 8 h, 24 h) aliquots with a volume of 10 µl were taken from the reaction mixtures and diluted in a ratio of 1:10 in 0.2% (v/v) formic acid/HPLC-grade Water.

Peptide Desalting by Solid Phase Extraction

The reaction products derived from the incubation of non-immobilized plasma proteins were desalted by solid phase extraction (Hydophilic lipophilic balance (HLB) µElution plate, Waters). The HLB material was equilibrated by three washing steps with 0.2% formic acid/HPLC-grade water by centrifugation of the plate for 1 min with 500×g. Next, the sample was applied to the HLB material and centrifuged for 2 min with 200×g. The flow-through was discarded and unbound molecules were removed by washing the HLB material 3 times with 0.2% formic acid/HPLC-grade water by centrifugation of the plate for 2 min with 500×g. Adsorbed molecules were eluted with 100 µl 60% MeOH followed by centrifugation for with 200×g. Eluates were collected in a 96 well plate and evaporated to complete dryness in a vacuum concentrator (RC 10, Thermo Scientific). Prior to LC/MS analysis samples were dissolved in 0,2% (v/v) formic acid/HPLC-grade Water.

Incubation of Recombinant Proteases with Angiotensin Peptides

Each angiotensin peptide (ANG-1-10, ANG-1-8, ANG-2-10, ANG-3-10, ANG-4-10 or ANG-5-10) was dissolved in HPLC-grade water to a final concentration of 10^{-5} M. Incubation was started by addition of 0.25 µg recombinant human aminopeptidase N or angiotensin converting enzyme-1 (ACE-1, R&D Systems) to 100 µl of each angiotensin solution. At defined incubation times aliquots (10 µl) were taken from the reaction mixtures. Reaction was stopped by addition of formic acid adjusting a final concentration of 0.2% (v/v) formic acid/HPLC-grade water. Samples were analyzed by MALDI-MS (Reflex IV, Bruker) using DHB as matrix.

Mass Spectrometric Peptide Identification and Quantification

Angiotensin peptides were identified by LC-ESI-IT-MS/MS (ion-trap, XCT, Agilent Technologies) or MALDI-MS (Reflex IV, Bruker) and quantified by SRM-coupled ESI-QQQ-MS (Triple quadrupole, 6430 Series, Agilent Technologies). ESI-IT-MS and ESI-QQQ-MS was coupled to an HPLC-chip-system (Agilent Technologies).

Details of the HPLC-chip–MS/MS-system used for analysis are described according to Trusch *et al.* [19]. Individual settings of the system are described in the supporting information part "mass spectrometric identification and quantification" (Methods S2) and in Table S1.

Detection of Angiotensin Peptides in Human Plasma

The immuno-fluorescence assay was performed as described in the instruction manual (Fluorescent EIA Kit, phoenix pharmaceuticals). For this assay calibration curves of the angiotensin peptides ANG1-10, ANG-4-10, ANG-5-10 and ANG-6-10 were measured using following concentrations (pg/ml): 1, 10, 1000, 10000.

Plasma peptide fractions were obtained by plasma protein precipitation of 15 ml plasma by addition of ACN/0.1% TFA in HPLC-grade water (v/v) in a ratio of 1:2 (plasma volume:ACN/TFA). The supernatant including the plasma peptides was subjected to a two step chromatographic purification. The first step included a purification by a cartridge filled with HLB material (Oasis HLB cartridge 6 g, Waters). All equilibration, washing and elution steps with this cartridge were carried out as already described for the desalting step of peptides by Oasis HLB-µelution plate in section "Peptide desalting by solid phase extraction". The eluate containing the desalted plasma peptide fraction was collected in a 50 ml tube. This sample was evaporated to complete dryness and redissolved in 500 µl 0.1% TFA (v/v)/HPLC-grade water. Afterwards a reversed phase chromatography (RP18e, 100 mm×4 mm, Chromolith®performance, Merck KGaA) of the plasma fraction was performed with a 1100 capillary pump (Agilent Technologies) working at 500 µl/min. HPLC-grade water with 0.1% TFA (solvent A) was used for sample loading and delivery. Peptides were eluted from the column using a gradient composed of solvent A and solvent B (acetonitrile) consisting of 3–21% solvent B in 5 min, 21–25% in 40 min, and 25–60% in 1 min.

Eluting fractions were collected with an ÄKTA prime fraction collector (GE Healthcare). Plasma fractions were evaporated to complete dryness prior to the immuno-fluorescence assay. Retention times of angiotensin peptides were determined using synthetic angiotensin peptides under the same chromatographic conditions as described above for the reversed phase chromatography. A volume of 500 µl of the synthetic peptides ANG-1-0, ANG-3-10, ANG-4-10, ANG-5-10 and ANG-6-10 in an equimolar concentration of 10^{-5} M dissolved in 0.1% TFA/HPLC-grade water was loaded onto the column. Separation performance was confirmed by MALDI-MS analysis of the eluted fractions.

Relative fluorescence intensity was measured by 3 flashes and 20 µs integration time with a microplate reader (Infinite m200, Tecan Group Ltd.) with an excitation wavelength of 325 nm (bandwidth 9 nm) and an emission wavelength of 420 nm (bandwidth 20 nm) using optimal gain. Data were processed with i-control software (Version 1.5.14.0, Tecan Group Ltd.) and plotted using Graph Pad Prism Software (Version 4.00).

Statistics

Data contained in figures with error bars are expressed as mean ± SEM. Statistical analysis was performed using Graph Pad Prism (Version 4.00). Statistical significance was assessed using a one sample 2-tailed student's *t* test. *P* values less than 0.05 were considered significant.

Results

Processing of ANG-1-10 by Immobilized Human and Murine Plasma Proteins

To investigate the processing of ANG-1-10 by human plasma proteases we used a mass spectrometry based enzyme screening (MES) system [20]. Therefore immobilized murine and human plasma proteins were incubated with ANG-1-10 and the reaction products were analyzed by mass spectrometry. Identification of the reaction products was done by comparison of the m/z values of the peaks in the mass spectra with the theoretical m/z values (Table S2) of possible angiotensin peptide products. ANG-1-10 (m/z 1296.5) was degraded by mouse and rat plasma proteins into the angiotensin peptides ANG-1-8 (m/z 1046.5), ANG-1-7 (m/z 899.5), ANG-2-10 (m/z 1181.7), ANG-3-10 (m/z 1025.6), ANG-4-10 (m/z 926.5), ANG-5-10 (m/z 763.4) (Figure S1+ Figure S2) in a time dependent manner. After 8 h of incubation the only reaction products left were ANG-1-8 after incubation with mouse plasma and ANG-1-7, ANG-1-8 and ANG-5-10 after incubation with rat plasma. All of these peptides were also generated from ANG-1-10 by human plasma proteins, but additionally ANG-6-10 (m/z 650.3) was detected (Figure 1). In addition it was investigated if the processing of ANG-1-10 by immobilized human male plasma proteins, which were used for this experiment, differs from the ANG-1-10-processing by immobilized human female plasma proteins (Figure S3). In comparison to the incubation experiments using male plasma the incubation with immobilized female plasma resulted in the same processing pattern after incubation with ANG-1-10. Neither the control sample containing glycine derivatized Sepharosebeads® without immobilized protein, nor the control sample with immobilized heat inactivated plasma proteins showed significant degradation of ANG-1-10. The identity of the angiotensin peptides generated by human plasma proteins was validated by LC-ESI-IT-MS/MS analysis (Figure S4).

Chromatographic Purification of Angiotensin Peptides and their Detection in Human Plasma

To prove whether the peptides bearing the same C-terminus as ANG-1-10 are present endogenously in human plasma we separated the plasma peptides from the plasma proteins by precipitation. Afterwards the plasma peptide fraction was purified by reversed phase high pressure liquid chromatography (RP-HPLC) (Figure 2 A). The separation efficiency and the retention time of the angiotensin peptides ANG-3-10, ANG-4-10, ANG-5-10, ANG-6-10 and ANG-1-10 was determined by RP-HPLC of synthetic angiotensin peptides (Figure 2 B) and subsequent analysis of the derived eluate fractions by MALDI-MS (Data not shown). The angiotensin peptides ANG-4-10, ANG-5-10 and ANG-6-10 were well separated by RP-HPLC and eluted after 32 min, 28 min and 22 min respectively. ANG-1-10 and ANG-3-10 co-eluted after 35 min.

The RP-HPLC plasma peptide fractionations corresponding to the retention times of the synthetic angiotensin peptides were analyzed by an immuno-fluorescence assay that was originally manufactured for the quantification of ANG-1-10. The calibration curves of the angiotensin peptides ANG-1-10, ANG-4-10, ANG-5-10 and ANG-6-10 were measured for the quantification of these peptides in human plasma. Using this assay calibration curves were generated for all angiotensin peptides (data not shown) demonstrating the ability of the anti ANG-1-10 antibody (rabbit polyclonal, no cross reactivity with ANG-1-8 or ANG-2-8) to bind angiotensin peptides bearing the same C-terminus. We determined a plasma concentration of about 8 pg/ml for ANG-4-10,

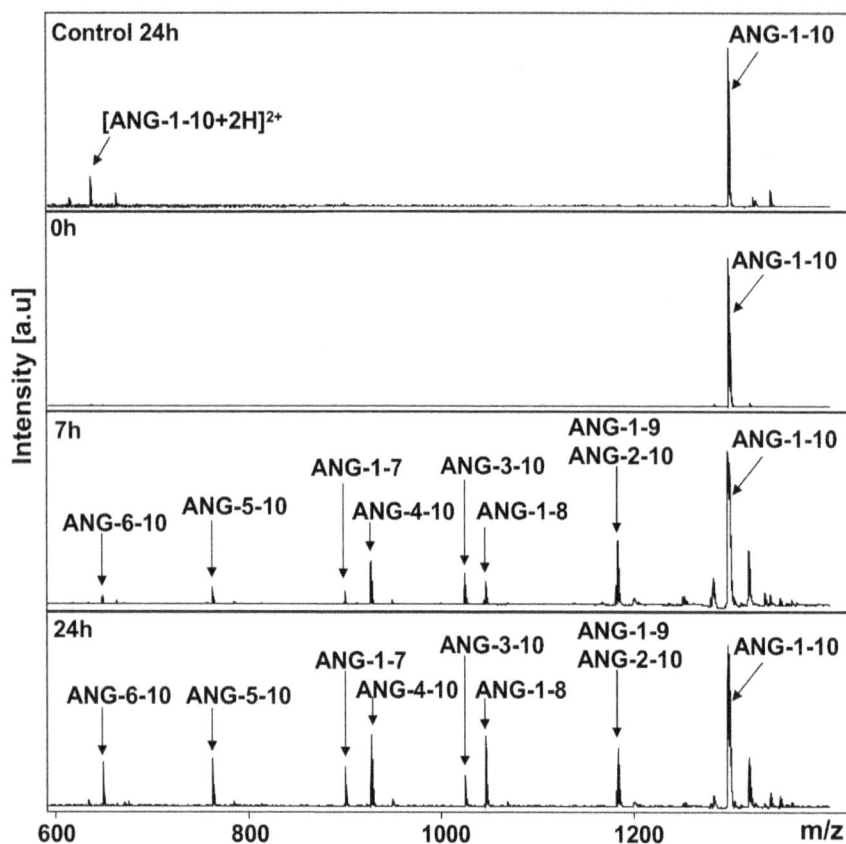

Figure 1. Processing of ANG-1-10 by immobilized human plasma proteins. ANG-1-10 (10^{-5} M) was incubated with immobilized human plasma proteins. Reaction products were detected by MALDI-MS after 0 h, 7 h and 24 h. MALDI-MS signals corresponding to angiotensin peptides are marked by arrows. Control: ANG-1-10 incubated for 24 h with Sepharosebeads® without immobilized proteins.

53 pg/ml for ANG-5-10 and 7 ng/ml for ANG-6-10 in the human plasma peptide fractions (Figure 3).

Processing of ANG-1-10 in Non-immobilized Human Plasma

Plasma protein immobilization is performed by coupling the plasma proteins covalently to CnBR activated Sepharosebeads® via free amino groups of the proteins. Immobilization of proteins stabilizes them because they are covalently fixed and therefore cannot proteolyze other proteins anymore. The hydrophilic environment of the sepharose matrix (agarose) additionally stabilizes proteins since it is comparable with physiological environments. Immobilization usually does not affect the protease activities significantly. In biotechnology immobilization of enzymes is a common approach to stabilize them and to retain their activities and specificities [21].

In some cases the *in vitro* generation of angiotensin peptides by immobilized plasma proteins might differ from generation with immobilized plasma proteases due to a change of protease activity caused by removal of protease inhibitors or the dissociation of cofactors through the process of immobilization. To address this question ANG-1-10 was directly incubated with non-immobilized undiluted and diluted human plasma (Figure 4). To compare the angiotensin generating activity of non-immobilized plasma with immobilized plasma, the data from the incubation experiments with ANG-1-10 and immobilized plasma proteins were included in the graphs.

In undiluted plasma the maximum amount of all angiotensin peptides except ANG-1-9 and ANG-2-10 was measured after an incubation time of 0.5 h. The amount of ANG-1-9 and ANG-2-10 showed its maximum after 0 h of incubation followed by a constant decrease, indicating that its generation in undiluted plasma starts within seconds after addition of ANG-1-10 followed by rapid proteolytic degradation. After 0.5 h the amount of the other angiotensin peptides also decreased, indicating their degradation. In diluted plasma samples the amplitude of ANG-2-10 generating activity appeared after 0.25 h. The highest amounts of ANG-3-10 and ANG-5-10 were measured after 8 h, whereas the generation of all other angiotensin peptides ANG-1-9, ANG-1-8, ANG-1-7, ANG-4-10 and ANG-6-10 seemed to be ongoing until the end of the incubation after 24 h. Hence as expected a higher proteolytic ANG-1-10-metabolizing activity was measured in undiluted plasma compared to diluted plasma. The angiotensin generating activity of the immobilized plasma was between the angiotensin generating activity of the non-immobilized undiluted plasma and the 1:100 diluted plasma. In summary all peptides which have been observed after incubation of ANG-1-10 with immobilized plasma proteins were also generated by incubation with non-immobilized undiluted and diluted plasma in a time dependent manner. It appears that endogenous plasma protease inhibitors do not play a crucial role concerning the proteolytic generation of these angiotensin peptides.

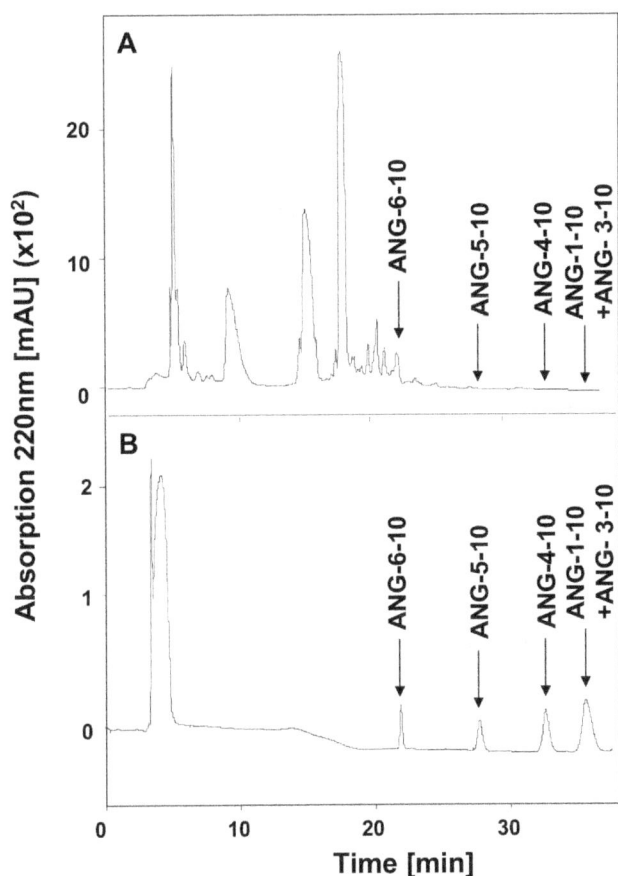

Figure 2. Reversed phase high pressure liquid chromatography (RP-HPLC) of angiotensin peptides. Chromatograms (Absorption at 220 nm plotted against time) after separation of human plasma peptides and synthetic angiotensin peptides by RP-HPLC are shown. A) Chromatogram of the plasma peptide fraction. Retention times of the synthetic angiotensin peptides determined by RP-HPLC of the synthetic angiotensin peptides are indicated by arrows. B) Chromatogram of the synthetic angiotensin peptide mixture including 10^{-5} M ANG-3-10, ANG-4-10, ANG-5-10 and ANG-6-10. Their retention times are indicated by arrows.

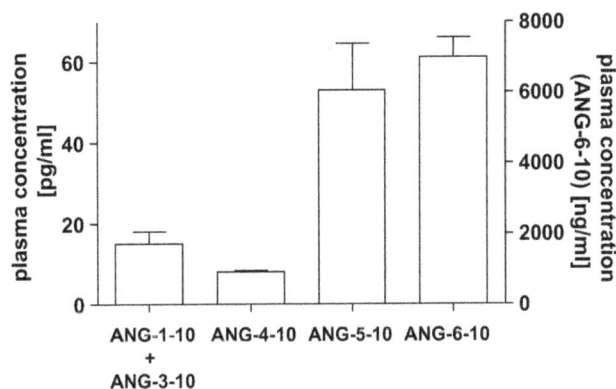

Figure 3. Detection of endogenous angiotensin peptides in human plasma. Concentrations of ANG-1-10+ANG-3-10, ANG-4-10, ANG-5-10 (shown on left y-axis) and ANG-6-10 (shown on right y-axis) were determined by an immuno-fluorescence assay.

was detected. After an incubation period of 8 h nearly no angiotensin peptides were detected in any of the reaction mixtures. These results demonstrate that, beside ANG-1-10, the intermediate angiotensin peptides ANG-3-10, ANG-4-10, ANG-5-10 serve as substrates for human plasma aminopeptidases.

Analysis of the reaction mixture by MALDI-MS showed, that ANG-2-10 was not only cleaved into the angiotensin peptides ANG-3-10, ANG-4-10, ANG-5-10 and ANG-6-10 but also into the peptides ANG-2-8 (m/z 931.5), ANG-3-8 (m/z 775.4) and ANG-4-8 (m/z 676.3) after 8 h of incubation (Figure S5). The signal for ANG-2-8 had a very low intensity, indicating fast further degradation into ANG-3-8 and ANG-4-8. These peptides were not observed by MALDI-MS after incubation of ANG-3-10 ANG-4-10 or ANG-5-10 (data not shown). This can be seen as an indication that the starting point for the generation of ANG-3-8 and ANG-4-8 likely is ANG-2-8. ANG-2-8 cannot be generated from angiotensin peptides smaller than ANG-2-10 and was also not generated from ANG-1-10 in human plasma. This would also explain why the generation of ANG-2-8, ANG-3-8 and ANG-4-10 was not observed in the reaction mixture after incubation of human plasma with ANG-1-10. Here the generated amount of ANG-2-10 was rapidly converted into the peptides ANG-3-10, ANG-4-10, ANG-5-10 and ANG-6-10 (Figure 1, Figure 4) thus minimizing the availability of ANG-2-10 for its hydrolysis into ANG-2-8. However, ANG-3-10 might also be processed into ANG-3-8 and ANG-4-8 by human plasma proteases. In the same way ANG-4-8 might be generated from ANG-4-10. The generation of these reaction products might be unrecognized due to a fast proteolytic degradation, resulting in low concentrations that might be below the detection limit of our instruments.

Influence of Protease Inhibitors on the Angiotensin-generating Activity of Immobilized Human Plasma Proteins

Proteases taking part in the aminopeptidase-dependent formation of angiotensin peptides were characterized by incubation of ANG-1-10 with immobilized human plasma proteins in presence of 200 μM AEBSF, 50 μM antipain, 150 μM bestatin, 10 μM captopril, 100 μM chymostatin, 100 μM EDTA and in absence of these inhibitors. Their protease specificity as well as their influences on the angiotensin-generating activity are shown in Table 1. The reaction products were analyzed by SRM-MS after 6 h (Figure S6).

Incubation of ANG-2-10, ANG-3-10, ANG-4-10 and ANG-5-10 with Immobilized Human Plasma Proteins

To investigate if the angiotensin peptides ANG-2-10, ANG-3-10, ANG-4-10, ANG-5-10 and ANG-6-10 are generated successively, each of these angiotensin peptides was incubated with immobilized human plasma proteins. Analysis of the reaction by SRM was performed on a LC-ESI-QQQ-MS system. Exoproteolytic processing of these angiotensin peptides requires the presence of plasma proteases which are able to cognate and cleave the intermediate angiotensin peptides. Incubation of the angiotensin peptide ANG-2-10 with immobilized human plasma proteins lead to the time dependent generation of ANG-3-10, ANG-4-10, ANG-5-10 and ANG-6-10 as was shown by SRM-MS-analysis (Figure 5). The highest amount of ANG-3-10 was generated after 3 h while the maximum amount of ANG-4-10, ANG-5-10 and ANG-6-10 was measured after 6 h. The substrate ANG-3-10 was processed into the peptides ANG-4-10, ANG-5-10 and ANG-6-10 and their maximum amount was generated after an incubation time of 2 h. When ANG-5-10 was incubated in presence of human immobilized plasma proteins time the dependent generation of ANG-6-10

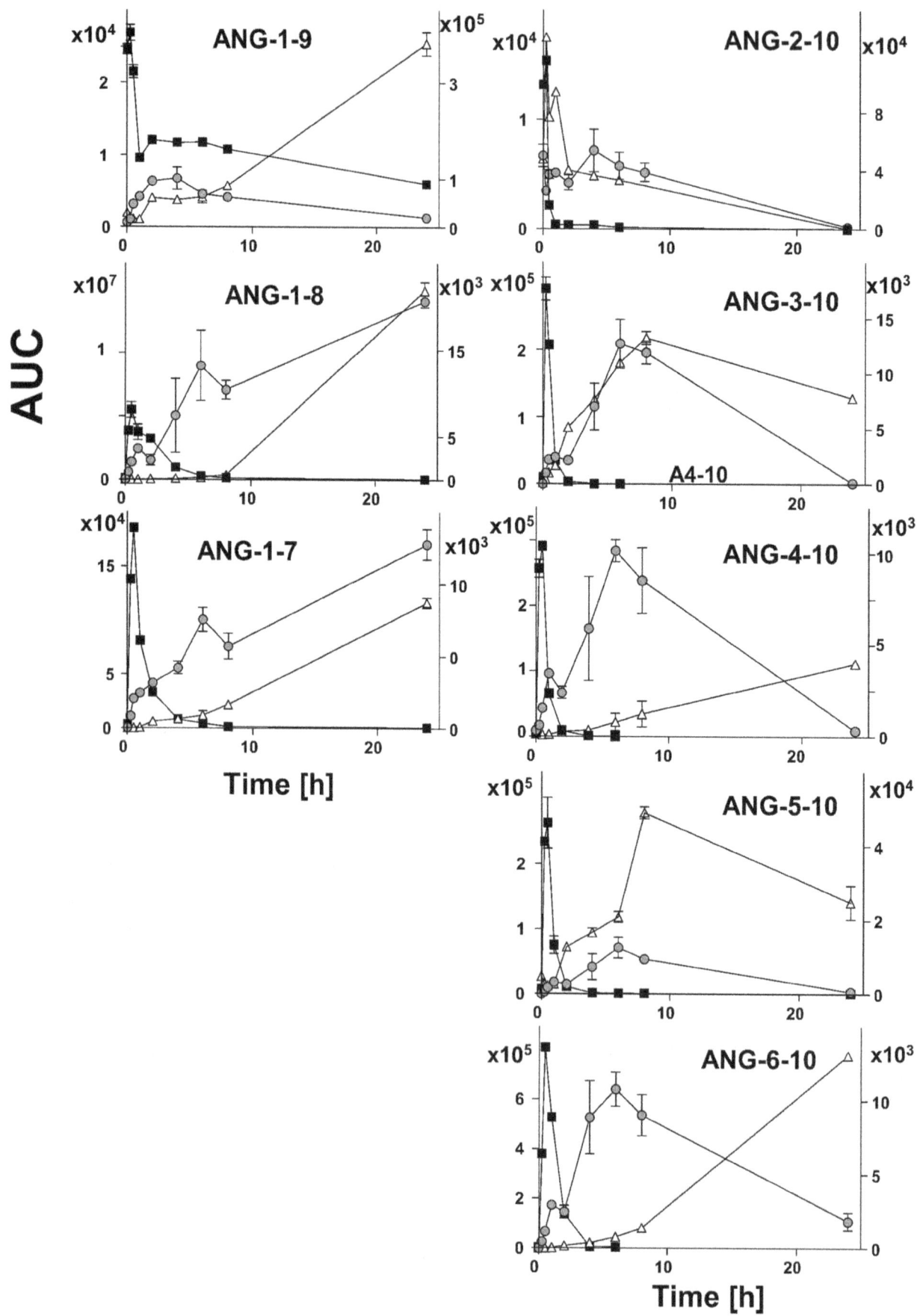

Figure 4. Processing of ANG-1-10 by immobilized, non-immoblized diluted and undiluted human plasma proteins. Immobilized plasma proteins (grey circle), non-immobilized undiluted (closed square) and 1:100 diluted plasma (open triangle) was incubated with ANG-1-10 (10^{-5} M). Reaction products were analyzed and relatively quantified by SRM-MS. The areas under the curves (AUC) of the reaction products ANG-1-9, ANG-1-8, ANG-1-7, ANG-2-10, ANG-3-10, ANG-4-10 ANG-5-10 and ANG-6-10 are shown. Left y-axis: AUCs of diluted non-immobilized plasma and immobilized plasma (ANG-3-10, ANG-1-9). Right y-axis: AUCs of undiluted non-immobilized plasma and immobilized plasma. Data are shown as mean $\pm$ SEM (n = 3, except for ANG-2-10 of the non-immobilized samples with n = 1).

The generation of ANG-2-10 was inhibited by antipain (-0.12 ± 0.023). The presence of EDTA and bestatin lead to higher ANG-2-10-generating activities. The ANG-3-10-generating activity was inhibited in the presence of AEBSF and antipain. Higher amounts of ANG-3-10 were detected in the presence of bestatin, but a significant raise of ANG-3-10 was only observed in the presence of EDTA (3.4 ± 0.8). The inhibition profile of ANG-4-10 showed many similarities compared to the inhibitor profiles of ANG-2-10 and ANG-3-10 and but a significant inhibition of ANG-4-10-generating activity was only observed in the presence of antipain (-0.21 ± 0.05). The generation of ANG-5-10 was inhibited by AEBSF, antipain and EDTA.

The inhibition profile of the ANG-6-10 generating activity shows clear differences compared to all other angiotensin peptides with the same C-terminus as ANG-1-10. The ANG-6-10-generating activity almost completely disappeared in presence of EDTA (-0.97 ± 0.006) and was slightly inhibited by antipain. Samples including AEBSF (2.80 ± 0.77), bestatin (0.76 ± 0.14), captopril (0.98 ± 0.32) and chymostatin (1.30 ± 0.14) lead to significantly higher amounts of ANG-6-10.

Aminopeptidase N (APN) as an Angiotensin-generating Protease

APN is a protease that is known to cleave ANG-2-8 between the Arg2-Val3 in vitro [22] and in vivo [23,24]. Thus it is a potential protease for the generation of ANG-3-10. To investigate if human plasma APN is able to generate ANG-3-10 and to analyze its substrate and cleavage site specificity recombinant APN (R&D Systems) was incubated with different angiotensin peptides. The reaction products were analyzed by MALDI-MS and the results are summarized in Table 2. After incubation of the APN with ANG-1-10 the reaction product ANG-6-10 was detected after 8 h of incubation. No reaction products were observed after incubation of ANG-1-8 with APN (data now shown).

When ANG-2-10 was used as a substrate for APN the signals of the reaction products ANG-3-10, ANG-4-10 were detected after 1 h of incubation, whereas ANG-5-10 and ANG-6-10 were detected after 4 h and 8 h.

ANG-3-10 was processed into ANG-4-10 and ANG-6-10 by APN within 1 h. The signal for ANG-4-10 was not present in the reaction mixture after longer incubation times. The signal for ANG-6-10 was still present after 24 h, indicating that no further degradation of ANG-6-10 occurred. When ANG-4-10 was incubated with APN, a signal for ANG-5-10 and ANG-6-10 was detected after 1 h of incubation. After an incubation period of 24 h almost no ANG-5-10 was detected, while ANG-6-10 was still present with a high signal to noise ratio.

Incubation of ANG-5-10 lead to the generation of ANG-6-10 whose signal remained stable until 24 h of incubation. No signals of smaller angiotensin reactions products were detected.

Conversion of Angiotensin Peptides by Recombinant ACE-1

It is known that ACE-1 generates ANG-1-8 by removal of the last two C-terminal amino acids (His9-Leu10) of ANG-1-10. To investigate if ACE-1 can also hydrolyze the angiotensin peptides

bearing the same C-terminus as ANG-1-10 we incubated recombinant human ACE-1 with the angiotensin peptides ANG-2-10, ANG-3-10, ANG-4-10, ANG-5-10 and ANG-6-10. The reaction products were analyzed by MALDI-MS (Figure S7). ACE-1 released the two C-terminal aminoacids His9-Leu10 from all angiotensin peptides, except for ANG-6-10. Thus ACE-1 generated the angiotensin peptides ANG-2-8, ANG-3-8, ANG-4-8 and ANG-5-8. This reaction was effectively inhibited by the ACE-1-inhibitor captopril. During incubation of ACE-1 with ANG-6-10 no cleavage product was observed. This is probably due to the small molecular weight (m/z 400.2) of the peptide ANG-6-8, as it consists of 3 amino acids. As a result the signal of ANG-6-8 can be suppressed by interfering signals of DHB-matrix photoproducts which generally occur in a mass range up to 500 Da.

Discussion

Beside the classical circulating RAS, many other localized RAS as in the brain [25] or the kidney [26] have been discovered. Recently Velez et al. investigated the proteolytic processing of ANG-1-10 in vitro by human glomerular endothelial cells and podocytes. They showed that ANG-1-10 was proteolytically degraded into ANG-1-8 but was also processed by aminopeptidases. The authors also proposed APA as an ANG-2-10-generating protease and APN as an ANG-3-10-generating protease [18], but still the question remained whether these peptides are also present in human plasma.

Here we show that ANG-1-10 is processed into A-1-9, ANG-1-8, ANG-1-7, ANG-2-10, ANG-3-10, ANG-4-10, ANG-5-10 and ANG-6-10 by immobilized and non-immobilized human plasma proteins (Figure 1, Figure 4, Figure S3 and Figure S4). Concerning the processing pattern of ANG-1-10 no difference could be observed between male (Figure 1) and female plasma (Figure S3). Some of these peptides have already been detected in other species. Velez et al. described, that in isolated rat glomeruli ANG-1-10 was mainly converted into ANG-2-10 and ANG-1-7 [27]. Furthermore ANG-2-10 is generated in the hypothalamic extract of rats [28]. ANG-3-10 has been described in the vascular bed of the cat [15,16], in rat brain and plasma [12] and, as well as ANG-4-10, in rat myocardial tissue [13]. ANG-5-10 has been shown to be generated by rat vascular tissues [17] and in rat peritoneal cell cultures [29]. We did not detect the generation of ANG-6-10 in mouse and rat plasma in contrast to ANG-2-10, ANG-3-10, ANG-4-10 and ANG-5-10 (Figure S1), pointing out that protease activities vary between different species. To the best of our knowledge, no report has described the generation of ANG-6-10 in any species before.

Many proteolytic activities of plasma proteases are strongly regulated by abundant circulating protease inhibitors like alpha-2-macroglobulin, which inhibits a variety of proteases like plasmin, kallikrein and thrombin [30] or C1-inactivator that inhibits proteases of the complement system [31]. In the RAS the ACE-2-activity in plasma was reported to be masked by an endogenous inhibitor that can be chromatographically removed from the protease [32]. With high probability we could exclude that the monitored proteolytic activities are masked by endogenous inhibitors in human plasma which might be removed or

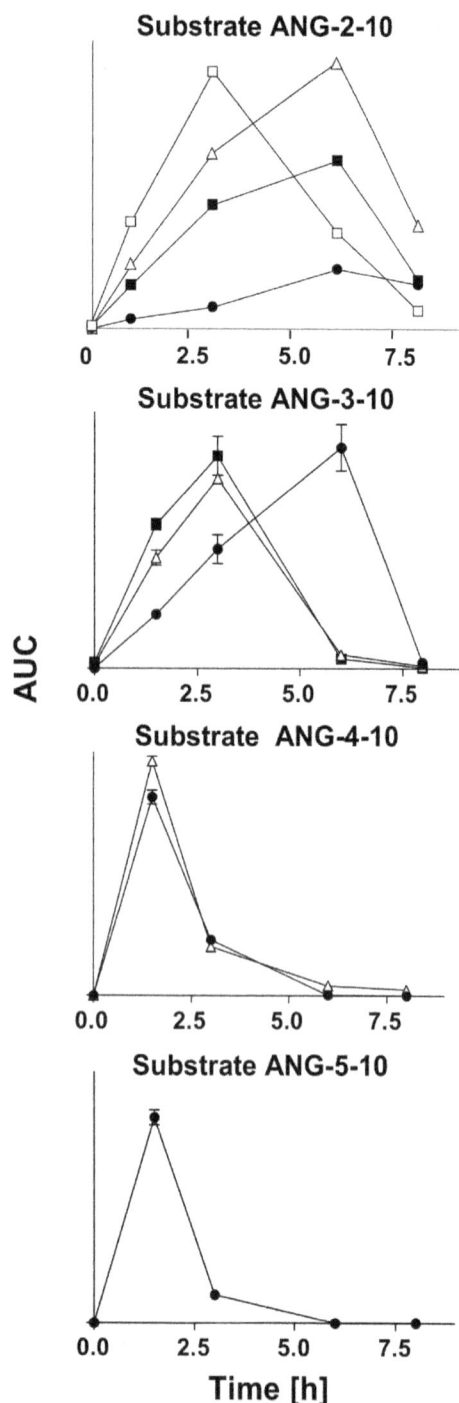

Figure 5. Processing of ANG-2-10, ANG-3-10, ANG-4-10 and ANG-5-10 by immobilized human plasma proteins. Each of these peptides was incubated with immobilized human plasma proteins. Reaction products were analyzed and relatively quantified by SRM-MS. The areas under the curves (AUC) of the generated angiotensin peptides ANG-3-10 (open square), ANG-4-10 (closed square), ANG-5-10 (open triangle), ANG-6-10 (closed circle) are shown (mean ± SEM.). Aliquots were analyzed after 0 h, 2 h, 3 h, 6 h and 8 h (n = 3, except for ANG-2-10 with n = 1).

immobilized human plasma the only observed effect of plasma protein immobilization was a reduced velocity of the angiotensin generation which was between the angiotensin generating velocity of non-immobilized undiluted and 1:100 diluted plasma.

We also investigated whether the peptides generated by aminopeptidase activities from ANG-1-10 are present endogenously in human plasma. Indeed, the angiotensin peptides ANG-4-10, ANG-5-10, ANG-6-10 were detected by an immuno-fluorescence assay in human plasma (Figure 3). The generated calibration curves of the angiotensin peptides ANG-1-10, ANG-4-10 and ANG-5-10 indicate cross-reactions of the immuno-fluorescence assay ANG-1-10-antibody with all peptides comprising the same C-terminus as ANG-1-10. Thus we confirm the finding of Velloso et al., who reported about 100% cross reaction of an ANG-1-10-antibody with ANG-3-10 and ANG-4-10 [33]. This finding is crucial concerning the direct measurement of ANG-1-10 in plasma by comparable antibody based assays, as the presence of peptides which derive from N-terminal cleavage of ANG-1-10 might lead to falsified quantification of the examined angiotensin peptide if the plasma peptides are not separated from each other prior to their quantification.

For the same reason we can not exclude that the quantity of ANG-1-10 that was determined by the immuno-fluorescence assay, was not affected by the presence of ANG-3-10 in human plasma, as the synthetic peptides ANG-3-10 and ANG-1-10 coeluted during reversed phase chromatography (Figure 2 A). Thus we prefer to refer to the sum of ANG-1-10 and ANG-3-10 which are present in human plasma with a concentration of 15 pg/ml. Therefore it might be estimated that ANG-3-10 itself is present in plasma in a picomolar concentration range, which is comparable to the finding of Chappell et al. who detected ANG-3-10 with a concentration of approximately 30 pg/ml in rat plasma [12]. With 53 pg/ml the ANG-5-10 plasma level we detected is comparable to the plasma levels of ANG-1-8 or ANG-1-7, which are in a range of 20–70 pg/ml [34,35,36,37]. The measured ANG-6-10 concentration of 7 ng/ml is much higher than the levels of ANG-1-8 or ANG-1-7 indicating that ANG-6-10 might be the end product of the catalytic cascade of aminopeptidase activities.

A hint for the successive formation of the angiotensin peptides with the same C-terminus as ANG-1-10 is given by the observation that human plasma aminopeptidase activities are not only able to hydrolyze ANG-1-10 but also ANG-2-10, ANG-3-10, ANG-4-10 and ANG-5-10 resulting in ANG-6-10 as the end product (Figure 5).

Various ANG-1-8-generating proteases like ACE-1, chymase and cathepsin G have been identified until today. In contrast, little is known about ANG-2-10, ANG-3-10, ANG-4-10, ANG-5-10 and ANG-6-10-generating proteases. Here, the proteases taking part in the aminopeptidase-dependent formation of angiotensin peptides were characterized by incubation of ANG-1-10 with plasma in the presence and absence of several protease inhibitors and the reaction products were analyzed by SRM-MS (Table 1, Figure S6).

One question of interest was, if only one protease could be responsible for the generation of all these angiotensin peptides. The similarity between the inhibition profiles of ANG-2-10, ANG-3-10, ANG-4-10 (Table 2, Figure S6) indicates that this might be the case. However, in comparison to the inhibition profiles of these peptides the inhibition profile of ANG-5-10 differed a little, in contrast to the profile of ANG-6-10, which showed remarkable divergence. Thus at least two proteases must be responsible for the generation of these angiotensin peptides. An antipain sensitive protease belonging to the serine or cysteine family must be

dissociated by the process of protein immobilization, since the peptides were also generated in non-immobilized undiluted and diluted human plasma (Figure 4). In comparison to non-

Table 1. List of different inhibitors used for characterization of angiotensin-generating proteases.

Inhibitor:	AEBSF	Antipain	Bestatin	Captopril	Chymostatin	EDTA
Protease specificity:	Serine	Serine+Cysteine	Amino-peptidases	ACE-1	Serine	Metallo-proteases
ANG-2-10	/	–	/	/	/	+
ANG-3-10	–	–	/	/	/	+
ANG-4-10	/	–	/	/	/	/
ANG-5-10	/	/	/	/	/	/
ANG-6-10	+	/	+	+	+	–

Influence of inhibitors on generating activity of different angiotensins is indicated by "+" (increase in angiotensin peptide amount) and "–"(decrease in angiotensin peptide amount). Backslash indicates no inhibitory effect.

responsible for the generation of ANG-2-10, ANG-3-10 and ANG-4-10.

APA, a metallopeptidase present in a variety of different tissues and in serum [38], theoretically accounts for the generation of ANG-2-10 as it cleaves ANG-1-8 between the N-terminal amino acids Asp1-Arg2 generating ANG-2-8. But as a metalloprotease EDTA should inhibit APA. Since ANG-2-10 generating activity in human plasma was increased in presence of EDTA APA cannot be responsible for the main part of ANG-2-10 generation. In contrast, EDTA almost completely inhibited the generation of ANG-6-10, whereas ANG-3-10- and ANG-4-10-generating activities increased. This observation clearly shows the existence of an ANG-6-10-generating metalloprotease that differs from the aminopeptidases generating ANG-2-10, ANG-3-10, ANG-4-10. In rat vascular tissue the generation of ANG-5-10 was reported to be completely inhibitable by chymostatin [17]. Formation of the angiotensin peptides ANG-5-10 and ANG-6-10 in human plasma was not inhibited by chymostatin suggesting that the ANG-5-10 generating protease in rat differs from the human one.

ANG-2-8 can be converted into ANG-3-8 by cleavage between the N-terminal amino acids Arg-Val by APN. APN can exist in a membrane bound or soluble state [38,39] since it was purified from human plasma [40] and urine [41]. Favoloro et al. found out that the soluble form of APN owns the predominant functional activity compared to surface associated form [39]. Palmieri et al. also showed that APN is able to process ANG-2-10 in cultured porcine aorta endothelium and smooth muscle cells [42]. Taken together APN is a protease that probably also accounts for the generation of ANG-3-10 from ANG-1-10 in plasma.

Incubation of APN with ANG-1-10 resulted in the generation of ANG-6-10 (Table 2). But since long incubation times were necessary to detect its generation ANG-1-10 does not seem to be the preferred angiotensin substrate of APN. ANG-1-8, when used as a substrate itself, was not cleaved by APN (Table 2). This shows

that generation of ANG-3-8 requires the initial cleavage of ANG-1-8 into ANG-2-8 by APA.

In contrast, the proteolytic APN-activity on ANG-2-10, ANG-3-10, ANG-4-10, ANG-5-10 and ANG-6-10 was much more distinctive and more effective. Hence we confirm the proposed scheme of Velez et al. [18] who suggested the generation of ANG-3-10 from ANG-2-10 by human APN. The same proteolytic pattern could be observed during incubation of ANG-2-10 with human plasma proteins (Figure S5).

Since ANG-6-10 is not further degraded by APN within 24 h it seems to be the stable end product. This might also be an explanation for the high concentration of ANG-6-10 in human plasma that was determined in this work by an immunofluorescence assay.

Taken together, APN as a zinc-dependent plasma protease presumably acts as an ANG-6-10-generating protease in human plasma. This would be substantiated by the fact that the ANG-6-10-generating activity was markedly reduced in the presence of EDTA (Table 1, Figure S6). On the other hand the ANG-6-10-generating activity in human plasma was not reduced in the presence of bestatin, a potent APN inhibitor [43], indicating the presence of an additional metalloprotease with ANG-6-10 generating activity.

The similarities between the generation of angiotensin peptides bearing the same C-terminus as ANG-1-10 in plasma and their generation by APN (Figure 6) also suggest its participation in the generation of ANG-3-10, ANG-4-10 and ANG-5-10 beside an antipain sensitive protease.

The angiotensins bearing the same C-terminus as ANG-1-10 might serve as alternative precursors for the generation of ANG-2-8, ANG-3-8, ANG-4-8, ANG-5-8 and ANG-6-8. Garcia et al. proposed that the generation of ANG-2-10 from ANG-1-10 in rat blood depicts a pathway for the proteolytic processing of ANG-1-10 that bypasses the generation of ANG-1-8 [44]. The authors also

Table 2. Generation of angiotensin peptides from different angiotensin substrates by recombinant aminopeptidase N (APN).

Substrate	ANG-1-10	ANG 1-8	ANG-2-10	ANG-3-10	ANG-4-10	ANG-5-10
Sequence	DRVYIHPFHL	DRVYIHPF	RVYIHPFHL	VYIHPFHL	YIHPFHL	IHPFHL
Cleavage products	HPFHL	none	VYIHPFHL	YIHPFHL	IHPFHL	
			YIHPFHL	HPFHL	HPFHL	HPFHL
			IHPFHL			
			HPFHL			

Sequences of the reaction products which were detected by MALDI-MS are shown.

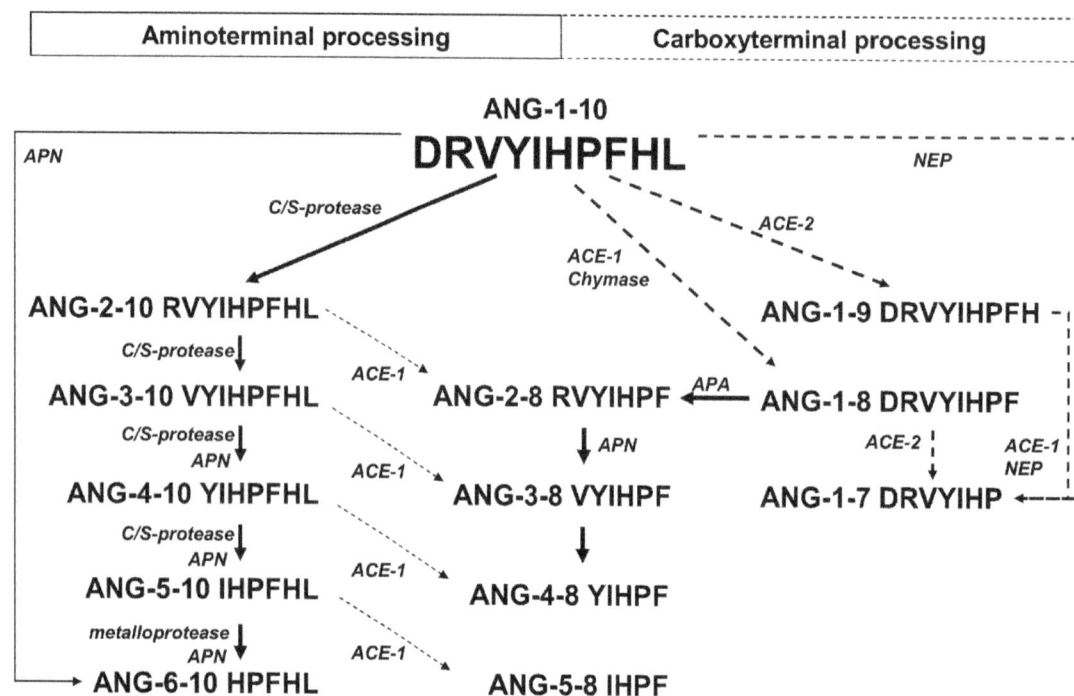

Figure 6. Overview of ANG-1-10 processing by human plasma proteins and by recombinant APN and ACE-1. Arrows with dashed lines indicate hydrolysis from C-terminus. N-terminal-processing is marked by arrows with continuous lines. APN: Aminopeptidase N. C/S-protease: cysteine/serine protease. ACE-1: Angiotensin converting enzyme-1, APA: Aminopeptidase A, NEP: Neprilysin.

stated that ACE-1 contributed to the degradation of ANG-2-10 into ANG-2-8 since this conversion was fully inhibited by the ACE-1-inhibitor captopril. It is likely that in humans ACE-1 also catalyzes the removal of the last two C-terminal amino acids His-Leu from ANG-2-10, ANG-3-10, ANG-4-10, ANG-5-10 and ANG-6-10 *in vivo*. This assumption is strongly supported by the results of our incubation experiments were these angiotensins were incubated with recombinant ACE-1 (Figure S7). In this work we show that recombinant human ACE-1 is able to hydrolyze these angiotensin peptides *in vitro*. The incubation of recombinant ACE-1 resulted in the captopril sensitive release of the last two C-terminal amino acids (His-Leu) from ANG-2-10, ANG-3-10, ANG-4-10 and ANG-5-10 finally leading to the generation of the peptides ANG-2-8, ANG-3-8, ANG-4-8 and ANG-5-8.

In Figure 6 an overview is given about the RAS system including the results of this study. Our results demonstrate for the first time the processing of ANG-1-10 by an aminopeptidase-dependent pathway in human plasma which exists in addition to the well known carboxypeptidase pathway. By the aminopeptidase-dependent pathway ANG-2-10, ANG-3-10, ANG-4-10, ANG-5-10 and ANG-6-10 are generated from ANG-1-10. This shows the efficient multi-parallel utilization of the peptide hormone precursor ANG-1-10.

For the generation of the biologically active angiotensin peptides ANG-2-8 and ANG-3-8 the peptides ANG-2-10 and ANG-3-10 can serve as additional substrates, thus bypassing Ang-1-8. Therefore this Ang-1-8 independent pathway can provide both of these physiologically important peptides.

Many physiological effects of ANG-2-8 and ANG-3-8 are similar to the effects of ANG-1-8, including blood pressure regulation by AT1-mediated vasoconstriction [45]. In patients with hypertension the inhibition of ACE-1 by captopril is a common treatment to lower the blood pressure by decreasing the

generation of ANG-1-8. Since ANG-2-8 and ANG-3-8 also require the action of ACE-1 captopril is also decreasing the generation of these two vasoconstrictors.

Padia et al. described that ANG-2-8 exerts natriuresis in rats via the AT2-receptor. This effect was not produced by ANG-1-8 [46]. The authors speculated that APN inhibitors might be used to treat diseases characterized by sodium and fluid retention, preventing the degradation of ANG-2-8 to ANG-3-8. This would result in an increase of ANG-2-10, which will not be converted to ANG-3-10 by APN. In addition, this should also increase the generation of ANG-2-8 in the presence of ACE-1. However, according to our results an inhibition of APN will presumably decrease the concentrations of ANG-3-10 and ANG-3-8. As a result side effects may occur, because at least ANG-3-8 [45] is known to exert physiological actions.

Before our study it was generally accepted that the generation of ANG-2-8 and ANG-3-8 goes hand in hand with the hydrolysis of ANG-1-8. Since the regulation of biological processes requires independent control circuits an ANG-1-8-independent pathway for the generation of ANG-2-8 and ANG-3-8 makes sense.

In conclusion our findings demonstrate that the RAS system is equipped with a large number of independently controllable regulator elements which in the future should be investigated in depth regarding the clinical relevance and their impact on drug actions.

Supporting Information

Figure S1 Processing of ANG-1-10 by immobilized mouse plasma proteins. ANG-1-10 (10^{-5} M) was incubated with immobilized mouse plasma proteins. Reaction products were detected by MALDI-MS after 0 h, 3 h, 4 h and 8 h. MALDI-MS signals corresponding to angiotensin peptides are marked by arrows. Control: ANG-1-10 incubated for 24 h with Sepharose-beads® without immobilized proteins.

Figure S2 Processing of ANG-1-10 by immobilized rat plasma proteins. ANG-1-10 (10^{-5} M) was incubated with immobilized rat plasma proteins. Reaction products were detected by MALDI-MS after 0 h, 1 h, 3 h and 8 h. MALDI-MS signals corresponding to angiotensin peptides are marked by arrows. Control: ANG-1-10 incubated for 24 h with Sepharosebeads® without immobilized proteins.

Figure S3 Processing of ANG-1-10 by immobilized human female plasma proteins. ANG-1-10 (10^{-5} M) was incubated with immobilized human female plasma proteins. Reaction products were detected by MALDI-MS after 0 h, 3 h, 4 h and 8 h. MALDI-MS signals corresponding to angiotensin peptides are marked by arrows. Control: ANG-1-10 incubated for 24 h with immobilized heat inactivated plasma proteins.

Figure S4 Identification of angiotensin peptides by tandem mass spectrometry. The reaction products A) ANG-1-9, B) ANG-1-8, C) ANG-1-7, D) ANG-2-10, E) ANG-3-10, F) ANG-4-10, G) ANG-5-10 and H) ANG-6-10 that were generated after incubation of ANG-1-10 with human immobilized plasma proteins were identified by LC/−ESI-IT-MS/MS. Relating MS/MS spectra with assigned b- and y-fragment ions (dashed lines) and the deduced peptide sequence are shown.

Figure S5 Processing of ANG-2-10 by immobilized human plasma proteins. ANG-2-10 (10^{-5} M) was incubated with immobilized human plasma proteins. Reaction products were detected by MALDI-MS after 8 h. MALDI-MS signals corresponding to angiotensin peptides are marked by arrows.

Figure S6 Processing of ANG-1-10 by immobilized human plasma proteases in the presence and abscence of protease inhibitors. ANG-1-10 (10^{-5} M) was incubated with human plasma proteins in presence of 200 μM AEBSF, 50 μM antipain, 150 μM bestatin, 10 μM captopril, 100 μM chymostatin, 100 μM EDTA and absence of any inhibitor. Reaction products were analyzed by SRM-MS after 6 h and the areas under the curves (AUC) of ANG-2-10, ANG-3-10, ANG-4-

10, ANG-5-10 and ANG-6-10 from the SRM-MES are shown (mean ± SEM, n = 3). Right y-axis: AUC of ANG-2-10 and ANG-3-10 after incubation in presence of EDTA.

Figure S7 Processing of angiotensin peptides by recombinant ACE-1. The angiotensin peptides (10^{-5} M) were incubated with 0.25 μg recombinant human ACE-1 in presence (left panel) and absence of captopril (right panel). ACE-1 was incubated with A) ANG-1-10, B) ANG-2-10, C) ANG-3-10, D) ANG-4-10, E) ANG-5-10. Reaction products were analyzed by MALDI-MS after an incubation time of 8 h. MALDI-MS-Signals which were assigned as angiotensin peptides are marked by arrows.

Table S1 SRM-transitions and settings for relative quantification of angiotensin peptides by SRM-coupled LC-ESI-QQQ-MS. Charge states of the precursor ions are denoted in brackets.

Table S2 Sequences and protonated monoisotopic masses of angiotensin peptides that were generated in human plasma.

Methods S1 Immobilization of plasma proteins.

Methods S2 Mass spectrometric peptide identification and quantification.

Acknowledgments

We thank Prof. Dr. Martin Tepel (Odense University Hospital and University of Southern Denmark, Institute of Molecular Medicine, Cardiovascular and Renal Research) for providing the murine plasma. In addition we thank Dr. Petra Henklein (Group of Prof. Dr. Peter-Michael Kloetzel, Charité Berlin, Institute for biochemistry) for the synthesis of the angiotensin peptides.

Author Contributions

Conceived and designed the experiments: DH HS. Performed the experiments: DH PM LFE. Analyzed the data: DH PM LFE. Contributed reagents/materials/analysis tools: DH HS. Wrote the paper: DH HS.

References

1. Fyhrquist F, Saijonmaa O (2008) Renin-angiotensin system revisited. J Intern Med 264: 224–236.
2. Capettini LS, Montecucco F, Mach F, Stergiopulos N, Santos RA, et al. (2012) Role of renin-angiotensin system in inflammation, immunity and aging. Curr Pharm Des 18: 963–970.
3. Kranzhofer R, Browatzki M, Schmidt J, Kubler W (1999) Angiotensin II activates the proinflammatory transcription factor nuclear factor-kappaB in human monocytes. Biochem Biophys Res Commun 257: 826–828.
4. Campbell-Boswell M, Robertson AL Jr (1981) Effects of angiotensin II and vasopressin on human smooth muscle cells in vitro. Exp Mol Pathol 35: 265–276.
5. Gard PR, Naylor C, Ali S, Partington C (2012) Blockade of pro-cognitive effects of angiotensin IV and physostigmine in mice by oxytocin antagonism. Eur J Pharmacol 683: 155–160.
6. Wright JW, Harding JW (2013) The brain renin-angiotensin system: a diversity of functions and implications for CNS diseases. Pflugers Arch 465: 133–51.
7. Urata H, Nishimura H, Ganten D (1996) Chymase-dependent angiotensin II forming systems in humans. Am J Hypertens 9: 277–284.
8. Welches WR, Brosnihan KB, Ferrario CM (1993) A comparison of the properties and enzymatic activities of three angiotensin processing enzymes: angiotensin converting enzyme, prolyl endopeptidase and neutral endopeptidase 24.11. Life Sci 52: 1461–1480.

9. Jackman HL, Massad MG, Sekosan M, Tan F, Brovkovych V, et al. (2002) Angiotensin 1–9 and 1–7 release in human heart: role of cathepsin A. Hypertension 39: 976–981.
10. Dupont AG, Brouwers S (2010) Brain angiotensin peptides regulate sympathetic tone and blood pressure. J Hypertens 28: 1599–1610.
11. von Bohlen und Halbach O (2003) Angiotensin IV in the central nervous system. Cell Tissue Res 311: 1–9.
12. Chappell MC, Brosnihan KB, Diz DI, Ferrario CM (1989) Identification of angiotensin-(1–7) in rat brain. Evidence for differential processing of angiotensin peptides. J Biol Chem 264: 16518–16523.
13. Neves LA, Almeida AP, Khosla MC, Santos RA (1995) Metabolism of angiotensin I in isolated rat hearts. Effect of angiotensin converting enzyme inhibitors. Biochem Pharmacol 50: 1451–1459.
14. Champion HC, Garrison EA, Estrada LS, Potter JM, Kadowitz PJ (1996) Analysis of responses to angiotensin I-(3–10) in the mesenteric vascular bed of the cat. Eur J Pharmacol 309: 251–259.
15. Garrison EA, Kadowitz PJ (1996) Analysis of responses to angiotensin I-(3–10) in the hindlimb vascular bed of the cat. Am J Physiol 270: H1172–1177.
16. Kaye AD, Nossaman BD, Smith DE, Ibrahim IN, Anwar M, et al. (1998) Analysis of responses to angiotensin I (3–10) and Leu3 angiotensin (3–8) in the pulmonary vascular bed of the cat. Am J Ther 5: 295–302.

17. Takai S, Sakaguchi M, Jin D, Yamada M, Kirimura K, et al. (2001) Different angiotensin II-forming pathways in human and rat vascular tissues. Clin Chim Acta 305: 191–195.

18. Velez JC, Ierardi JL, Bland AM, Morinelli TA, Arthur JM, et al. (2012) Enzymatic processing of angiotensin peptides by human glomerular endothelial cell. Am J Physiol Renal Physiol 302: F1583–1594.

19. Trusch M, Bohlick A, Hildebrand D, Lichtner B, Bertsch A, et al. (2010) Application of displacement chromatography for the analysis of a lipid raft proteome. J Chromatogr B Analyt Technol Biomed Life Sci 878: 309–314.

20. Schluter H, Jankowski J, Rykl J, Thiemann J, Belgardt S, et al. (2003) Detection of protease activities with the mass-spectrometry-assisted enzyme-screening (MES) system. Anal Bioanal Chem 377: 1102–1107.

21. Brena BM, Batista-Viera F (2006) Immobilization of Enzymes and Cells (Methods in Biotechnology). New Jersey: © Humana Press Inc. 464 p.

22. Ward PE, Benter IF, Dick L, Wilk S (1990) Metabolism of vasoactive peptides by plasma and purified renal aminopeptidase M. Biochem Pharmacol 40: 1725–1732.

23. Reaux A, Fournie-Zaluski MC, David C, Zini S, Roques BP, et al. (1999) Aminopeptidase A inhibitors as potential central antihypertensive agents. Proc Natl Acad Sci USA 96: 13415–13420.

24. Zini S, Fournie-Zaluski MC, Chauvel E, Roques BP, Corvol P, et al. (1996) Identification of metabolic pathways of brain angiotensin II and III using specific aminopeptidase inhibitors: predominant role of angiotensin III in the control of vasopressin release. Proc Natl Acad Sci USA 93: 11968–11973.

25. Wright JW, Harding JW (2011) Brain renin-angiotensin–a new look at an old system. Prog Neurobiol 95: 49–67.

26. Bader M, Ganten D (2008) Update on tissue renin-angiotensin systems. J Mol Med (Berl) 86: 615–621.

27. Velez JC, Ryan KJ, Harbeson CE, Bland AM, Budisavljevic MN, et al. (2009) Angiotensin I is largely converted to angiotensin (1–7) and angiotensin (2–10) by isolated rat glomeruli. Hypertension 53: 790–797.

28. Sim MK, Qiu XS (1994) Formation of des-Asp-angiotensin I in the hypothalamic extract of normo- and hypertensive rats. Blood Press 3: 260–264.

29. Cole KR, Kumar S, Trong HL, Woodbury RG, Walsh KA, et al. (1991) Rat mast cell carboxypeptidase: amino acid sequence and evidence of enzyme activity within mast cell granules. Biochemistry 30: 648–655.

30. Barrett AJ, Starkey PM (1973) The interaction of alpha 2-macroglobulin with proteinases. Characteristics and specificity of the reaction, and a hypothesis concerning its molecular mechanism. Biochem J 133: 709–724.

31. Heeb MJ, Espana F (1998) alpha2-macroglobulin and C1-inactivator are plasma inhibitors of human glandular kallikrein. Blood Cells Mol Dis 24: 412–419.

32. Lew RA, Warner FJ, Hanchapola I, Yarski MA, Manohar J, et al. (2008) Angiotensin-converting enzyme 2 catalytic activity in human plasma is masked by an endogenous inhibitor. Exp Physiol 93: 685–693.

33. Velloso EP, Vieira R, Cabral AC, Kalapothakis E, Santos RA (2007) Reduced plasma levels of angiotensin-(1–7) and renin activity in preeclamptic patients are associated with the angiotensin I- converting enzyme deletion/deletion genotype. Braz J Med Biol Res. 40: 583–590.

34. Bluher M, Kratzsch J, Paschke R (2001) Plasma levels of tumor necrosis factor-alpha, angiotensin II, growth hormone, and IGF-I are not elevated in insulin-resistant obese individuals with impaired glucose tolerance. Diabetes Care 24: 328–334.

35. Jalil JE, Palomera C, Ocaranza MP, Godoy I, Roman M, et al. (2003) Levels of plasma angiotensin-(1–7) in patients with hypertension who have the angiotensin-I-converting enzyme deletion/deletion genotype. Am J Cardiol 92: 749–751.

36. Kappelgaard AM, Nielsen MD, Giese J (1976) Measurement of angiotensin II in human plasma: technical modifications and practical experience. Clin Chim Acta 67: 299–306.

37. Reyes-Engel A, Morcillo L, Aranda FJ, Ruiz M, Gaitan MJ, et al. (2006) Influence of gender and genetic variability on plasma angiotensin peptides. J Renin Angiotensin Aldosterone Syst 7: 92–97.

38. Lalu K, Lampelo S, Nummelin-Kortelainen M, Vanha-Perttula T (1984) Purification and partial characterization of aminopeptidase A from the serum of pregnant and non-pregnant women. Biochim Biophys Acta 789: 324–333.

39. Favaloro EJ, Browning T, Facey D (1993) CD13 (GP150; aminopeptidase-N): predominant functional activity in blood is localized to plasma and is not cell-surface associated. Exp Hematol 21: 1695–1701.

40. Tokioka-Terao M, Hiwada K, Kokubu T (1984) Purification and characterization of aminopeptidase N from human plasma. Enzyme 32: 65–75.

41. Scherberich JE, Wiemer J, Herzig C, Fischer P, Schoeppe W (1990) Isolation and partial characterization of angiotensinase A and aminopeptidase M from urine and human kidney by lectin affinity chromatography and high-performance liquid chromatography. J Chromatogr 521: 279–289.

42. Palmieri FE, Ward PE (1989) Dipeptidyl(amino)peptidase IV and post proline cleaving enzyme in cultured endothelial and smooth muscle cells. Adv Exp Med Biol 247A: 305–311.

43. Inoue T, Kanzaki H, Imai K, Narukawa S, Higuchi T, et al. (1994) Bestatin, a potent aminopeptidase-N inhibitor, inhibits in vitro decidualization of human endometrial stromal cells. J Clin Endocrinol Metab 79: 171–175.

44. Garcia Del Rio C, Smellie WS, Morton JJ (1981) des-Asp-angiotensin I: its identification in rat blood and confirmation as a substrate for converting enzyme. Endocrinology 108: 406–412.

45. Li Q, Feenstra M, Pfaffendorf M, Eijsman L, van Zwieten PA (1997) Comparative vasoconstrictor effects of angiotensin II, III, and IV in human isolated saphenous vein. J Cardiovasc Pharmacol 29: 451–456.

46. Padia SH, Kemp BA, Howell NL, Siragy HM, Fournie-Zaluski MC, et al. (2007) Intrarenal aminopeptidase N inhibition augments natriuretic responses to angiotensin III in angiotensin type 1 receptor-blocked rats. Hypertension 49: 625–630.

13

Neural Circuit Interactions between the Dorsal Raphe Nucleus and the Lateral Hypothalamus: An Experimental and Computational Study

Jaishree Jalewa[1☉], **Alok Joshi**[2☉], **T. Martin McGinnity**[2], **Girijesh Prasad**[2], **KongFatt Wong-Lin**[2*], **Christian Hölscher**[3*]

1 School of Biomedical Sciences, University of Ulster, Coleraine, Northern Ireland, United Kingdom, **2** Intelligent Systems Research Centre, University of Ulster, Magee Campus, Londonderry, Northern Ireland, United Kingdom, **3** Division of Biomedical and Life Sciences, Faculty of Health and Medicine, Lancaster University, Lancaster, United Kingdom

Abstract

Orexinergic/hypocretinergic (Ox) neurotransmission plays an important role in regulating sleep, as well as in anxiety and depression, for which the serotonergic (5-HT) system is also involved in. However, little is known regarding the direct and indirect interactions between 5-HT in the dorsal raphe nucleus (DRN) and Ox neurons in the lateral hypothalamus (LHA). In this study, we report the additional presence of $5\text{-HT}_{1B}R$, $5\text{-HT}_{2A}R$, $5\text{-HT}_{2C}R$ and fast ligand-gated $5\text{-HT}_{3A}R$ subtypes on the Ox neurons of transgenic Ox-enhanced green fluorescent protein (Ox-EGFP) and wild type C57Bl/6 mice using single and double immunofluorescence (IF) staining, respectively, and quantify the colocalization for each 5-HT receptor subtype. We further reveal the presence of $5\text{-HT}_{3A}R$ and $5\text{-HT}_{1A}R$ on GABAergic neurons in LHA. We also identify NMDAR1, OX_1R and OX_2R on Ox neurons, but none on adjacent GABAergic neurons. This suggests a one-way relationship between LHA's GABAergic and Ox neurons, wherein GABAergic neurons exerts an inhibitory effect on Ox neurons under partial DRN's 5-HT control. We also show that Ox axonal projections receive glutamatergic (PSD-95 immunopositive) and GABAergic (Gephyrin immunopositive) inputs in the DRN. We consider these and other available findings into our computational model to explore possible effects of neural circuit connection types and timescales on the DRN-LHA system's dynamics. We find that if the connections from 5-HT to LHA's GABAergic neurons are weakly excitatory or inhibitory, the network exhibits slow oscillations; not observed when the connection is strongly excitatory. Furthermore, if Ox directly excites 5-HT neurons at a fast timescale, phasic Ox activation can lead to an increase in 5-HT activity; no significant effect with slower timescale. Overall, our experimental and computational approaches provide insights towards a more complete understanding of the complex relationship between 5-HT in the DRN and Ox in the LHA.

Editor: Maurice J. Chacron, McGill University, Canada

Funding: The work has been supported by the Centre of Excellence in Intelligent Systems Project, ISRC, University of Ulster, Magee, Northern Ireland. The funders had no role in study design, data collection and analysis, decision to publish, or preparation of the manuscript.

* E-mail: c.holscher@lancaster.ac.uk (CH); k.wong-lin@ulster.ac.uk (KW)

☉ These authors contributed equally to this work.

Introduction

Mood and neuropsychiatric disorders such as depression have a close relationship with sleep disturbances, and are instantiated by the overlap of emotional processing and the sleep-wake regulation neuronal circuitries [1]. The neuropeptide hormone orexin/hypocretin (Ox) has been known to regulate sleep and its deregulation is related to narcolepsy, and novel drugs to facilitate sleep induction by activating Ox receptors are currently under development [2,3,4]. Recent studies also suggest a role for Ox in depression, emotional processing, reward seeking behaviour and in the regulation of endocrine functions [5,6,7,8,9,10]. Ox neurons comprising of neuropeptides Ox A and Ox B are found predominantly in the lateral hypothalamus (LHA) [11,12] and are known to function through OX_1R and OX_2R G-protein coupled receptors, respectively [13,14,15,16].

The neurotransmitter/neuromodulator 5-hydroxytryptamine (5-HT) released by serotonergic neurons, substantially located in the midbrain's dorsal raphe nucleus (DRN) is often associated with mood and emotional processing, and its dysfunction is related to mood and neuropsychiatric disorders [17,18,19]. In addition, perturbations of 5-HT have also been found to influence sleep [18,20,21,22]. Numerous drugs to treat depression are already on the market that target 5-HT neurotransmission [23,24].

It is important to understand how drugs that target Ox and/or 5-HT systems alter neuronal activity and signal transmission in order to understand the underlying mechanisms of antidepressive and sleep inducing effects. Therefore, we set out to map what subtypes of 5-HT receptors are expressed by Ox neurons, and how neuronal transmission and signal transduction in neuronal circuits may be controlled by these receptors.

Up till now, only the (inhibitory) 5-HT$_{1A}$R has so far been found in LHA's Ox neurons [25,26,27]. In addition to their inhibitory (5-HT$_{1A}$) autoreceptors [28], 5-HT also excites the GABAergic inhibitory neurons within the DRN for self-regulation [29,30,31,32,33]. Within the LHA, in addition to their self-excitatory Ox autoreceptors [4], Ox neurons can send direct long-range excitation to 5-HT neurons, and the GABAergic neurons in the DRN [31], mediated by both OX$_2$R and OX$_1$R [30,32]. Thus, Ox can have both direct and indirect influences on the DRN's 5-HT neurons. However, it remains unknown whether 5-HT can reciprocally indirectly influence LHA's Ox neurons by influencing the LHA's GABAergic neurons, and whether this connection is effectively excitatory or inhibitory. It is also not known whether Ox can innervate its local GABAergic neurons similar to how 5-HT neurons excite their local GABAergic neurons [34].

Knowledge of direct and indirect circuit connections is important to provide a more complete understanding of diversified neural circuit dynamics and regulations within the DRN-LHA system [35,36]. Furthermore, it is not known how the interplay between fast synaptic transmission and slow currents induced by 5-HT and Ox can affect the relay of information in these circuits. In this work, primarily through immunofluorescence (IF) staining, we attempt to map out a more complete neural circuit between the principal (5-HT and Ox) and non-principal (GABAergic and glutamatergic) neurons. Based on some of these findings, we present a computational model to investigate possible neural circuit dynamics between the DRN and LHA. As our goal in the modelling is to understand how the various neural circuit architecture and connectivity timescale affect the DRN-LHA activity, we keep the model as simple as possible. Hence we make use of population-based or "mean-field" firing-rate type approach, which compromises between previous biophysical models [37,38,39,40] and more abstract mathematical models [41,42,43].

Our experimental results reveal various other 5-HT receptor subtypes expressed in Ox and GABAergic neurons in the LHA, provide more evidence to support a unidirectional relationship between these LHA's neurons, and suggest that Ox can project to DRN's 5-HT neurons indirectly through local non-5-HT neurons. Our computational modelling results show that if 5-HT is weakly excitatory or inhibitory on LHA's GABAergic neurons, the network can exhibit slow oscillation. This is not observed if the connection is strongly excitatory. Furthermore, we show the importance of the timescale for the Ox-to-DRN connection during transient behaviour.

Materials and Methods

Animals

Twelve-week-old C57BL/6 male mice (n = 4) were used for each qualitative experiment and n = 6 mice for each quantification experiment described in this study. An orexin/enhanced green fluorescent protein (Ox-EGFP) breeder pair was a kind gift from Prof. Takeshi Sakurai (Kanazawa University, Japan). Brain sections from these mice show green fluorescence in the Ox neurons when excited at 488 nm wavelength. Breeding was set up in-house and the male pups were aged to ten-twelve weeks before the start of the experiments. Animals were maintained on a 12/12 h light/dark cycle (lights on at 8:00 A.M., off at 8:00 P.M.), in a temperature-controlled room (21.5±1°C). Animals received food and water *ad libitum*. All animal experiments were licenced by the UK Home Office in accordance with the Animals (Scientific Procedures) Act of 1986 and in agreement with UK and EU laws.

Perfusion, Fixation and Sectioning

Mice were anaesthetised with pentobarbitone (0.3 ml; Euthanal, Bayer AG, Leverkusen, Germany) and perfused transcardially with 0.1 M PBS (pH 7.4) buffer followed by ice-cold 4% paraformaldehyde in PBS. The brains were removed and fixed in 4% paraformaldehyde for at least 24 hr and cryoprotected in a 30% sucrose solution in PBS overnight at 4°C. Brains were then snap frozen using Envirofreez, and coronal sections of 45 μm thickness were cut using a Leica cryostat. According to the mouse brain atlas by Paxinos and Franklin (2004), LHA and DRN sections were cut at a depth of −0.34 mm to −2.80 mm bregma and −4.04 mm to −5.20 mm bregma respectively. Sections were chosen according to the stereological rules, with the first section taken at random and every sixth section afterward. In the case of LHA, by taking every 6th section (in total 54 sections per LHA per brain half), at least 10–11 sections were taken per immunostaining experiment (n = 4). In the case of DRN, by taking every 3rd section (in total 25 sections per DRN per full brain), at least 11–12 sections were taken per immunostaining experiment (n = 4). 10–11 sections from 4 mice brain halves (in case of LHA) and full brain (in case of DRN) were processed so, at least 40 sections were considered in total for every immunostaining experiment.

Immunohistochemistry

Single, double or triple immunofluorescence (IF) staining experiments were performed on 45 μm free-floating sections using primary antibodies: i) affinity purified goat polyclonal Ox-A (C-19) IgG (1:400 dilution, Santa Cruz, sc-8070), raised against a peptide mapping at the C-terminus of Orexin-A of human origin; ii) rabbit polyclonal anti-5HT$_{1B}$ receptor IgG (1:500 dilution, Abcam, ab102700) raised against a synthetic peptide taken from within the region 230–280 (designed to the 3rd cytoplasmic domain) of the human 5HT$_{1B}$ receptor conjugated to an immunogenic carrier protein; iii) rabbit polyclonal anti-5HT$_{1A}$ receptor IgG (1:500 dilution, Abcam, ab79230) raised against a synthetic peptide from the 3rd cytoplasmic domain of mouse 5HT$_{1A}$ receptor, conjugated to an immunogenic carrier protein [44]; iv) rabbit polyclonal anti-5HT$_{2A}$ receptor IgG (1:500 dilution, Abcam, ab16028) raised against a synthetic peptide conjugated to KLH derived from within residues 1–100 of rat 5HT$_{2A}$ receptor [45,46,47,48,49]; v) rabbit polyclonal anti-5HT$_{2C}$ receptor IgG (1:500 dilution, Abcam, ab32172) raised against a synthetic peptide derived from within residues 400 to the C-terminus of rat 5HT2C receptor [48,50,51,52]; vi) rabbit polyclonal anti-HTR3A receptor IgG (1:500 dilution, Sigma-Aldrich, AV13046) raised against a synthetic peptide corresponding to a region of human HTR3A with an internal ID of P01375. The immunogen for anti-HTR3A antibody was a synthetic peptide directed towards the N-terminal of human HTR3A, with the following sequence: LLWVQQAL-LALLLPTLLAQGEARRSRNTTRPALLRLSDYLLTNYRKVG-RP; vii) rabbit monoclonal anti-NMDAR, clone 1.17.2.6 IgG (1:500 dilution, Millipore, AB9864R) raised against linear peptide corresponding to human NMDAR1; viii) guinea pig polyclonal anti-GABA IgG (1:400 dilution, Millipore, AB175) raised against GABA coupled to KLH via glutaraldehyde; ix) rabbit polyclonal anti-Orexin receptor 1 IgG (1:500 dilution, Abcam, ab83960) raised against a synthetic peptide from the internal region (250–300) of mouse Orexin receptor 1 conjugated to an immunogenic carrier; x) rabbit polyclonal anti-Orexin receptor 2 IgG (1:500 dilution, Abcam, ab85899) raised against a synthetic peptide from the C terminal region (350–415) of human Orexin receptor 2 conjugated to an immunogenic carrier; xi) rabbit monoclonal PSD-95 (D27E11) XP IgG (1:500 dilution, Cell Signaling Technology, 3450S) raised against a synthetic peptide corresponding to residues surrounding Gly99 of human PSD95; xii) rabbit polyclonal Anti-Gephyrin IgG (1:500 dilution, Abcam, ab32206) raised against a synthetic peptide conjugated to KLH derived from within residues 700 to the C-terminus of mouse Gephyrin.

All primary antibodies used in this study were previously characterized and their specificity was verified according to the respective manufacturer. After blocking in 1% BSA and 5% donkey normal serum in TBS buffer (pH 8, Sigma Aldrich) to avoid nonspecific antibody binding, sections were incubated in the primary antibody overnight at 4°C. The following day, sections were incubated in the secondary antibody for an hour at room temperature and mounted using Vectashield mounting medium (Vector Laboratories) on the slides coated with 3-aminopropyl triethoxy silane (Sigma Aldrich). For double IF stainings, a simultaneous method was used where sections were incubated with two primary antibodies together for 48 hrs at 4°C. In the case of triple IF stainings, three primary antibodies were added together and sections were incubated for 72 hrs at 4°C. Labeled donkey IgG (H+L) anti-goat Alexa Fluor 488 (1:800 dilution, Cat. # A11055, Molecular Probes), anti-rabbit Alexa Fluor 546 (1:1000 dilution, Cat. # A11056, Molecular Probes), anti-chicken CF 594 (1:1000 dilution, Cat. # BTIU20167, Biotium) and anti-guinea pig CF 633 (1:1000 dilution, Cat. # BTIU20171, Biotium) secondary antibodies were used in the study. Negative controls were performed for single IF staining by omitting the primary antibody and for double/triple IF staining by omitting the primary and secondary antibody.

Antigen Retrieval

Antigen retrieval was done while performing NMDAR1 IF staining. Sections were incubated in 10 mM sodium citrate (pH 6) at 80°C for 30 min, before blocking the sections.

Microscopy

Fluorescence microscopy. All single IF stainings in EGFP brain sections for 5-HT receptors on the Ox neurons in the LHA were visualized using Qimaging (Chromaphor). Microscopy was performed using an Olympus BX51 (Surveyor version 5.5.5.30, automated specimen scanning for the OASIS automation control system).

Confocal microscopy. Imaging was performed using a confocal microscope (Leica Microsystems; SP5 LAS IF Software). For quantification experiments, three sections of similar density of Ox neurons in the LHA were analyzed per brain (n = 6). 4–5 images were obtained from each section thus, 70–90 images were analyzed for each quantification experiment.

Co-localization Quantification

For quantification of the 5HT receptor subtype on the Ox neurons in LHA, images were acquired using 63× objective in a Leica SP5 confocal microscope. Once the conditions such as photomultiplier gain for each channel and pinhole settings were adjusted to minimize background noise and saturated pixels, parameters were kept constant for all acquisitions. Triple-stained images were obtained by sequential scanning for each channel to eliminate the cross talk of chromophores and to ensure reliable quantification of co-localization. Ambiguity and inconsistency are the two major issues affecting colocalization analysis. In the context of digital imaging, colocalization means the colours emitted by the fluorescent molecules occupy the same pixel in the image [53,54]. Therefore, we have used the JACoP (Just Another Colocalization Plug-in) tool of ImageJ for colocalization analysis. The degree of Ox neuron (Alexa488, green) signal

colocalizing with 5HT receptor (Alexa546, red) signal was quantified on single-plane 8-bit color images using the JACoP plugin [55]. A simple way of measuring the dependency of pixels in dual-channel images is to plot the pixel grey values of two images against each other. The intensity of a given pixel in the green image is used as the x-coordinate of the scatter plot and the intensity of the corresponding pixel in the red image as the y-coordinate. Results in JACoP are displayed in a pixel distribution diagram called a scatter plot or fluorogram in addition to the calculated co-localization coefficients such as Pearson's and Overlap coefficient.

Pearson's correlation coefficient (Rr) is the most quantitative estimate of colocalization that depends on the amount of colocalized signals in both channels in a nonlinear manner and is a well-defined and commonly accepted means for describing the extent of overlap between image pairs [56]. It is used for describing the correlation of the intensity distributions between channels. It takes into consideration only similarity between shapes, while ignoring the intensities of signals. The values of Pearson's coefficient range from -1 to 1, with values from 0.5 to 1.0 indicating colocalization and -1.0 to 0.5 indicating no-colocalization. As Pearson's Correlation does some averaging of pixel information and can return negative values another method, the Overlap Coefficient, is simultaneously used to describe overlap. Manders' overlap coefficient (R) is based on the Pearson's correlation coefficient with average intensity values being taken out of the mathematical expression. This new coefficient varies from 0 to 1, with values from 0.6 to 1.0 indicating colocalization and 0 to 0.6 indicating no colocalization. Overlap coefficient according to Manders indicates an overlap of the signals and thus represents the true degree of Colocalization. Costes randomization (number of randomization rounds = 1000) was used to exclude any co-localization of pixels that might have occurred due to chance [57]. P-value for each image pair was 100.0% (calculated from the fitted data).

Statistics

Statistical analyses were performed using Prism 5 (GraphPad software Inc. USA) with the level of probability set at 95% and the results are expressed as means±SEM. Data for 5-HT receptor quantification was analysed by two-tailed unpaired t-test.

Computational Model

To investigate the consequences of the DRN-LHA circuit architecture on systems dynamics, we implement neural network model that is an extension and modification of our previous model [58]. The aim of the model is to understand how the circuit connectivity and timescale affect the DRN-LHA activity.

Neural units. Our neural network model consists of 4 populations of neurons, namely, Ox neurons in the LHA, local LHA inhibitory GABAergic neurons, 5-HT neurons in the DRN, and local DRN inhibitory GABAergic neurons. Glutamatergic neurons will be ignored in this work primarily due to the evidence showing that glutamatergic effects in DRN are locally weaker when compared to local GABAergic influence [59], and that we can implicitly encompass the effects of the LHA's glutamatergic neurons on Ox neurons [4] with self-excitatory Ox connections. Furthermore, incorporating two additional (glutamatergic) neural populations can lead to more free parameters in the model.

The chosen model is of the population-averaged or "mean-field" firing rate type model [60,61,62,63]. This simplifies a population of neurons into its representative unit. The 4 neural populations to be considered are the LHA's Ox and GABAergic neural populations, and the DRN's 5-HT and GABAergic neural populations. With support from electrophysiological data, the input-output or current-frequency relationship (f-I curve) for each neural population can be described by threshold-linear functions [64,65]:

$$f_j(I_j) = g_j\left[I_{local,j} - I_{0,j} - I_{Background,j}\right]_+ \qquad (1)$$

where f_j and $(I_{Local,j} - I_{Background,j})$ denote the population-averaged firing rate activity and total afferent input of the jth neural population, respectively. $I_{Background,j}$ is the background current, consisting of inputs from the rest of the brain areas. g_j and $I_{0,j}$ determine the input-output slope and the current threshold, respectively. The threshold-linear function $[z]_+ = z$ if $z>0$, and 0 otherwise. Based on their known neuronal electrophysiological properties [4,27,66]), we determine and constrain the values of the g_j's (i.e. g_{5-HT} and g_{GABA} for DRN; g_{GABA} and g_{Ox} for LHA) and the $I_{0,j}$'s, while consider $I_{Background,j}$ as a free parameter. Note that we have assumed the dynamics of the neural population (neuronal membrane time constant $\sim$ 10 ms) to be relatively instantaneous and slaved to the much slower timescale of the connections ($\sim$ s) [63]. This simplification reduces 4 model parameters (neuronal membrane time constants) and 4 dynamical (differential) equations for the 4 neuronal types [60].

2.8.2. Inputs and connections: Since each neural unit receives inputs from self-feedback and from 2 other neural units (e.g. 5-HT neuronal population receives both afferent inputs from DRN's GABAergic neurons and longer range connection from Ox neurons), the local afferent input (minus the background input) can be described by:

$$I_j = I_{j,self} + I_{j,1} + I_{j,2} \qquad (2)$$

where the subscript "self" denotes a self-feedback connection (e.g. due to autoreceptors in Ox and 5-HT neural populations, or GABAergic synapses in the two inhibitory neural populations), and the subscript "1" or "2" denotes the afferent inputs due to the other 2 neural populations. For local GABAergic neurons, they receive self-inhibition, and projections from their local principal neural population (5-HT or Ox if from DRN or LHA, respectively), and long-range projection from the other brain region (LHA or DRN). If the net effect of the ith neural population on the jth neural population is inhibitory or excitatory, the coefficient in front of $I_{j,i}$ will be -1 or $+1$, respectively. For example, suppose the 5-HT neural population receives inhibitory autoreceptor influence, direct projection from Ox neurons, and local GABAergic influence. Then,

$$I_{5-HT} = -I_{5-HT-auto} + I_{Ox-on-5-HT} - I_{GABA(DRN)-on-5-T} \quad (3)$$

To simulate the phasic response of the circuit, an extra term I_{stim} is added to the 5-HT or/and Ox inputs

$$I_{5-HT} = -I_{5-HT-auto} + I_{Ox-on-5-HT} -$$
$$I_{GABA(DRN)-on-5-T} + I_{stim} \qquad (4)$$

The dynamics of each input are filtered by its (synaptic/effective) time constant ($\tau_{syn, j, i}$) as follows:

$$T_{syn,j,i}\frac{dI_{j,i}}{dt} = -I_{j,i} + J_{syn,j,i}f_i \qquad (5)$$

In principle, there are 10 such similar dynamical equations describing the currents caused by ionotropic (4 equations) and metabotropic (6 equations) receptors. The synaptic time constants associated with the ionotropic receptors are obtained from electrophysiological data while the effective metabotropic time constants are deduced from the associated G protein-coupled inwardly-rectifying potassium (GIRK) current, or if unavailable, the temporal change in firing rate due to the injection of 5-HT or Ox [67]. For simplicity, a simple linear relationship is assumed between the input current $I_{j,i}$ and activity level f_i. The other important parameter, $J_{syn,j,i}$ is the connection strength within or between the neuronal populations. Under steady state condition ($dI_{j,i}/dt = 0$), $J_{syn,j,i}$ is defined as the ratio of the current $I_{j,i}$ and the associated (presynaptic) activity f_i. These currents are obtained from various experiments [4,26,27,31,68,69]. The in vivo baseline firing rates (f_i) for the Ox and GABAergic populations in LHA are $\sim$5 Hz (3–8 Hz in experiments) [70,71,72]. In DRN, the baseline neuronal firing rate of 5-HT and GABAergic neuronal population is $\sim$5 and $\sim$15 Hz, respectively [73]. The relationship among these baseline activities will be used to constrain our model parameters, namely the J_i, g_i and $I_{j,0}$ (Table 1). Consistent with our assumption on ignoring the relatively much faster neuronal membrane dynamics ($\sim$10 s ms), we shall also ignore the dynamics for the relatively fast GABAergic synapses ($\sim$4 ms), assuming they attain instantaneous steady states. This further reduces 4 dynamical equations in describing the associated currents.

Thus the free parameters in the model are: $J_{GABA(DRN)-to-GABA(DRN)}$, $J_{GABA(LHA)-to-GABA(LHA)}$ connection strengths of the GABAergic self-inhibition in DRN and LHA neuronal groups; $J_{5-HT-to-GABA(LHA)}$ connection strengths of the effect of 5-HT on GABAergic neurons in LHA; $\tau_{5-HT-on-GABA(LHA)}$ time constants of the 5-HT effects on GABAergic neurons in LHA; $\tau_{Ox-on-GABA(LHA)}$ time constant of the effect of Ox on GABAergic neurons in LHA; and afferent background currents to the neuronal groups, $I_{Background-5-HT}$, $I_{Background-Ox}$, $I_{Background-GABA(DRN)}$, and $I_{Background-GABA(LHA)}$.

In addition we make the following further constraints on the values of J's and τ's:

(i)   Time constant of 5-HT effect on LHA's GABAergic neurons is equivalent to that of 5-HT on LHA's Ox neurons, i.e. $\tau_{5-HT-on-GABA(LHA)} \sim \tau_{5-HT-on-Ox}$

(ii)   Time constants of Ox (Ox$_2$) effect on GABAergic neurons in LHA is equivalent to the time constant of self-excitation of Ox (Ox$_2$) autoreceptors in LHA, i.e. $\tau_{Ox-on-GABA(LHA)} \sim \tau_{Ox-auto}$

(iii)   Connection strengths of 5-HT on GABAergic neurons (LHA) are equivalent to the connection strengths of 5-HT on Ox neurons in LHA, i.e. $J_{5-HT-to-GABA(LHA)} \sim J_{5-HT-to-Ox}$

(iv)   Connection strength of Ox on GABAergic neurons (LHA) is equivalent to the connection strength of the Ox autoreceptors in LHA, i.e. $J_{Ox-to-GABA(LHA)} \sim J_{Ox-auto}$

(v)   Connection strengths of GABAergic (GABA$_A$) self-inhibition in DRN or LHA is equivalent, i.e. $J_{GABA(LHA)-to-GABA(LHA)} \sim J_{GABA(DRN)-to-GABA(DRN)}$.

Further details of the model parameter values, justifications, and their related references are summarized in Table 1.

Table 1. Time constants, currents of all the neuronal groups, values are deduced from experiments, and by using above constraints and Connection strength (C. strength), and background current of all the neuronal groups.

A. Time constants, currents of all the neuronal groups, values are deduced from experiments, and by using above constraints.

Parameter	Description	Value	Reference, remarks
$\tau_{5\text{-HT-auto}}$	5-HT$_{1A}$ autoreceptor time constant	1 s	[90,91]
$I_{5\text{-HT-auto}}$	5-HT$_{1A}$ autoreceptor induced current amplitude	80 pA	[68]
$I_{GABA(DRN)\text{-on-}5\text{-HT}}$	GABA$_A$ mediated current amplitude in 5-HT neurons	70 pA	[31]
$\tau_{Ox\text{-on-}5\text{-HT}}$	Ox$_{1,2}$ induced current time constant on 5-HT neurons	60 s	[31]
$I_{Ox\text{-on-}5\text{-HT}}$	Ox$_{1,2}$ induced current amplitude on 5-HT neurons	75 pA	[31]
$I_{GABA(DRN)\text{-on-}GABA(DRN)}$	GABA$_A$ mediated current amplitude in DRN's GABAergic neurons	70 pA	$I_{GABA(DRN)\text{-on-}GABA(DRN)} \sim I_{GABA(DRN)\text{-on-}5\text{-HT}}$
$\tau_{5\text{-HT-on-}GABA(DRN)}$	5-HT induced current time constant in GABAergic neurons in DRN	60 s	[92]
$I_{5\text{-HT-on-}GABA(DRN)}$	5-HT induced current amplitude in GABAergic neurons in DRN	50 pA	[92]
$\tau_{Ox\text{-on-}GABA(DRN)}$	Ox induced current timescale on GABAergic neurons in DRN	5 s	[31]
$I_{Ox\text{-on-}GABA(DRN)}$	Ox induced current amplitude in GABAergic neurons in DRN	25 pA	[31]
$\tau_{Ox\text{-auto}}$	Ox$_2$ autoreceptor time constant	10 s	[4]
$I_{Ox\text{-auto}}$	Ox$_2$ autoreceptor induced current amplitude	30 pA	[4]
$I_{GABA(LHA)\text{-on-}Ox}$	GABAergic induced current amplitude in Ox neurons	590 pA	[93]
$\tau_{5\text{-HT-on-}Ox}$	5-HT$_{1A}$ induced current time constant on Ox neurons	2 s	[26]
$I_{5\text{-HT-on-}Ox}$	5-HT$_{1A}$ induced current amplitude in LHA	32 pA	[26]
$I_{GABA(LHA)\text{-on-}GABA(LHA)}$	GABA mediated current amplitude in LHA's GABAergic neurons	70 pA	$I_{GABA(LHA)\text{-on-}GABA(LHA)} \sim I_{GABA(DRN)\text{-on-}Ox}$
$\tau_{Ox\text{-on-}GABA(LHA)}$	Ox induced current time constant on LHA's GABAergic neurons	10 s	$\tau_{Ox\text{-on-}GABA(LHA)} \sim \tau_{Ox\text{-auto}}$
$I_{Ox\text{-on-}GABA(LHA)}$	Ox induced current amplitude in LHA's GABAergic neurons	30 pA	$I_{Ox\text{-on-}GABA(LHA)} \sim I_{Ox\text{-auto}}$
$\tau_{5\text{-HT-on-}GABA(LHA)}$	5-HT induced current time constant in LHA's GABAergic neurons	2 s	$\tau_{5\text{-HT-on-}GABA(LHA)} \sim \tau_{5\text{-HT-on-}Ox}$
$I_{5\text{-HT-on-}GABA(LHA)}$	5-HT induced current amplitude inLHA's GABAergic neurons	32 pA	$I_{5\text{-HT-on-}GABA(LHA)} \sim I_{5\text{-HT-on-}Ox}$
$g_{5\text{-HT}}$	Slope of input-output function of 5-HT neurons	0.033 Hz/pA	[64]
g_{Ox}	Slope of input-output function of LHA's GABAergic neurons	0.205 Hz/pA	[65]
$g_{GABA(DRN)}$	Slope of input-output function of DRN's GABAergic neurons	0.061 Hz/pA	[64]
$g_{GABA(LHA)}$	Slope of input-output function of Ox neurons	0.195 Hz/pA	[65]
$I_{5\text{-HT,0}}$	Current threshold of 5-HT neurons	0.13 pA	[64]
$I_{Ox,0}$	Current threshold of Ox neurons	0 pA	[65]
$I_{GABA(DRN),0}$	Current threshold of DRN's GABAergic neurons	0 pA	[65]
$I_{GABA(LHA),0}$	Current threshold of LHA's GABAergic neurons	0 pA	[64]
a	Stimulus current amplitude applied to the 5-HT or Ox neurons	150 pA	2-fold activity increase in the 5-HT neurons
–	Duration of the applied stimulus	0.5 s	Behavioural timescale

B. Connection strength (C. strength), and background current of all the neuronal groups.

Parameter	Description	Value	Reference, remarks
$J_{5\text{-HT-auto}}$	C. strength of 5-HT$_{1A}$ autoreceptors	16 pA/Hz	*
$J_{GABA(DRN)\text{-to-}5\text{-HT}}$	C. strength of GABAergic on 5-HT neurons in DRN	5 pA/Hz	*
$J_{Ox\text{-to-}5\text{-HT}}$	C. strength of Ox neurons on 5-HT neurons	15 pA/Hz	*
$J_{GABA(DRN)\text{-to-}GABA(DRN)}$	C. strength of GABAergic neurons in DRN	5 pA/Hz	*
$J_{5\text{-HT-to-}GABA(DRN)}$	C. strength of 5-HT on GABAergic neurons in DRN	10 pA/Hz	*
$J_{Ox\text{-to-}GABA(DRN)}$	C. strength of Ox on GABAergic neurons in DRN	5 pA/Hz	*
$J_{Ox\text{-auto}}$	C. strength of Ox$_2$ autoreceptors	6 pA/Hz	*
$J_{GABA(LHA)\text{-to-}Ox}$	C. strength of GABA on Ox neurons	118 pA/Hz	*
$J_{5\text{-HT-to-}Ox}$	C. strength of 5-HT neurons on Ox neurons	6 pA/Hz	*
$J_{GABA(LHA)\text{-to-}GABA(LHA)}$	C. strength of GABAergic neurons in LHA	5 pA/Hz	$J_{GABA(LHA)\text{-on-}GABA(LHA)} \sim J_{GABA(DRN)\text{-on-}GABA(DRN)}$
$J_{Ox\text{-to-}GABA(LHA)}$	C. strength of Ox on GABAergic neurons in LHA	~0	Very weak connection, [4]

Table 1. Cont.

A. Time constants, currents of all the neuronal groups, values are deduced from experiments, and by using above constraints.			
Parameter	Description	Value	Reference, remarks
$J_{5-HT-to-GABA(LHA)}$	C. strength of 5-HT on LHA's GABAergic neurons	6 pA/Hz	$J_{5-HT(DRN)-on-GABA(LHA)} \sim$ $J_{5-HT(DRN)-on-Ox(LHA)}$
$I_{Background-5-HT}$	Background current amplitude in 5-HT neurons	231.47 pA	**
$I_{Background-Ox}$	Background current amplitude in Ox neurons	617.6 pA	**
$I_{Background-GABA(DRN)}$	Background current amplitude in DRN's GABAergic neurons	246 pA	**
$I_{Background-GABA(LHA)}$	Background current amplitude in LHA's GABAergic neurons	20.9 pA	**

*Values of $J_{syn,j,i}$ are calculated from the current $I_{j,i}$ and the associated firing rate activity f_i and also by using the above defined constraints.
**Parameter values are chosen such that the baseline activities and basic electrophysiological properties of the neurons are similar to those in experiments [4,27,35,66,71,72,73].

Simulations and analysis. We use XPPAUT [74] for neural circuit dynamics simulations and stability analysis. The Runge-Kutta 2 numerical integration algorithm with a time step of 10 ms is used. Smaller time steps do not affect our results.

Results

5-HT$_{3A}$R, 5-HT$_{1B}$R, 5-HT$_{2A}$R and 5-HT$_{2C}$R Receptors on Ox Neurons in the LHA

Muraki et al. (2004) have shown the presence of 5-HT$_{1A}$ receptors on Ox neurons in the LHA. To elucidate the presence of additional 5-HT receptor subtypes on Ox neurons, we performed

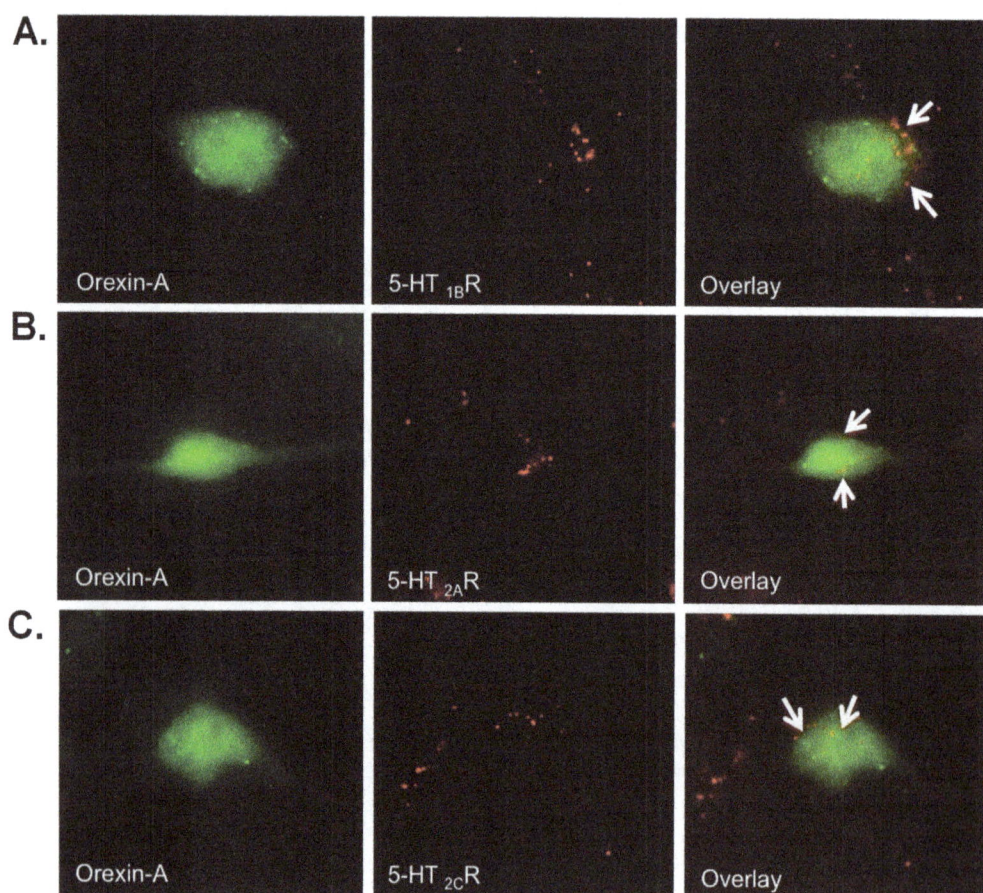

Figure 1. Single-label immunofluorescence analysis showing Ox neurons expressing additional 5HT receptor subtypes. (A) 5- HT$_{1B}$R, (B) 5-HT$_{2A}$R and (C) 5-HT$_{2C}$R in the LHA of EGFP transgenic mice expressing GFP (green) exclusively in the Ox neurons. Immunoreactivity for 5-HT receptors (Alexa 546, red), 100× magnification.

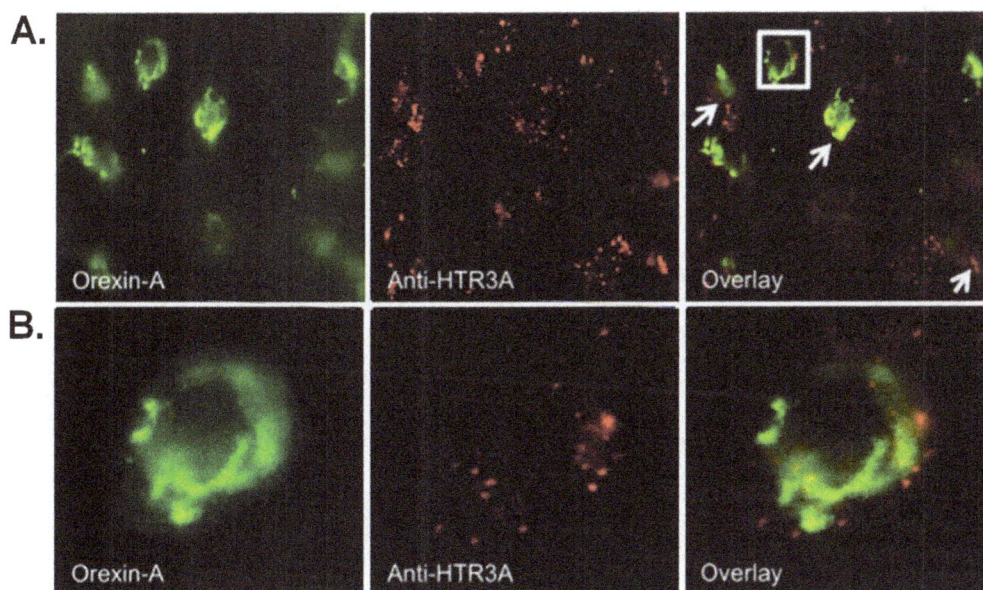

Figure 2. Double-label immunofluorescence analysis showing 5-HT$_{3A}$R (anti-HTR3A) immunoreactivity of Ox (Orexin-A) neurons in the LHA. Confocal microscopy: Chromogens were Alexa 488 (green) and Alexa 546 (red), 100× magnification. (A) Arrows indicate the presence of 5-HT$_{3A}$R on Ox neurons in the LHA. (B) Enlarged image of the selected square in A.

single IF staining for 5-HT$_{1B}$R, 5-HT$_{2A}$R and 5-HT$_{2C}$R on Ox-EGFP transgenic mice brains (Figure 1). Co-localization of the receptor and the Ox neuron immunoreactivity (Figure 1A–C) can be observed in the respective overlay images. Using double IF labeling, we found fast ligand-based 5-HT$_{3A}$ receptors on Ox neurons in the LHA of wild type C57BL/6 mice (Figure 2). This suggests that the DRN-to-LHA connection may transmit signals fast. Figure 3 shows representative z-stack$_{max}$ (30 µm thickness) confocal images showing colocalization of Orexin A with 5HT$_{1A}$R (Figure 3A) and 5HT$_{3A}$R (Figure 3B), respectively.

As we demonstrated the additional 5HT receptor subtypes on the Ox neurons in LHA for the first time, the next obvious step was to quantify the double-labeled Ox neurons for each 5HT receptor subtype. Using the JACoP plug-in tool, we quantified the

degree of co-localization by calculating Pearson's correlation coefficient (Figure 4B) and overlap coefficient by Manders (Figure 4C). It is important to note, that the two coefficients, namely Pearson's Correlation Coefficient and Overlap Coefficient, showed similar pattern of changes while revealing the different aspects of the colocalization process, proving the applicability of the calculations to investigate the degree of colocalization of serotonin receptor subtypes and orexin A (Ox neurons).

5-HT$_{3A}$R and 5-HT$_{1A}$R Receptors on Ox as well as GABAergic Interneurons in the LHA

To examine the serotonergic long-range connections from DRN in the midbrain to the Ox and GABAergic neurons in the LHA, triple-label IF staining was performed. We searched for 5-HT$_{3A}$R

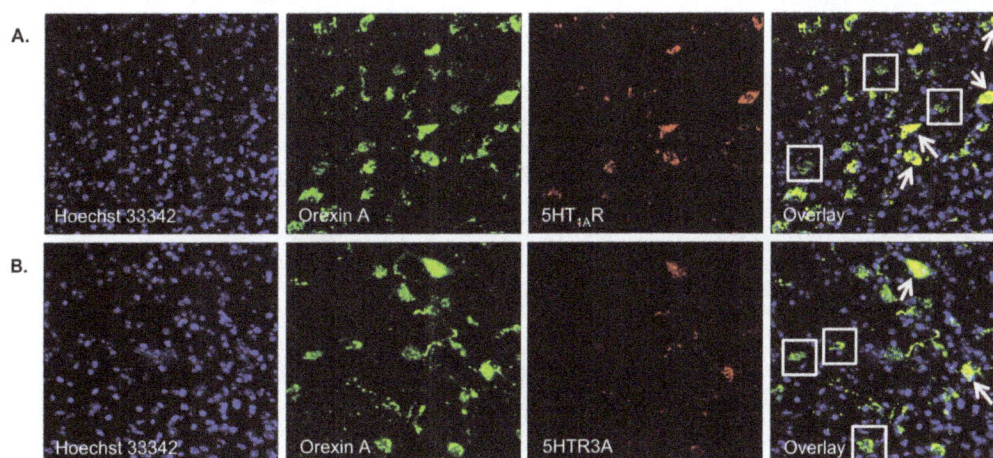

Figure 3. Representative z-stack$_{max}$ (30 µm thickness) confocal images. Colocalization of Orexin A with (A) 5HT$_{1A}$R, (B) 5HT$_{3A}$R, 63× magnification. White arrows indicate orexin neuron expressing the respective 5HT receptor and white boxes indicate orexin neuron not expressing the receptor.

A.

B.

C.

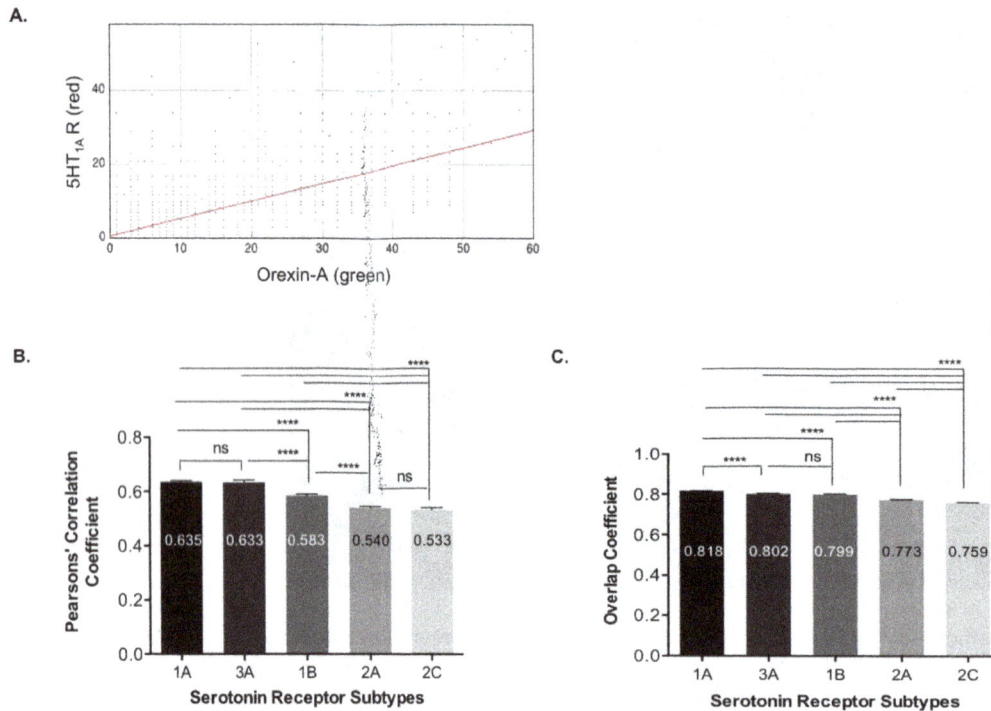

Figure 4. Quantitative analysis of the degree of colocalization of serotonin receptor subtypes and Ox neurons in the LHA. (A) Representative Scatter plot (cytofluorogram) indicating colocalization of $5HT_{1A}R$ and Orexin A, Pearson's coefficient = 0.641, (B) Pearson's correlation coefficient calculated for the estimation of degree of colocalization of serotonin reseptor subtypes ($5HT_{1A}R$, $5HT_{3A}R$, $5HT_{1B}R$, $5HT_{2A}R$ and $5HT_{2C}R$) and Ox neurons in the LHA, (C) Overlap coefficient by Manders calculated for the estimation of extent of overlap of serotonin receptor subtypes ($5HT_{1A}R$, $5HT_{3A}R$, $5HT_{1B}R$, $5HT_{2A}R$ and $5HT_{2C}R$) and Ox neurons in the LHA. Data was analysed by two tailed unpaired t-test (Prism 5 software). Level of significance ****($p < 0.0001$).

(Figure 5A) and $5\text{-}HT_{1A}R$ (Figure 5B) on the Ox and GABAergic neurons and found that these 5-HT receptors may be present on Ox as well as local GABAergic neurons. This may suggest both direct and indirect (via LHA's GABAergic neurons) 5-HT influence on the Ox neurons.

OX_1R, OX_2R and NMDAR1 Receptors on Ox Neurons but not on GABAergic Neurons in the LHA

To identify the role of Ox neurons in regulating the activity of local GABAergic neurons, we carried out a triple IF labeling in the

LHA. To study this inter-relationship, we searched for OX_1R and OX_2R receptors on the Ox and GABAergic neurons. We identified OX_1R (Figure 6A) and OX_2R (Figure 6B) on the Ox neurons but found these to be absent on the GABAergic interneurons. This sheds light on the self-regulatory mechanism of Ox neurons via OX_1R and OX_2R auto-receptors. Previous studies have demonstrated the presence of inhibitory $GABA_B$ receptors on the Ox neurons [33,75,76,77] and there is evidence that $GABA_A$ receptors are also present on Ox neurons (Backberg et al., 2004, Kokare et al., 2006). This could suggest a one-way

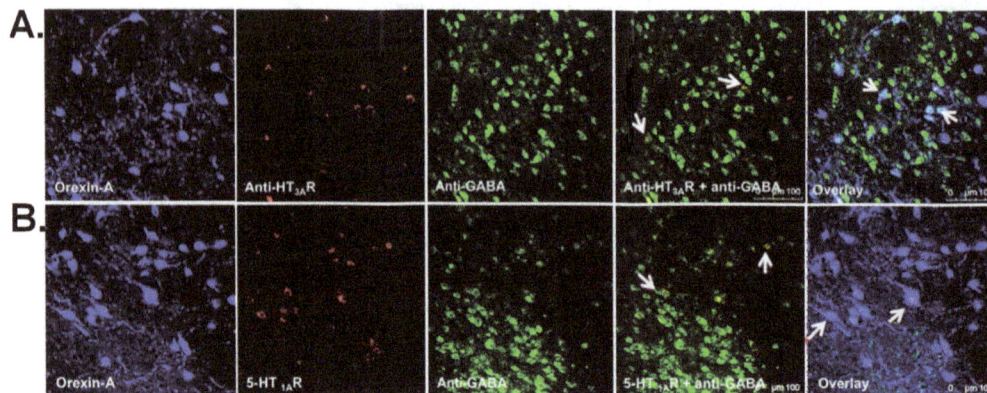

Figure 5. Representative photomicrographs of triple-label IF staining in LHA for Ox (Orexin-A) and GABAergic (anti-GABA) neurons. (A) $5\text{-}HT_{3A}R$, overlay image indicates an overlay of Orexin-A, $5\text{-}HT_{3A}R$ and anti-GABA, (B) $5\text{-}HT_{1A}R$, overlay image indicates an overlay of Orexin-A, $5\text{-}HT_{1A}R$ and anti-GABA. Confocal microscopy: Chromogens were Alexa 488 (blue), Alexa 546 (red) and CF633 (green), 40× magnification.

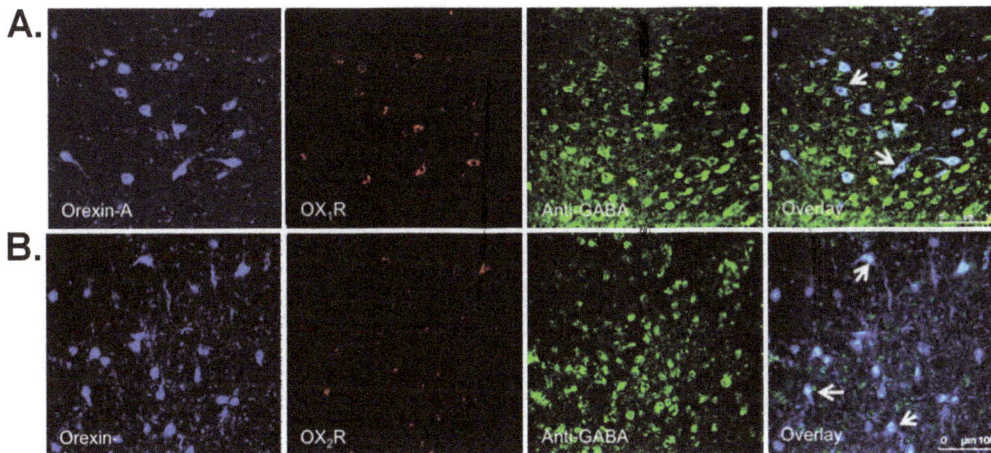

Figure 6. Representative photomicrographs of triple-label IF staining for Ox and GABAergic neurons in LHA. (A) OX_1R, overlay image indicates an overlay of Orexin-A, OX_1R and anti-GABA, (B) OX_2R, overlay image indicates an overlay of Orexin-A, OX_2R and anti-GABA. Arrows in A and B overlay images indicate presence of OX_1R and OX_2R on Ox neurons, respectively. Confocal microscopy: Chromogens were Alexa 488 (blue), Alexa 546 (red) and CF633 (green), 40×.

relationship between GABAergic and Ox neurons wherein GABAergic neurons exerts an inhibitory effect on the Ox neurons under the partial DRN's serotonergic control, consistent with Yamanaka et al. (2010).

Ox neurons form a distinct group of a hypothalamic neuronal population that project to multiple brain regions and coordinate many physiological functions [9,78]. Also, there is a convergence of signals that regulate the activity of Ox neurons by neurotransmitters, hormones, etc. Glutamate is an important neurotransmitter for Ox neurons and excitatory AMPA receptors have been shown to mediate the miniature EPSC in Ox neurons [79]. Also, NMDA receptors activate Ox neurons in the perifornical region of LHA [80]. To verify whether NMDA receptors are present on the overall Ox neurons and not confined solely to the perifornical area, we did triple-label IF staining. Here, we show the presence of N-methyl D-aspartate receptor 1 (NMDAR1) on the Ox neurons

(Figure 7) and their absence on the GABAergic interneurons (Figure 7A, overlay image).

Ox Axonal Projections Receive Glutamatergic and GABAergic Post-synaptic Inputs in the DRN

To study the long-range connections of Ox neurons in the DRN, we performed IF labeling in the DRN region for PSD-95 (a marker for glutamatergic synapses) and gephyrin (a marker for GABAergic synapses) post-synaptic proteins [81]. Firstly, we carried out double-label IF staining for PSD-95 and Ox in the DRN and observed PSD-95 immunopositive Ox axonal terminals (Figure 8) indicating glutamatergic input to the Ox axonal projections. In another set of experiments, we examined GABAergic post-synaptic input to the Ox terminals as it has been shown that Ox fibres project to both 5-HT and GABAergic neurons in the DRN [31]. A double-label IF analysis for gephyrin and Ox in the DRN (Figure 9) showed gephyrin immunopositive

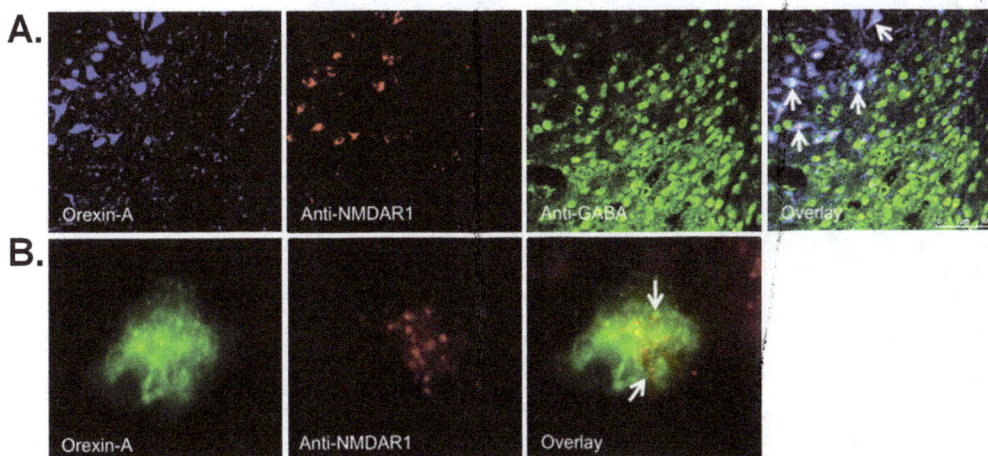

Figure 7. Representative photomicrographs of triple-label IF staining for NMDAR1 on Ox (Orexin-A) and GABAergic (anti-GABA) neurons. (A) NMDAR1 (anti-NMDAR1) in the LHA. (B) 100× magnification in A. Magenta colour in the overlay image in A and yellow/orange in the overlay image in B indicates the presence of NMDAR1 on Ox neurons. Confocal microscopy: Chromogenes were Alexa 488 (blue in A and green in B), Alexa 546 (red) and CF633 (green), 40×.

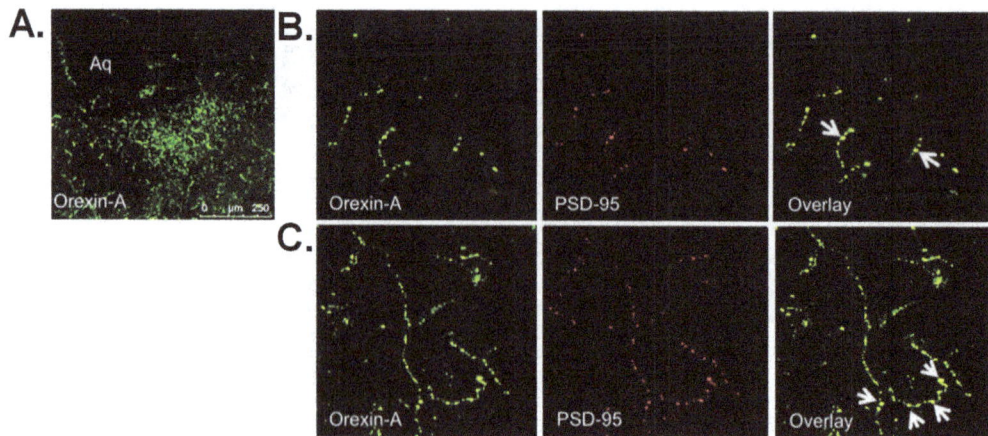

Figure 8. Ox axonal projections receive glutamatergic post-synaptic inputs in the DRN. (A) Ox (Orexin-A) axonal projections in the DRN, *Aq* indicates aqueduct, $20\times$ magnification. (B) Double IF staining was done for Ox axons and PSD-95 (post-synaptic glutamatergic marker) in the DRN, $100\times$. Confocal microscopy: magnification $100\times$. (C) Z-stack$_{max}$ image of 30 μm thickness selected from 45 μm section.

Ox terminals, further indicating GABAergic input to the Ox projections. This result is in agreement with the findings of Liu et al. (2002) where they showed Ox axons in proximity to GABA/GABA-transporter immunoreactive neurons.

Map of the DRN-LHA Circuit

Based on the above results and in other previous studies (see below), a tentative map of the DRN and LHA is illustrated in Figure 10A. Connection (i) in the figure involves $5\text{-HT}_{1A}R$ [26], $5\text{-HT}_{1B}R$, $5\text{-HT}_{2A}R$, $5\text{-HT}_{2C}R$, $5\text{-HT}_{3A}R$ (found in our current study). Although connection (ii), found in this study, consists of $5\text{-HT}_{1A}R$ and $5\text{-HT}_{3A}R$, the specific types of connection (e.g. effectively excitatory or inhibitory) and their strengths are not known. Connection (iii) and its receptor types are not known yet. Similarly, the receptor types for connection (iv) (Ox_1R and Ox_2R

found in this study are consistent with previous findings [30,32]), and connections (v) and (vi) (found in this study) are unknown. Connection (vii) receptors are not known and specifically Ox_1R and Ox_2R are not found in our study.

Other connections in the circuit based on other previous work: (viii) excitatory connection and receptor types are not known [82]; (ix) $GABA_{A/B}$ [33,75,76,77]; (x) AMPAR [79] and NMDAR1 (found in the current study); (xi) $GABA_B$ [77]; (xiii) Ox_2R [4], Ox_1R and Ox_2R (found in the current studies); (xv) $5\text{-HT}_{1B/1D}$ [34]; $5\text{-HT}_{1A/2A/2C}$ [69]; (xvi) 5-HT_7 [34]; (xvii) $GABA_{A/B}$ [34]; (xviii) AMPAR and NMDAR [34]; (xix) AMPAR and NMDAR [34]; (xxi) $5\text{-HT}_{1A}R$ [83]. For the connections (xii), (xiv), (xx) and (xxii) (shown by black dotted circle/arrow in Figure 10A), we implicitly assume that non-principal neurons (GABAergic and glutamatergic neurons in LHA and DRN) are self-coupled.

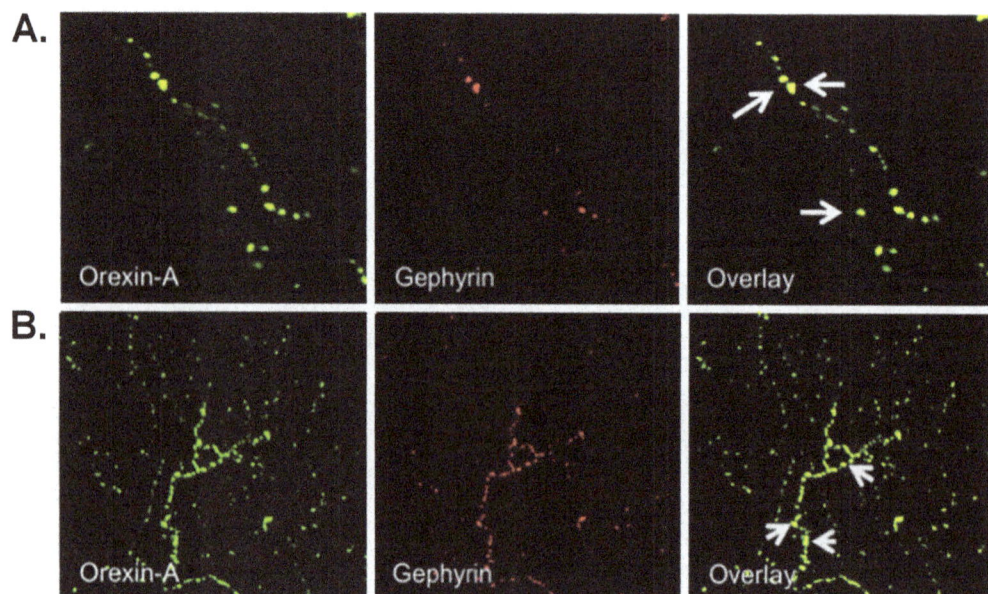

Figure 9. Ox axonal projections receive GABAergic post-synaptic inputs in the DRN. (A) Double IF staining was performed for Ox axons (Orexin-A) and Gephyrin (post-synaptic GABAergic marker) in the DRN. Confocal microscopy: magnification $100\times$. (B) Z-stack$_{max}$ image of 30 μm thickness selected from 45 μm section.

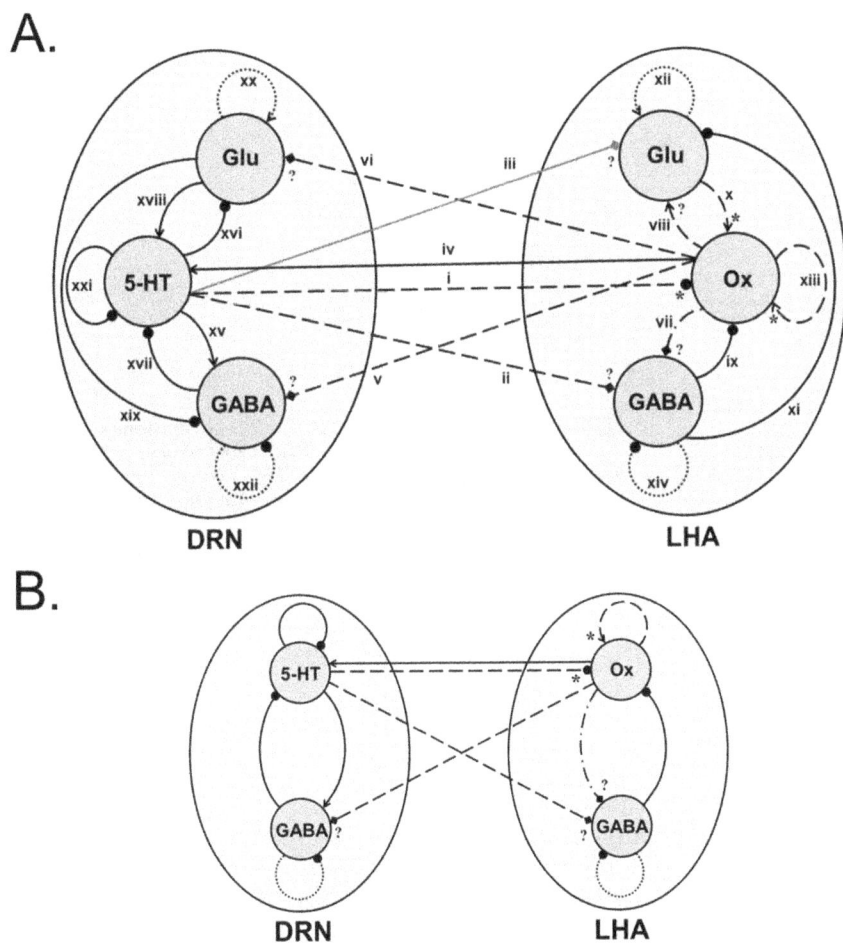

Figure 10. Neural circuit map of the LHA-DRN system. (A) Diagrammatic representation of the receptor mapping suggests a complex bi-directional relationship between DRN and LHA. Black circle/arrow represents effective inhibitory/excitatory connections between LHA and DRN, solid black line represents various receptor types derived from previous experimental studies, dashed black line shows the recent experimental findings, dashed arrows with * (asterisk) indicates that some of these receptors are known in the previous studies, dashed-dotted line signifies that receptors (in the target neurons) are not identified in this current study, dotted lines shows the hypothesized connection, diamond arrows indicate that connection types (excitatory/inhibitory) are not known, ? (question mark) means that receptors are not known and gray line signifies that the connection is not studied. All these receptor types are numbered on the figure and details are provided below. See text for more information. (B) Reduced LHA-DRN circuit used in the computational model. Label as in (A).

Neural Circuit Modelling and Analysis of the Direct and Indirect Interactions within the DRN-LHA System

The experimental findings in this study and in other previous work now provide sufficient information to build a computational model to investigate how the DRN-LHA neural architecture can influence the systems dynamics. We purposely simplified the implementation of the neural population units and connections in order to better illustrate the effects of network topology on dynamics (see Materials and Methods section). We also omit modelling the glutamatergic neurons in the DRN and LHA because there is evidence that glutamate has a weaker effect than GABA in the DRN [59]; and that indirect excitation from LHA's glutamatergic neurons on Ox neurons may be represented implicitly by the Ox autoreceptors. This significantly reduces the number of model parameters and dynamical equations involved. In our model, we find that the various receptor subtypes will not affect the steady-state activities of the system. We shall henceforth not elaborate on the specific receptor subtype till we investigate the transient activation part of the results (section 3.6.3) when we study

the influence of various timescales induced by the various receptors.

Next we investigate how the unknown model parameters and different timescales affect the system behaviour. Specifically, the focus is to understand the effects of: (i) 5-HT on LHA's GABAergic neurons; (ii) local Ox and GABAergic interactions; and (iii) how connection timescales affect phasic 5-HT or Ox activations.

Oscillatory DRN-LHA Behaviour if 5-HT Weakly Excites or Inhibits LHA's GABAergic Neurons

It is not yet known whether the connection from 5-HT neurons to LHA's GABAergic neurons ($J_{5\text{-}HT\text{-}to\text{-}GABA(LHA)}$) is effectively excitatory or inhibitory. Here, we shall explore these possibilities. If this connection is excitatory, we find that as its connection strength $J_{5\text{-}HT\text{-}to\text{-}GABA(LHA)}$ is increased, the steady-state firing rate activities for most of the neural populations decrease (Figure 11). Note that LHA's GABAergic neural population barely increases. This slight change is due to the self-inhibition within these GABAergic neurons. In contrast, the relatively larger changes for

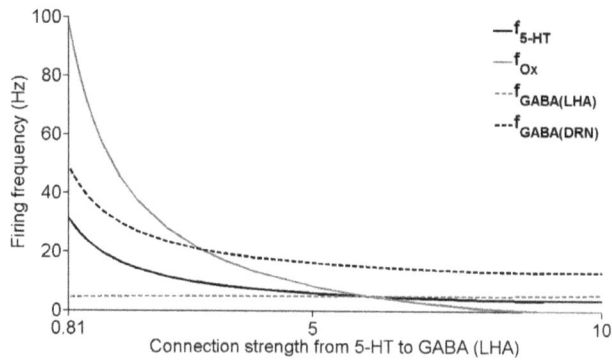

Figure 11. Neural circuit responses to change in connection strength from 5-HT to GABAergic neurons (LHA). Change in the steady-state values of the neural firing frequencies of the neuronal groups with varying connection strength factor $J_{5\text{-}HT\text{-}to\text{-}GABA(LHA)}$ (in pA/Hz) from 5-HT neurons to the GABAergic neurons (LHA) for excitatory connection. Steady states are obtained after simulating for a sufficiently long time. $f_{5\text{-}HT}$, f_{Ox}, $g_{GABA(LHA)}$, and $g_{GABA(DRN)}$: population firing frequencies of 5-HT, Ox, LHA's GABAergic and DRN's GABAergic neurons, respectively.

Figure 12. LHA-DRN system can exhibit oscillations with weak excitatory from 5-HT to GABAergic (LHA) neurons. Steady-state values of the firing rate activity of 5-HT neurons are plotted as a function of the connection strength $J_{5\text{-}HT\text{-}to\text{-}GABA(LHA)}$ (in pA/Hz). Oscillatory region (left of dashed) is bounded by the values of $J_{5\text{-}HT\text{-}to\text{-}GABA(LHA)}$ below 0.804 pA/Hz. Max (Min): maximum (minimum) firing rates during oscillation. Label as in Figure 11.

the other neural populations (especially when the connection strength is weaker) are due to the strong inhibitory projection from the LHA's GABAergic neuron onto the Ox neurons (see Table 1), which subsequently affect the neurons in the DRN.

When the connection from 5-HT to LHA's GABAergic neurons becomes very weak ($J_{5\text{-}HT\text{-}to\text{-}GABA\ (LHA)}$ close to 0), the DRN-LHA circuit will begin to exhibit slow oscillatory behaviour. Figure 12 shows such transition towards oscillatory behaviour for one sample neural population (the rest of the neural populations look similar). The oscillatory period can be as slow as a few minutes (Figure 12, inset). It can also be observed that as the connection strength decreases, the amplitude of oscillation (bounded by the top and bottom lines in grey region) increases. When this particular connection becomes inhibitory ($J_{5\text{-}HT\text{-}to\text{-}GABA(LHA)} < 0$), the oscillation amplitude can become very large (not shown). It is well-known that excitatory-inhibitory network can easily create oscillation [60,61]. Similarly, the observed oscillation phenomenon can be explained by the interplay between strong inhibition and self-excitatory auto-regulation in the Ox neural population; the oscillation disappears when Ox autoreceptors are removed from the model e.g. when $J_{Ox\text{-}auto} = 0$ (not shown). As far as we know, there has yet to be any observable slow oscillation of Ox in the timescale of minutes. Hence, our model suggests that strong excitatory connection from 5-HT to LHA's GABAergic neurons is more plausible, and we shall proceed with this assumption from here on.

DRN-LHA System is Resilient to Changes in the Ox-to-GABAergic Neurons in the LHA

One of our experimental findings in this study shows that the LHA's GABAergic neurons do not have Ox receptors, consistent with indirect results from previous works [4]. Using our model, we check for the significance of such Ox receptors' absence. In our model, we gradually increase the connection strength of Ox to LHA GABAergic neurons ($J_{Ox\text{-}to\text{-}GABA(LHA)}$). This only marginally decreases the DRN-LHA steady-state activities (Figure 13). A similar explanation as that for the results in Figure 11 can be used to account for this phenomenon.

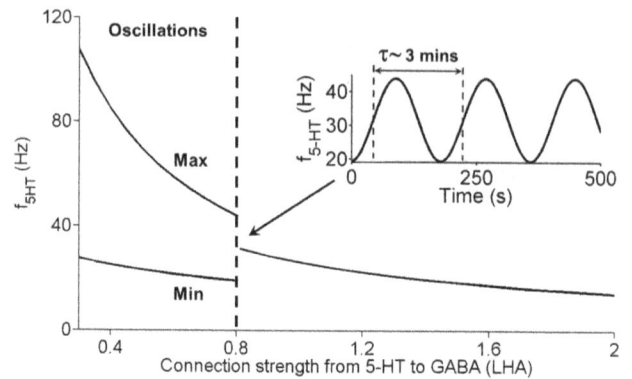

Transient Ox and 5-HT Activities can be Affected by Different Receptor Timescales

Having studied the tonic activities (stable steady states) of the LHA-DRN system, we shall now proceed to investigate how transient or phasic activations of 5-HT or Ox can affect the system. The motivation for this is that it has been known that 5-HT and Ox can phasically activate in the presence of behaviourally relevant stimuli [84,85,86]. Furthermore, our current experimental finding suggests that 5-HT can influence LHA neurons over multiple timescales, through both the slow G-protein-coupled (non-5-HT$_{3A}$) and fast ligand-gated (5-HT$_{3A}$) receptors. To simulate this, the (more plausible) excitatory connection $J_{5\text{-}HT\text{-}to\text{-}GABA(LHA)}$ is considered, and a pulse stimulus current of the duration of 0.5 s with an amplitude of 150 pA is applied to either the 5-HT or Ox neural populations. To understand the individual role of the slow and fast timescales, and minimize any confounding effect, we simulate the two timescales separately.

Figure 14A (left) shows that a 0.5 s stimulation of 5-HT neurons rapidly increase its activity, followed by a slower decay back towards baseline. The 5-HT activity decay is due to feedback inhibition from its autoreceptors and the local GABAergic neurons (hence the undershoot below baseline). The Ox activity responses are generally suppressed by the strong 5-HT-mediated inhibition (Figure 14A, right). When the 5-HT-to-Ox and 5-HT-to-GABA (LHA) connections are fast (mimicking 5-HT$_{3A}$R), the rebound upon removal of stimulus ($\sim$ after the 2 s mark in Figure 14A, right) is higher, suggesting the faster disinhibition than that for the slower connections. Varying these timescales do not affect the transient 5-HT activity in DRN (overlapping curves, Figure 14A, left).

When a similar stimulus is applied to the Ox neurons, the activity of the Ox neurons increases considerably throughout the stimulus duration (Figure 14B, right), which is due to the self-amplifying effect of Ox autoreceptors dominating over the local GABAergic inhibition (feedback inhibition is not strong due to the weak excitatory connection from Ox neurons to their local GABAergic neurons). These Ox neurons, affect 5-HT neurons either directly or indirectly (through the DRN's GABAergic neurons). From our simulations, we find that when the Ox-to-5-HT connection acts on a fast timescale (5 s), they can transiently excite the 5-HT neurons (Figure 14B, left). If the direct timescale is

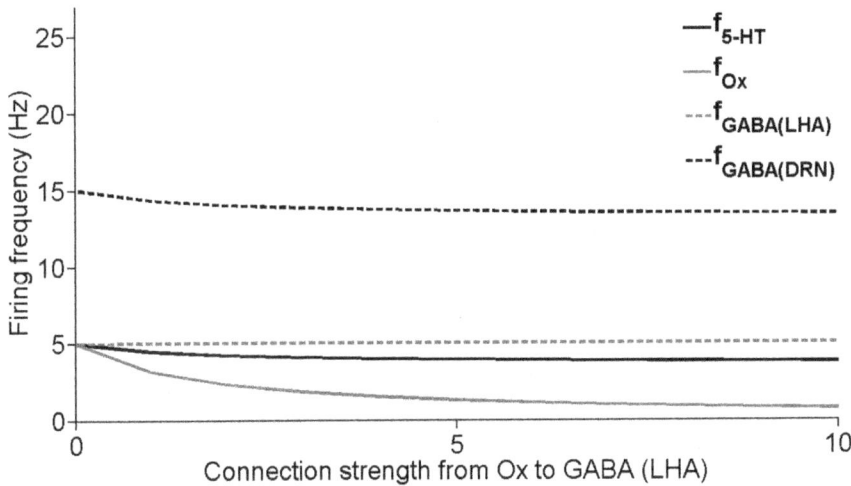

Figure 13. Neural circuit responses to change in connection strength from Ox to GABAergic neurons (LHA). Steady-state firing frequencies of the neural populations as functions of the connection strength (in pA/Hz). Label as in Figure 11.

much slower (e.g. 60 s), then 5-HT neurons can hardly be activated. Varying Ox-to-GABA(DRN) connection does not affect the system significantly. Note the post-stimulus undershoot due to recurrent inhibition, suppressing the Ox and 5-HT transiently below baseline. Taken together, when Ox-to-5-HT acts on a faster timescale, the phasic influence of Ox on 5-HT activity is much larger.

Discussion

We have mapped out the direct and indirect connections between and within the DRN and LHA brain regions using IF staining to identify the receptors and trace long-range connections. To consider indirect connections, non-principal neurons have to

be involved. In our study, the non-principal neurons are the inhibitory GABAergic and excitatory glutamatergic neurons, although there exist other neuronal types (e.g. neurons containing neuropeptide Y, etc) [87].

We have confirmed a previously identified 5-HT_{1A} receptor on Ox neurons, and have also identified multiple major 5-HT receptors in the LHA. They include, on Ox neurons, 5-HT_{1B}R, 5-HT_{2A}R, and 5-HT_{2C}R and also fast ligand-gated 5-HT_{3A}R. It is interesting to note that Muraki et al. (2004) has shown that the 5-HT_{1A} receptor antagonist WAY-100635 can completely block the 5-HT hyperpolarizing effect on Ox neurons. It could perhaps be that the 5-HT_{1A} receptors have the highest affinity as compared to the other 5-HT receptor subtypes found in this study. Further

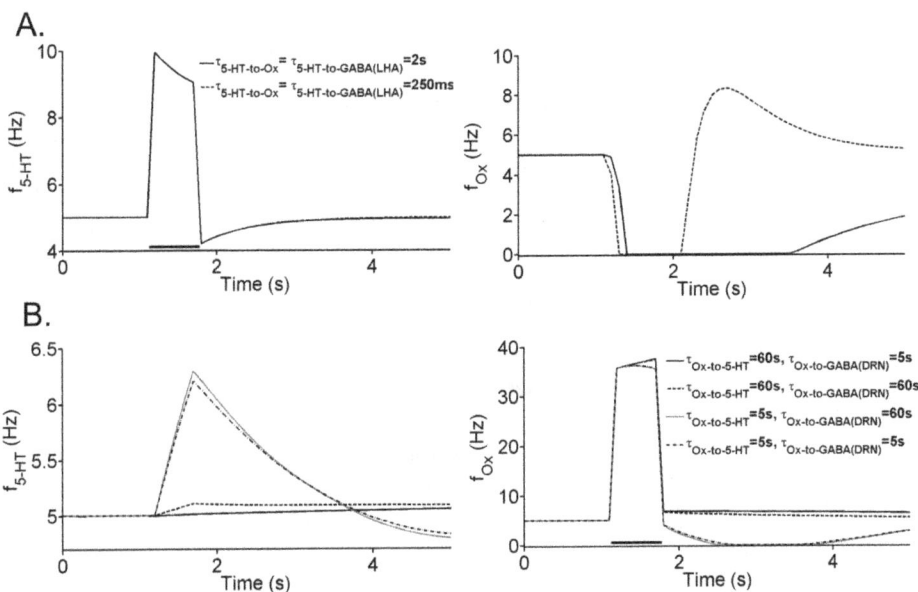

Figure 14. Transient activities of 5-HT and Ox neuronal groups under phasic stimulation. (A) Firing frequency of 5-HT neurons for different slow and fast 5-HT receptor timescales (left) and for the Ox neurons (right). (B) Firing frequency of 5-HT neurons for different slow and fast Ox receptor time scales (left) and for the Ox neurons (right). Solid horizontal black lines denote presence of stimulus with duration of 500 ms, and amplitude of 150 pA. Label as in Figure 11.

experimental work would help to clarify this issue. In addition, we found both fast and slow 5-HT receptors (5-HT$_{3A}$R and 5-HT$_{1A}$R) on LHA's GABAergic neurons. Hence, these 5-HT (especially 1A) receptors could directly and indirectly influence Ox neurons, and further studies would be required to compare their relative affinities, and hence relative influences on Ox neurons. Within the LHA, we have also found the presence of OX$_1$R, OX$_2$R and NMDAR1 receptors on Ox neurons but not on GABAergic neurons. For LHA-to-DRN projections, we have identified Ox neurons receiving glutamatergic and GABAergic post-synaptic inputs in the DRN. The results from our experimental work and from previous work are summarized in Figure 10A. Our results also suggest that LHA's GABAergic neurons could be isolated from direct (excitatory) afferent influences from local glutamatergic and Ox neurons, but could be influenced directly by 5-HT neurons, or through a unidirectional closed loop that involves the Ox and 5-HT neurons (Figure 10A).

From Figure 10B, the local LHA architecture looks similar in some respects to the well-studied cortical column architecture, with the Ox neurons acting as pyramidal neurons, with excitatory feedback among themselves [88]. However, our present work and in a previous study [4] have shown Ox neurons to have very weak influence on their local GABAergic neurons. This differs from the stereotypical excitatory-inhibitory feedback loop in a cortical column model. The DRN also looks similar to that of a cortical column if we include glutamatergic neurons (or glutamate-containing 5-HT neurons) to provide excitatory feedback. But excitation in the DRN is known to be weaker than GABA-mediated inhibition [59]. To provide a systemic understanding of the identified DRN-LHA architecture, we incorporated some of our current findings and previous published data into a computational model, which is an extension of our previous model [58].

We purposely simplified the neural unit and neuronal interaction implementations to focus primarily on this unique neural architecture of the DRN-LHA system. As a first step, glutamatergic neurons were justifiably omitted in our model simulations and analyses. We have included various constraints to the model, specifically on the relative values of the input-output slope, total input currents, and time constants and relative strengths of the connections. More importantly, we intentionally constructed a model to demonstrate the complex consequences of the neural circuit architecture we have established from our current experimental study.

We first use our model to explore the consequences of 5-HT's effect on LHA's GABAergic neurons. We found that the system becomes oscillatory when the connection strength is weak or inhibitory (Figure 12). This oscillatory behaviour has so far not been observed in experiments. Thus, based on these results, we hypothesize that this connection is excitatory. It would be interesting to test the strength of this connection using 5-HT$_{3A}$R agonists/antagonists on LHA's GABAergic neurons. This hypothesis would also imply that DRN may send inhibition to Ox neurons both directly and indirectly (through the GABAergic neurons), i.e. no balanced projections. This is in contrast with the long-range projection of Ox to the DRN, which excites both 5-HT neurons and GABAergic neurons in the DRN (Figure 10B).

Another interesting experimental finding in our work is the indication of an absence of Ox receptors on LHA's GABAergic neurons. Our model simulations show that this particular connection does not affect the coupled DRN-LHA system significantly (Figure 13). This means that Ox receptor on LHA's GABAergic neurons will have little influence on the circuit's dynamics. It is interesting to speculate that the absence of these Ox receptors could be due to a lack of significant functional roles at the circuit level.

From our experimental work, we have identified both slow 5-HT G-protein-coupled and fast ligand-based receptors on both Ox and GABAergic neurons in the LHA. Our model attempted to mimic these different 5-HT receptor mediated timescales separately to investigate how the DRN-LHA circuit as a whole can be affected. We found that the 5-HT timescales do not change the tonic (steady-state) activities of the system, but can greatly affect the transient activations (Figure 14). In general, a faster transient 5-HT influence on Ox neurons does not affect the suppression much but can result in a faster disinhibition of Ox neurons. This could mean that the faster 5-HT$_{3A}$R could be useful for quickly resetting the Ox neurons back to baseline after phasic 5-HT activation. More interestingly, a faster transient Ox influence can excite phasically 5-HT activity while slower timescale does not.

In summary, we have established aspects of the neurobiological circuitry function between the levels of 5-HT and Ox through direct and indirect pathways between the DRN and LHA. This work could have important implications in clinical neuroscience and neuropsychopharmacology as this DRN-LHA loop has been interpreted in two ways. It has been hypothesized that lower levels of 5-HT (common in depression) provide weaker inhibition to Ox neurons. Taking into account the effects of the circadian regulation of Ox and its influence on other neurotransmitters or neuromodulators, the Ox level may increase by this change in 5-HT. This increase in Ox levels may then generate a state similar to insomnia and other mood alterations. Similarly, in the reverse direction, it has been argued that lower levels of 5-HT are due to either a weaker excitatory connection from the Ox neurons (LHA) or because of lower levels of Ox (LHA), a situation commonly seen in hypersomnia or in narcolepsy [3,89]. To explore the potential effects of Ox knock-out mice on 5-HT activity, our model can simulate such an effect by removing all the Ox effects in the circuit, and we observed that 5-HT neurons can still fire at ~3.5 Hz (not shown). Furthermore, drug studies often do not consider integrating multiple targeted and non-targeted but connected brain areas. For example, in the DRN-LHA circuit considered in this study, an administration of 5-HT$_{1A}$R agonist can directly affect not only 5-HT autoreceptors, but also all the 5-HT$_{1A}$R in the DRN's GABAergic neurons, and LHA's Ox and GABAergic neurons. Other brain regions without 5-HT or Ox receptors, but connected to the affected DRN and LHA, will also be indirectly affected. Thus, the overall effect is complex, and this could be one important reason underlying serious side effects of various neuropharmacological drugs. A promising approach to gain a holistic understanding of such complex neurobiological systems is to perform more intensive computational modelling, simulations and analyses.

Acknowledgments

We thank Prof. Takeshi Sakurai, Department of Molecular Neuroscience and Integrative Physiology, Graduate School of Medical Science, Kanazawa University, Japan, and Mr. Mohit K. Sharma for providing the Ox-EGFP and C57Bl/6 mice, respectively. We also thank Dr. Nicolas J. Penington, SUNY Downstate Medical Center, USA, for advice on some of the model parameters.

Author Contributions

Conceived and designed the experiments: JJ AJ KW CH. Performed the experiments: JJ AJ. Analyzed the data: JJ CH KW. Contributed reagents/materials/analysis tools: CH GP TMM. Wrote the paper: JJ KW CH. Grant holder: TMM.

References

1. Saper CB, Scammell TE, Lu J (2005) Hypothalamic regulation of sleep and circadian rhythms. Nature 437: 1257–1263.

2. Palasz A, Lapray D, Peyron C, Rojczyk-Golebiewska E, Skowronek R, et al. (2013) Dual orexin receptor antagonists - promising agents in the treatment of sleep disorders. Int J Neuropsychopharmacol: 1–12.

3. Mignot E (2001) A commentary on the neurobiology of the hypocretin/orexin system. Neuropsychopharmacology 25: S5–13.

4. Yamanaka A, Tabuchi S, Tsunematsu T, Fukazawa Y, Tominaga M (2010) Orexin directly excites orexin neurons through orexin 2 receptor. J Neurosci 30: 12642–12652.

5. Feng P, Vurbic D, Wu Z, Hu Y, Strohl KP (2008) Changes in brain orexin levels in a rat model of depression induced by neonatal administration of clomipramine. J Psychopharmacol 22: 784–791.

6. Borgland SL, Labouebe G (2010) Orexin/hypocretin in psychiatric disorders: present state of knowledge and future potential. Neuropsychopharmacology 35: 353–354.

7. Brundin L, Bjorkqvist M, Petersen A, Traskman-Bendz L (2007) Reduced orexin levels in the cerebrospinal fluid of suicidal patients with major depressive disorder. Eur Neuropsychopharmacol 17: 573–579.

8. Salomon RM, Ripley B, Kennedy JS, Johnson B, Schmidt D, et al. (2003) Diurnal variation of cerebrospinal fluid hypocretin-1 (Orexin-A) levels in control and depressed subjects. Biol Psychiatry 54: 96–104.

9. Lopez M, Tena-Sempere M, Dieguez C (2010) Cross-talk between orexins (hypocretins) and the neuroendocrine axes (hypothalamic-pituitary axes). Front Neuroendocrinol 31: 113–127.

10. Sakurai T, Mieda M, Tsujino N (2010) The orexin system: roles in sleep/wake regulation. Ann N Y Acad Sci 1200: 149–161.

11. de Lecea L, Kilduff TS, Peyron C, Gao X, Foye PE, et al. (1998) The hypocretins: hypothalamus-specific peptides with neuroexcitatory activity. Proc Natl Acad Sci U S A 95: 322–327.

12. Sakurai T, Amemiya A, Ishii M, Matsuzaki I, Chemelli RM, et al. (1998) Orexins and orexin receptors: a family of hypothalamic neuropeptides and G protein-coupled receptors that regulate feeding behavior. Cell 92: 573–585.

13. Urbanska A, Sokolowska P, Woldan-Tambor A, Bieganska K, Brix B, et al. (2012) Orexins/hypocretins acting at Gi protein-coupled OX 2 receptors inhibit cyclic AMP synthesis in the primary neuronal cultures. J Mol Neurosci 46: 10–17.

14. Sakurai T (2005) Reverse pharmacology of orexin: from an orphan GPCR to integrative physiology. Regul Pept 126: 3–10.

15. Xu TR, Ward RJ, Pediani JD, Milligan G (2012) Intramolecular fluorescence resonance energy transfer (FRET) sensors of the orexin OX1 and OX2 receptors identify slow kinetics of agonist activation. J Biol Chem.

16. Scammell TE, Winrow CJ (2011) Orexin receptors: pharmacology and therapeutic opportunities. Annu Rev Pharmacol Toxicol 51: 243–266.

17. Lanni C, Govoni S, Lucchelli A, Boselli C (2009) Depression and antidepressants: molecular and cellular aspects. Cell Mol Life Sci 66: 2985–3008.

18. Adrien J (2002) Neurobiological bases for the relation between sleep and depression. Sleep Med Rev 6: 341–351.

19. Wisor JP, Wurts SW, Hall FS, Lesch KP, Murphy DL, et al. (2003) Altered rapid eye movement sleep timing in serotonin transporter knockout mice. Neuroreport 14: 233–238.

20. Catena-Dell'osso M, Marazziti D, Rotella F, Bellantuono C (2012) Emerging targets for the pharmacological treatment of depression: focus on melatonergic system. Curr Med Chem 19: 428–437.

21. Grace KP, Liu H, Horner RL (2012) 5-HT1A Receptor-Responsive Pedunculopontine Tegmental Neurons Suppress REM Sleep and Respiratory Motor Activity. J Neurosci 32: 1622–1633.

22. de Carvalho TB, Suman M, Molina FD, Piatto VB, Maniglia JV (2012) Relationship of obstructive sleep apnea syndrome with the 5-HT2A receptor gene in Brazilian patients. Sleep Breath.

23. Artigas F (2013) Developments in the field of antidepressants, where do we go now? Eur Neuropsychopharmacol.

24. Artigas F (2013) Serotonin receptors involved in antidepressant effects. Pharmacol Ther 137: 119–131.

25. Li Y, van den Pol AN (2005) Direct and indirect inhibition by catecholamines of hypocretin/orexin neurons. J Neurosci 25: 173–183.

26. Muraki Y, Yamanaka A, Tsujino N, Kilduff TS, Goto K, et al. (2004) Serotonergic regulation of the orexin/hypocretin neurons through the 5-HT1A receptor. J Neurosci 24: 7159–7166.

27. Yamanaka A, Muraki Y, Tsujino N, Goto K, Sakurai T (2003) Regulation of orexin neurons by the monoaminergic and cholinergic systems. Biochem Biophys Res Commun 303: 120–129.

28. Cooper MA, McIntyre KE, Huhman KL (2008) Activation of 5-HT1A autoreceptors in the dorsal raphe nucleus reduces the behavioral consequences of social defeat. Psychoneuroendocrinology 33: 1236–1247.

29. Brown RE, Sergeeva O, Eriksson KS, Haas HL (2001) Orexin A excites serotonergic neurons in the dorsal raphe nucleus of the rat. Neuropharmacology 40: 457–459.

30. Brown RE, Sergeeva OA, Eriksson KS, Haas HL (2002) Convergent excitation of dorsal raphe serotonin neurons by multiple arousal systems (orexin/hypocretin, histamine and noradrenaline). J Neurosci 22: 8850–8859.

31. Liu RJ, van den Pol AN, Aghajanian GK (2002) Hypocretins (orexins) regulate serotonin neurons in the dorsal raphe nucleus by excitatory direct and inhibitory indirect actions. J Neurosci 22: 9453–9464.

32. Soffin EM, Gill CH, Brough SJ, Jerman JC, Davies CH (2004) Pharmacological characterisation of the orexin receptor subtype mediating postsynaptic excitation in the rat dorsal raphe nucleus. Neuropharmacology 46: 1168–1176.

33. Matsuki T, Nomiyama M, Takahira H, Hirashima N, Kunita S, et al. (2009) Selective loss of GABA(B) receptors in orexin-producing neurons results in disrupted sleep/wakefulness architecture. Proc Natl Acad Sci U S A 106: 4459–4464.

34. Harsing LG, Jr., Prauda I, Barkoczy J, Matyus P, Juranyi Z (2004) A 5-HT7 heteroreceptor-mediated inhibition of [3H]serotonin release in raphe nuclei slices of the rat: evidence for a serotonergic-glutamatergic interaction. Neurochem Res 29: 1487–1497.

35. Lee HS, Park SH, Song WC, Waterhouse BD (2005) Retrograde study of hypocretin-1 (orexin-A) projections to subdivisions of the dorsal raphe nucleus in the rat. Brain Res 1059: 35–45.

36. Kumar S, Szymusiak R, Bashir T, Rai S, McGinty D, et al. (2007) Effects of serotonin on perifornical-lateral hypothalamic area neurons in rat. Eur J Neurosci 25: 201–212.

37. Postnova S, Voigt K, Braun HA (2009) A mathematical model of homeostatic regulation of sleep-wake cycles by hypocretin/orexin. J Biol Rhythms 24: 523–535.

38. Williams KS, Behn CG (2011) Dynamic interactions between orexin and dynorphin may delay onset of functional orexin effects: a modeling study. J Biol Rhythms 26: 171–181.

39. Patriarca M, Postnova S, Braun HA, Hernandez-Garcia E, Toral R (2012) Diversity and noise effects in a model of homeostatic regulation of the sleep-wake cycle. PLoS Comput Biol 8: e1002650.

40. Carter ME, Brill J, Bonnavion P, Huguenard JR, Huerta R, et al. (2012) Mechanism for Hypocretin-mediated sleep-to-wake transitions. Proc Natl Acad Sci U S A 109: E2635–2644.

41. Diniz Behn CG, Kopell N, Brown EN, Mochizuki T, Scammell TE (2008) Delayed orexin signaling consolidates wakefulness and sleep: physiology and modeling. J Neurophysiol 99: 3090–3103.

42. Rempe MJ, Best J, Terman D (2010) A mathematical model of the sleep/wake cycle. J Math Biol 60: 615–644.

43. Kumar R, Bose A, Mallick BN (2012) A mathematical model towards understanding the mechanism of neuronal regulation of wake-NREMS-REMS states. PLoS One 7: e42059.

44. Omenetti A, Yang L, Gainetdinov RR, Guy CD, Choi SS, et al. (2011) Paracrine modulation of cholangiocyte serotonin synthesis orchestrates biliary remodeling in adults. Am J Physiol Gastrointest Liver Physiol 300: G303–315.

45. Arenkiel BR, Peca J, Davison IG, Feliciano C, Deisseroth K, et al. (2007) In vivo light-induced activation of neural circuitry in transgenic mice expressing channelrhodopsin-2. Neuron 54: 205–218.

46. Hu Z, Rudd JA, Fang M (2012) Development of the human corpus striatum and the presence of nNOS and 5-HT2A receptors. Anat Rec (Hoboken) 295: 127–131.

47. Johansson S, Povlsen GK, Edvinsson L (2012) Expressional changes in cerebrovascular receptors after experimental transient forebrain ischemia. PLoS One 7: e41852.

48. Wai MS, Lorke DE, Kwong WH, Zhang L, Yew DT (2011) Profiles of serotonin receptors in the developing human thalamus. Psychiatry Res 185: 238–242.

49. Yeung LY, Kung HF, Yew DT (2010) Localization of 5-HT1A and 5-HT2A positive cells in the brainstems of control age-matched and Alzheimer individuals. Age (Dordr) 32: 483–495.

50. Ren LQ, Wienecke J, Chen M, Moller M, Hultborn H, et al. (2013) The time course of serotonin 2C receptor expression after spinal transection of rats: an immunohistochemical study. Neuroscience.

51. Weber M, Schmitt A, Wischmeyer E, Doring F (2008) Excitability of pontine startle processing neurones is regulated by the two-pore-domain K+ channel TASK-3 coupled to 5-HT2C receptors. Eur J Neurosci 28: 931–940.

52. Rivera HM, Santollo J, Nikonova LV, Eckel LA (2012) Estradiol increases the anorexia associated with increased 5-HT(2C) receptor activation in ovariectomized rats. Physiol Behav 105: 188–194.

53. Zinchuk V, Grossenbacher-Zinchuk O (2009) Recent advances in quantitative colocalization analysis: focus on neuroscience. Prog Histochem Cytochem 44: 125–172.

54. Zinchuk V, Zinchuk O, Okada T (2007) Quantitative colocalization analysis of multicolor confocal immunofluorescence microscopy images: pushing pixels to explore biological phenomena. Acta Histochem Cytochem 40: 101–111.

55. Bolte S, Cordelieres FP (2006) A guided tour into subcellular colocalization analysis in light microscopy. J Microsc 224: 213–232.

56. Adler J, Parmryd I (2010) Quantifying colocalization by correlation: the Pearson correlation coefficient is superior to the Mander's overlap coefficient. Cytometry A 77: 733–742.

57. Costes SV, Daelemans D, Cho EH, Dobbin Z, Pavlakis G, et al. (2004) Automatic and quantitative measurement of protein-protein colocalization in live cells. Biophys J 86: 3993–4003.

58. Joshi A, Wong-Lin K, McGinnity TM, Prasad G (2011) A mathematical model to explore the interdependence between the serotonin and orexin/hypocretin systems. Conf Proc IEEE Eng Med Biol Soc 2011: 7270–7273.

59. Tao R, Auerbach SB (2003) Influence of inhibitory and excitatory inputs on serotonin efflux differs in the dorsal and median raphe nuclei. Brain Res 961: 109–120.

60. Wilson HR, Cowan JD (1972) Excitatory and inhibitory interactions in localized populations of model neurons. Biophys J 12: 1–24.

61. Dayan P, Abbott L (2011) Theoretical Neuroscience: Computational and Mathematical Modeling of Neural Systems. The MIT Press.

62. Shriki O, Hansel D, Sompolinsky H (2003) Rate models for conductance-based cortical neuronal networks. Neural Comput 15: 1809–1841.

63. Wong KF, Wang XJ (2006) A recurrent network mechanism of time integration in perceptual decisions. J Neurosci 26: 1314–1328.

64. Crawford LK, Craige CP, Beck SG (2010) Increased intrinsic excitability of lateral wing serotonin neurons of the dorsal raphe: a mechanism for selective activation in stress circuits. J Neurophysiol 103: 2652–2663.

65. Karnani MM, Szabo G, Erdelyi F, Burdakov D (2013) Lateral hypothalamic GAD65 neurons are spontaneously firing and distinct from orexin- and melanin-concentrating hormone neurons. J Physiol 591: 933–953.

66. Kirby LG, Pernar L, Valentino RJ, Beck SG (2003) Distinguishing characteristics of serotonin and non-serotonin-containing cells in the dorsal raphe nucleus: electrophysiological and immunohistochemical studies. Neuroscience 116: 669–683.

67. Katayama J, Yakushiji T, Akaike N (1997) Characterization of the K+ current mediated by 5-HT1A receptor in the acutely dissociated rat dorsal raphe neurons. Brain Res 745: 283–292.

68. Williams JT, Colmers WF, Pan ZZ (1988) Voltage- and ligand-activated inwardly rectifying currents in dorsal raphe neurons in vitro. J Neurosci 8: 3499–3506.

69. Liu R, Jolas T, Aghajanian G (2000) Serotonin 5-HT(2) receptors activate local GABA inhibitory inputs to serotonergic neurons of the dorsal raphe nucleus. Brain Res 873: 34–45.

70. Lee MG, Hassani OK, Jones BE (2005) Discharge of identified orexin/hypocretin neurons across the sleep-waking cycle. J Neurosci 25: 6716–6720.

71. Mileykovskiy BY, Kiyashchenko LI, Siegel JM (2005) Behavioral correlates of activity in identified hypocretin/orexin neurons. Neuron 46: 787–798.

72. Takahashi K, Lin JS, Sakai K (2008) Neuronal activity of orexin and non-orexin waking-active neurons during wake-sleep states in the mouse. Neuroscience 153: 860–870.

73. Sakai K (2011) Sleep-waking discharge profiles of dorsal raphe nucleus neurons in mice. Neuroscience 197: 200–224.

74. Ermentrout GB, Kopell N (1990) Oscillator death in systems of coupled neural oscillators. SIAM Journal on Applied Mathematics 50: 125–146.

75. Xie X, Crowder TL, Yamanaka A, Morairty SR, Lewinter RD, et al. (2006) GABA(B) receptor-mediated modulation of hypocretin/orexin neurones in mouse hypothalamus. J Physiol 574: 399–414.

76. Backberg M, Ultenius C, Fritschy JM, Meister B (2004) Cellular localization of GABA receptor alpha subunit immunoreactivity in the rat hypothalamus: relationship with neurones containing orexigenic or anorexigenic peptides. J Neuroendocrinol 16: 589–604.

77. Kokare DM, Patole AM, Carta A, Chopde CT, Subhedar NK (2006) GABA(A) receptors mediate orexin-A induced stimulation of food intake. Neuropharmacology 50: 16–24.

78. Peyron C, Tighe DK, van den Pol AN, de Lecea L, Heller HC, et al. (1998) Neurons containing hypocretin (orexin) project to multiple neuronal systems. J Neurosci 18: 9996–10015.

79. Alberto CO, Hirasawa M (2010) AMPA receptor-mediated miniature EPSCs have heterogeneous time courses in orexin neurons. Biochem Biophys Res Commun 400: 707–712.

80. Doane DF, Lawson MA, Meade JR, Kotz CM, Beverly JL (2007) Orexin-induced feeding requires NMDA receptor activation in the perifornical region of the lateral hypothalamus. Am J Physiol Regul Integr Comp Physiol 293: R1022–1026.

81. Henny P, Jones BE (2006) Innervation of orexin/hypocretin neurons by GABAergic, glutamatergic or cholinergic basal forebrain terminals evidenced by immunostaining for presynaptic vesicular transporter and postsynaptic scaffolding proteins. J Comp Neurol 499: 645–661.

82. Li Y, Gao XB, Sakurai T, van den Pol AN (2002) Hypocretin/Orexin excites hypocretin neurons via a local glutamate neuron-A potential mechanism for orchestrating the hypothalamic arousal system. Neuron 36: 1169–1181.

83. Verge D, Daval G, Patey A, Gozlan H, el Mestikawy S, et al. (1985) Presynaptic 5-HT autoreceptors on serotonergic cell bodies and/or dendrites but not terminals are of the 5-HT1A subtype. Eur J Pharmacol 113: 463–464.

84. Ranade SP, Mainen ZF (2009) Transient firing of dorsal raphe neurons encodes diverse and specific sensory, motor, and reward events. J Neurophysiol 102: 3026–3037.

85. Bromberg-Martin ES, Hikosaka O, Nakamura K (2010) Coding of task reward value in the dorsal raphe nucleus. J Neurosci 30: 6262–6272.

86. Wong-Lin K, Joshi A, Prasad G, McGinnity TM (2012) Network properties of a computational model of the dorsal raphe nucleus. Neural Netw 32: 15–25.

87. Diaz-Cabiale Z, Parrado C, Narvaez M, Puigcerver A, Millon C, et al. (2011) Galanin receptor/Neuropeptide Y receptor interactions in the dorsal raphe nucleus of the rat. Neuropharmacology 61: 80–86.

88. Mountcastle VB (1997) The columnar organization of the neocortex. Brain 120 (Pt 4): 701–722.

89. Ebrahim IO, Sharief MK, de Lacy S, Semra YK, Howard RS, et al. (2003) Hypocretin (orexin) deficiency in narcolepsy and primary hypersomnia. J Neurol Neurosurg Psychiatry 74: 127–130.

90. Morikawa H, Manzoni OJ, Crabbe JC, Williams JT (2000) Regulation of central synaptic transmission by 5-HT(1B) auto- and heteroreceptors. Mol Pharmacol 58: 1271–1278.

91. Ishibashi H, Kuwano K, Takahama K (2000) Inhibition of the 5-HT(1A) receptor-mediated inwardly rectifying K(+) current by dextromethorphan in rat dorsal raphe neurones. Neuropharmacology 39: 2302–2308.

92. Gocho Y, Sakai A, Yanagawa Y, Suzuki H, Saitow F (2013) Electrophysiological and pharmacological properties of GABAergic cells in the dorsal raphe nucleus. J Physiol Sci 63: 147–154.

93. Yamanaka A, Muraki Y, Ichiki K, Tsujino N, Kilduff TS, et al. (2006) Orexin neurons are directly and indirectly regulated by catecholamines in a complex manner. J Neurophysiol 96: 284–298.

14

Functional Epitope Core Motif of the *Anaplasma marginale* Major Surface Protein 1a and Its Incorporation onto Bioelectrodes for Antibody Detection

Paula S. Santos[1,℘], **Rafael Nascimento**[1,℘], **Luciano P. Rodrigues**[2], **Fabiana A. A. Santos**[1], **Paula C. B. Faria**[3], **João R. S. Martins**[4], **Ana G. Brito-Madurro**[2], **João M. Madurro**[2], **Luiz R. Goulart**[1]*

1 Laboratório de Nanobiotecnologia, Instituto de Genética e Bioquímica, Universidade Federal de Uberlândia, Uberlândia, Brazil, 2 Laboratório de Filmes Poliméricos e Nanotecnologia, Instituto de Química, Universidade Federal de Uberlândia, Uberlândia, Brazil, 3 Laboratório de Leishmanioses, Universidade Federal de Minas Gerais, Belo Horizonte, Brazil, 4 Laboratório de Parasitologia, Instituto de Pesquisas Veterinárias Desidério Finamor, Eldorado do Sul, Brazil

Abstract

Anaplasmosis, a persistent intraerythrocytic infection of cattle by *Anaplasma marginale*, causes severe anemia and a higher rate of abortion, resulting in significant loss to both dairy and beef industries. Clinical diagnosis is based on symptoms and confirmatory laboratory tests are required. Currently, all the diagnostic assays have been developed with whole antigens with indirect ELISA based on multiple epitopes. In a pioneer investigation we demonstrated the use of critical motifs of an epitope as biomarkers for immunosensor applications. Mimotopes of the MSP1a protein functional epitope were obtained through Phage Display after three cycles of selection of a 12-mer random peptide library against the neutralizing monoclonal antibody 15D2. Thirty-nine clones were randomly selected, sequenced, translated and aligned with the native sequence. The consensus sequence SxSSQSEASTSSQLGA was obtained, which is located in C-terminal end of the 28-aa repetitive motif of the MSP1a protein, but the alignment and sequences' variation among mimotopes allowed us to map the critical motif STSSxL within the consensus sequence. Based on these results, two peptides were chemically synthesized: one based on the critical motif (STSSQL, Am1) and the other based on the consensus sequence aligned with the native epitope (SEASTSSQLGA, Am2). Sera from 24 infected and 52 healthy animals were tested by ELISA for reactivity against Am1 and Am2, which presented sensitivities of 96% and 100%, respectively. The Am1 peptide was incorporated onto a bioelectrode (graphite modified with poly-3-hydroxyphenylacetic acid) and direct serum detection was demonstrated by impedance, differential pulse voltammetry, and atomic force microscopy. The electrochemical sensor system proved to be highly effective in discriminating sera from positive and negative animals. These immunosensors were highly sensitive and selective for positive IgG, contaminants did not affect measurements, and were based on a simple, fast and reproducible electrochemical system.

Editor: Bernhard Kaltenboeck, Auburn University, United States of America

Funding: The work was supported by Universidade Federal de Uberlândia, Conselho Nacional de Desenvolvimento Científico e Tecnológico (CNPq), Fundação de Amparo à Pesquisa do estado de Minas Gerais (FAPEMIG) and Coordenação de Aperfeiçoamento de Pessoal de Nível Superior (CAPES). The funders had no role in study design, data collection and analysis, decision to publish, or preparation of the manuscript.

Competing Interests: The authors have declared that no competing interests exist.

* E-mail: lrgoulart@ufu.br

℘ These authors contributed equally to this work.

Introduction

The tick-borne intracellular pathogen *Anaplasma marginale* (Rickettsiales: Anaplasmataceae) is the causal agent of anaplasmosis, a hemoparasitic disease of cattle. *A. marginale* is distributed worldwide in tropical and subtropical regions of the world [1], resulting in considerable economic loss to both dairy and beef industries. *A. marginale* resides within erythrocytes of ruminants, and induces pyrexia, anemia, weight loss, abortion, lethargy, icterus, spleno and hepatomegaly, and often death [2]. *A. marginale* is transmitted horizontally by ixodid ticks, while mechanical transmission occurs when infected blood is transferred to susceptible cattle by fly bites or blood-contaminated fomites [3].

Six major surface proteins (MSPs) have been characterized on the erythrocytic stage of *A. marginale*. Among them, the major surface protein 1 (MSP1) has been extensively studied [4]. MSP1 is a complex of two covalently linked unrelated polypeptides, MSP1a and MSP1b. The MSP1a has been shown that is involved in the adhesion, infection and tick transmission of *A. marginale*, as well as to contribute to protective immunity in cattle [5,6]. MSP1a contains a variable number of tandemly repeated peptides in the amino-terminal region, which are exposed extracellularly for interaction with host cell receptors [7,8].

Many different methods have been reported for the diagnosis of *A. marginale* in cattle. The clinical diagnosis is usually based on the observation of clinical signs, necropsy findings, and the geographic region [9]. In order to confirm the diagnosis, laboratory tests such as light microscopy evaluation of Giemsa-stained blood smears or serological/molecular diagnostic procedures are required. In carrier animals, microscopy-based diagnosis can be difficult, due to variable parasitemia, and thus, a variety of serologic tests for detection of specific antibodies [10,11] or PCR based assays [12] are necessary. However, these methods that claim high sensitivity also require greater technical skills as well as expensive instrumentation. In such a scenario, rapid identification methods using simple immunological assays for laboratory use, such as ELISA, and field portable biosensors could be more useful.

In general all antibody detection assays are based on whole antigens with multiple epitopes, which show greater sensitivity, but cross-reactions are often observed. On the other hand, epitope-specific antibody response assays are not commonly used, because it is well established that genetic background can influence the specificity of B-cell responses [13]; therefore, simple epitopes are rarely used as markers because of the difficulty in selecting common motifs that recognize broad immune responses of animals. However, the development of novel epitopes through Phage Display (PD) technology [14] has become possible, especially because selected mimotopes that mimic natural antigenic determinants are mainly originated from dominant responses, and selection favors highly reactive motifs, due to their optimized structure or functional properties [15]. Importantly, selected stable short peptide sequences assessed for tight binding to antibodies, receptors or proteins may present potential applications in diagnostics, therapeutics and vaccines [16,17].

Because of the importance of the carrier animal in disease transmission, and also due to the difficulty in producing total purified antigens from infected erythrocyte cultures, an effective diagnostic test with synthetic peptides may be an interesting alternative tool to reduce disease transmission and economic losses. Therefore, in this present study, we have selected peptides through PD against a monoclonal antibody that targets the major surface protein 1a (MSP1a) in order to map its epitope and to develop new mimotopes that are more effective than the native epitope in detecting antibody responses in cattle against *A. marginale*. We have also proposed a bioeletrode conjugated to the epitope to detect antibodies in crude serum by electrochemistry, which may become the basis of novel biosensors based on a specific-epitope antibody response detection that can be used in field conditions due to its flexibility, easiness, fastness, and low cost.

Results

Epitope mapping of MSP1a by Phage Display

Thirty-nine randomly selected MSP1a mimotopes were obtained after three rounds of biopanning using a phage displayed 12-mer random peptide library against the anti-MSP1a monoclonal antibody 15D2 (Figure 1A). Alignment analysis revealed the consensus sequence SxSSQSEASTSSQLGA, which is depicted as a sequence logo in Fig. 1B. This sequence corresponds to tandem repeats located in the amino-terminal region of MSP1a, and may be considered part of the antigenic determinant region.

All 39 selected peptide sequences were different, but with the presence of the critical motif STSSxL in 43.6% of the clones (17/39), in which 17 of them presented the highest scores in ELISA. Alignment has also shown that 29 clones presented the full or partial sequence of the critical motif at the C-terminal end of the peptides, and all of them presented at least 4 matches with the original sequence.

Based on the frequency of residues in the selected clones and the original MSP1a sequence, we have chosen two motifs for chemical synthesis, STSSQL and SEASTSSQLGA, for additional analysis.

Immunoreactivity of selected phagotopes for the anti-MSP1a mAb and pooled IgG from infected animals

Phage-ELISA assays were performed to validate the selected phage-fused peptide clones (phagotopes) and a successful reactivity

was demonstrated for both anti-MSP1 mAb (Figure 1C) and IgG from *A. marginale* infected animals (Figure 1D). The wild type M13 phage vector (no peptide) was used as negative control to confirm the selection efficiency. The reactivities of phagotopes to the mAb were similar, except for clones C_{12} and H_{01} that presented low reactivities; however all phagotopes recognized IgG from serum of *A. marginale* infected bovines, demonstrating the ability of phagotopes to discriminate infected from non-infected animals.

To confirm the surface exposure probability of the consensus epitope sequence, we have performed a simulation to generate a 3D structure of the MSP1a protein, because its PDB structure is not available, and the putative localization of the epitope within the structure was shown in Figure 1E, corroborating the possible antibody binding region in the external sequences of the predicted protein.

Immunoreactivity of synthetic peptides against IgG from *A. marginale* infected animals and negative controls

Two peptides were chemically synthesized representing the most repetitive motif (STSSQL, Am1) and the putative natural epitope (SEASTSSQLGA, Am2) based on the consensus sequence. Both synthetic molecules were able to discriminate sera from infected animals and healthy controls (p<0.0001) (Figure 2). The ROC curve analysis were significant for both peptides Am1 (AUC = 0.8906) and Am2 (AUC = 0.8938), and based on cut-off values they presented sensitivities of 95.83% and 100%, and specificities and 53.85% and 57.69%, respectively.

Testing specificity for anaplasmosis

Both synthetic peptides Am1 and Am2 presented high reactivity against sera of *A. marginale* infected animals; however, when both were tested (ELISA) for reactivity to other diseases, the Am1 specifically reacted with IgG antibodies from anaplasmosis (p<0.05), while the Am2 presented cross-reactivity with bovine brucellosis (Figure 3).

Bioelectrode functionalization and electrochemical detection of peptide-antibody complexes

Differential pulse voltammograms of a bioelectrode functionalized with the peptide Am1 were carried out aiming to evaluate the interaction process between the graphite electrode/poly(3-HPA)/Am1 (probe) and the target IgG (Figure 4). After immersion of the functionalized bioelectrode in a positive pooled serum sample (IgG+), it was observed a significant decrease in the amplitude of the current signal in relation to the negative serum (IgG−) with an approximate reduction of 140 µA after antibody binding.

The impedance response of the graphite electrode (Figure 5) demonstrated significant changes in the surface resistivity, as shown by experimental curves for the polymeric film alone (poly(3-HPA)), the functionalized bioelectrode (poly(3-HPA)/Am1) without sera and with IgG+ and IgG− sera, generating two more curves for the bioelectrode peptide-antibody complex test. The positive serum (poly(3-HPA)/Am1:IgG+) presented a significant difference in resistivity in comparison to the three controls (poly(3-HPA)), poly(3-HPA)/Am1 and poly(3-HPA)/Am1:IgG-). The polymeric film (poly(3-HPA)) alone was different from the other two controls, which presented curves with similar behaviors.

The equivalent circuit used to fit the experimental data was: $R_s[(W\ R_{ct,1})Q_{dl,1}]Q_{dl,2}$, ($R_S$: solution resistance, Q_{dl}: double-layer capacitance, R_{ct}: charge transfer resistance, W: Warburg impedance).

Comparison of the double layer capacitance in Table 1 showed that the poly(3-HPA)/Am1:IgG+ complex presented a greater

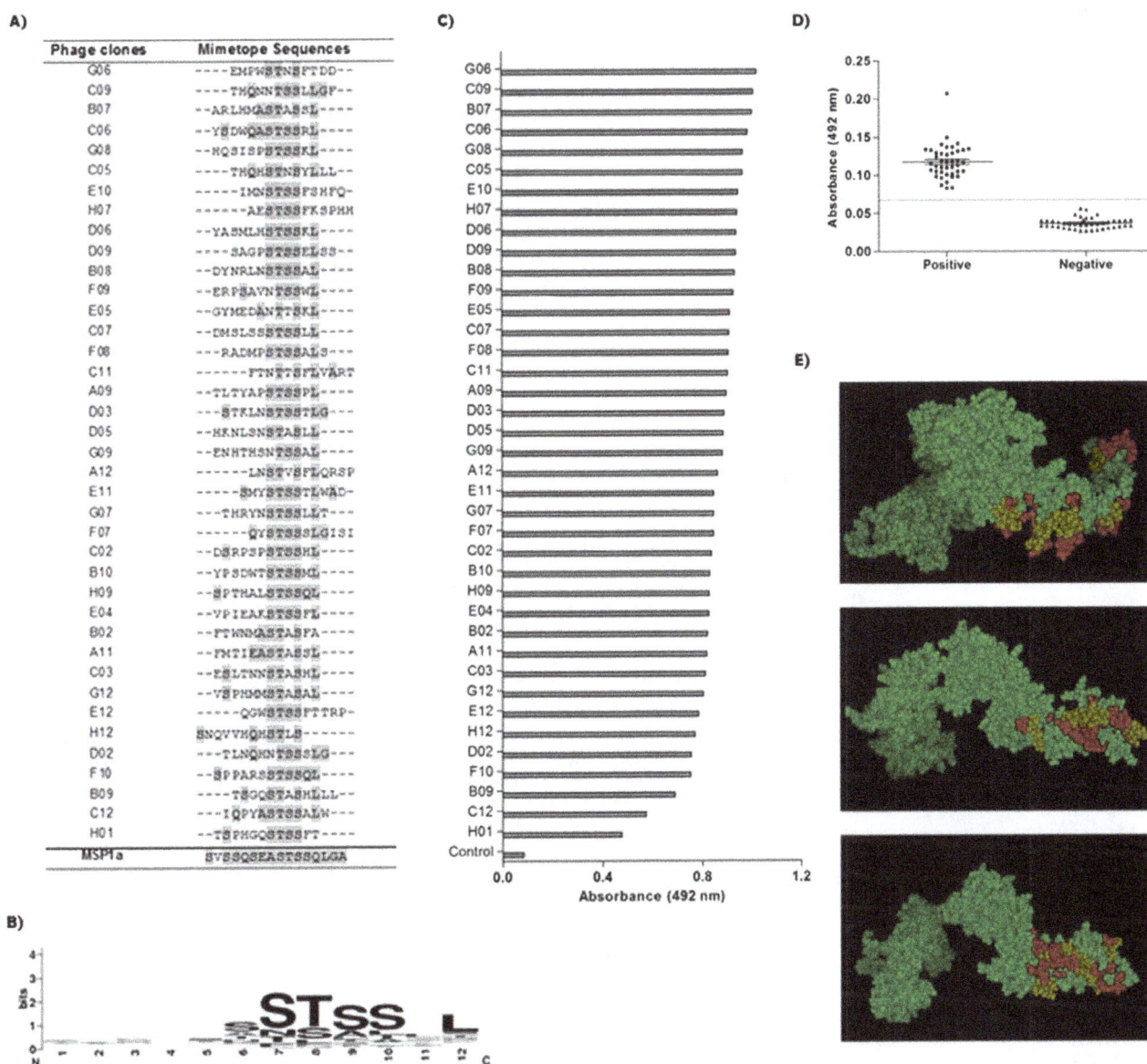

Figure 1. Phages selected by Phage Display and its performance. (A) Peptides sequences of 39 phage clones and their consensus sequence, according to the original sequence determined for the N-terminal region of the MSP1a protein. (B) Graphical representation of the sequence logo of MSP1a-binding motifs. The conserved sequence pattern was generated using WebLogo3 (http://weblogo.berkeley.edu/). Bits represent the relative frequency of amino acids. (C) Interaction of phage clones with the commercially available anti-MSP1 monoclonal antibody 15D2, and (D) binding specificity of each clone to pooled sera from *Anaplasma marginale* infected animals and non-infected. M13 wild type phage (without peptide) was used as a negative control. (E) Models of 3D structures predicted for the MSP1a protein and its putative epitope localization. In red, the consensus motif SxSSQSEASTSSQLGA, and in yellow, the critical motif STSSQL.

resistivity (267 $\Omega.cm^2$) than the controls poly(3-HPA)/Am1 (158 $\Omega.cm^2$) and poly(3-HPA)/Am1:IgG− (153 $\Omega.cm^2$). Therefore, the presence of the target positive IgG in the bioelectrode's interface presented nearly 2-fold increase in the double layer capacitance. The chi-square values (χ^2) of the Kramers-Kronig was in the order of 10^{-2}–10^{-3}.

The atomic force microscopic experiments (Figure 6) demonstrated that after immobilization of the Am1 (probe) on the modified electrode (Figure 6B), the bioelectrode presented a more irregular surface, but after interaction with the peptide:IgG+ (Figure 6C), the roughness of the bioelectrode increased, as observed with the formation of numerous clusters. The root-mean-square roughness values of the graphite electrode modified with

poly(3-HPA), poly(3-HPA):Am1, poly(3-HPA):Am1:IgG+ were 29.7 nm, 37.0 nm and 45.9 nm, respectively.

Discussion

In this investigation, we have used PD to select immunodominant epitopes against the neutralizing monoclonal antibody 15D2 anti-MSP1 that recognizes all geographical *A. marginale* isolates, for which two epitopes were previously characterized (QASTSS and EASTSS) [18,19]. However, evidences have also demonstrated that the full epitope sequence consists of a larger repetitive motif of 28 or 29 amino acids (ADSSSAGGQQQESSVSSQSD-QASTSSQLG) with changes in only seven residues [8]. In our

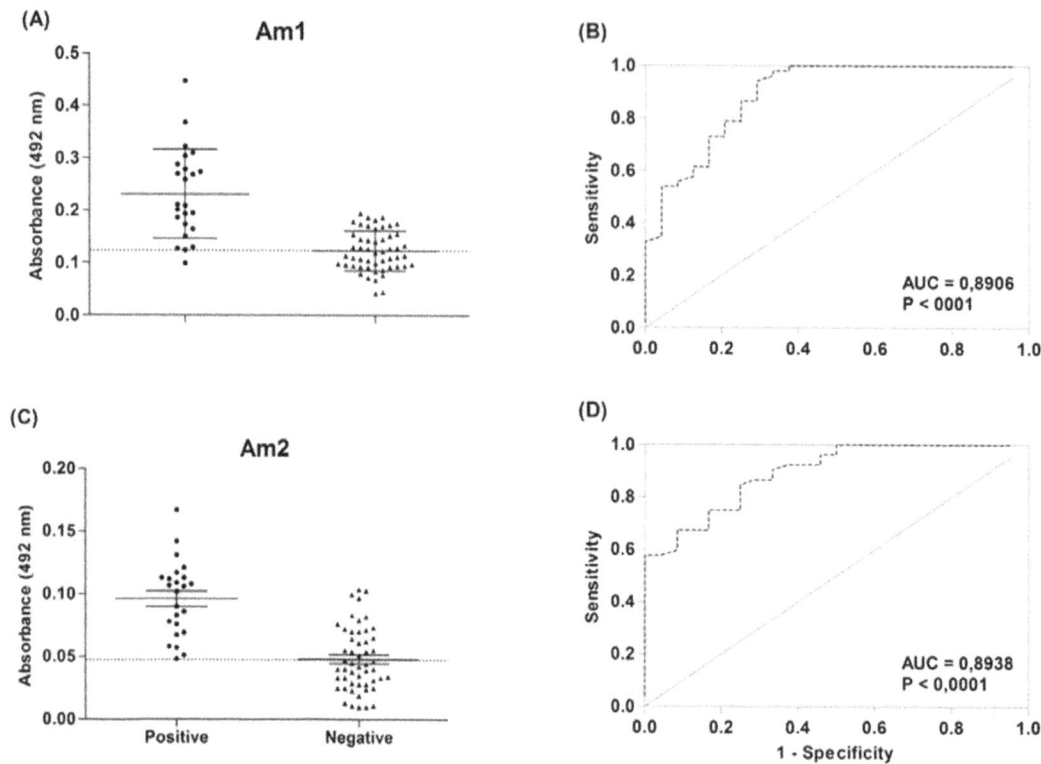

Figure 2. Antibody detection by ELISA. Detection of immunoglobulin G antibodies anti-*Anaplasma marginale* MSP1a in serum samples from infected and non-infected bovine with a definitive diagnosis of anaplasmosis (n = 24), and apparently healthy individuals (n = 52) by enzyme-linked immunosorbent assay using the Am1 (A) and Am2 (B), and the respectively ROC curve.

mimotopes' selection of 12-mer ligands against the paratope region of the monoclonal antibody anti-MSP1a, we have revealed a shorter consensus motif (SxSSQSEASTSSQLGA) that is located within the 28-aa tandemly repeat peptide at the C-terminal side. Due to the high frequency of the critical motif at the C-terminal end, it is possible that the N-terminal end of the peptide may favor specificity and antibody binding affinity, however it is not critical due to the very large variation of residues in that region without contributing to the improved sensitivity, and apparently the alanine insertion in the critical motif (ASTSSxL) may slightly improve the reaction, although not significantly.

Interestingly, we have also demonstrated that a critical motif, STSSxL, resulted from the alignment of all selected mimotopes with the native epitope, is important for the antibody recognition and appears to play a predominant role in dictating the formation of the antigen-antibody complex and may possibly be used as a vaccine immunogen. This is corroborated by predictions of 3D structures of the MSP1a protein, regardless which model is correct, and the critical core epitope was mapped in the most exposed region of the protein in all three models, located in a hydrophilic region, and with adjacent residues in both sides showing less surface exposure, which also explain why the STSSxL was mainly selected.

This result corroborates the powerful application of the PD technology in selecting peptide ligands [20], especially against monoclonal antibodies [21], which has also proven to be an effective strategy for vaccine development [22] and drug discovery [23]. The mimotopes have strongly bound to the mAb anti-MSP1 with reactivities that were superior to the peptide that were synthetized based on the native MSP1a sequence as demonstrated by ELISA assays. Similarly, the mimotopes fused to phage

particles have also recognized IgG from *A. marginale* infected bovine sera, and efficiently discriminated infected from non-infected animals. This indicates that the surface characteristics of these mimetic peptides are not only equivalent to the epitope, but PD selections may have even improved their affinity by optimizing their structure.

The final aim of this investigation was to use these short epitope peptides in bovine anaplasmosis diagnostics, but immunoassays based on single epitopes usually present low sensitivity and specificity. It is well established that the polyclonal B cell response to specific antigens is an adaptive immune system of mammals that ensures the recognition of multiple epitopes of an antigen. Because antigens can be large and complex substances, any single antibody can bind only to a single epitope, which are often specific, and highly dependent on the genetic background [13], consequently, an effective immune response frequently involves the production of many different antibodies. This means that a diagnostic tool based on a single epitope may only be effective if this antigenic determinant is highly dominant and presents a broad recognition by antibodies, which explains the rare number of publications with unique epitopes as biomarkers.

Our protein target, MSP1a, was chosen because of its occurrence in all *A. marginale* strains [18], the presence of repetitive motifs, its surface exposure, and the conserved nature across rickettsia species [19], and also because we believe that a reduced critical epitope sequence could be identified based on its dominant affinity to the antibody target, which could only be accomplished by selection and enrichment processes obtained through the PD technology.

Although peptide sequences were similar due to the critical motif STSSxL, they all differ from the native MSP1a epitope

A

Am1

B

Am2

Figure 4. Differential pulse voltammograms of graphite electrode modified with poly(3-HPA). Synthetic peptide Am1 (A) before or after addition of negative serum (IgG−) (B) or positive serum (IgG+) (C), in aqueous solution containing $K_3Fe(CN)_6/K_4Fe(CN)_6$ (0.33 mmol.L^{-1}) and KCl (0.1 mol.L^{-1}). Modulation amplitude: 25 mV; 16 mVs^{-1}.

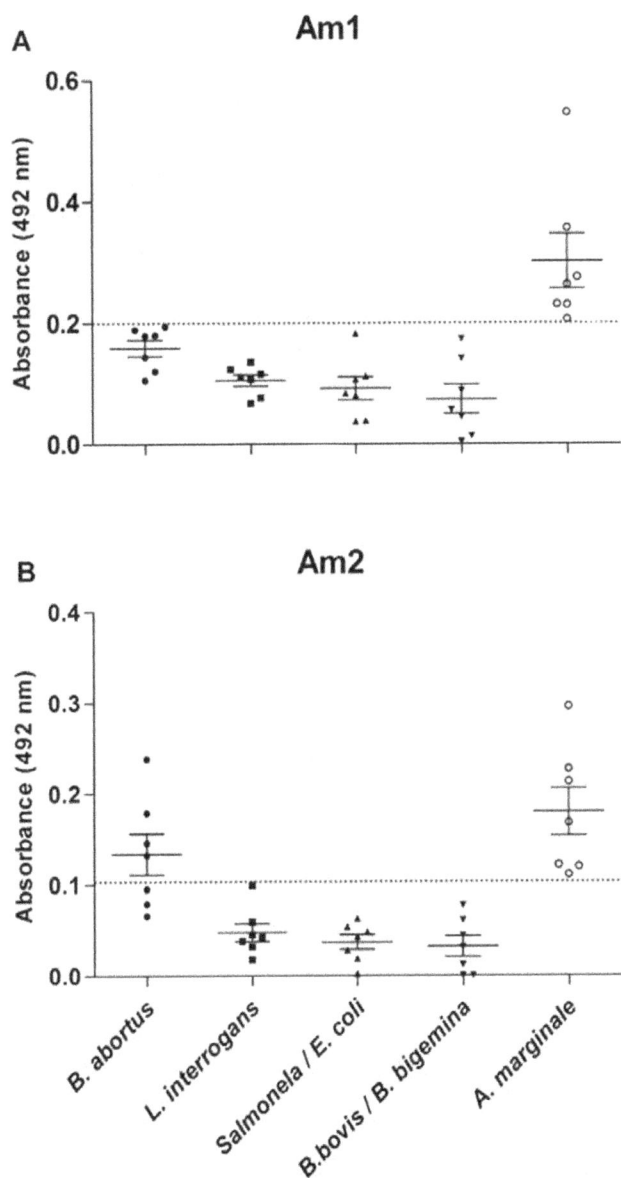

Figure 3. Synthetic peptides binding specificity analysis. Reactivity of synthetic peptides, Am1 (A) and Am2 (B), to bovine sera infected with *Anaplasma marginale* (n = 7), *Brucella abortus* (n = 7), *Leptospira interrogans* (n = 7), *Salmonella typhimurium* (n = 7), *Escherichia coli* (n = 7), *Babesia bovis* (n = 7), *Babesia bigemina* (n = 7); dog sera infected with *Anaplasma phagocytophilum* (n = 7), *Ehrlichia canis* (n = 7), *Ehrlichia ewingii* (n = 1); and a rabbit serum infected with *Ehrlichia sennetsu* (n = 1) by enzyme-linked immunosorbent assays.

sequence, which makes them true mimotopes, but for the proof-of-principle that a minimum dominant sequence would be required for antibody recognition, we have chemically synthesized two peptides with restricted sizes based on the critical motif (STSSQL, Am1) and the consensus sequence (SEASTSSQLGA, Am2) for affinity studies. Importantly, the synthetic peptides have also presented a high reactivity against infected animal sera, which is discordant from other reports that affirm that mimotopes fused to the phage can only be detected when anchored in the phage capsid [24–26].

Classification of animals in the target population as infected or uninfected is conditional upon how well the reference animal

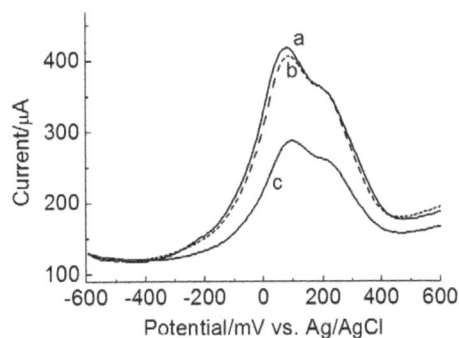

population is used to validate the assay [27]. Therefore, we have used two well classified groups of animals (infected and uninfected) in order to compare with sensitivity and specificity indices reported with whole antigens [28–31]. The immunoassays for Am1 and Am2 synthetic peptides presented sensitivity and specificity around 96%/53% and 100%/57%, respectively. Interestingly, another study using recombinant MSP1a and MSP2 as targets reached sensitivity and specificity of 99% and 100%, respectively [32]. The differences in specificity between the epitope-based assay and the whole-antigen assay, is probably associated with other epitopes that are not recognized by the monoclonal antibody, demonstrating the differential B cell response or because short epitopes may present a higher probability of cross reactions with other proteins. However, the sensitivity remains similar, which indicates that the two restricted epitope sequences are immunodominant.

Similar to the MSP1, the whole MSP5 protein has also been successfully used for antibodies detection, with sensitivities that vary from 96 to 100% and specificities that varied from 89% to 100% [28–31]. Again, the slightly inferior sensitivity of Am1 compared to the whole antigens may be due to the greater number of epitopes, a hypothesis that is corroborated by another report

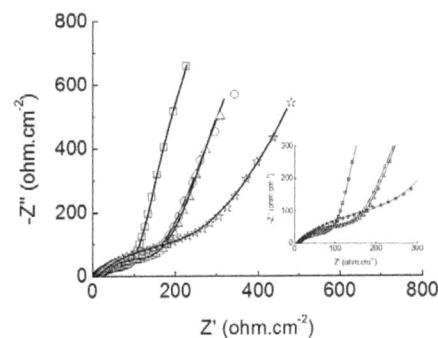

Figure 5. Impedance response of graphite electrode. Nyquist diagrams of a the polymeric film poly(3-HPA) (-□-), poly(3-HPA)/Am1 (-Δ-), poly(3-HPA)/Am1:IgG+ (-☆-), and poly(3-HPA)/Am1:IgG- (-○-) obtained in aqueous solution containing $K_3Fe(CN)_6/K_4Fe(CN)_6$ (5 mmol.L^{-1}) and KCl (0.1 mol.L^{-1}), recorded from an applied potential of +0.24 V, amplitude of 10 mV, and frequency range from 100 KHz to 10 Hz. The continuous lines represent the fitting curve to the equivalent circuit. Inset: amplification of high frequencies region.

Table 1. Values to double layer capacitance and charge transfer resistance of the polymer-biomolecules/electrolyte interface of the bioelectrode, prepared with AC impedance analysis from Nyquist plots.

	poly(3-HPA):Am1	poly(3-HPA):Am1:(IgG+)	poly(3-HPA):Am1:(IgG-)
$Q_{dl,1}$	0.018	0.033	0.013
$R_{ct,1}$	158	267	153

$R(\Omega.cm^2)$, $Q(mF.cm^{-2})$.

that demonstrated the cross-reactivity of the MSP5 with multiple species of *Anaplasma* [33].

Importantly, the five additional residues in the Am2 peptide improved the sensitivity to 100%, indicating the immunodominance of the epitope, but it did not significantly differ from the assay with the critical motif (STSSxL), suggesting that this six-amino acid core epitope may have an important role in the antibody recognition. The lower specificity of ELISA tests with both synthetic peptides (Am1 and Am2) may also be due to the presence of these motifs in other pathogen species or because of cross-reactions with erythrocyte proteins demonstrated elsewhere [34].

Because of the low specificity within uninfected samples, we have also investigated cross-reactions with other ehrlichiosis (*Anaplasma phagocytophilum*, *Ehrlichia canis*, *Ehrlichia ewingii* and *Ehrlichia sennetsu*) and other bovine diseases (brucellosis, leptospirosis, salmonellosis and babesiosis), and only the Am2 peptide cross-reacted with sera of animals with brucellosis. This may suggest that the full epitope may have conformational changes in the whole antigen, which prevents the cross-reaction with other sera, but this hypothesis remains to be demonstrated, although a recombinant MSP1a has been demonstrated to be 100% specific [32]. Interestingly the critical core peptide did not react with other pathogens, which may be due to its short sequence that favors a linear recognition.

Our long-term goal is to develop peptide-based bioelectronic sensor to accurately detect animal diseases. Synthetic peptides are attractive for high precision diagnostics because they are easily produced and are free of contaminants. Therefore, once the critical core epitope was identified and validated, the next step was to incorporate this functional epitope onto bioelectrodes for electrochemical detection of infected sera without any extra manipulation and without labeling.

After functionalizing the bioelectrode with the Am1 peptide, differential pulse voltammograms were used to differentiate samples into positive and negative sera, and we have successfully demonstrated that the system was very selective for the positive sera, and although the evaluated sera contains many interfering substances, such as urea, uric acid, glutamate and albumin, none of them have disturbed the detection process.

Differently from the voltammetric sensor, an impedimetric immunosensor was also developed, which is based on the antigen–antibody complex that is formed in the interface, affecting the capacitance and the charge transfer resistance in the interface [35]. The complex plane plot, known as the Nyquist plot, representative of the impedance response of a graphite electrode, demonstrated significant electrical changes (resistance) in the presence of specific (IgG+), when the poly(3-HPA) polymer was functionalized with the Am1 peptide, which was different from the non-specific antibodies (negative sera) that were similar to the

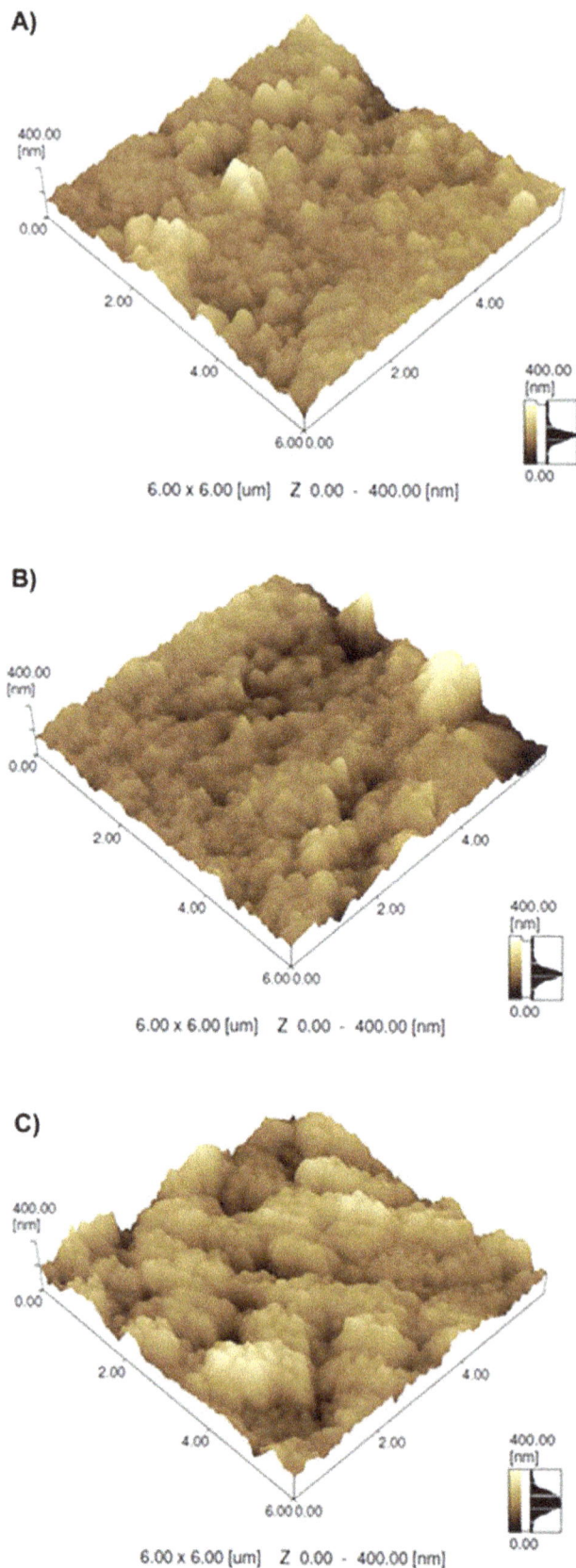

Figure 6. AFM topographical images of modified graphite electrode with poly(3-HPA). (A) Without biomolecules; (B) polymeric film after immobilization of the synthetic peptide Am1, and (C) agglutination process (Am1:IgG+).

curves obtained for the polymeric film alone [poly(3-HPA)] and for the polymeric film conjugated to the peptide, confirming the high specificity of the bioelectrode. These results are compatible with our previous experiments with differential pulse voltammetry.

The model for the morphology of the electroactive layer of the bioelectrode (equivalent circuit) supports the existence of two regions: a more internal region, formed by the graphite/polymer interface, described by $Q_{dl,2}$, and a external region, formed by the polymer-biomolecules/electrolyte interface, represented by $R_{ct,1}$ and $Q_{dl,1}$. The comparison of the double layer capacitance showed that the bioelectrode interface, when the peptide interacts with the specific antibody (positive sera), produced a 2-fold increase due to the electrical charges near the bioelectrode's surface. This charge transfer resistance is utilized as a main indicator in the faradaic EIS (Electrochemical impedance spectroscopy) detection for the electrode kinetics at the interface, which is modified by probes that are capable of selectively capturing a given target on the electrode's surface [36]. Comparisons of this parameter indicated that the complex poly(3-HPA)/Am1:IgG+, presented a significant increase in the R_{ct} value, resulting in a 1.7-fold decrease in the charge transfer, which was significantly different from the poly(3-HPA)/Am1 and poly(3-HPA)/peptide:IgG−. This bioelectrode corroborates other studies using impedimetric systems [37,38].

Additional experiments with atomic force microscopy were carried out to evaluate the morphological changes in the electrode's surface, and the interaction of the peptide with the positive sera was evidenced by the formation supramolecular assemblies, with large clusters and significant conformational changes in the topographical view. The polymeric film and the conjugated film did not differ significantly from each other in the film formation, with a polymer height that varied from 100 nm to <200 nm, while for the agglutination process the range of film formation was from 200 to >300 nm, with greater roughness, which are in accordance with the voltammetric and impedimetric studies. Our results are in agreement with a report published elsewhere [39] that demonstrate that peptides obtained from PD selections can be readily used as sensing probes in biosensor development.

Interestingly, an optical immunosensor based on the anti-MSP5 antibodies detection was developed to improve diagnosis of naturally infected animals with anaplasmosis [40], but surprisingly, its sensitivity and specificity have significantly decreased (93% and 70%, respectively), probably because larger molecules may suffer important structural changes due to its interaction with polymeric surfaces.

In conclusion, we have demonstrated that highly reactive peptides selected by PD against the mAb anti-MSP1a have successfully generated the most reduced and dominant epitope motif (STSSxL) that could recognize circulating antibodies of *A. marginale* infected animals with high sensitivity, and this peptide was effectively incorporated onto a bioelectrode surface based on the polymeric film poly(3-HPA) functionalized with the critical core epitope, and both impedimetric and differential pulse voltammetric immunosensors presented a very sensitive detection of sera from *A. marginale* infected animals, resulting in a simple, fast and reproducible technique.

Materials and Methods

Targeted-monoclonal antibody

Monoclonal antibody against the outer membrane protein MSP1 (mAb 15D2, IgG3 isotype) used in this study was acquired Veterinary Medical Research & Development Inc., VMRD (Pullman, WA, USA).

Peptide selection through phage display

For the peptide screening, a PhD-12 phage library (New England Biolabs, Beverly, MA, USA) was used. This is a 12-mer random peptide library fused to the minor coat protein (pIII) of the M13 bacteriophage, with a peptide diversity of 1.9×10^9. A sample of the library containing 2×10^{11} infectious phage particles was subjected to three rounds of selection and amplification. The selection was carried out using 50 µL of Recombinant Protein G Agarose (Invitrogen) previously washed with 1 mL of TBS-T 0,1% (Tris Buffered Saline plus 0.1% of Tween 20). The Protein G Agarose was blocked with TBS-BSA 3% at 8°C for 1 h and washed four times with TBS-T 0.1%. Meanwhile, 300 ng of the monoclonal antibody anti-MSP1a was incubated with 2×10^{11} phage particles from the PhD-12 library in 200 µL of TBS-T 0,1% solution at room temperature for 20 min. Thereafter, the mAbs-phage solution was incubated with blocked agarose for 15 min at room temperature. After incubation, the resin was washed ten times with TBS-T 0,1% and the unbound phage particles discarded, followed by elution of bound phages with 1 mL of elution buffer (0.2 M Glicine-HCl, pH 2.2 and BSA 1 mg/mL) for 10 min at room temperature. After elution, the solution was centrifuged at 4000 rpm and 4°C for 1 min and the supernatant transferred to a new microtube containing 150 µL of 1 M Tris-HCl (pH 9.1) for neutralization. The eluted phages were amplified in *E. coli* ER2738 strain (New England Biolabs, Beverly, MA, USA), purified using PEG-NaCl precipitation and after each of the three rounds of biopanning, individual bacterial colonies containing amplified phage clones were grown in a microtiter plate and titrated essentially as described [15].

Bioinformatic analysis

Phagemid DNA was isolated from 1 mL overnight cultures, and the sequencing reactions were carried out by using the DyEnamic ET Dye Terminator Cycle Sequencing Kit (GE Healthcare) with the primer −96 M13 (5′-CCCTCATTAGTTAGCGCGTAA-CG-3′), according to the manufacturer's instructions, and detection was performed in a MegaBace 1000 Genetic Analyzer (Amersham Biosciences) automatic capillary sequencer. Amino acid sequences were deduced according to the nucleotide sequences and analyzed using DNA2PRO2 software from Relic Program [17,41]. The similarity of selected peptides with *A. marginale* MSP1a was performed using BLAST search followed by sequence alignment with ClustalW2 software (http://www.ebi.ac.uk/Tools/ msa/clustalw2/). A graphical representation of the conserved sequence patterns within a multiple sequence alignment was generated using WebLogo3 (http://weblogo.berkeley.edu/) [42].

The three-dimensional structure predictions of the MSP1a protein were performed with the I-TASSER server [43] and the analysis was performed using PyMOL (http://www.pymol.org).

Phage-ELISA and reactivity to the mAb 15D2

To test specific binding of these peptides to the target molecule, we performed duplicate phage-ELISA experiments. A ninety-six-well Maxisorp™ microtiter plate (NUNC, NY, USA) was coated with 1 µg/well of anti-MSP1 mAb in 50 µL of carbonate buffer (0.1 M NaHCO₃, pH 8.6) overnight at 4°C. The wells were washed with PBS-T (phosphate-buffered saline plus 0.1% Tween 20) and then blocked for 1 h at 37°C with 3% BSA in PBS (BSA/PBS). The plate was washed twice with PBS-T and incubated with culture supernatant containing amplified phage particles ($\sim 10^{10}$ pfu/mL) for 2 h at 37°C. The wells were washed four times with PBS-T followed by incubation with HRP-conjugated anti-M13 (Roche Applied Science) diluted (1:5000) in BSA/PBS for 1 h at 37°C. The plate was washed four times in PBS-T, revealed with OPD Sigma*Fast*[TM] (Sigma-Aldrich) and read at 492 nm. M13 phage without displaying any peptide was used as negative control.

Phage-ELISA with bovine serum

Serum samples from healthy and *A. marginale*-infected animals used in this study were kindly provided by The Institute Desidério Finamor of Veterinary Research (Rio Grande do Sul, Brazil).

To test the reactivity of selected clones against bovine sera from infected and non-infected animals, another phage-ELISA was carried out. Briefly, a ninety-six-well Maxisorp[TM] microtiter plate (NUNC, NY, USA) was coated in duplicates with phages (10^{10} pfu/mL), diluted in carbonate buffer (0.1 M NaHCO$_3$, pH 8.6) overnight at 4°C. The wells were washed with PBS-T 0.05% and then blocked for 1 h at 37°C with PBS-BSA 5%. The plate was incubated for 1 h at 37°C with a pool of serum (1:200 in PBS-BSA 5%) from animals known to be infected and non-infected with *A. marginale*. The wells were washed 3 times with PBS-T 0.05% followed by incubation with HRP-conjugated goat anti-bovine IgG (Sigma-Aldrich) diluted (1:2000) in PBS-BSA 5% for 1 h at 37°C. The plate was washed 3 times in PBS-T 0.05%, revealed with OPD Sigma*Fast*[TM] (Sigma-Aldrich) and read at 492 nm. All samples were tested in duplicate. Each serum sample was tested against M13 phage without displaying any peptide as negative control.

Peptide design and synthesis

After bioinformatics analysis of selected clones, two peptide sequences were designed and chemically synthesized by GenScript USA Inc. To increase immunogenicity, peptides were coupled to Bovine Serum Albumin. The peptide Am1 (STSSQLGGGS-STSSQLGGGSSTSSQL) as well the peptide Am2 (SEASTSSQL-GAGGGSSEASTSSQLGA) were constructed with 26 residues, both containing repeats of a MSP1a motif sequence (underlined) separated by a 4-aa spacer, GGGS.

Antibody detection by ELISA

To determine the peptides Am1 and Am2 reactivity to serum from infected and non-infected animals, specific ELISA test was carried out. High affinity microtiter plates were coated with the peptides (1 μg/weel) in carbonate bicarbonate buffer, pH 9.6, and incubated overnight at 4°C. Microplates were washed with PBS-T 0.05%. After blocking with 5% BSA in PBS at 37°C for 1 h, 100 μL/well of 24 infected and 52 non-infected bovine sera diluted in PBS-BSA 5% (1:50 to Am1; 1:250 to Am2) were added and incubated for 1 h at 37°C. After washing, conjugated goat anti-bovine IgG (Sigma-Aldrich) was added in a dilution of 1:5000 in PBS-BSA and incubated for 1 h at 37°C. All samples were tested in duplicates, and the assay was developed was determined as described above.

Specificity tests for synthetic peptides against sera from other bovine diseases

The binding specificity of synthetic peptides were analyzed by ELISA using sera from bovines infected with *Brucella abortus* (n = 7), *Leptospira interrogans* (n = 7), *Salmonella typhimurium* (n = 7), *Escherichia coli* (n = 7), *Babesia bovis* (n = 7), *Babesia bigemina* (n = 7) and *Anaplasma marginale* (n = 7); dogs infected with *Anaplasma phagocytophilum* (n = 7), *Ehrlichia canis* (n = 7), *Ehrlichia ewingii* (n = 1); and a rabbit infected with *Ehrlichia sennetsu* (n = 1),

following the same protocol as previously described. Sera used in this step were kindly provided by The Institute Desidério Finamor of Veterinary Research and by the State University of Londrina.

Construction and analysis of the bioelectrode

All reagents used were of analytical grade. Ultra high pure water (Millipore Milli-Q system) was used in the preparation of solutions. Monomer solutions, 3-hydroxyphenylacetic acid, were prepared in 0.5 mol.L^{-1} HClO$_4$ solution, immediately before their use. The electrochemical experiments were conducted at room temperature (25 ± 1°C).

The electropolymerizations were performed in three-compartment electrochemical cell connected to a potentiostat (CH Instruments, 420A-model, Austin, USA). The working electrode was graphite (99.9995%) from Alfa Aesar, in disk form, 6.18 mm of diameter. A platinum plate and electrodes of Ag/AgCl, KCl (3 M) were used as auxiliary and reference electrodes, respectively. Electrochemical impedance spectroscopy (EIS) was performed in an Autolab Electrochemical System (PGSTAT302N and FRA2 module, Eco Chemie, Utrecht, The Netherlands), using aqueous solution containing K$_3$Fe(CN)$_6$/K$_4$Fe(CN)$_6$ (5 mmol.L^{-1}) and KCl (0.1 mol.L^{-1}). The frequency range was from 100 KHz to 10 Hz using the open-circuit potential system, +0.24 V. The voltage amplitude was 10 mV. Film morphology and roughness values were assessed by atomic force microscopy (Shimadzu, model SPM-9600).

Graphite carbon electrodes were modified with polymer derived from 3-hydroxyphenylacetic acid [poly(3-HPA)] as described elsewhere [44].

The modified electrode with poly(3-HPA) was pre-treated by applying a potential of −0.2 V in PBS buffer, pH 7.3 for 2 minutes. After, 1 μg of synthetic peptide Am1 was diluted in the acetate buffer, pH 4.3, added on the modified electrodes and incubated for 30 min at 25°C. Graphite electrode/poly(3-HPA)/Am1 was immersed for 6 seconds in PBS buffer, pH 7.3 and dried with N$_2$.

For specific *A. marginale* infected sera detection, 1 μL of positive serum in 17 μL of PBS was added to the bioelectrode (graphite electrode/poly(3-HPA)/Am1) for 15 minutes. Negative serum (1 μL) solubilized in PBS (17 μL) was used as negative control.

Statistical analysis

Unpaired t test with Welch's correction was used to determine differences among groups for phage clones and peptides reactivity. A value of $p < 0.05$ was considered statistically significant. Sensitivity and specificity parameters were calculated based on the ROC curve analysis. One-way analysis of variance and Tukey's Multiple Comparison test was used to determine differences among other diseases.

Acknowledgments

We thank Dr. Odilon Vidotto (State University of Londrina) for providing dog and rabbit infected samples.

Author Contributions

Conceived and designed the experiments: PSS RN LRG. Performed the experiments: PSS RN LPR FAAS PCBF. Analyzed the data: PSS AGBM JMM LRG. Contributed reagents/materials/analysis tools: JRSM AGBM JMM LRG. Wrote the paper: PSS RN AGBM LRG.

References

1. Kocan KM, de la Fuente J, Blouin EF, Garcia-Garcia JC (2004) Anaplasma marginale (Rickettsiales: Anaplasmataceae): recent advances in defining host-pathogen adaptations of a tick-borne rickettsia. Parasitology 129 Suppl: S285–300.

2. Ajayi SA, Wilson AJ, Campbell RS (1978) Experimental bovine anaplasmosis: clinico-pathological and nutritional studies. Res Vet Sci 25: 76–81.

3. Kocan KM, de la Fuente J, Blouin EF, Coetzee JF, Ewing SA (2010) The natural history of Anaplasma marginale. Vet Parasitol 167: 95–107.

4. Oberle SM, Palmer GH, Barbet AF, Mcguire TC (1988) Molecular-Size Variations in an Immunoprotective Protein Complex among Isolates of Anaplasma-Marginale. Infection and Immunity 56: 1567–1573.

5. McGarey DJ, Barbet AF, Palmer GH, McGuire TC, Allred DR (1994) Putative adhesins of Anaplasma marginale: major surface polypeptides 1a and 1b. Infection and Immunity 62: 4594–4601.

6. de la Fuente J, Lew A, Lutz H, Meli ML, Hofmann-Lehmann R, et al. (2005) Genetic diversity of anaplasma species major surface proteins and implications for anaplasmosis serodiagnosis and vaccine development. Anim Health Res Rev 6: 75–89.

7. de la Fuente J, Garcia-Garcia JC, Blouin EF, Kocan KM (2001) Differential adhesion of major surface proteins 1a and 1b of the ehrlichial cattle pathogen Anaplasma marginale to bovine erythrocytes and tick cells. Int J Parasitol 31: 145–153.

8. de la Fuente J, Garcia-Garcia JC, Blouin EF, Kocan KM (2003) Character-ization of the functional domain of major surface protein 1a involved in adhesion of the rickettsia Anaplasma marginale to host cells. Vet Microbiol 91: 265–283.

9. Jones EW, Brock WE (1966) Bovine Anaplasmosis - Its Diagnosis Treatment and Control. Journal of the American Veterinary Medical Association 149: 1624-&.

10. Nakamura Y, Shimizu S, Minami T, Ito S (1988) Enzyme-Linked Immunosor-bent-Assay Using Solubilized Antigen for Detection of Antibodies to Anaplasma marginale. Tropical Animal Health and Production 20: 259–266.

11. Ekici OD, Sevinc F (2011) Comparison of cELISA and IFA tests in the serodiagnosis of anaplasmosis in cattle. African Journal of Microbiology Research 5: 1188–1191.

12. Corona B, Martínez S (2011) Detección de Anaplasma marginale em bovinos, mediante la amplificación por PCR Del gen msp5. Rev Salud Anim 33: 24–31.

13. Kennedy MW, Mcintosh AE, Blair AJ, Mclaughlin D (1990) Mhc (Rt1) restriction of the antibody repertoire to Infection with the nematode Nippostrongylus-Brasiliensis in the Rat. Immunology 71: 317–322.

14. Smith GP (1985) Filamentous fusion phage - novel expression vectors that display cloned antigens on the virion surface. Science 228: 1315–1317.

15. Barbas CF, Burton DR, Scott JK, Silverman GJ (2001) Phage display: a laboratory manual. Cold Spring Harbor, NY: Cold Spring Harbor Laboratory Press.

16. Scott JK, Smith GP (1990) Searching for peptide ligands with an epitope library. Science 249: 386–390.

17. Huang J, Ru B, Dai P (2011) Bioinformatics resources and tools for phage display. Molecules 16: 694–709.

18. Palmer GH, Waghela SD, Barbet AF, Davis WC, McGuire TC (1987) Characterization of a neutralization-sensitive epitope on the Am 105 surface protein of Anaplasma marginale. Int J Parasitol 17: 1279–1285.

19. Allred DR, Mcguire TC, Palmer GH, Leib SR, Harkins TM, et al. (1990) Molecular-basis for surface-antigen size polymorphisms and conservation of a neutralization-sensitive epitope in Anaplasma-Marginale. Proceedings of the National Academy of Sciences of the United States of America 87: 3220–3224.

20. Ehrlich GK, Bailon P (2001) Identification of model peptides as affinity ligands for the purification of humanized monoclonal antibodies by means of phage display. Journal of Biochemical and Biophysical Methods 49: 443–454.

21. Smith GP, Petrenko VA (1997) Phage Display. Chem Rev 97: 391–410.

22. Westerink MAJ, Lesinski GB (2001) Novel vaccine strategies to T-independent antigens. Journal of Microbiological Methods 47: 135–149.

23. Kay BK, Kurakin AV, Hyde-DeRuyscher R (1998) From peptides to drugs via phage display. Drug Discovery Today 3: 370–378.

24. Felici F, Luzzago A, Folgori A, Cortese R (1993) Mimicking of discontinuous epitopes by phage-displayed peptides, II. Selection of clones recognized by a protective monoclonal antibody against the Bordetella pertussis toxin from phage peptide libraries. Gene 128: 21–27.

25. Murthy KK, Ekiel I, Shen SH, Banville D (1999) Fusion proteins could generate false positives in peptide phage display. Biotechniques 26: 142–149.

26. Schillberg S, Zhang MY, Zimmermann S, Liao YC, Breuer G, et al. (2001) GST fusion proteins cause false positives during selection of viral movement protein specific single chain antibodies. Journal of Virological Methods 91: 139–147.

27. Jacobson RH (1998) Validation of serological assays for diagnosis of infectious diseases. Revue Scientifique Et Technique De L Office International Des Epizooties 17: 469–486.

28. Fosgate GT, Urdaz-Rodriguez JH, Dunbar MD, Rae DO, Donovan GA, et al. (2010) Diagnostic accuracy of methods for detecting Anaplasma marginale infection in lactating dairy cattle of Puerto Rico. Journal of Veterinary Diagnostic Investigation 22: 192–199.

29. Vidotto O, Marana ERM, Kano FS, Vicentini JC, Spurio RS, et al. (2009) Cloning, expression, molecular characterization of the MSP5 protein from PR1 strain of Anaplasma marginale and its application in a competitive enzyme-linked immunosorbent test. Revista Brasileira De Parasitologia Veterinaria 18: 5–12.

30. Araujo FR, Melo ESP, Ramos CAN, Soares CO, Rosinha GMS, et al. (2007) ELISA based on recombinant truncated MSP5 for detection of antibodies against Anaplasma marginale in cattle. Pesquisa Veterinaria Brasileira 27: 301–306.

31. Madruga CR, Marques APC, Leal CRB, Carvalho CME, Araujo FR, et al. (2000) Evaluation of an enzyme-linked immunosorbent assay to detect antibodies against Anaplasma marginale. Pesquisa Veterinaria Brasileira 20: 109–112.

32. Araujo FR, Melo VSP, Ramos CAN, Madruga CR, Soares CO, et al. (2005) Development of enzyme-linked immunosorbent assays based on recombinant MSP1a and MSP2 of Anaplasma marginale. Memorias Do Instituto Oswaldo Cruz 100: 765–769.

33. Munodzana D, McElwain TF, Knowles DP, Palmer GH (1998) Conformational dependence of Anaplasma marginale major surface protein 5 surface-exposed B-cell epitopes. Infection and Immunity 66: 2619–2624.

34. Barry DN, Parker RJ, Devos AJ, Dunster P, Rodwell BJ (1986) A microplate enzyme-linked-immunosorbent-assay for measuring antibody to Anaplasma marginale in cattle serum. Australian Veterinary Journal 63: 76–79.

35. Lindholm-Sethson B, Nystrom J, Malmsten M, Ringstad L, Nelson A, et al. (2010) Electrochemical impedance spectroscopy in label-free biosensor applica-tions: multivariate data analysis for an objective interpretation. Analytical and Bioanalytical Chemistry 398: 2341–2349.

36. Park SM, Park JY (2009) DNA Hybridization Sensors Based on Electrochemical Impedance Spectroscopy as a Detection Tool. Sensors 9: 9513–9532.

37. Ouerghi O, Touhami A, Jaffrezic-Renault N, Martelet C, Ouada HB, et al. (2002) Impedimetric immunosensor using avidin-biotin for antibody immobili-zation. Bioelectrochemistry 56: 131–133.

38. Radi AE, Munoz-Berbel X, Lates V, Marty JL (2009) Label-free impedimetric immunosensor for sensitive detection of ochratoxin A. Biosens Bioelectron 24: 1888–1892.

39. Banta S, Wu J, Cropek DM, West AC (2010) Development of a Troponin I Biosensor Using a Peptide Obtained through Phage Display. Analytical Chemistry 82: 8235–8243.

40. Oliva A, Silva M, Wilkowsky S, De Echaide ST, Farber M (2006) Development of an immunosensor for the diagnosis of bovine anaplasmosis. Impact of Emerging Zoonotic Diseases on Animal Health 1081: 379–381.

41. Rodi DJ, Mandava S, Makowski L, Devarapalli S, Uzubell J (2004) RELIC - A bioinformatics server for combinatorial peptide analysis and identification of protein-ligand interaction sites. Proteomics 4: 1439–1460.

42. Crooks GE, Hon G, Chandonia JM, Brenner SE (2004) WebLogo: A sequence logo generator. Genome Research 14: 1188–1190.

43. Roy A, Kucukural A, Zhang Y (2010) I-TASSER: a unified platform for automated protein structure and function prediction. Nat Protoc 5: 725–738.

44. Madurro JM, Oliveira RML, Vieira SN, Alves HC, Franca EG, et al. (2010) Electrochemical and morphological studies of an electroactive material derived from 3-hydroxyphenylacetic acid: a new matrix for oligonucleotide hybridiza-tion. Journal of Materials Science 45: 475–482.

15

Functional Divergence in Shrimp Anti-Lipopolysaccharide Factors (ALFs): From Recognition of Cell Wall Components to Antimicrobial Activity

Rafael Diego Rosa[1][¤a], Agnès Vergnes[1], Julien de Lorgeril[1], Priscila Goncalves[2][¤b], Luciane Maria Perazzolo[2], Laure Sauné[1], Bernard Romestand[1], Julie Fievet[1][¤c], Yannick Gueguen[1][¤c], Evelyne Bachère[1], Delphine Destoumieux-Garzón[1]*

1 Ecologie des Systèmes Marins Côtiers, UMR5119, Centre National de la Recherche Scientifique, Institut Français de Recherche pour l'Exploitation de la Mer, Institut de la Recherche pour le Développement, Université Montpellier 1, Université Montpellier 2, Montpellier, France, 2 Laboratory of Immunology Applied to Aquaculture, Department of Cell Biology, Embryology and Genetics, Federal University of Santa Catarina, Florianópolis SC, Brazil

Abstract

Antilipopolysaccharide factors (ALFs) have been described as highly cationic polypeptides with a broad spectrum of potent antimicrobial activities. In addition, ALFs have been shown to recognize LPS, a major component of the Gram-negative bacteria cell wall, through conserved amino acid residues exposed in the four-stranded β-sheet of their three dimensional structure. In penaeid shrimp, ALFs form a diverse family of antimicrobial peptides composed by three main variants, classified as ALF Groups A to C. Here, we identified a novel group of ALFs in shrimp (Group D ALFs), which corresponds to anionic polypeptides in which many residues of the LPS binding site are lacking. Both Group B (cationic) and Group D (anionic) shrimp ALFs were produced in a heterologous expression system. Group D ALFs were found to have impaired LPS-binding activities and only limited antimicrobial activity compared to Group B ALFs. Interestingly, all four ALF groups were shown to be simultaneously expressed in an individual shrimp and to follow different patterns of gene expression in response to a microbial infection. Group B was by far the more expressed of the ALF genes. From our results, nucleotide sequence variations in shrimp ALFs result in functional divergence, with significant differences in LPS-binding and antimicrobial activities. To our knowledge, this is the first functional characterization of the sequence diversity found in the ALF family.

Editor: Jérôme Nigou, French National Centre for Scientific Research - Université de Toulouse, France

Funding: This study received financial support from IFREMER and CNRS. RDR was supported by a doctoral fellowship from CNPq-Brazil and PG by a master fellowship provided by CAPES Brazil. The funders had no role in study design, data collection and analysis, decision to publish, or preparation of the manuscript.

Competing Interests: The authors have declared that no competing interests exist.

* E-mail: ddestoum@ifremer.fr

¤a Current address: Laboratory of Biochemistry and Immunology of Arthropods, Department of Parasitology, Institute of Biomedical Sciences, University of São Paulo, São Paulo SP, Brazil

¤b Current address: Department of Biological Sciences, Faculty of Science, Macquarie University and Sydney Institute of Marine Science, Sydney, New South Wales, Australia

¤c Current address: Ifremer, Centre Océanologique du Pacifique, Taravao, Tahiti, French Polynesia

Introduction

Anti-lipopolysaccharide factors (ALFs) are antimicrobial peptides (AMPs) only found in marine chelicerates (horseshoe crabs) and crustaceans, which exhibit a potent antimicrobial activity against a broad range of microorganisms. The spectrum of the antimicrobial activity of ALFs covers a large number of Gram-positive and Gram-negative bacteria, filamentous fungi as well as enveloped viruses [1–5]. Initially isolated from the hemolymph of the horseshoe crabs *Tachypleus tridentatus* (ALF-T) and *Limulus polyphemus* (ALF-L) [6], ALFs were later identified in penaeid shrimps by transcriptomic-based approaches [7–9]. ALFs are known as highly cationic polypeptides of about 100 residues with a hydrophobic N-terminal region. The horseshoe crab ALF-L and the shrimp ALFPm3, from *Penaeus monodon*, share a similar three-dimensional structure, consisting in three α-helices packed against a four-stranded β-sheet [10,11]. Two conserved cysteine residues are involved in an intramolecular disulfide bridge which delimits the central β-hairpin.

ALF homologues have been described in many crustacean species, especially in penaeid shrimp (for review see [12,13]). In *P. monodon*, for which studies in ALF characterization are relatively advanced, six different ALF variants (ALFPm1 to 6) have been identified [14,15]. Based on amino acid sequence comparisons, these variants were classified into three main groups: Group A (ALFPm1-2), Group B (ALFPm3-5) and Group C (ALFPm6), which are encoded by distinct genomic loci [14,16]. Beyond differences in gene organization (two or three introns), shrimp ALF Groups A to C are differentially transcribed among shrimp tissues. Indeed, ALFPm3 and ALFFc from *Fenneropenaeus chinensis* (Group B) are expressed in hemocytes, while *Mj*ALF2 from *Marsupenaeus japonicus* and ALFPm6 from *P. monodon* (Group C) are expressed in different shrimp tissues [14,17–19].

As observed for other cationic AMPs from amphibians and fishes [20,21], both horseshoe crab and shrimp ALFs from Groups B and C bind and neutralize LPS [6,10,19,22], a major component of the outer membrane of Gram-negative bacteria. In horseshoe crab ALFs, positively-charged residues recognizing the lipid A moiety of LPS would be located in the β-hairpin stabilized by the disulfide bridge [10]. Such a β-hairpin structure is preserved in cyclic synthetic peptides designed on the central-most residues of the β-hairpin, hydrophobic residues at position 44 and 46 of the horseshoe crab ALF sequence playing a crucial role in stabilizing the hydrophobic face of the β-hairpin [23]. In such cyclic synthetic peptides, the positively-charged residues bind in an exothermic reaction with the negative charges of LPS and Lipid A, which further shows the crucial role of the β-hairpin in its interaction with the LPS acyl chains [24]. Like the native horseshoe crab ALF, the recombinant shrimp ALFPm3 binds to Lipid A and the most probable LPS-binding site involves six positively-charged residues and one negatively-charged amino acid located in the cysteine-stabilized β-hairpin and in the two neighboring β-strands (**Figure 1**) [11]. Recombinant ALFPm3 was also proved to bind to lipoteichoic acid (LTA), a major cell wall component of Gram-positive bacteria [19]. However, the amino acid residues involved in LTA-binding are still unknown. Interestingly, the ability of shrimp ALFs to bind to bacterial cell wall components, such as LPS and LTA, has been suggested to be associated with their antibacterial activities [19].

Studies on ALFs have mainly focused on the highly active and cationic ALFs belonging to Group B (for review see [12,13]), which have been shown to be essential in the protection of shrimp against different microbial infections [7,14,25,26]. However, while many sequences of shrimp ALFs have been described, little attention has been paid to the functional consequences of the ALF sequence diversity in terms of biochemical properties and biological activities.

Herein, taking advantage of the identification of a novel group of shrimp ALFs with unique anionic properties and displaying an incomplete LPS binding site (Group D ALFs), we have undertaken a study of shrimp ALF functional diversity. Our phylogenetic and functional data show that shrimp ALFs have evolved as four functionally diverse groups, among which some have shaped important motifs for interacting with the cell wall components of bacteria and inhibiting bacterial growth. Those groups show different patterns of expression in response to infection. We provide here the first evidence of functional divergence in shrimp ALFs. In order to standardize the current classification of this diverse family of AMPs, we propose here a common nomenclature for shrimp ALFs.

Materials and Methods

Molecular Cloning and Sequence Analysis

For the molecular cloning of a cationic ALF member (Group B) in the blue shrimp *Litopenaeus stylirostris*, specific primers were designed based on consensus nucleotide sequences of ALFs from different shrimp species (**Table 1**). Total RNA was extracted from hemocytes by homogenization with Trizol reagent (Invitrogen) following the manufacturer's instructions. Following heat denaturation (70°C for 5 min), reverse transcription was performed using 1 μg of total RNA with 50 ng/μl oligo(dT)$_{12-18}$ in a 20-μl reaction volume containing 1 mM dNTPs, 1 unit/μl of RNAse-OUT Ribonuclease and 200 units/μl M-MLV reverse transcriptase in reverse transcriptase buffer according to the manufacturer's instructions (Invitrogen). PCR reactions were conducted in a 25-μl reaction volume using 1 μl of synthesized complementary DNA (cDNA) as template. PCR conditions were as follows: 30 cycles of 94°C for 1 min, 55°C for 1 min, 72°C for 1 min and a final elongation step of 72°C for 10 min. The amplification products were cloned into a pCR 2.1 TA cloning vector using a TA cloning

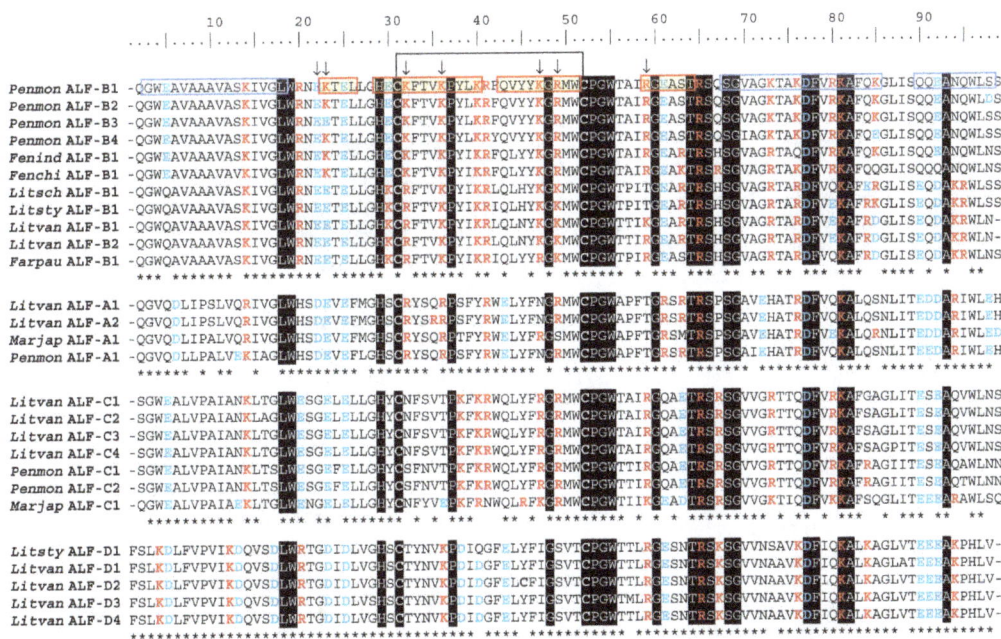

Figure 1. Amino acid sequence alignments of mature polypeptides of the four ALF groups (Group A, B, C and D) found in penaeids. Asterisks (*) indicate residues conserved in each specific group of ALF sequences. Residues conserved in all ALF sequences are highlighted in black. The conserved cysteine bridge is indicated. Residues involved in LPS-binding of Group B ALFs [11] are indicated by arrows. α-helices and β-strands identified in the three-dimensional structure of *Penmon* ALF-B1 (PDB entry 2JOB) [11] are indicated by blue and red boxes, respectively.

Table 1. Nucleotide sequence of primers used in this study.

Primer name	Forward primer (5′–3′)	Reverse primer (5′–3′)
Primers for cDNA amplification		
LstyB1	AGTAACTTTCCTAGTTTAGA	CTGGCGCGGGAAAGGCCTA
Primers for protein expression		
LstyD1rp	GCGCGAATTCATGTTTTCGCTAAAAGACCTTTTTG	ATATATGTCGACTTATACAAGGTGTGGTTTGGC
Primers for quantitative Real-Time PCR (RT-qPCR)		
Lvanrpl40qt	GAGAATGTGAAGGCCAAGATC	TCAGAGAGAGTGCGACCATC
LvanAqt	CTGATTGCTCTTGTGCCACG	TGACCCATGAACTCCACCTC
LvanBqt	GTGTCTCCGTGTTGACAAGC	ACAGCCCAACGATCTTGCTG
LvanCqt	ATGCGAGTGTCTGTCCTCAG	TGAGTTTGTTCGCGATGGCC
LvanDqt	TGTGTTGGTTGTGGCACTGG	CAACGAGGTCAATGTCACCG

kit (Invitrogen). The positive recombinant clones were identified by colony PCR and were sequenced in both directions.

Annotated ALF sequences from penaeid shrimp were methodically collected from publicly accessible databases (GenBank, EMBL) and used for the search of homologous sequences among EST sequences from the NCBI database. Homology searches were performed using BLAST at NCBI. The multiple alignments were generated using the MAFFT alignment program (http://align. bmr.kyushu-u.ac.jp/mafft/online/server/). Prediction of signal peptide was performed with the SignalP program (http://www. cbs.dtu.dk/services/SignalP/). The phylogenetic analysis based on the amino acid sequences of shrimp and chelicerate ALFs and scygonadins (outgroup) was performed using the Neighbour-Joining method with the software MEGA version 4.0 [27]. Bootstrap sampling was reiterated 1,000 times. The uncovered sequence phylogenies were used to define distinct groups of ALF sequences.

Synthetic and Recombinant Peptides

Peptide amino acid sequences are shown in **Table 2**. Synthetic peptides Penmon ALF-B1 β-hairpin, Litsty ALF-B1 β-hairpin and Litsty ALF-D1 β-hairpin were obtained by Fmoc chemistry and purchased from Genepep S.A. (Montpellier, France). Recombinant Penmon ALF-B1 (also referred to as rALFPm3) was overexpressed and purified as previously described [4]. Recombinant Litsty ALF-D1 was expressed in Escherichia coli Rosetta (DE3) as an N-terminal His6-tagged fusion protein using the pET-28a system (Novagen). The open reading frame of interest was PCR-amplified from cDNA samples derived from hemocytes of L. stylirostris using specific primers (**Table 1**), a Met-coding trideoxynucleotide was incorporated 5′ of each cDNA and cloned in-frame with the N-terminal His6 in the EcoRI/SalI sites of pET-28a. Expression of recombinant Litsty ALF-D1 was performed as described previously for oyster defensins [28,29]. The purification procedure started with affinity chromatography by incubating bacterial cell lysates with TALON® metal affinity resin (Clontech) at a ratio of 25:1 (v/v) in 6 M guanidine HCl, 50 mM sodium phosphate, 300 mM NaCl, 5 mM imidazole, (pH 8.5) for 4 h at 4°C with gently agitation. Then, resin was washed twice by decantation in 6 M guanidine HCl, 50 mM sodium phosphate, 300 mM NaCl (pH 8.5), and fusion proteins were eluted by decantation with two column volumes of 6 M guanidine HCl, 50 mM sodium phosphate and 1 M imidazole (pH 6.4). The eluate was desalted on a reversed phase Sep-pak C-18 cartridge (2O cc). Separation was performed with a step gradient of 10% and 80% of

acetonitrile (ACN) containing 0.05% Trifluoroacetic acid (TFA). The 80% ACN fraction containing the peptide mixture was then frozen and lyophilized. The methionine residue introduced at the peptide N-terminus was subjected to CNBr cleavage as described previously [28]. The cleaved fusion peptide mixture was then directly folded at pH 8.1 in a refolding solution containing 0.1 M NaHCO3, 3 mM reduced glutathione and 0.3 mM oxidized glutathione in the presence of 2 M urea and 25% N,N-dimethylformamide, at room temperature for 72 h. The peptide mixture containing Litsty ALF-D1 was then purified on a preparative reversed-phase Zorbax 300 SB-C8 column with a biphasic gradient of 0–32% ACN over 10 min and 32–47% ACN over 50 min at a flow rate of 3 ml/min. Peptide purity was assessed by MALDI-TOF-MS and SDS-PAGE analysis.

Antimicrobial Assays

The antimicrobial activity of peptides was assayed against different microorganisms: the Gram-positive bacteria Aerococcus viridans (CIP 104 074), Micrococcus luteus (CIP 5345 and IBMC collection), Bacillus megaterium (CIP 6620), Brevibacterium stationis (CIP 101 282) and Staphylococcus aureus (ATCC 65 38), the Gram-negative bacteria E. coli SBS363, Salmonella enterica (CIP 5858), Vibrio alginolyticus (CIP 103336 T), V. harveyi E22, V. anguillarum V62, V. penaeicidae AM101 and V. nigripulchritudo SFn1 and the fungi Candida albicans, Fusarium oxysporum, Rhizopus stolonifer, Septoria nodorum and Botrytis cinerea. Minimum inhibitory concentrations (MICs) were determined in triplicate by the liquid growth inhibition assay based on the procedure described in [30]. MIC values are expressed as the lowest concentration tested that causes 100% of growth inhibition (μM). Poor Broth (PB, 1% bacto-tryptone, 0.5% NaCl w/v, pH 7.5) was used as a culture medium for bacteria. It was supplemented with 0.5 M NaCl for marine bacteria (B. stationis and Vibrio strains). When needed, sea salts (20 μM KCl, 5 μM MgSO4, 1.5 μM CaCl2 final concentration) were also added to the culture medium. Brain heart infusion (BHI) broth (Becton Dickinson) was used for Aerococcus viridans. Potato Dextrose Broth (Difco) at half strength was used for cultures of filamentous fungi, while Sabouraud Dextrose Broth (Difco) was used for yeast (C. albicans) cultures. Growth was monitored spectrophotometrically at 620 nm on a Multiscan microplate reader (Labsystems).

Table 2. Peptide sequences.

Peptide	Amino acid sequence	Mass (kDa)	p*I*
Penmon **ALF-B1** (GenBank: ABP73289)	QGWEAVAAAVASKIVGLWRNEKTELLGHE**C**KFTVKPYLKRFQVYYKGR MW**C**PGWTAIRGEASTRSQSGVAGKTAKDFVRKAFQKGLISQQEANQWLSS	11.05	9.95
Litsty **ALF-D1** (GenBank: AAY33769)	FSLKDLFVPVIKDQVSDLWRTGDIDLVGHS**C**TYNVKPDIQGFELYFIGSV T**C**PGWTTLRGESNTRSKSGVVNSAVKDFIQKALKAGLVTEEEAKPHLV	10.81	6.10
Penmon **ALF-B1** (Synthetic β-hairpin)	G**C**KFTVKPYLKRFQVYYKGRMW**C**G	2.96	9.93
Litsty **ALF-D1** (Synthetic β-hairpin)	G**C**TYNVKPDIQGFELYFIGSVT**C**G	2.61	4.37
Litsty **ALF-B1** (Synthetic β-hairpin)	G**C**RFTVKPYIKRIQLHYKGKMW**C**G	2.91	10.04

Sequences of *Penmon* ALF-B1 and *Litsty* ALF-D1 are shown without signal peptide, in their expected mature form. Conserved cysteine residues are in bold face. Molecular weight (kDa) and predicted isoelectric point (p*I*) of the recombinant peptides and synthetic ALF β-hairpins are indicated on the right.

LPS-binding Assay

The LPS-binding properties of recombinant ALFs were tested using the QCL-1000 *Limulus* amoebocyte lysate (LAL, Lonza). Briefly, *E. coli* LPS at 1 EU/ml was incubated in a 1:1 (v/v) ratio with 0.001 to 5 µM recombinant ALFs or polymixine B (Sigma), used as a control. After 15 min at 37°C, 1 vol of *Limulus* amoebocyte lysate was added to the reaction. After 10 min at 37°C, the chromogenic LAL substrate was added. The reaction was stopped after a 6 min-incubation by adding 1 vol of 25% acetic acid. Absorbance was read at 405 nm.

Animals, Tissue Collection and Immune Challenge

Litopenaeus vannamei adult shrimp (10±2 g) were obtained from the Laboratory of Marine Shrimp (Federal University of Santa Catarina, Brazil). Shrimp were acclimated at 23°C for at least 72 h before experimental infection. Then, shrimp received in intramuscular injection (50 µl) of *Fusarium solani* (5×10^6 spores/animal) or sterile sea water (SSW), as injury control. The standardization of the experimental infections and the preparation of the fungal inoculum were performed as previously described [31]. Three groups of three shrimp were used in each experimental condition. Hemolymph from unchallenged and challenged shrimp was collected at 24 h and 48 h from the ventral sinus into a precooled modified Alsever solution (MAS: 27 mM sodium citrate, 336 mM NaCl, 115 mM glucose, 9 mM EDTA, pH 7.0). Hemocytes were isolated by centrifugation to discard plasma (700×*g* for 10 min at 4°C) and directly processed for RNA extraction.

Real-time Quantitative PCR (RT-qPCR) Analysis of Gene Expression

Total RNA was extracted from hemocytes using Trizol reagent according to manufacturer instructions (Invitrogen). RNA was then treated with DNase I (Invitrogen) for 15 min at room temperature and inactivated by heat, 10 min at 65°C. After sodium acetate precipitation, quantity and quality of total RNA were determined using a NanoDrop spectrophotometer (*NanoDrop* Technologies) and agarose gel electrophoresis, respectively. Following heat denaturation of 1 µg of total RNA (65°C for 5 min), first strand synthesis was carried out using ImProm-II™ reverse transcriptase following the manufacturer protocol (Promega).

Gene expression of ALF members was analyzed by RT-qPCR on a LightCycler 480 System (Roche) in a final volume of 6 µl containing 5 mM MgCl$_2$, 0.5 µM of each primer, 3 µl of reaction mix (LightCycler 480 SYBR Green I Master 2X) and 1 µl of each reverse transcribed RNA (diluted at 1:19 in DNase/RNase-free distilled water). Primer sequences are listed in **Table 1**. RT-qPCR assays were submitted to an initial denaturation step of 10 min at 95°C followed by 40 cycles of denaturation at 95°C for 10 s, annealing at 57°C for 20 s and extension time at 72°C for 25 s. RT-qPCR assays were performed in triplicate, and standard curves were generated using six two-fold serial dilutions from a pool of all cDNAs (in DNase/RNase-free distilled water) to determine primer pair efficiencies (E) according to the equation: $E = 10^{[-1/\text{slope}]}$. Only primer pairs with efficiencies of 2±0.2 were selected for RT-qPCR relative quantification using the $2^{-\Delta\Delta Cq}$ method [32] with the *L. vannamei* ribosomal protein L40 (*Litvanrpl40*, GenBank: FE077602) as reference gene. Data were analyzed using the LightCycler 480 software version 1.5.0.39 and the 2nd derivative max algorithm. Statistical significance was determined using Student's *t*-test between conditions and differences were considered when *p*<0.05.

Results

Identification of Novel Members of the ALF Antimicrobial Peptide Family in Shrimp

All annotated ALF sequences available for penaeid shrimp in publicly accessible databases were collected (**Table S1**) and used to screen non-annotated nucleotide sequences from EST projects using BLAST at NCBI. New EST sequences homologous to ALFs were identified, which included 28 sequences from the Pacific white leg shrimp *L. vannamei* (GenBank: FE109538; FE087264; FE088625; FE105941; FE153599; FE052210; FE176556; FE176555; FE058235; FE155445; FE079082; FE088301; FE090668; FE078559; FE079755; FE151634; FE152534; FE110967; FE115964; FE098450; FE156649; FE155982; FE152063; FE116643; FE115660; FE183080; FE092417; FE109539) and 4 sequences from the black tiger shrimp *P. monodon* (GenBank: DW678002; HO000126; GO080476; GW993385). Only full-length coding sequences (CDS) were kept for subsequent analyses. These data were completed by PCR amplification and sequencing of a novel ALF sequence obtained from the blue shrimp *L. stylirostris* (GenBank: KC346373).

From those 63 sequences, shrimp ALFs appear to be encoded as precursors starting with a 22–26 residue hydrophobic signal peptide followed by a 69–98 residue mature polypeptide containing two conserved cysteine residues (**Table S1**; **Figure 1**). Mature ALFs differed in terms of size with calculated molecular weights ranging from 7.69 to 11.52 kDa (**Table S1**; **Figure 1**), and more remarkably, in terms of electrostatic properties. Indeed, in addition to the known highly cationic ALFs (calculated pI ≥ 9.5), we found here a series of anionic ALFs (calculated pI ≤ 6) which differed from the previously described Group A ALFs [16,33]. Interestingly, both cationic and anionic ALFs were shown to be present in a same shrimp species (**Table S1**).

Shrimp ALFs Cluster into Four Distinct Groups with Contrasted Overall Charges

Phylogenetic analyses were performed with ALF sequences from penaeid shrimp and from the horseshoe crabs *T. tridentatus* (TACTR_ALF: P07087; TACTR2_ALF: AAK00651), *L. polyphemus* (LIMPO_ALF: P07086) and *Carcinoscorpius rotundicauda* (CARRO_ALF: CK086627). A total of 42 sequences were analyzed. A phylogenetic tree was generated from the sequence alignments with scygonadins [34], another two cysteine-containing AMP family found in crustaceans, used here as an outgroup.

Sequences of shrimp ALFs clustered into four groups distinct from horseshoe crab ALFs (**Figure 2**). A first group named Group A [16], which was the more distant from the other three groups, corresponded to sequences similar to ALF*Pm*2 from *P. monodon* (referred to here as *Penmon* ALF-A1). Groups B [16] and C [14] corresponded to sequences similar to ALF*Pm*3 (*Penmon* ALF-B1) and ALF*Pm*6 (*Penmon* ALF-C1), respectively. Finally, a new group of shrimp ALFs, termed here Group D, clustered separately and contained sequences similar to those found *L. setiferus* [8] and *L. stylirostris* [7] (**Figure 2**). Interestingly, Group A to C gather sequences from occidental and Asian shrimp species, whereas Group D gathers sequences from occidental shrimp only. Scygonadins (outgroup) showed no clear relationship with shrimp ALFs and clustered in a distinct group from both shrimp and horseshoe crab ALFs.

Remarkably, the four ALF groups showed contrasted calculated isoelectric points (pI). Peptides from Group C and B were cationic (calculated pI = 7.77–9.77) and highly cationic (calculated pI = 9.70–10.29), respectively. Peptides from Group A ranged from anionic to cationic (calculated pI = 5.88–8.06). Finally, peptides from Group D were very anionic (calculated pI = 5.58–6.10).

Based on both our sequence and phylogenetic analyses (**Figures 1 and 2**; **Table S1**), we propose here a uniform nomenclature for shrimp ALFs, similar to that earlier proposed for penaeidins, another family of antimicrobial peptides from penaeid shrimp [35]. Like the penaeidin (PEN) nomenclature, it includes: (i) the name of the penaeid shrimp in six letters and in italics (three for the genus and three for the species) followed by a space, (ii) the abbreviation "ALF" (for anti-lipopolysaccharide factor) followed by a hyphen and (iii) a number for the identification inside the group (e.g. ALF*Pm*3 is now referred to as *Penmon* ALF-B1). The determination of shrimp ALF Groups is based on the amino acid sequences of the mature peptide (**Figure 1**; **Table S1**).

Group D ALFs Display an Incomplete LPS-binding Site and have Impaired LPS-binding Activities

Alignment of the amino acid sequences of the most anionic shrimp ALFs (Group D, pI ≤ 6.10) with the most cationic ones (Group B, pI ≥ 9.95). showed that 18% of the amino acids were identical. In particular, the two cysteines delimiting the central β-hairpin are conserved in all sequences. Other residues were group specific (**Figure 1**). Interestingly, we observed that the Group D

ALFs were lacking most of the residues proposed to be involved in LPS-binding of Group B ALFs [11] (**Figure 1**).

To investigate the functional consequences of this incomplete LPS-binding site, the Group D ALF from *L. stylirostris*, termed here *Litsty* ALF-D1, was over-expressed in *E. coli* and purified by reversed-phase HPLC. The molecular mass of the recombinant *Litsty* ALF-D1 determined by MALDI-TOF-MS (10816.9 Da) (**Figure 3**) was compatible with its calculated molecular mass (10812.3 Da). We compared *Litsty* ALF-D1, as a representative form of Group D ALFs, to the previously characterized *Penmon* ALF-B1 (also known as ALF*Pm*3) [4,11], which belongs to the extensively studied cationic Group B ALFs. Both peptides presented highly contrasted LPS-binding properties in the *Limulus* amoebocyte lysate assay (**Figure 4**). Indeed, *Penmon* ALF-B1 bound LPS almost as efficiently as polymixin B (used as a control) with a 50% inhibition of the limulus amoebocyte lysate assay at 0.1 μM instead of 0.05 μM for PmB. In contrast, *Litsty* ALF-D1, which displayed an incomplete LPS-binding site, was 20 times less active, 2 μM being required to achieve an equivalent inhibition (**Figure 4**). Therefore, unlike the Group B *Penmon* ALF-B1, the Group D *Litsty* ALF-D1, which has an incomplete LPS-binding site, is also deficient in binding LPS.

The Group D *Litsty* ALF-D1 has Strongly Weakened Antimicrobial Properties

Antimicrobial activities of *Litsty* ALF-D1 and *Penmon* ALF-B1 were determined by the liquid growth inhibition assay against a series of Gram-positive and Gram-negative bacterial strains as well as against filamentous fungal strains. Interestingly, *Litsty* ALF-D1 was almost inactive below 10 μM. Only one strain of *Bacillus megaterium* and one strain of *E. coli* were inhibited at 2.5 μM (**Table 3**). No activity was recorded against fungal strains. On the contrary, *Penmon* ALF-B1 was active against both bacteria and fungi with MICs as low as 0.15 μM (**Table 3**). Comparison of MICs against *E. coli* SBS363 showed that *Litsty* ALF-D1 is at least 16-fold less active that *Penmon* ALF-B1. None of the ALF tested were active against *Vibrio* strains. However, *Penmon* ALF-B1 was active against the marine strain *Aerococcus viridans* CIP 104074 at 1.25 μM. Because anionic antibacterial peptides such as *Litsty* ALF-D1 may require divalent cations for antimicrobial activity, we performed a few antimicrobial assays in the presence of sea salts (1.5 μM Ca^{2+}, 5 μM Mg^{2+}, 20 μM K^+). However, *Litsty* ALF-D1 remained inactive below 10 μM against both *V. alginolyticus* and *V. harveyi* under such conditions (data not shown).

The Antimicrobial Activity Carried by Central β-hairpin of Group B ALFs is Lacking in *Litsty* ALF-D1 (Group D)

The central β-hairpin of ALFs, which carries LPS-binding properties in many species [22–24], is one region poorly conserved between Group B and D compared to other groups. Indeed, 17 over the 22 residues (77.2%) of the β-hairpin sequence are identical or have conserved functions between groups B and C, 14 (63.6%) between groups B and A, and only 9 (40.9%) between groups B and D. To determine whether the central β-hairpin, could be responsible for the weak activity of *Litsty* ALF-D1, we compared the antimicrobial activity of synthetic peptides corresponding to this region in both ALF sequences. Synthetic peptides were named *Litsty* ALF-D1 β-hairpin and *Penmon* ALF-B1 β-hairpin for the anionic *Litsty* ALF-D1 and the cationic *Penmon* ALF-B1, respectively. An additional peptide was used, *Litsty* ALF-B1 β-hairpin, which corresponds to a cationic Group B ALF from *L. stylirostris*. In the liquid growth inhibition assay, no activity was recorded for *Litsty* ALF-D1 β-hairpin up to 10 μM (**Table 3**). On

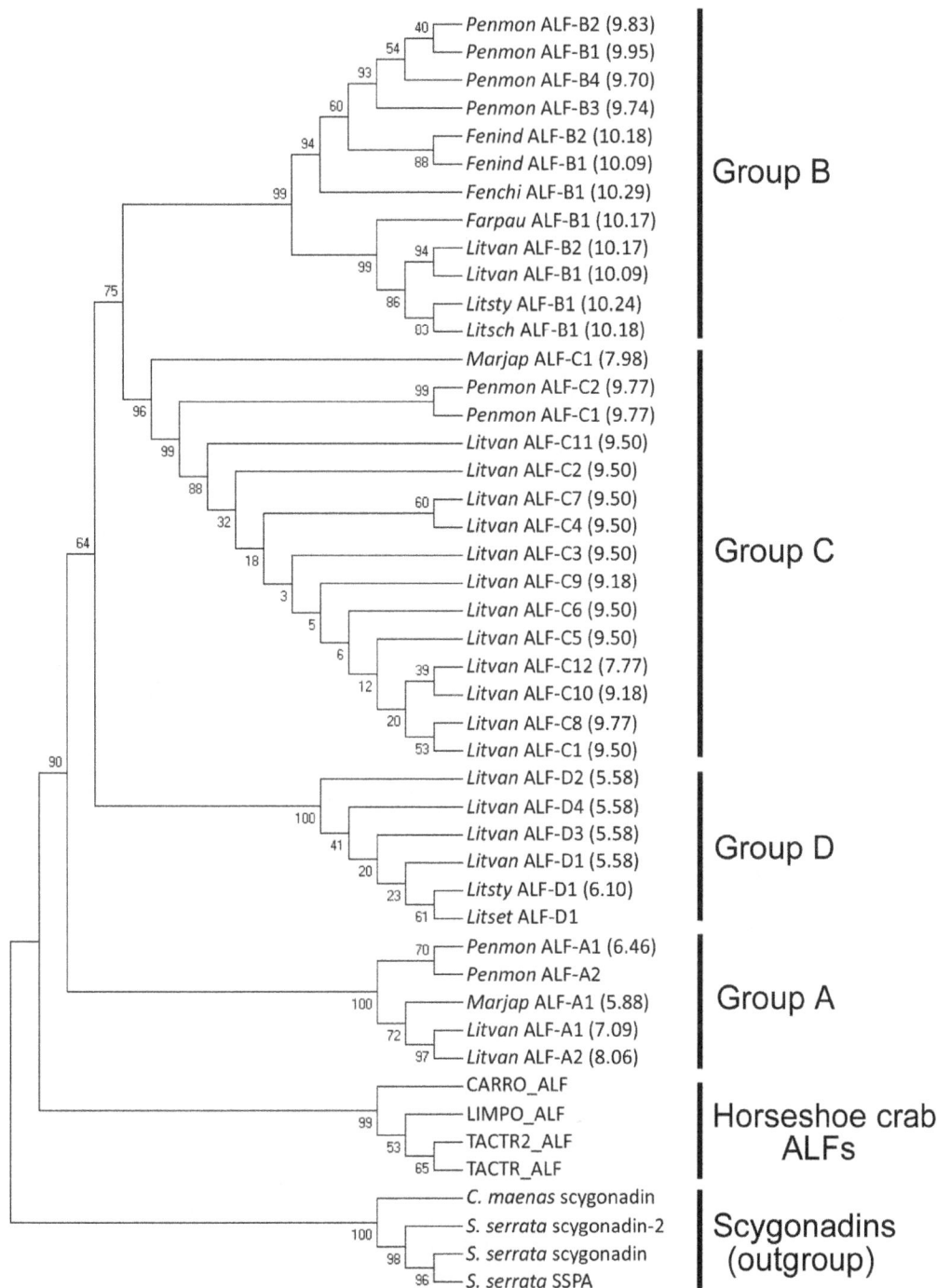

Figure 2. Penaeid shrimp ALFs cluster into four groups. ALF sequences from penaeid shrimps and horseshoe crabs, and scygonadins from crabs (outgroup) were aligned with MAFFT alignment program prior to phylogenetic analysis. The tree was constructed using the Neighbour-Joining method (Pairwise deletion) in MEGA 4. Bootstrap sampling was reiterated 1,000 times. Sequences included in analyses were the following: (i) <u>Shrimp ALFs</u>: black tiger shrimp *Penaeus monodon* (*Penmon* ALF-A1 or ALF*Pm*2: ABP73291; *Penmon* ALF-A1 or ALF*Pm*1: ABP73290; *Penmon* ALF-B1 or ALF*Pm*3: ABP73289; *Penmon* ALF-B2 and -B3 [47]; *Penmon* ALF-B4: ADC32520; *Penmon* ALF-C1 or ALF*Pm*6: ADM21460; *Penmon* ALF-C2 or ALF*Pm*6: AER45468), kuruma prawn (*Marjap* ALF-A1 or *Mj*ALF2: BAH22585; *Marjap* ALF-C1 or M-ALF: BAE92940), fleshy prawn *Fenneropenaeus chinensis* (*Fenchi* ALF-B1 or ALF*Fc*: AAX63831), Indian prawn *F. indicus* (*Fenind* ALF-B1: ADE27980; *Fenind* ALF-B2: ADK94454), pink shrimp *Farfantepenaeus paulensis* (*Farpau* ALF-B1 or ALF*Fpau*: ABQ96193), Atlantic white shrimp *Litopenaeus setiferus* (*Litset* ALF-D1: BE846661), Pacific white leg shrimp *L. vannamei* (*Litvan* ALF-A1 or *Lv*ALF1: EW713395; *Litvan* ALF-A2: FE087264; *Litvan* ALF-B1 or ALF*Lv*3: ABB22833; *Litvan* ALF-B2 or ALF*Lv*2: ABB22832; *Litvan* ALF-C1: FE153599; *Litvan* ALF-C2: FE176556; *Litvan* ALF-C3: FE058235; *Litvan* ALF-C4: FE079082; *Litvan* ALF-C5: FE088301; *Litvan* ALF-C6: FE078559; *Litvan* ALF-C7: FE079755; *Litvan* ALF-C8: FE105941; *Litvan* ALF-C9: FE090668; *Litvan* ALF-C10: FE052210; *Litvan* ALF-C11: FE088625; *Litvan* ALF-C12 or: *Lv*ALF2: EW713396; *Litvan* ALF-D1: FE152534; *Litvan* ALF-D2: FE151634; *Litvan* ALF-D3: FE110967; *Litvan* ALF-D4: FE115964), Southern white shrimp *L. schmitti* (*Litsch* ALF-B1 or ALF*Lsch*: ABJ90465) and blue shrimp *L. stylirostris* (*Litsty* ALF-B1: AGH32549; *Litsty* ALF-D1: AAY33769); (ii) <u>Horseshoe crab ALFs</u>: Chinese horseshoe crab *Tachypleus tridentatus* (TACTR_ALF: P07087; TACTR2_ALF: AAK00651), Atlantic horseshoe crab *Limulus polyphemus* (LIMPO_ALF: P07086) and Southeast Asian horseshoe crab *Carcinoscorpius rotundicauda* (CARRO_ALF: CK086627); (iii) <u>Scygonadins (outgroup)</u>: giant

mud crab *Scylla serrata* (*S serrata* scygonadin: AAW57403; *S serrata* scygonadin-2: ABI96918; *S serrata* SSAP: ABM05493) and green crab *Carcinus maenas* (*C. maenas* scygonadin: DY307310).

the contrary, both *Penmon* ALF-B1 β-hairpin and *Litsty* ALF-B1 β-hairpin, which derive from ALF Group B sequences, were active against both bacteria and fungi. They had very similar potencies and spectra of activity, the more potent activities being recorded against *B. megaterium* CIP 6620 at 0.6 and 1.25 μM, respectively (**Table 3**). Altogether, these data show that unlike the central β-hairpin of Group B ALFs, the central β-hairpin of Group D *Litsty* ALF-D1 is devoid of antimicrobial activity.

All Four ALF Groups are Simultaneously Expressed in a Single shrimp

Because sequences of the four ALF groups were identified in *L. vannamei* (**Figure 1**), we asked whether they could be expressed simultaneously in a single shrimp. Their expression was monitored by quantitative Real-Time PCR (RT-qPCR) in circulating hemocytes of 15 non-stimulated (naïve) *L. vannamei* shrimps. Specific primers sets were used for each ALF group (**Table 1**). Results showed that all four ALF groups are constitutively transcribed in the circulating hemocytes of a single *L. vannamei* shrimp, but at significantly different levels. *Litvan ALF-B* was the most transcribed gene, followed by *Litvan ALF-A*, *-D* and *-C* (**Figure 5**). The basal gene expression levels of *Litvan ALF-B* was 49-fold ($p<0.0001$), 253-fold ($p<0.0001$) and 361-fold ($p<0.0001$) higher than *Litvan ALF-A*, *-D* and *-C*, respectively. Comparatively, the basal mRNA expression of *Litvan ALF-A* was 5.2-fold ($p<0.0001$) and 7.4-fold ($p<0.0001$) higher than *Litvan ALF-D* and *-C*, respectively (**Figure 5**). Although we found a significant difference between the basal mRNA expression of *Litvan ALF-C* and *Litvan ALF-D* ($p<0.002$), it was not possible to determine which of these two genes is the most expressed due to the small differences in gene expression (<2-fold change) and the use of primer pairs with different efficiencies ($E = 2 \pm 0.2$).

ALF Groups are Differentially Regulated in Response to a Microbial Challenge

The functional diversity of shrimp ALFs evidenced here, led us ask how these genes respond to an injury or to a microbial infection. For that, we analyzed by RT-qPCR the expression profile of ALF groups in circulating hemocytes of shrimp injected with sterile sea water (SSW) or infected with the fungal pathogen *Fusarium solani* [31]. Infections were performed in the shrimp *L. vannamei* for which ALF sequences from all four groups were available in the databases. Every group of ALFs responded differently to the microbial challenge (**Figure 6**). The transcript abundance of *Litvan ALF-A* did not differ between shrimp injected with *Fusarium* or SSW and non-injected shrimp. On the contrary, the transcript abundance of *Litvan ALF-B*, *-C*, and *-D* followed different expression profiles. Thus, after an injection of SSW or *Fusarium*, the abundance of *Litvan ALF-B* transcripts was 2.4 to 3.1 fold higher than in non-injected shrimp (48 h after *Fusarium* injection and 24 h after SSW injection, respectively). No significant difference was observed between 24 h and 48 h. Conversely, the transcript abundance of *Litvan ALF-C* varied with the type of injection. Thus, after 24 h, only shrimp injected with *Fusarium* showed a *Litvan ALF-C* transcript abundance 8.1 fold higher than non-injected shrimp. However, after 48 h, it was 4.8 fold and a 16 fold for SSW and *Fusarium*-injected shrimp, respectively. Finally, for *Litvan ALF-D*, only shrimp injected with SSW had significantly higher transcript abundance, from 3.7 fold (24 h) to 3.1 fold (48 h), than non-injected shrimp.

Discussion

Results showed that shrimp ALFs cluster into four groups diverse in terms of primary sequence, biochemical properties, biological activities and gene expression. We provide here the first evidence of functional divergence in shrimp ALFs. The novel

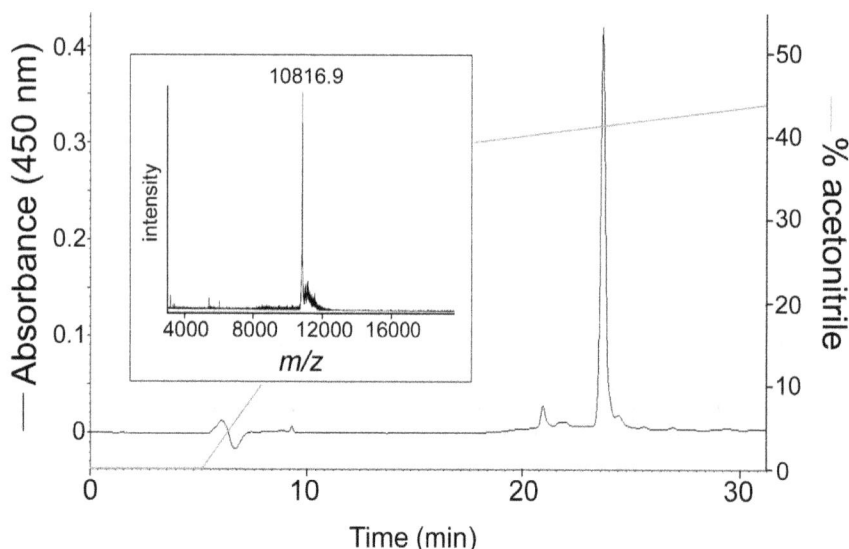

Figure 3. Purification of recombinant *Litsty* ALF-D1. HPLC profile of recombinant *Litsty* ALF-D1 on a C18 UP5NEC 25QS column. The acetonitrile percentage corresponding to the biphasic gradient is shown as a *grey line*. Recombinant *Litsty* ALF-D1 was observed as an absorbance peak (black line) eluted at 42% acetonitrile. The MALDI-TOF-MS spectrum of the collected peak (*inset*) showed a single mass at *m/z* = 10816.9.

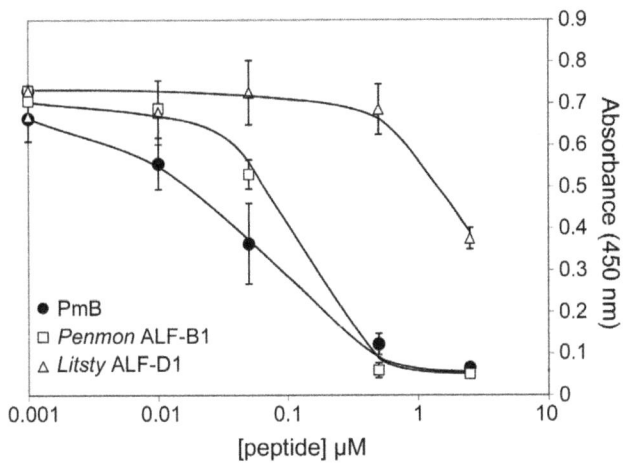

Figure 4. Differential LPS-binding properties of shrimp ALFs from Groups B and D. The ability of *Litsty* ALF-D1 (Group D, *open triangles*) to bind LPS was compared to that of *Penmon* ALF-B1 (Group B, *open squares*) in the *Limulus* amoebocyte lysate (LAL) assay. Polymixin B (PmB, *black circles*) was used as a positive control. Absorbance (405 nm) is indicative of LAL assay activation by LPS. LPS neutralization by LPS-binding peptides prevents LAL assay activation. It takes here 20-fold more *Litsty* ALF-D1 than *Penmon* ALF-B1 to neutralize an equivalent amount of LPS.

members of the ALF family identified in this study form a

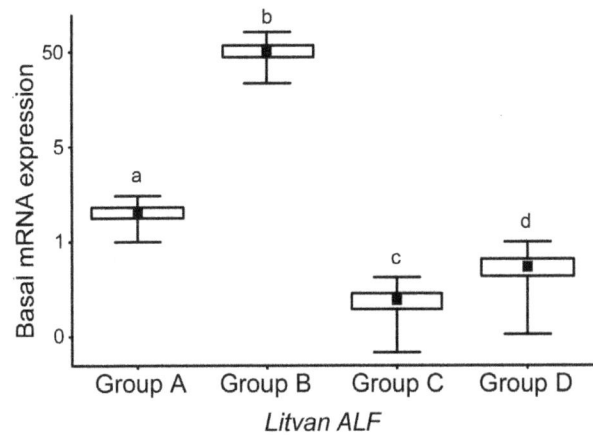

Figure 5. All four ALF groups are simultaneously transcribed in a single shrimp. The basal gene expression levels of the four ALF Groups identified in *L. vannamei* shrimp (*Litvan ALF-A* to *-D*) were determined by RT-qPCR on circulating hemocytes of individual shrimps (n = 15) according to the $2^{-\Delta\Delta Cq}$ method [32]. The ribosomal protein L40 (*Litvan-rpl40*, GenBank: FE077602) was used as a reference gene. Results are expressed as mean values (central black squares) ±SE (boxes) and ±SD (whiskers) of relative expressions on a logarithmic scale. Significant differences between the basal RNA expression levels of the four ALF Groups (Student's t-test, $p < 0.05$) are indicated by different lowercase letters (a, b, c and d). The use of a same letter indicates the absence of significant difference, while the use of different letters indicates significant difference.

Table 3. Spectrum of antimicrobial activities of recombinant ALFs and synthetic ALF β-hairpins.

MIC (µM)	Recombinant ALFs		Synthetic peptides		
	Litsty **ALF-D1**	*Penmon* **ALF-B1**	*Litsty* **ALF-D1** **β-hairpin**	*Penmon* **ALF-B1** **β-hairpin**	*Litsty* **ALF-B1** **β-hairpin**
Gram-positive bacteria					
Aerococcus viridans CIP 104 074	>10	**1.25**	>10	**1.25**	**2.50**
Bacillus megaterium (IBMC collection)	**2.50**	**0.15**	>10	**1.25**	**2.5**
Bacillus megaterium CIP 6620	>10	**1.25**	>10	**0.60**	**1.25**
Brevibacterium stationis CIP 101 282	>10	>10	>10	**10**	>10
Micrococcus luteus CIP 5345	>10	**10**	>10	**1.25**	**1.25**
Staphylococcus aureus ATCC 65 38	>10	>10	>10	>10	**10**
Gram-negative bacteria					
Escherichia coli SBS363	**2.50**	**0.15**	>10	**5**	**2.50**
Salmonella enterica CIP 5858	>10	**2.50**	>10	**10**	**5**
Vibrio alginolyticus CIP 103336 T	>10	**0.60**	>10	>10	>10
Vibrio harveyi E22	>10	**10**	>10	>10	>10
Vibrio anguillarum V62	>10	>10	>10	>10	>10
Vibrio penaeicidae AM101	>10	>10	>10	>10	>10
Vibrio nigripulchritudo SFn1	>10	>10	>10	>10	>10
Fungi					
Candida albicans	>10	**10**	>10	>10	**10**
Fusarium oxysporum	>10	**10**	>10	**10**	**2.50**
Rhizopus stolonifer	>10	>10	>10	>10	**10**
Septoria nodorum	NT	NT	>10	**10**	**2.50**
Botrytis cinerea	>10	**10**	>10	>10	**10**

MIC (µM) values refer to the minimal inhibitory concentration required to achieve 100% growth inhibition. NT: not tested.

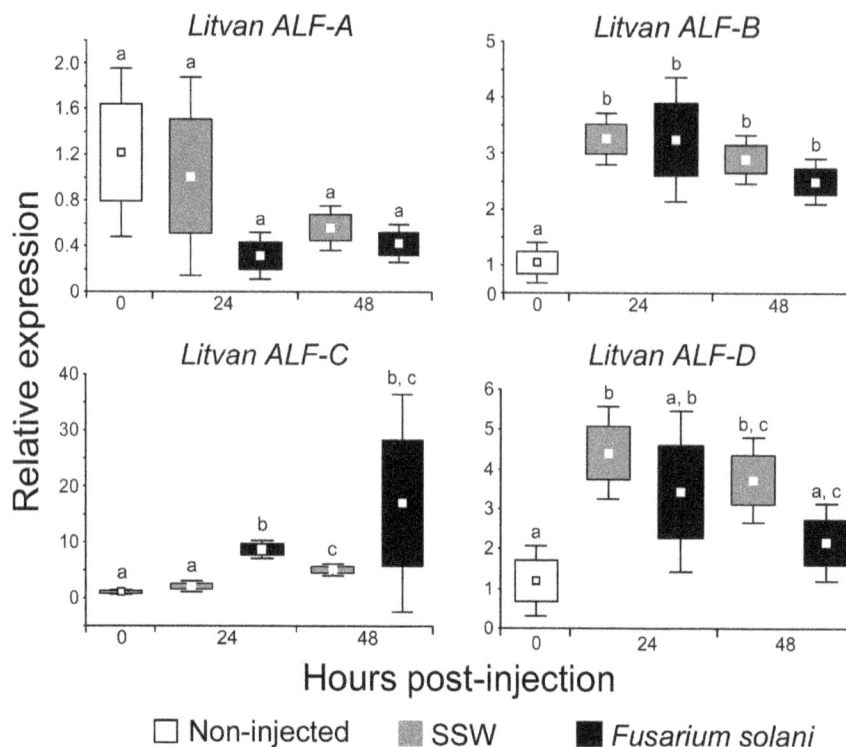

Figure 6. Differential expression of ALF groups in response to an experimental infection. The transcript abundance of the four ALF Groups (*Litvan ALF-A, -B, -C,* and *-D*) was measured by RT-qPCR on circulating hemocytes of shrimp (three groups of three shrimps per condition) injected with the shrimp pathogen *Fusarium solani* (black boxes), sterile sea water (SSW, grey boxes) or non-injected (white boxes). Analyses were performed 24 and 48 hours after injection. Relative abundances of ALF transcripts are calculated according the $2^{-\Delta\Delta Cq}$ method [32], with the *L. vannamei* ribosomal protein L40 (*Litvan-rpl40*, GenBank: FE077602) used as reference gene. Results are expressed as mean values (central squares) $\pm$SE (boxes) and $\pm$SD (whiskers) of relative expressions. Significant differences between conditions for each ALF groups (Student's t-test, $p<0.05$) are indicated by different lowercase letters (a, b, and c). The use of a same letter indicates the absence of significant difference, while the use of different letters indicates significant difference.

phylogenetic group termed Group D distinct from the previously characterized Groups A, B and C. Group D ALFs have a negative net charge (pI ≤6.2) which contrasts with the most common properties of eukaryotic antimicrobial peptides, whose positive net charge mediates interaction with the electronegative membranes of bacteria [36]. In addition, Group D ALFs display only two conserved residues out of seven charged residues [Glu^{21}, Lys^{22}, (Lys/Arg)^{31}, Lys^{35}, (Lys/Arg)^{46}, (Lys/Arg)^{48}, (Lys/Arg)^{58}] from the ALF β-sheet proposed to be involved in the LPS-binding site of *Penmon* ALF-B1 [11]. The other five residues are substituted by uncharged amino acids (**Figure 1**). We showed here that such substitutions impaired the interaction of Group D ALFs with LPS. Indeed, Group D *Litsty* ALF-D1 produced here as a recombinant peptide was 20-fold less effective in binding LPS than our reference recombinant Group B ALF, *Penmon* ALF-B1 (**Figure 4**). Together with previous biochemical data on the horseshoe crab ALF [24] and our previous molecular modeling data based on the shrimp *Penmon* ALF-B1 NMR structure [11], this strongly supports the role of the charged residues carried by the four-stranded β-sheet of ALFs in their interaction with the hydrophilic and phosphate groups of lipid A. Importantly, *Litsty* ALF-D1, which displays a poor conservation of the putative LPS-binding residues, was also poorly active against Gram-negative bacteria with MIC values at least 16-fold higher than the Group B *Penmon* ALF-B1 (**Table 3**). The same was true for the synthetic β-harpin region of *Litsty* ALF-D1, which lacked antibacterial activity, as opposed to the synthetic β-harpin regions of *Litsty* ALF-B1 and *Penmon* ALF-

B1 (**Table 3**). Therefore, LPS-binding appears essential for the anti-Gram-negative activity of shrimp ALFs.

Altogether, our phylogenetic and functional data indicate that ALFs have evolved as functionally diverse groups, among which some have selected important motifs for interacting with LPS. Our present study demonstrates the very intimate link existing between such recognition functions and the subsequent antimicrobial properties of ALFs. Together with our previous study on oyster defensin functional diversity [29], this shows how far sequence diversity affecting amino acids involved in the recognition of microbe cell wall components has shaped the activity of marine invertebrate antimicrobial peptides and led to functional divergence.

Because *Litsty* ALF-D1 was only poorly active against Gram-negatives, we asked whether it had evolved towards other antibacterial functions. However, *Litsty* ALF-D1 was not active against filamentous fungi and only poorly active against Gram-positive bacteria as compared to our broad-spectrum Group B reference (**Table 3**). Similarly, synthetic peptides corresponding to the central β-hairpin of Group D ALFs were devoid of anti-Gram-positive activity, as opposed to peptides designed on the same region of Group B ALFs (**Table 3**). This is in agreement with the important role of the central β-hairpin in the anti-Gram-positive activity of Group B ALFs [4]. It also tells that the major reduction of the positive net charge in this functional region may not only impair the electrostatic interaction with the LPS of Gram-negative bacteria but also with LTA, a major cell wall component of Gram-positive bacteria previously reported to bind to Group B ALFs [19]. Because some Group A ALFs also display a negative net charge [13], but have a more conserved β-harpin sequence, investigations on their antibacterial properties should help understanding whether conserved amino acids interacting with specific cell wall components or a global positive charge are required for the antibacterial activity of shrimp ALFs.

From our *in silico* analysis (**Figure 2**), Group D ALFs would be only found in occidental shrimp, such as *L. vannamei*, *L. setiferus* and *L. stylirostris*. This is reminiscent of penaeidins, another major AMP family of penaeid shrimp. Indeed, while *PEN2*, *PEN3* and *PEN4* are found in occidental shrimp, only *PEN3* and *PEN5* are found in Asian shrimps [13]. Therefore, for those two main families of shrimp AMPs, genes encoding AMP groups appear to be differentially distributed among Asian and occidental shrimp species.

By using *L. vannamei* to conduct our expression study, we showed here that all four members of the ALF family (A, B, C and D) are simultaneously expressed in a single shrimp. This was also shown for penaeidin genes (*PEN2*, *PEN3* and *PEN4*) from the same species [37]. Interestingly, as in *P. monodon* [16], Group B ALFs were by far the most transcribed ALF genes in *L. vannamei* shrimp (by more than 50 fold; **Figure 5**). The high expression and the highly potent antimicrobial activity of Group B ALFs (**Table 3**) suggest they likely play an important protective role in the shrimp antimicrobial defense. Whereas there is no evidence of ALF expression at the protein level, RNA interference data strongly support the major role of *Penmon* ALF-B1 in shrimp antibacterial and antiviral defense [14,26].

Group A and C ALFs, which were shown here to be expressed at much lower levels than Group B ALFs (**Figure 5**), would also display important functions in shrimp survival to both bacterial and fungal infections [14,25]. Since all groups of antimicrobial ALFs are expressed simultaneously in a single shrimp (**Figure 5**), it can be hypothesized that the high sequence diversity of shrimp ALFs contributes to create synergism improving shrimp antimicrobial defenses [12]. Such a synergism has been previously demonstrated for variants of oyster defensins [29,38].

The role of Group D ALFs in shrimp defense remains much more ambiguous. On the one hand, *Litsty* ALF-D1 was shown to be poorly antimicrobial (**Table 3**). On the other hand, expression of *Litsty* ALF-D1 is not correlated with the capacity of *L. stylirostris* shrimp to survive a pathogenic *Vibrio* infection [7]. Therefore, unlike other ALF groups, Group D ALFs would not play a direct antibacterial/antifungal functions in shrimp defense but could have evolved towards uncharacterized functions (neofunctionalization). Indeed, it is now well established that AMPs are multifunctional peptides whose defense functions go far beyond their antimicrobial properties [39]. One interesting function for shrimp defense that remains to be investigated is the antiviral activity of this ALF group.

Finally, while the gene expression profile of ALFs has been mainly investigated in response to viruses and bacteria challenges [13], we analyzed here the transcript abundance of the four *Litvan*

ALF genes in shrimp infected with the fungal pathogen *F. solani*. Interestingly, shrimp ALFs were shown to follow different gene regulation patterns in response to a fungal infection. Thus, while *Litvan ALF-A* is not modulated, *Litvan ALF-B*, *-C*, and *-D* are induced in circulating hemocytes in response to *F. solani* challenge and to injury (**Figure 6**). Such a differential regulation has been earlier observed among variants of other AMP families such as the oyster big defensins (*Cg*-BigDefs) and human β-defensins (HBD) [40,41]. Together with important sequence variations between ALF groups, this suggests that different genes encode ALFs from different groups in *L. vannamei*. This is in agreement with data on *P. monodon* showing that each of the three ALFs (Group A to C) is encoded by a distinct genomic sequence [14,16]. The existence of different genetic locations for ALF groups supports further the functional divergence of this family of AMPs, as also proposed for oyster defensins [42].

To conclude, taking into account the high diversity of shrimp ALFs explored here, we propose a uniform nomenclature for shrimp ALFs, as adopted for other AMP families, such as avian β-defensins [43], amphibian skin peptides [44], and shrimp penaeidins [35]. Like the penaeidin (PEN) nomenclature, the shrimp ALF nomenclature is based on the conserved amino acid residues found in the mature polypeptides of each ALF Group (**Figure 1**). For instance, in *L. vannamei* shrimp, the described ALF variants were named as ALF*Lv*1 to -3 by Jiménez-Vega and Vargas-Albores [45] or *Lv*ALF1 and -2 by de la Vega *et al.* [25]. The aim of this recommended nomenclature is to standardize the confusing terminology currently employed for this highly diverse family of shrimp antimicrobial peptides [13,17,18,22,46].

Supporting Information

Table S1 Sequences and biochemical properties of shrimp ALFs. The names of ALF sequences (groups A to D) are given in the new nomenclature. The corresponding ancient names are also provided together with existing GenBank numbers. Amino acid sequences of signal peptides and mature peptides are displayed with the number of amino acids and theorical pI of the polypeptides.

Acknowledgments

The authors are grateful to Marc Leroy for technical assistant and to Dr. Claudia R. B. de Souza to provide the cultures of *Fusarium solani*. We have benefited from molecular genetic analysis technical facilities of the SFR "Montpellier Environnement Biodiversité" and the "Platform qPHD UM2/Montpellier GenomiX".

Author Contributions

Conceived and designed the experiments: YG EB DDG. Performed the experiments: RDR AV JdL PG LS BR JF YG DDG. Analyzed the data: RDR JdL LMP YG DDG. Wrote the paper: RDR JdL YG DDG.

References

1. Carriel-Gomes MC, Kratz JM, Barracco MA, Bachère E, Barardi CR, et al. (2007) *In vitro* antiviral activity of antimicrobial peptides against herpes simplex virus 1, adenovirus, and rotavirus. Mem Inst Oswaldo Cruz 102: 469–472.

2. Liu Y, Cui Z, Li X, Song C, Li Q, et al. (2012) A new anti-lipopolysaccharide factor isoform (PtALF4) from the swimming crab *Portunus trituberculatus* exhibited structural and functional diversity of ALFs. Fish & shellfish immunology 32: 724–731.

3. Morita T, Ohtsubo S, Nakamura T, Tanaka S, Iwanaga S, et al. (1985) Isolation and biological activities of *Limulus* anticoagulant (anti-LPS factor) which interacts with lipopolysaccharides (LPS). J Biochem 97: 1611–1620.

4. Somboonwiwat K, Marcos M, Tassanakajon A, Klinbunga S, Aumelas A, et al. (2005) Recombinant expression and anti-microbial activity of anti-lipopolysaccharide factor (ALF) from the black tiger shrimp Penaeus monodon. Developmental and comparative immunology 29: 841–851.

5. Yedery RD, Reddy KV (2009) Identification, cloning, characterization and recombinant expression of an anti-lipopolysaccharide factor from the hemocytes of Indian mud crab, *Scylla serrata*. Fish & shellfish immunology 27: 275–284.

6. Tanaka S, Nakamura T, Morita T, Iwanaga S (1982) *Limulus* anti-LPS factor: an anticoagulant which inhibits the endotoxin mediated activation of *Limulus*

coagulation system. Biochemical and biophysical research communications 105: 717–723.

7. de Lorgeril J, Gueguen Y, Goarant C, Goyard E, Mugnier C, et al. (2008) A relationship between antimicrobial peptide gene expression and capacity of a selected shrimp line to survive a *Vibrio* infection. Molecular immunology 45: 3438–3445.

8. Gross PS, Bartlett TC, Browdy CL, Chapman RW, Warr GW (2001) Immune gene discovery by expressed sequence tag analysis of hemocytes and hepatopancreas in the Pacific White Shrimp, *Litopenaeus vannamei*, and the Atlantic White Shrimp, *L. setiferus*. Dev Comp Immunol 25: 565–577.

9. Supungul P, Klinbunga S, Pichyangkura R, Jitrapakdee S, Hirono I, et al. (2002) Identification of immune-related genes in hemocytes of black tiger shrimp (*Penaeus monodon*). Mar Biotechnol 4: 487–494.

10. Hoess A, Watson S, Siber GR, Liddington R (1993) Crystal structure of an endotoxin-neutralizing protein from the horseshoe crab, *Limulus* anti-LPS factor, at 1.5 A resolution. EMBO J 12: 3351–3356.

11. Yang Y, Boze H, Chemardin P, Padilla A, Moulin G, et al. (2009) NMR structure of rALF-*Pm3*, an anti-lipopolysaccharide factor from shrimp: model of the possible lipid A-binding site. Biopolymers 91: 207–220.

12. Rosa RD, Barracco MA (2010) Antimicrobial peptides in crustaceans. Inv Surv J 7: 262–284.

13. Tassanakajon A, Amparyup P, Somboonwiwat K, Supungul P (2011) Cationic antimicrobial peptides in penaeid shrimp. Mar Biotechnol (NY) 13: 639–657.

14. Ponprateep S, Tharntada S, Somboonwiwat K, Tassanakajon A (2012) Gene silencing reveals a crucial role for anti-lipopolysaccharide factors from *Penaeus monodon* in the protection against microbial infections. Fish & shellfish immunology 32: 26–34.

15. Supungul P, Klinbunga S, Pichyangkura R, Hirono I, Aoki T, et al. (2004) Antimicrobial peptides discovered in the black tiger shrimp *Penaeus monodon* using the EST approach. Dis Aquat Organ 61: 123–135.

16. Tharntada S, Somboonwiwat K, Rimphanitchayakit V, Tassanakajon A (2008) Anti-lipopolysaccharide factors from the black tiger shrimp, *Penaeus monodon*, are encoded by two genomic loci. Fish Shellfish Immunol 24: 46–54.

17. Liu F, Liu Y, Li F, Dong B, Xiang J (2005) Molecular cloning and expression profile of putative antilipopolysaccharide factor in Chinese shrimp (*Fenneropenaeus chinensis*). Mar Biotechnol (NY) 7: 600–608.

18. Mekata T, Sudhakaran R, Okugawa S, Kono T, Sakai M, et al. (2010) Molecular cloning and transcriptional analysis of a newly identified anti-lipopolysaccharide factor gene in kuruma shrimp, *Marsupenaeus japonicus*. Letters in applied microbiology 50: 112–119.

19. Somboonwiwat K, Bachère E, Rimphanitchayakit V, Tassanakajon A (2008) Localization of anti-lipopolysaccharide factor (ALFPm3) in tissues of the black tiger shrimp, *Penaeus monodon*, and characterization of its binding properties. Developmental and comparative immunology 32: 1170–1176.

20. Bhunia A, Domadia PN, Torres J, Hallock KJ, Ramamoorthy A, et al. (2010) NMR structure of pardaxin, a pore-forming antimicrobial peptide, in lipopolysaccharide micelles: mechanism of outer membrane permeabilization. J Biol Chem 285: 3883–3895.

21. Bhunia A, Saravanan R, Mohanram H, Mangoni ML, Bhattacharjya S (2011) NMR structures and interactions of temporin-1Tl and temporin-1Tb with lipopolysaccharide micelles: mechanistic insights into outer membrane permeabilization and synergistic activity. J Biol Chem 286: 24394–24406.

22. Nagoshi H, Inagawa H, Morii K, Harada H, Kohchi C, et al. (2006) Cloning and characterization of a LPS-regulatory gene having an LPS binding domain in kuruma prawn *Marsupenaeus japonicus*. Molecular immunology 43: 2061–2069.

23. Mora P, De La Paz ML, Pérez-Payá E (2008) Bioactive peptides derived from the *Limulus* anti-lipopolysaccharide factor: structure-activity relationships and formation of mixed peptide/lipid complexes. J Pept Sci 14: 963–971.

24. Andrä J, Howe J, Garidel P, Rössle M, Richter W, et al. (2007) Mechanism of interaction of optimized *Limulus*-derived cyclic peptides with endotoxins: thermodynamic, biophysical and microbiological analysis. Biochem J 406: 297–307.

25. de la Vega E, O'Leary NA, Shockey JE, Robalino J, Payne C, et al. (2008) Anti-lipopolysaccharide factor in *Litopenaeus vannamei* (LvALF): a broad spectrum antimicrobial peptide essential for shrimp immunity against bacterial and fungal infection. Molecular immunology 45: 1916–1925.

26. Tharntada S, Ponprateep S, Somboonwiwat K, Liu H, Soderhall I, et al. (2009) Role of anti-lipopolysaccharide factor from the black tiger shrimp, *Penaeus*

monodon, in protection from white spot syndrome virus infection. J Gen Virol 90: 1491–1498.

27. Tamura K, Dudley J, Nei M, Kumar S (2007) MEGA4: Molecular Evolutionary Genetics Analysis (MEGA) software version 4.0. Mol Biol Evol 24: 1596–1599.

28. Gueguen Y, Herpin A, Aumelas A, Garnier J, Fievet J, et al. (2006) Characterization of a defensin from the oyster *Crassostrea gigas*. Recombinant production, folding, solution structure, antimicrobial activities, and gene expression. J Biol Chem 281: 313–323.

29. Schmitt P, Wilmes M, Pugnière M, Aumelas A, Bachère E, et al. (2010) Insight into invertebrate defensin mechanism of action: oyster defensins inhibit peptidoglycan biosynthesis by binding to lipid II. J Biol Chem 285: 29208–29216.

30. Hetru C, Bulet P (1997) Strategies for the isolation and characterization of antimicrobial peptides of invertebrates. Methods Mol Biol 78: 35–49.

31. Goncalves P, Vernal J, Rosa RD, Yepiz-Plascencia G, de Souza CR, et al. (2012) Evidence for a novel biological role for the multifunctional beta-1,3-glucan binding protein in shrimp. Molecular immunology 51: 363–367.

32. Livak KJ, Schmittgen TD (2001) Analysis of relative gene expression data using real-time quantitative PCR and the $2^{-\Delta\Delta CT}$ Method. Methods 25: 402–408.

33. Supungul P, Klinbunga S, Pichyangkura R, Hirono I, Aoki T, et al. (2004) Antimicrobial peptides discovered in the black tiger shrimp *Penaeus monodon* using the EST approach. Dis Aquat Organ 61: 123–135.

34. Wang KJ, Huang WS, Yang M, Chen HY, Bo J, et al. (2007) A male-specific expression gene, encodes a novel anionic antimicrobial peptide, scygonadin, in *Scylla serrata*. Molecular immunology 44: 1961–1968.

35. Gueguen Y, Garnier J, Robert L, Lefranc MP, Mougenot I, et al. (2006) PenBase, the shrimp antimicrobial peptide penaeidin database: sequence-based classification and recommended nomenclature. Developmental and comparative immunology 30: 283–288.

36. Yeaman MR, Yount NY (2003) Mechanisms of antimicrobial peptide action and resistance. Pharmacol Rev 55: 27–55.

37. O'Leary NA, Gross PS (2006) Genomic structure and transcriptional regulation of the penaeidin gene family from *Litopenaeus vannamei*. Gene 371: 75–83.

38. Schmitt P, Lorgeril J, Gueguen Y, Destoumieux-Garzón D, Bachère E (2012) Expression, tissue localization and synergy of antimicrobial peptides and proteins in the immune response of the oyster *Crassostrea gigas*. Developmental and comparative immunology.

39. Lai Y, Gallo RL (2009) AMPed up immunity: how antimicrobial peptides have multiple roles in immune defense. Trends Immunol 30: 131–141.

40. O'Neil DA, Porter EM, Elewaut D, Anderson GM, Eckmann L, et al. (1999) Expression and regulation of the human beta-defensins hBD-1 and hBD-2 in intestinal epithelium. J Immunol 163: 6718–6724.

41. Rosa RD, Santini A, Fievet J, Bulet P, Destoumieux-Garzón D, et al. (2011) Big defensins, a diverse family of antimicrobial peptides that follows different patterns of expression in hemocytes of the oyster *Crassostrea gigas*. PLoS ONE 6: e25594.

42. Schmitt P, Gueguen Y, Desmarais E, Bachère E, de Lorgeril J (2010) Molecular diversity of antimicrobial effectors in the oyster *Crassostrea gigas*. BMC evolutionary biology 10: 23.

43. Lynn DJ, Higgs R, Lloyd AT, O'Farrelly C, Herve-Grepinet V, et al. (2007) Avian beta-defensin nomenclature: a community proposed update. Immunology letters 110: 86–89.

44. Thomas P, Kumar TV, Reshmy V, Kumar KS, George S (2012) A mini review on the antimicrobial peptides isolated from the genus *Hylarana* (Amphibia: Anura) with a proposed nomenclature for amphibian skin peptides. Mol Biol Rep 39: 6943–6947.

45. Jiménez-Vega F, Vargas-Albores F (2007) Isoforms of *Litopenaeus vannamei* anti-lipopolysaccharide and its expression by bacterial challenge. J Shell Res 26: 1169–1175.

46. Rosa RD, Stoco PH, Barracco MA (2008) Cloning and characterisation of cDNA sequences encoding for anti-lipopolysaccharide factors (ALFs) in Brazilian palaemonid and penaeid shrimps. Fish & shellfish immunology 25: 693–696.

47. Somboonwiwat K, Supungul P, Rimphanitchayakit V, Aoki T, Hirono I, et al. (2006) Differentially expressed genes in hemocytes of Vibrio harveyi-challenged shrimp Penaeus monodon. Journal of biochemistry and molecular biology 39: 26–36.

16

Charged and Hydrophobic Surfaces on the A Chain of Shiga-Like Toxin 1 Recognize the C-Terminal Domain of Ribosomal Stalk Proteins

Andrew J. McCluskey[1], Eleonora Bolewska-Pedyczak[2], Nick Jarvik[3], Gang Chen[3], Sachdev S. Sidhu[3], Jean Gariépy[1,2,4]*

1 Department of Pharmaceutical Sciences, University of Toronto, Toronto, Ontario, Canada, **2** Sunnybrook Research Institute, Toronto, Ontario, Canada, **3** Banting and Best Department of Medical Research, Terrence Donnelly Center for Cellular and Biomolecular Research, University of Toronto, Toronto, Ontario, Canada, **4** Department of Medical Biophysics, University of Toronto, Toronto, Ontario, Canada

Abstract

Shiga-like toxins are ribosome-inactivating proteins (RIP) produced by pathogenic *E. coli* strains that are responsible for hemorrhagic colitis and hemolytic uremic syndrome. The catalytic A_1 chain of Shiga-like toxin 1 (SLT-1), a representative RIP, first docks onto a conserved peptide SD[D/E]DMGFGLFD located at the C-terminus of all three eukaryotic ribosomal stalk proteins and halts protein synthesis through the depurination of an adenine base in the sarcin-ricin loop of 28S rRNA. Here, we report that the A_1 chain of SLT-1 rapidly binds to and dissociates from the C-terminal peptide with a monomeric dissociation constant of 13 µM. An alanine scan performed on the conserved peptide revealed that the SLT-1 A_1 chain interacts with the anionic tripeptide DDD and the hydrophobic tetrapeptide motif FGLF within its sequence. Based on these 2 peptide motifs, SLT-1 A_1 variants were generated that displayed decreased affinities for the stalk protein C-terminus and also correlated with reduced ribosome-inactivating activities in relation to the wild-type A_1 chain. The toxin-peptide interaction and subsequent toxicity were shown to be mediated by cationic and hydrophobic docking surfaces on the SLT-1 catalytic domain. These docking surfaces are located on the opposite face of the catalytic cleft and suggest that the docking of the A_1 chain to SDDDMGFGLFD may reorient its catalytic domain to face its RNA substrate. More importantly, both the delineated A_1 chain ribosomal docking surfaces and the ribosomal peptide itself represent a target and a scaffold, respectively, for the design of generic inhibitors to block the action of RIPs.

Editor: Ludger Johannes, Institut Curie, France

Funding: This project was supported by an operation grant to J.G. from the Canadian Institutes of Health Research. The funders had no role in study design, data collection and analysis, decision to publish, or preparation of the manuscript.

Competing Interests: The authors have declared that no competing interests exist.

* E-mail: gariepy@sri.utoronto.ca

Introduction

Shiga toxins such as Shiga-like toxin 1 (SLT-1) are produced by enteropathogenic *Escherichia coli* strains and represent the major cause of hemorrhagic colitis and hemolytic uremic syndrome [1,2]. SLT-1 is a type II ribosome-inactivating protein (RIP) composed of a catalytically active A subunit non-covalently associated with a pentamer of B-subunits [3,4]. This pentamer binds to the glycolipid globotriaosylceramide (CD77,Gb3), an event that leads to its internalization [5,6,7]. SLT-1 then traffics in a retrograde manner through the Golgi apparatus where it is proteolytically cleaved into an N-terminal catalytic A_1 domain and a C-terminal A_2 fragment non-covalently associated with its B-pentamer. Both A chain fragments remain linked by a single disulfide bond which is thought to be reduced in the ER lumen [8,9,10]. The A_1 domain is then retrotranslocated to the cytosol by virtue of its newly exposed hydrophobic C-terminus, where it eventually docks onto ribosomes and subsequently depurinates a single adenine base (A^{4324}) in the sarcin-ricin loop (SRL) of 28S rRNA [11,12,13,14,15]. This depurination event creates an apurinic site that prevents elongation factor 1 (EF-1)-dependent amino-acyl tRNA from binding to the ribosome and EF-2-catalysed translocation during elongation, leading to an inhibition of protein synthesis [16,17,18].

The protein component of the ribosome was first shown to contribute to the toxicity of RIPs when a 10^5 fold increase in depurination rate was observed for ricin on native ribosomes when compared to protein-depleted ribosomes [19]. SLT-1 as well as other structurally and functionally related RIPs, require their docking to ribosomal proteins in addition to rRNA to maintain their optimal depurination rate and cytotoxic function [15,19,20,21]. More recently, it has been revealed that the ribosomal protein components required for interacting with either type I (trichosanthin (TCS)) or type II (SLT-1 and ricin) RIPs are the ribosomal proteins RPP0, RPLP1 and RPLP2 (P0, P1, and P2) [15,20,22,23]. These three proteins form the ribosomal stalk which is required for the binding of elongation factors leading to protein translation [24,25,26]. The eukaryotic stalk structure is composed of two heterodimers of the P1 and P2 proteins [27,28,29], which interact by virtue of the N-terminus of the P1 protein, at two specific locations on the P0 protein [30,31,32,33], which subsequently binds to rRNA [34].

We have previously shown that the A_1 chain of SLT-1 interacts with the ribosomal stalk proteins P0, P1, and P2 via a conserved C-terminal peptide (SDXDMGFGLFD, where X = D or E) [15]. In the present study, we demonstrate by yeast-2-hybrid (Y2H) and surface plasmon resonance (SPR) that the A_1 chain of SLT-1 interacts with the C-terminal ribosomal stalk peptide with a micromolar dissociation constant. Specifically, the interaction of the A_1 chain with the conserved C-terminal peptide SDDD MGFGLFD common to all three ribosomal stalk proteins exhibits a modest binding constant (K_d 13 μM), towards the monovalent peptide, with rapid *on* and *off* rates. This transient interaction is mediated by distinct charged and hydrophobic surfaces on the SLT-1 A_1 chain, which are also essential for its full catalytic activity. Moreover, alanine-scanning mutagenesis revealed that anionic tripeptide and hydrophobic tetrapeptide motifs within the sequence SDDDMGFGLFD represent key anchor residues recognized by the A_1 chain. These findings suggest that the nature of these interactions may play a guiding role in properly orienting RIP catalytic domains towards their substrate, the sarcin-ricin loop, and may represent a scaffold for the generation of RIP-specific antidotes.

Methods

Protein expression and purification

The wild-type SLT-1 was expressed as an N-terminal His$_8$-tagged fusion construct in the *E. coli* strain JM101 (Agilent Technologies, Mississauga, ON), and purified as an AB$_5$ holotoxin on nickel-NTA resin (Sigma-Aldrich, St. Louis, MO). The A_1 chain was further purified from the holotoxin by first treating the purified AB$_5$ variants with the protease furin (New England BioLabs, Ipswich, MA) and reducing the disulfide bond with 10 mM DTT. The A_1 chain was then recovered on Nickel-NTA resin in the presence of a guanidine-HCl gradient to remove the untagged A_2 chain and B subunits, followed by a re-folding step in PBS.

SLT-1 mutant variants corresponding to those that exhibited a lack of interaction as defined by Y2H screens (R172A, R176A, R179A, R188A, R176/179/188A, V191A, F226A, L233A, and S235A) were created by multi-step PCR using Taq polymerase. The first PCR step consisted of two reactions: (1) using sense primer 2 and one of the mutagenic antisense primers and (2) using an antisense primer 2 and one of the mutagenic sense primers (Table S1). The second step consisted of a single reaction using the previous PCR reactions as templates with sense and antisense primer 2 which allowed for the amplification of the entire mutated sequence including the incorporation of unique restriction endonuclease sites. These mutant gene sequences were digested with the appropriate restriction endonuclease, cloned into the *NheI* and *XhoI* sites of the pECHE10a vector (Molecular Templates Inc., Austin, Texas). The resulting SLT-1 A_1 mutants were expressed and purified in the same manner as the wild-type SLT-1 A_1 chain. The above-mentioned SLT-1 A_1 variants used in subsequent experiments were judged by densitometry to be ≥85% pure (Figure S1).

Peptide Synthesis

Synthetic peptides corresponding to the final 17, 11, and 7 residues of the C-terminal domain of ribosomal proteins P1 and P2, a control peptide, as well as all alanine-containing peptide variants of the final 11 residue peptide SDDDMGFGLFD used to measure binding affinities were assembled using the 9-fluorenyl-methoxycarbonyl (Fmoc) method and Wang resin on a PS3 Peptide Synthesizer (Protein Technologies Inc., Tucson, AZ).

Fmoc-Asp(OMpe)-OH (NovaBiochem, Gibbstown, NJ) and Fmoc-Glu(OBt)-Ser(ΨMe, MePro)-OH (NovaBiochem) dipeptide were used to avoid aspartimide formation. The Fmoc protecting groups were removed using 20% Piperidine/0.1 M HOBt in DMF during synthesis. The N-α-amino group (N-terminus) of the peptides was labeled overnight with biotin using a ten-fold excess of biotin (Molecular Probes, Burlington, ON) in the presence of HCTU/HOBt and DIPEA. Biotinylated peptides were cleaved from their support using 3 mL of a TFA/TIS/EDT/Water (92.5:2.5:2.5:2.5%) mixture for 4 hrs at RT and purified by HPLC on a C$_{18}$ semi-preparative column using an acetonitrile gradient from 5% to 100% in 20 min. Peptide masses were then confirmed by mass spectrometry.

Yeast-2-Hybrid

Yeast ribosomes are rapidly inactivated by the A chains of RIPs such as SLT-1. A catalytically inactive form of the A_1 chain (CIA$_1$) was thus generated by introducing two mutations, namely E167A and R170A, within its catalytic region [11,15]. These two point mutations decrease the toxicity of the A_1 chain by 10,000 fold, therefore enabling yeast to grow during the expression of the A_1 chain of SLT-1 [35,36]. The yeast-2-hybrid (Y2H) technique [37], takes advantage of the GAL4 transcription factor that can be spliced into 2 complementary domains: a DNA binding domain (DNA-BD) and a transcription activation domain (AD). When pairs of bait/prey proteins interact, the GAL4 DNA-BD and transcription AD modules are assembled leading to the activation of survival genes (i.e.: HIS3).

The CIA$_1$ SLT-1 gene sequence was used as a template to construct several charged and hydrophobic single point mutations to alanine by multistep PCR using sets of mutagenic primers (Table S1). The CIA$_1$ as well as the point mutants were cloned into the bait vector pGBKT7 (Clontech, Mountain View, CA) between the *NdeI* and *BamHI* sites and expressed as fusions to the C-terminus of the GAL4 DNA-binding domain (GAL4 DNA-BD). The human gene sequence corresponding to RPLP2 (P2) was cloned into the prey vector pGADT7 (Clontech) and expressed as a fusion construct to the GAL4 activation domain (GAL4-AD). The pGBKT7-CIA$_1$ or one of the charged or hydrophobic point mutants were co-transformed with pGADT7-P2 separately into the yeast strain AH109 [*MATa, trp1-901, leu2-3, 112, ura3-52, his3-200, gal4Δ, gal80Δ, LYS2::GAL1_{UAS}-GAL1_{TATA}-HIS3, GAL2_{UAS}-GAL2_{TATA}-ADE2, URA3::MEL1_{UAS}-MEL1_{TATA}-lacZ, MEL1*] (Clontech).

Transformed cells were plated onto SD agar lacking Trp and Leu (−Trp/−Leu), to select for the presence of both plasmids, and incubated for 72 h at 30°C. A single colony of each transformation was then inoculated into SD −Trp/−Leu broth and shaken at 30°C overnight. Overnight cultures were centrifuged at 4000 rpm for five minutes followed by washing and equilibration to OD$_{600}$ of 1.0 in phosphate-buffered saline (PBS). Samples were then serially diluted 10-fold followed by spotting onto SD agar −Trp/−Leu to select for the presence of both plasmids and SD −Trp/−Leu/−His to select for an interaction between the two proteins. Plates were incubated for 72 h at 30°C.

Surface Plasmon Resonance Measurements

Toxin-peptide binding affinities were assessed by surface plasmon resonance (SPR) using a ProteOnXPR36 array biosensor [38] (Bio-Rad) in PBS buffer at 25°C. A neutravidin (NLC) sensor chip (Bio-Rad) was preconditioned by four pulses of 1 M NaCl, 50 mM NaOH, and 100 mM HCl, which was followed by an equilibration step in PBS. Biotinylated peptides corresponding to the final 17, 11, and 7 residues of the conserved peptide and a control peptide with no sequence homology were immobilized on the NLC sensor chip followed by a wash step in 1 M NaCl, to remove unbound peptide and a PBS wash to equilibrate the chip. Purified SLT-1 wild-type A_1 chain was tested at 10 different concentrations in quadruplicate beginning with 30 μM and diluting 2-fold in PBS to 60 nM, followed by an injection of PBS alone. The wild-type A_1 chain was exposed to all ribosomal peptide variants. Each concentration of SLT-1 A_1 chain variant was exposed to the NLC chip harboring the ribosomal peptides at a flow rate of 50 μl/min for 60 s with a dissociation time of 120 s. A 1 M NaCl wash step was performed for 18 s at a flow rate of 100 μl/min following each protein concentration. The chip was then equilibrated with PBS at a flow rate of 100 μl/min for 120 s.

Additional SPR experiments measuring (i) the affinities of charge and hydrophobic A_1 chain mutants, (ii) the contributions of electrostatic interactions by increasing salt concentrations, and (iii) to establish the anchor residues within the conserved ribosomal stalk peptide, were performed in triplicate using the same protocol described above. Only the final 11 residues of the conserved peptide KEESEESDDDMGFGLFD were used, as the removal of the six N-terminal residues had no effect on binding and therefore do not contribute significantly to the interaction (Figure 1).

Equilibrium data was used to calculate peptide binding constants to the A_1 chain in light of fast on- and off-rates. In all cases, the response data given from the ribosomal peptides was subtracted from the control peptide as well as PBS alone to eliminate non-specific binding. The resulting data for each concentration was averaged and plotted as percent response units (% RU) versus A_1 chain concentration using GraphPad Prism® (Figure 1). In the cases where binding affinities could not be calculated, due to a decrease or loss of binding, raw data was plotted as relative units (RU) compared to the same concentration (15 μM) of wild-type toxin A_1 chain in the same buffer (Figures 2B, 2C, 3B, and 3C).

Protein Synthesis Assay

T7-coupled transcription-translation (TnT) reticulocyte lysate assays (Promega, Madison, WI) were performed in order to determine if a decrease in ribosomal stalk binding altered the toxicity profiles of SLT-1 A_1 chain point mutants when compared to the wild-type toxin. TnT assays were performed, according to the manufacturer's instructions, using eight 10-fold serial dilutions of the wild-type or SLT-1 A_1 variants in PBS (starting with 1 μM). Protein synthesis was measured using a luciferase reporter plasmid (500 μg) through the incorporation of [^{35}S]-methionine (10 μCi; GE Healthcare, Piscataway, NJ) after a 90 min incubation at 30°C. Samples (20 μl) were loaded on a gradient (4–12%) SDS PAGE gel, and labeled protein bands revealed using a Storm® Phosphorimager (GE Healthcare). The addition of PBS alone was used as a control.

Results

The catalytic A_1 chain of SLT-1 binds to the conserved C-terminal ribosomal peptide

We have previously shown that the A_1 chain of SLT-1 binds to the conserved peptide KEESEESD(D/E)DMGFGLFD found at the C-terminus of ribosomal stalk proteins P0, P1 and P2 [15]. However, the molecular details of this interaction are unknown and could provide new evidence surrounding the mechanism of ribosome-inactivation by RIPs. The binding of the conserved C-terminal domain of the ribosomal proteins P1 and P2 to the wild-type SLT-1 A_1 chain was thus analyzed by surface plasmon resonance (SPR). Biotinylated peptides corresponding to the final 17, 11 and 7 residues of ribosomal stalk proteins P1 and P2 were immobilized on an sensor chip (NLC; Bio-Rad, Hercules, CA) and each peptide ligand was exposed to increasing concentrations of the wild-type SLT-1 A_1 chain. In light of the fast association and dissociation times observed for this binding event, equilibrium data from four independent experiments were instead used to calculate the dissociation constants of peptides binding to the A_1 chain. It was determined that the A_1 chain of SLT-1 interacted with the C-terminal 17 (KEESEESDDDMGFGLFD) and 11 residues (SDDDMGFGLFD) of P1 and P2 with comparable binding affinities of 11±5 μM and 13±2 μM respectively (Figure 1). The final 7 residues (MGFGLFD) did not bind to wild-type SLT-1 A_1 chain (data not shown), as previously reported by pull-down experiments [15].

Delineating key residues within the ribosomal peptide recognized by the SLT-1 A_1 chain

The SLT-1 A_1 chain docks within the ribosomal stalk by binding to specific residues within the peptide sequence SDDDMGFGLFD. A series of synthetic peptides containing alanine substitutions were generated by solid-phase peptide synthesis, to establish which residues within the peptide motif were important for its interaction with the A_1 chain of SLT-1. Specifically, alanine was introduced at each position within the

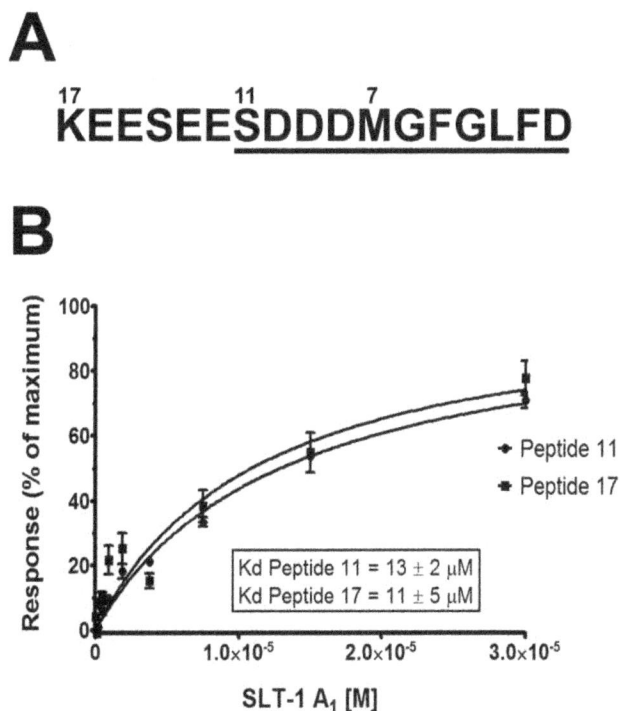

A

17   11   7
KEESEESDDDMGFGLFD

B

Kd Peptide 11 = 13 ± 2 μM
Kd Peptide 17 = 11 ± 5 μM

Figure 1. The A_1 chain of SLT-1 binds to a conserved C-terminal ribosomal peptide. (A) Amino acid sequence representing the 17-residue C-terminus common to ribosomal stalk proteins P1 and P2. The last 11 amino acids (underlined) delimit the shortest peptide element shown to interact with the A_1 chain [15]. (B) Relative surface plasmon resonance (SPR) signals for the A_1 chain of SLT-1 binding to immobilized, biotinylated monomeric synthetic peptides were plotted as a function of SLT-1 A_1 chain concentration. The calculated dissociation constants (K_d) suggest that both monomeric peptides have similar affinities for the A_1 chain. Each point on the curve represents the average relative SPR signals from experiments performed in quadruplicate.

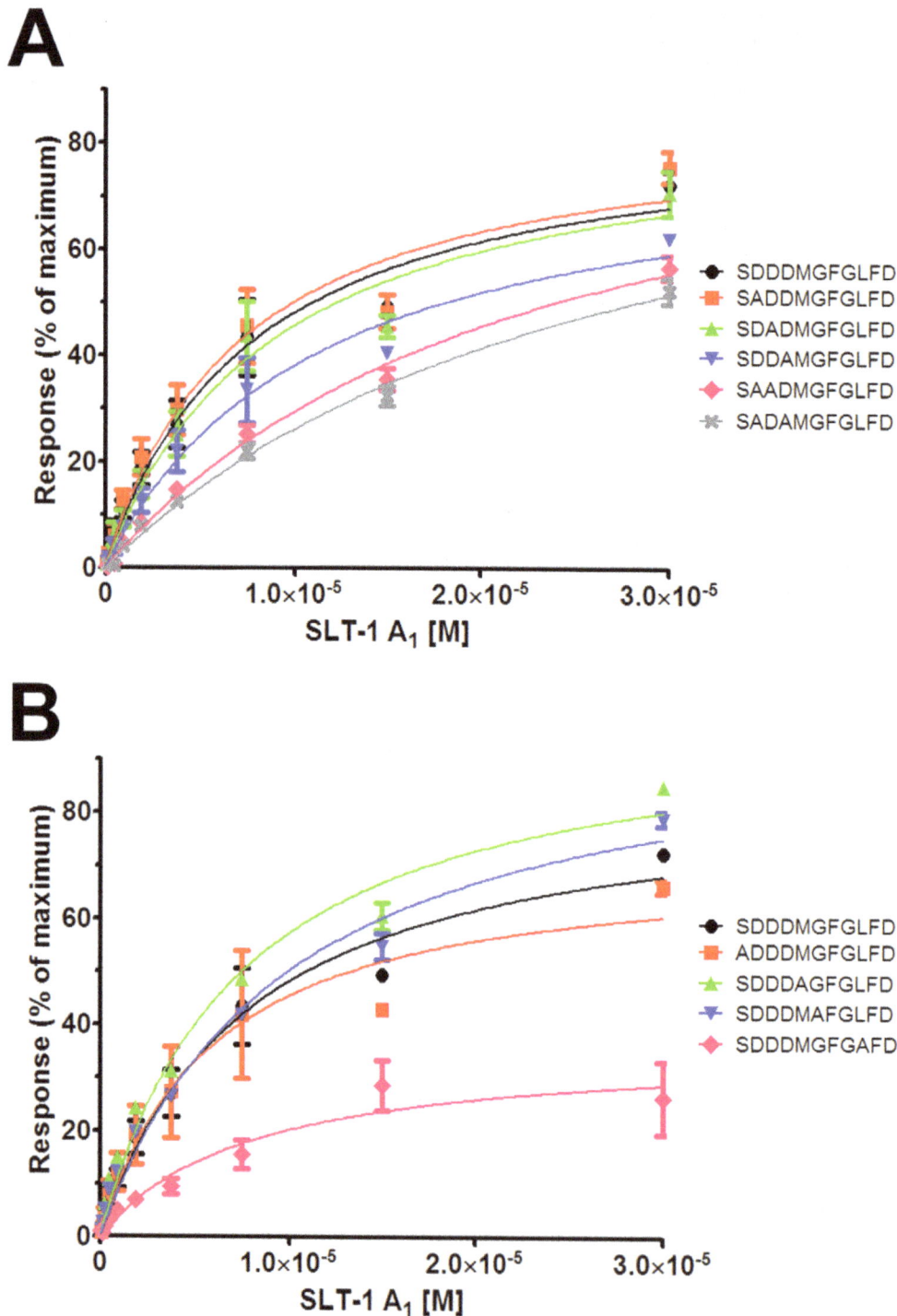

Figure 2. Surface plasmon resonance analysis of alanine-containing peptide variants of the conserved C-terminal ribosomal stalk peptide SDDDMGFGLFD confirms that the interaction with the A₁ chain of SLT-1 requires both electrostatic and hydrophobic contacts. The peptide sequence corresponding to the final 11 residues of the conserved C-terminal peptide (SDDDMGFGLFD) was substituted at each position for an alanine residue. Individual peptides corresponding to a substitution of charged (Panel A) or other residues (Panel B) were biotinylated and immobilized on an NLC SPR sensor chip. Each monomeric peptide was exposed to ten 2-fold serial dilutions of the A₁ chain of SLT-1 in triplicate and the responses were subtracted from buffer alone and a control peptide. The SPR responses for the single and double/triple alanine variants were graphed and compared to the control natural peptide. Amino acid substitutions that resulted in a peptide that lacked an interaction with the A₁ chain of SLT-1 could not be plotted. Calculated dissociation constants are reported in Table 1.

peptide SDDDMGFGLFD. These peptides were modified at their N-terminus with biotin, immobilized on an NLC sensor chip (BioRad), and exposed to graded concentrations of the SLT-1 A₁

chain. The SPR equilibrium data was collected, in triplicate, and responses were plotted as a function of A₁ chain concentration to calculate dissociation constants. It was determined that mutations

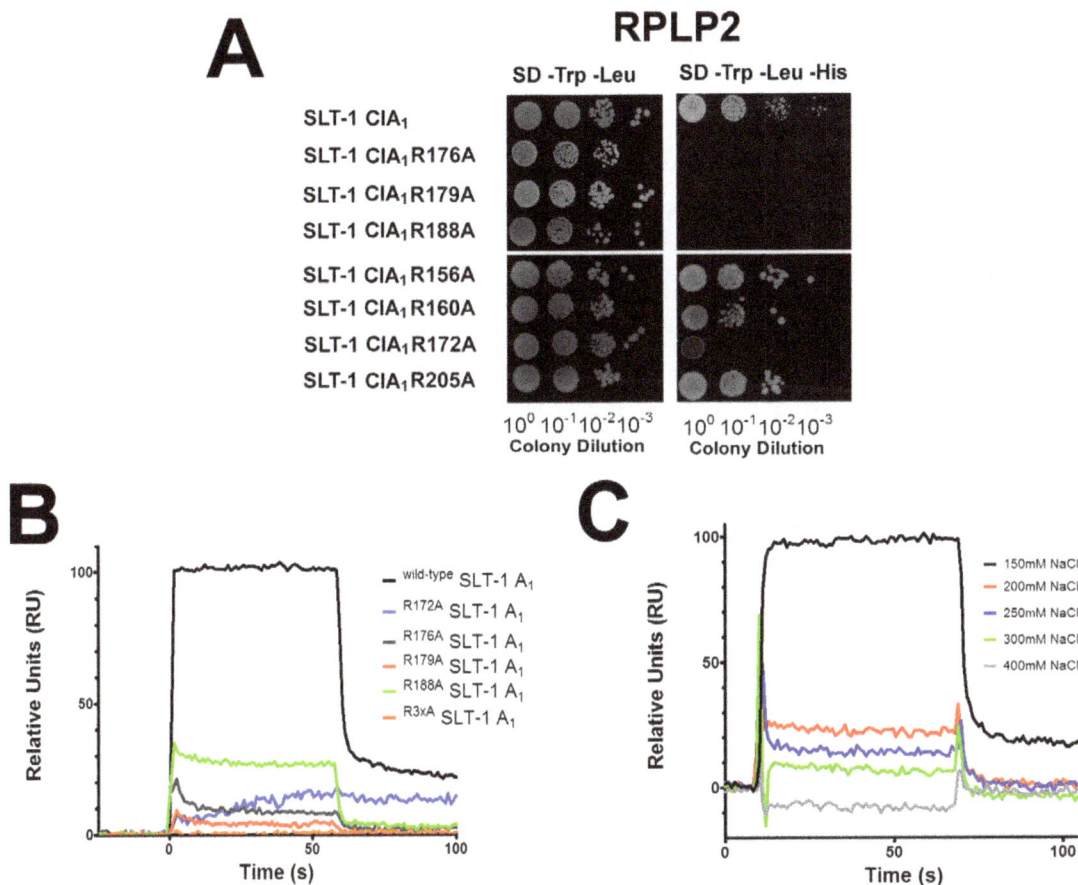

Figure 3. The A₁ chain of SLT-1 harbors a cationic surface composed of a cluster of arginine residues that interact with the ribosomal stalk protein P2 and the conserved C-terminal peptide. (A) A vector expressing a catalytically inactive variant of the SLT-1 A₁ domain (CIA₁) or one of the arginine-to-alanine point mutants as fusion partners with the GAL4 DNA-BD domain were co-transformed in the yeast strain AH109 with a vector expressing ribosomal protein P2 as a fusion construct to the GAL4-AD. The transformed yeast cells were plated on SD agar −Trp/−Leu. The resulting yeast colonies were grown overnight, and spotted (10 µl) as 10-fold serial dilutions onto SD medium lacking Trp and Leu to select for the presence of each plasmid followed by spotting on SD media lacking Trp, Leu, and His to select for interacting partners leading to colony growth. (B) SPR profiles illustrating the decrease in relative units for the arginine-to-alanine SLT-1 A₁ chain variants in relation to the wild-type A₁ chain, at a concentration of 15 µM, when presented to the immobilized peptide SDDDMGFGLFD. (C) Increasing salt concentrations led to a decrease or loss of binding of wild-type SLT-1 A₁ chain when exposed to the peptide SDDDMGFGLFD. SPR traces were plotted for the wild-type SLT-1 A₁ chain (15 µM) as a function of increasing salt concentrations.

of any of the five C-terminal residues to alanine (SDDDMG<u>FGLFD</u>) resulted in a decrease or complete loss of binding to the A₁ chain. (Table 1 and Figure 2). Interestingly, individual mutations of the N-terminal, negatively charged aspartyl residues to alanine did not show any significant effect on the affinity suggesting that such interaction with the A₁ chain may require the removal of more than one aspartic acid residues within the DDD tripeptide regardless of their position (Figure 2A). Therefore, the contribution of the charged aspartic acid residues within the peptide was further assessed through the generation of double and triple aspartate-to-alanine SLT-1 A₁ chain mutants. It was observed that mutations to any two aspartic acid residues to alanines, particularly involving residues at positions 3 and 4 caused at least a two-fold decrease in binding (Table 1 and Figure 2A). This comprehensive binding analysis confirmed that both the charge and hydrophobic elements (Figures 2A and B, respectively) of the conserved peptide SDDDMGFGLFD are required for the optimal binding of the A₁ chain of SLT-1 to the ribosomal stalk.

The binding between the SLT-1 A₁ chain and the conserved peptide SDDDMGFGLFD involves electrostatic interactions

In view of the negatively charged nature of the conserved ribosomal C-terminal peptide (aspartic acid [D] residues), we investigated whether a complementary positively charged surface was present on the A₁ chain of SLT-1 that would promote electrostatic interactions with this ribosomal peptide. Surface-exposed arginine residues were mutated to alanine (R-A) in order to assess their importance as anchoring residues for the conserved ribosomal stalk peptide. SLT-1 A₁ domain constructs with point mutations at arginine residues were generated by PCR using a catalytically-inactive SLT-1 A₁ chain (CIA₁) gene for their non-lethal expression in yeast. These R-A mutants were assessed for their ability to bind the full-length P2 protein in a yeast two-hybrid (Y2H) assay where co-transformation of each mutant with P2 were then serially diluted on medium which selected for an interaction between the two proteins (SD −Trp/−Leu/−His). Alanine mutations at residues R172, R176, R179, and R188 in the A₁

Table 1. Binding of synthetic, alanine-containing peptide variants of the ribosomal stalk peptide SDDDMGFGLFD to the A_1 chain of SLT-1 as measured by surface plasmon resonance.

Peptide	K_d (μM)
SDDDMGFGLFD	10.6±2.4
ADDDMGFGLFD	12.7±4.3
SADDMGFGLFD	9.75±2.4
SDADMGFGLFD	11.6±2.8
SDDAMGFGLFD	15.9±3.5
SDDDAGFGLFD	7.5±0.2
SDDDMAFGLFD	9.9±0.5
SDDDMGAGLFD	≥50
SDDDMGFALFD	≥50
SDDDMGFGAFD	41±11
SDDDMGFGLAD	≥50
SDDDMGFGLFA	≥50
SAADMGFGLFD	23.9±1.6
SADAMGFGLFD	28.2±1.9
SDAAMGFGLFD	≥50
SAAAMGFGLFD	≥50

Dissociation constants (K_d) were calculated from curves relating changes in relative surface plasmon resonance signal observed as a function of SLT-1 A_1 concentration (Figure 2). Each K_d value was derived from an average of three experiments. A K_d value of ≥50 μM was assigned to peptides displaying no measurable binding to the A_1 chain of SLT-1.

domain led to a loss of growth, implying that these arginines interact with the ribosomal peptide (Figure 3A). Alanine mutations at other arginine sites within the A_1 domain did not affect yeast survival suggesting that they are not involved in ribosomal docking (Figure 3A). Importantly, residues R172, R176, R179, and R188 are clustered on the surface of the A_1 chain suggesting the existence of a complementary positively charged surface within the A_1 chain.

To further define the effect of charged residues on the binding strength of A_1 chain to the peptide SDDDMGFGLFD, we expressed and purified recombinant SLT-1 A_1 chain variants, which showed a loss of interaction with the P2 protein by Y2H (R172A, R176A, R179A, R188A, and R176/179/188A), to directly assess their binding affinities towards this peptide by SPR. As expected, the binding affinity of the SLT-1 A_1 chain R-A mutants for the peptide (≥50 μM, no measurable binding) was lower than the affinity observed for the wild-type toxin (K_d = 13 μM). The SLT-1 R188A A_1 chain mutant was the only variant with a measurable binding constant (approximately 50 μM), while the other SLT-1 A_1 chain variants (R172A, R176A, R179A, and R176/179/188A) lacked detectable SPR responses even at high concentrations and were determined to be weaker than 50 μM (Figure 3B), In addition, when the concentration of salt in the running buffer (PBS) was increased, the interaction between the wild-type SLT-1 A_1 chain and the peptide SDDDMGFGLFD was decreased, further confirming the importance of electrostatic interactions (Figure 3C).

Hydrophobic interactions also contribute to binding the conserved C-terminal domain of ribosomal stalk proteins

The presence of a hydrophobic tetrapeptide motif (SDDDMG<u>FGLF</u>D) led us to hypothesize that a hydrophobic

patch on the A_1 chain of SLT-1 may also contribute to its interaction with the ribosomal stalk peptide. To test this hypothesis, several hydrophobic and serine residues namely, L185, V191, I224, S225, F226, L233, and S235 on the surface of the A_1 chain that are in the vicinity of the previously defined arginine cluster were mutated to alanine residues. These point mutants were generated by PCR using a catalytically-inactive SLT-1 A_1 chain (CIA$_1$) gene serving as a template in order to subsequently express them as non-toxic variants in yeast. Their binding to the ribosomal protein P2 was examined by Y2H (Figure 4A). Yeast transformations were serially diluted on SD-media lacking Trp, Leu, and His, which selects for the interaction between the two proteins. It was determined that residues V191, F226, L233, and S235 are important for the A_1 chain – ribosomal peptide interaction, while residues L185, I224 and S225 do not participate in these interactions (Figure 4A). The contribution of the hydrophobic and serine residues located on the surface of the A_1 chain was investigated by SPR to confirm the interactions of recombinant A_1 chain variants with the C-terminal peptide SDDDMGFGLFD. It was found that the V191A and L233A variants exhibited a large decrease in binding while variants harboring either a F226A or a S235A mutation displayed different binding kinetics and binding affinities of 3 and 5 μM, respectively. These affinities are comparable to that of wild type A_1 chain and may be attributed to the altered binding kinetics (Figure 4B and Figure S2).

Altering the cationic or hydrophobic surfaces of the catalytic domain of SLT-1 is sufficient to lower its ability to inhibit protein synthesis

We have shown by both Y2H and SPR that positively charged and hydrophobic residues located on the surface of the A_1 chain form cationic and hydrophobic patches allowing for its binding to ribosomal stalk proteins via the C-terminal peptide SDDDMGFGLFD. These interactions were measured as monomeric events, meaning that such interactions and affinity constants reflect complexes involving a single ribosomal peptide binding to a single SLT-1 A_1 chain. In reality, the peptide SDDDMGFGLFD is repeated five times within the context of the ribosomal stalk suggesting that even a dissociation constant in the low μM range could be significant due to the higher concentration of low affinity targets that could maintain the docking of the A_1 chain to the stalk.

To determine the effects of avidity on the binding and subsequent cytotoxicity of the A_1 chain of SLT-1, we further tested whether the charged or hydrophobic residues that are responsible for the interaction with the stalk peptide were also required for its full effect on protein synthesis inhibition. It was thus hypothesized that decreasing the affinity of the A_1 chain to the ribosomal stalk proteins by introducing arginine-to-alanine (R-A) or hydrophobic-to-alanine mutations, which have previously been shown to perturb the interaction, would result in a decrease in its ability to inhibit protein synthesis when compared to the wild-type A_1 chain. To test this hypothesis, ten-fold serial dilutions of each A_1 chain variant (R172A, R176A, R179A, R188A, R176/179/188A, V191A, F226A, L233A, and S235A) were added to a T7-coupled rabbit reticulocyte lysate transcription-translation system to quantify their ability to block the biosynthesis of [35S]-methionine-labeled luciferase. As predicted, A_1 chain mutants that exhibited a drastic decrease in affinity for the ribosomal stalk peptide (namely R172A, R176A, R179A, R188A, R176/179/188A, V191A, and L233A) displayed an increase in protein expression as compared to the wild-type toxin (Figure 5). These findings suggest that the binding of the A_1 chain to the ribosomal stalk, in the context of a functional eukaryotic ribosome, correlates

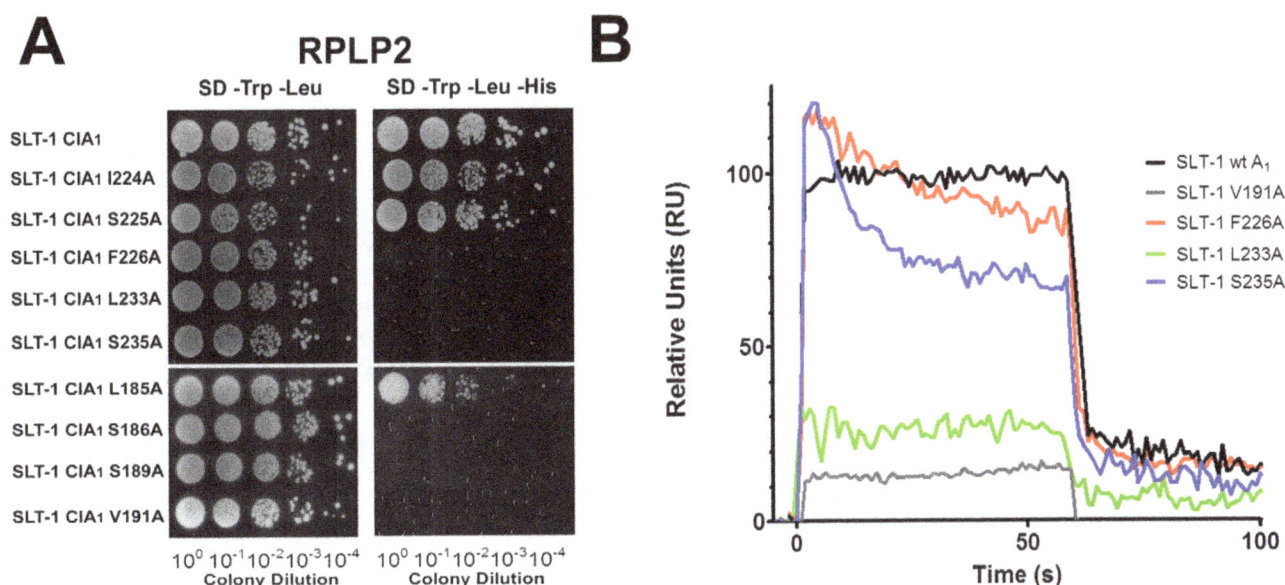

Figure 4. The interaction of the A_1 chain of SLT-1 with the ribosomal stalk protein P2 and the C-terminal peptide SDDDMGFGLFD also involves hydrophobic residues within the A_1 chain. (A) Bait vectors expressing either a catalytically inactive variant of the wild-type SLT-1 A_1 domain (CIA_1) or one of the hydrophobic mutants were co-transformed in the yeast strain AH109 with a prey vector expressing ribosomal protein P2. The transformed yeast cells were plated on SD agar −Trp/−Leu. The resulting yeast colonies were grown overnight, and spotted (10 μl) as 10-fold serial dilutions onto SD medium lacking Trp and Leu to select for the presence of each plasmid followed by spotting on SD media lacking Trp, Leu, and His to select for interacting partners. (B) SPR profiles (plotted at 15 μM) demonstrate that hydrophobic mutants F226A and S235A in the SLT-1 A_1 chain have a minor effect on the binding to the conserved peptide SDDDMGFGLFD and the SLT-1 V191A and L233A A_1 chain mutants cause a drastic decrease in binding. Experiments were performed in triplicate.

with its ability to inhibit protein synthesis with all SLT-1 A_1 chain mutants able to block protein synthesis at high concentrations. As expected, the S235A mutant did not significantly alter cytotoxicity profiles when compared to the wild type A_1 chain. The translation inhibition effects were also smaller for the R188A mutant, which can be attributed to the retention of approximately 30% of the binding compared to the wild-type toxin (Figures 3B and 5). Interestingly, the F226A mutation displayed an increase in catalytic activity as measured by TnT, an observation that is currently under investigation (Figure 5).

Discussion

The catalytic domains of ribosome-inactivating proteins (RIPs) such as SLT-1, ricin, and TCS [15,20,22,23,39] have been shown to interact with the eukaryotic ribosomal stalk, a heteropentamer composed of two heterodimers of the proteins RPLP1 (P1) and RPLP2 (P2) non-covalently associated to the protein RPP0 (P0) [27,28,29,30,31,32,33]. One known docking site for these catalytic chains on ribosomes is through their binding to a C-terminal peptide KEESEESDXDMGFGLFD (where X is either D or E) encoded by all three stalk proteins P0, P1, and P2. The pentavalent display of this peptide sequence within the stalk provides in theory up to five docking sites for RIP catalytic chains such as the A_1 chain of SLT-1 to orient themselves on ribosomes, an event that may facilitate ribosome depurination leading to the inhibition of protein synthesis and apoptosis. The presence of five copies of this peptide motif suggests that valency is a critical factor and that the peptide-RIP interaction may be transient and of low affinity. In the present study, we have established by surface plasmon resonance that the A_1 chain of SLT-1 binds to the conserved monomeric 17-residue long C-terminal peptide (KEE-SEESDDDMGFGLFD) and its truncated 11-residue form

(SDDDMGFGLFD) with comparable modest affinities (K_d of 11±5 μM and 13±2 μM, respectively) (Figure 1). Interestingly, the *on-* and *off-*rates could not be measured for the A_1 chain interacting with the monomeric peptide and such fast association and dissociation rates were expected since the binding of other RIPs namely saporin, restrictocin, and ricin to ribosome also show unusually fast kinetics and catalytic efficiencies [39,40,41,42]. The rapid *on-* and *off-*rates are plausible in light of the fact that the interaction between RIPs and the ribosome must be a transient event in order for one ricin molecule to depurinate ~2000 mammalian ribosomes/min and for a single SLT-1 A_1 toxin molecule being required to enter the cytosol to elicit cell death [43,44].

A previous study by our group had indicated that the C-terminal peptide inhibits the catalytic activity of the SLT-1 A_1 chain in blocking protein synthesis in an *in vitro* transcription-translation assay (McCluskey *et al.*, 2008). This finding suggested that such a peptide may serve as a template in designing a new class of inhibitors able to block the action of RIPs. We thus further analyzed the key components of this peptide involved in its interaction with SLT-1 A_1 chain. Specifically, synthetic peptide analogues of this sequence were used to define that the anionic tri-aspartyl sequence DDD and the four C-terminal residues FGLF within the C-terminal peptide (S<u>DDD</u>MG<u>FGLF</u>D) represent the two major anchors that interact with complementary cationic (Figure 2A) and hydrophobic (Figure 2B) surfaces on SLT-1 A_1 chain. As in the case of most ER-routed toxins, the A_1 chain of SLT-1 contains very few lysines (only 2 lysines located at its N-terminus) [45]. Therefore, arginine residues were projected to contribute to the creation of a positively charged surface on the A_1 chain that may interact with the anionic aspartic acid residues of SDDDMGFGLFD. Arginines at positions 172, 176, 179, and 188 were mutated to alanines leading to A_1 chain variants that were

[SLT-1 A₁] (M)

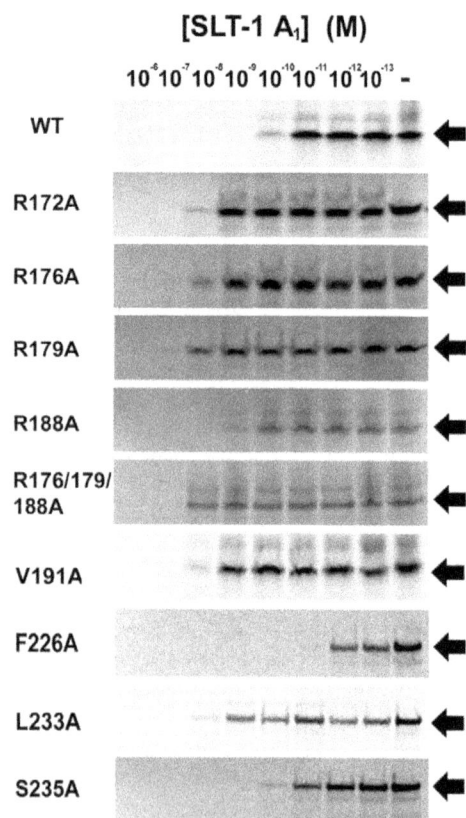

Figure 5. Arginine-to-alanine and hydrophobic variants of SLT-1 A₁ that bind weakly to the monomeric conserved C-terminal motif display altered ribosome-inactivating activities when compared to the wild-type A₁ chain. Eight ten-fold serial dilutions of the wild-type and each charge and hydrophobic A₁ chain variant was dispensed into an *in vitro* transcription and translation-coupled rabbit reticulocyte lysate system to monitor their ability to block protein synthesis (methods section). The level of *in vitro* protein synthesis was assessed by measuring the incorporation of [^{35}S]-methionine into the reporter protein luciferase during its synthesis. The expression of radiolabeled luciferase (arrow) was then resolved by SDS-PAGE and quantified using a phosphorimager. The addition of PBS alone (- lane) was used as a control.

unable to interact in a yeast-2-hybrid experiment with SDDDMGFGLFD presented in the context of P2 (Figure 3A). These same SLT-1 A₁ chain variants were also expressed and the resulting purified recombinant proteins were shown to have a decreased affinity or loss of binding to the conserved peptide SDDDMGFGLFD as determined by SPR (Figure 3B). In addition, increasing salt concentrations inversely correlated with binding of the wild-type A₁ chain to the peptide as measured by SPR (Figure 3C). These findings coincide with recent evidence that shows that electrostatic interactions are critical for the binding of ricin to whole ribosomes and for the targeting of restrictocin to the sarcin-ricin loop [39,40,46]. Moreover, several surface-exposed hydrophobic or serine residues on the A₁ chain in close proximity to the previously defined arginine cluster were mutated to alanine. Residues V191, F226, L233, and S235 were identified by Y2H as critical in maintaining the interaction with the conserved peptide SDDDMGFGLFD, the only docking site on the full length ribosomal protein P2 (Figure 4A). It was confirmed by SPR that residues V191, and L233 serve important roles in the interaction, whereas alanine mutations to residues F226 and S235 had no effect on binding (Figure 4B).

Structural data of a type I RIP, trichosanthin (TCS), and a type III RIP from maize supports our previous hypothesis and experimental data that charged and hydrophobic residues are important for the interaction with the C-terminus to occur [47,48]. When the structure of SLT-1 A₁ chain is compared to that of TCS it reveals the presence of a similar complementary surface-exposed groove theoretically predicted to involve residues R176, R179, and S235 of the A₁ chain of SLT-1 [47]. Two of the three residues (R176 and R179) have been identified in this study as being essential for the interaction with the C-terminal peptide. Recent evidence has also been published highlighting the importance of R176 on the depurination activity of SLT-1 [49]. In addition, we have observed that V191, R172, R188, and L233 also contribute to the binding of the A₁ chain to the conserved ribosomal peptide. However, the predicted A₁ chain-peptide interaction appears to be unique when compared to the known structure of TCS in complex with the conserved peptide [47].

The SLT-1 A₁ chain variants harboring R-A or hydrophobic mutations, which display striking decreased affinities for the ribosomal stalk peptide also show a reduction in their abilities to inhibit protein synthesis as measured by *in vitro* protein translation assays (Figure 5). These variants still retain their catalytic activity at high concentrations suggesting that a single substitution does not affect the overall three-dimensional fold of the A₁ chain (Figure 5). These results suggest that depurination may still occur at high concentrations due to rRNA binding or that the pentavalent presentation of the conserved ribosomal peptide in the context of the intact ribosomal stalk still favors A₁ chain docking to the stalk. Specifically, it may be sufficient for the A₁ chain of SLT-1, and other related RIPs cytotoxic domains, to bind directly to rRNA since the depurination of the sarcin-ricin loop has been observed in protein-depleted ribosomes, although the depurination rate is remarkably reduced [19].

This conserved ribosomal peptide represents a docking site for RIPs as it has been previously shown to interact with the catalytic domains of SLT-1, ricin, TCS, and maize RIP [15,20,47,48]. It is therefore highly likely that SLT-2, an SLT-1 homologue that is produced by clinically more severe bacterial strains [50], would possess similar docking residues as we have shown for SLT-1. Even though the two catalytic domains share only a 55% homology based on amino acid sequence [51], it was observed that most of the surface residues important for the SLT-1-ribosomal peptide interaction were conserved on the surface of SLT-2 when the two toxins were aligned based on their respective tertiary structures (Figure 6). Specifically, five of the six residues shown to be important for the SLT-1 interaction (R172, R176, R179, V191, and L233) were conserved on SLT-2 (R172, R176, R179, Y189, and L232) (Figure 6, panels B and C). The only missing A₁ chain interacting residue is arginine 188 of SLT-1, which is absent in SLT-2 (Figure 6, panels B and C). However, the A₁ chain variant R188A displayed a catalytic activity comparable to that of wild type A₁ chain (Figure 5).

This surface pocket and its affinity for the conserved C-terminal peptide may help explain why different RIP family members have the ability to bind the stalk conserved peptide and also why some RIPs such as pokeweed antiviral protein (PAP) do not require the stalk proteins for ribosome inactivation [52]. For example, the RIP saporin has a homologous tertiary structure to that of TCS, the A chains of SLT-1 and ricin, and was predicted to bind to the C-terminal tail of ribosomal stalk proteins [47]. Yet, we did not observe any measurable interactions between the C-terminal peptide SDDDMGFGLFD and the RIP saporin or the ribonuclease alpha-sarcin, which both target the sarcin-ricin loop, by either pull-down experiments, SPR, or by isothermal titration

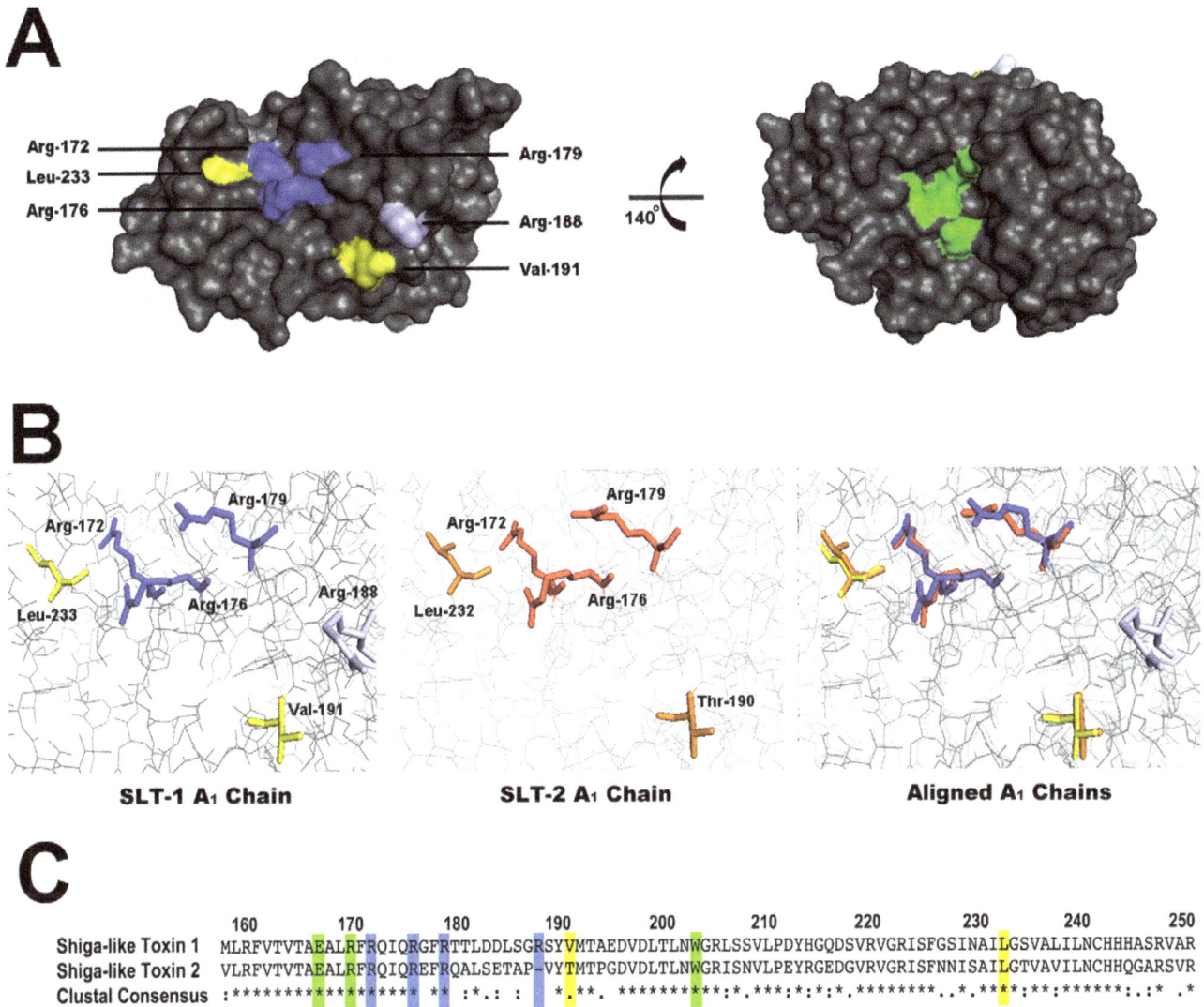

Figure 6. Primary and tertiary structural comparisons between SLT-1 and SLT-2 highlighting the conservation of important ribosomal stalk peptide contact sites. (A) *Left Panel* - Surface rendering of the SLT-1 A₁ chain (PDB# 1DM0) depicting the cationic (blue) and hydrophobic (yellow) residues essential for optimal binding to the conserved stalk peptide SDDDMGFGLFD as well as Arg-188 (light blue) which has a modest effect on peptide binding. *Right Panel* – Structure as shown in the left panel rotated by 140°, highlighting the catalytic residues in green. (B) Three-dimensional stick structures of SLT-1 (left panel), SLT-2 (PDB# 1R4P; middle panel), and the structural alignment of the two toxins (right panel). Cationic residues are labeled in blue and red, while hydrophobic residues are labeled in yellow and orange for SLT-1 and SLT-2 respectively. (C) Primary amino acid sequence alignment of SLT-1 and SLT-2 within residues 158 and 250. Catalytic residues are highlighted in green and cationic and hydrophobic residues in blue and yellow, respectively. Surface and stick renderings and alignments were performed using the The PyMOL Molecular Graphics System (Version 1.3, Schrödinger, LLC), whereas amino acid sequences were aligned using BioEdit software [59].

calorimetry (Figure S3). The A₁ chain of SLT-1 was confirmed to interact with the C-terminal peptide by both isothermal calorimetry and pull-down experiments (Figure S3). Although the *on* and *off* rates for the toxin-peptide interaction could not be measured by SPR, the HiCaM-tethered peptide was able to interact with the toxin most likely due to the excess (10:1) of the HiCaM-peptide used in the pull-down experiments in relation to the SLT-1 A₁ chain. Thus, the identified surface on the SLT-1 A₁ chain may therefore represent only one docking element that may not be generalized to the docking process of all RIPs.

Gastrointestinal disease outbreaks caused by infection with Shiga toxin-producing *E. coli* strains remain very common and suggest the need for post-infection therapeutic inhibitors [53,54,55]. The charged and hydrophobic surfaces mapped in this study provide a binding interface that is distinct from the catalytic site and are required for the full toxicity of SLT-1 *in vitro*. This structural information can be used for the generation of therapeutic inhibitors for both SLT-1 and SLT-2. For example, it has been shown that small molecule virtual-docking can be exploited to generate chemical compounds directed towards the catalytic domains of ricin, shiga toxins, and *Clostridium* botulinum neurotoxin [56,57,58]. These studies may provide molecular "leads" that may be more suited in terms of cell-permeability, affinities and *in vivo* stability than ribosomal stalk C-terminal peptide mimics as RIP antidotes.

In summary, the A₁ chain of SLT-1 interacts transiently ($K_d \sim 13$ μM with rapid *on* and *off* rates) with a short 11- amino acid conserved peptide located at the C-terminus of three ribosomal stalk proteins (P0, P1, and P2). The interaction involves both electrostatic and hydrophobic surfaces on both the A₁ chain of SLT-1 and the ribosomal peptide SDDDMGFGLFD. Conversely, a cluster of positively charged arginine residues within the A₁ chain in spatial proximity to a series of hydrophobic residues were defined as being critical for A₁ chain binding to the peptide SDDDMGFGLFD and to inhibit protein synthesis *in vitro*. Thus, the catalytic A chain of Shiga toxins and other related RIPs may have evolved to interact rapidly and with low affinity to proteins constituting the ribosomal stalk in order to properly orient themselves towards their substrate, the sarcin-ricin loop.

Supporting Information

Figure S1   SDS-PAGE gel showing the relative purities of recombinantly expressed and purified SLT-1 A₁ chain mutants. Each SLT-1 variant was expressed and purified as described in the methods section. Purified wild-type SLT-1 A₁ and point mutants were analyzed by SDS-PAGE and protein bands visualised by Coomassie blue staining. Numbers below each lane correspond to the purity of the major protein band (as a percentage) in relation to minor contaminating proteins as derived from densitometry measurements using the ImageJ software package.

Figure S2   The SLT-1 A₁ chain mutants F226A and S235A bind to the conserved C-terminal ribosomal peptide with similar affinity to wild-type SLT-1 A₁. Relative surface plasmon resonance (SPR) signals for the F226A and S235A SLT-1 A₁ chain variants binding to immobilized synthetic SDDDMGFGLFD peptide were plotted as a function of SLT-1 A₁ chain concentration. The calculated dissociation constants (K_d) suggest that the F226A and S235A mutations in the A₁ chain do not affect their affinity for the ribosomal stalk peptide SDDDMGFGLFD. Each point on the curve represents the average relative SPR signals from experiments performed in quadruplicate.

Figure S3   The conserved peptide SDDDMGFGLFD interacts with the A₁ chain of SLT-1 but may not be a generic contact site for all ribotoxins. (Top Panel) Phenyl Sepharose bound HiCaM [60] fusion constructs (100 μg) displaying the C-terminal 7 amino acids (Lanes 4–5), 11 amino acids (Lanes 6–7), 17 amino acids (Lanes 8–9) of P1 and P2, or HiCaM alone (Lanes 2–3) were incubated briefly with 10 μg of SLT-1 A₁ chain (Lane 1; Panel A), 20 μg saporin (Lane 1; Panel B), or 20 μg sarcin (Lane 1; Panel C) and separated on SDS-PAGE followed by Coomassie blue staining, as described previously [15]. The presence of a protein band in the thrombin cleavage (TC) lanes indicates an interaction and is only seen when the RIP A chain interacts with the final 11 or 17 residues of the conserved peptide. Legend: FT, column flow-through (unbound RIP); TC, thrombin-cleaved peptide. **(Lower Panel)** Synthetic peptide (starting with 500 μM) was titrated into a sample cell containing a 25 μM solution of degassed recombinant RIP and heat changes were measured using a VP-ITC (MicroCal Inc., Northampton, MA). The resulting calorimetric titration curves, minus the first injection of only 2 μl, were fitted using a single site binding model using the ORIGIN® software.

Table S1   Primers used to construct expression vectors. Restriction endonuclease sites are underlined and amino acid substitutions are in bold.

Acknowledgments

We are grateful to Dr. Christopher Bachran for supplying saporin used in the SPR experiments as well as Dr. Mohammed Yousef for SPR data analyses.

Author Contributions

Conceived and designed the experiments: AJM JG SSS. Performed the experiments: AJM NJ GC. Analyzed the data: AJM NJ JG. Contributed reagents/materials/analysis tools: EBP JG SSS. Wrote the paper: AJM JG.

References

1. Karmali MA (1989) Infection by verocytotoxin-producing Escherichia coli. Clin Microbiol Rev 2: 15–38.
2. Riley LW (1987) The epidemiologic, clinical, and microbiologic features of hemorrhagic colitis. Annu Rev Microbiol 41: 383–407.
3. Fraser ME, Chernaia MM, Kozlov YV, James MN (1994) Crystal structure of the holotoxin from Shigella dysenteriae at 2.5 A resolution. Nat Struct Biol 1: 59–64.
4. Kozlov YV, Chernaia MM, Fraser ME, James MN (1993) Purification and crystallization of Shiga toxin from Shigella dysenteriae. J Mol Biol 232: 704–706.
5. Lindberg AA, Brown JE, Stromberg N, Westling-Ryd M, Schultz JE, et al. (1987) Identification of the carbohydrate receptor for Shiga toxin produced by Shigella dysenteriae type 1. J Biol Chem 262: 1779–1785.
6. Lingwood CA, Law H, Richardson S, Petric M, Brunton JL, et al. (1987) Glycolipid binding of purified and recombinant Escherichia coli produced verotoxin in vitro. J Biol Chem 262: 8834–8839.
7. Jacewicz M, Clausen H, Nudelman E, Donohue-Rolfe A, Keusch GT (1986) Pathogenesis of shigella diarrhea. XI. Isolation of a shigella toxin-binding glycolipid from rabbit jejunum and HeLa cells and its identification as globotriaosylceramide. J Exp Med 163: 1391–1404.
8. Johannes L, Goud B (1998) Surfing on a retrograde wave: how does Shiga toxin reach the endoplasmic reticulum? Trends Cell Biol 8: 158–162.
9. Garred O, van Deurs B, Sandvig K (1995) Furin-induced cleavage and activation of Shiga toxin. J Biol Chem 270: 10817–10821.
10. Lea N, Lord JM, Roberts LM (1999) Proteolytic cleavage of the A subunit is essential for maximal cytotoxicity of Escherichia coli O157:H7 Shiga-like toxin-1. Microbiology 145(Pt 5): 999–1004.
11. LaPointe P, Wei X, Gariepy J (2005) A role for the protease-sensitive loop region of Shiga-like toxin 1 in the retrotranslocation of its A1 domain from the endoplasmic reticulum lumen. J Biol Chem 280: 23310–23318.
12. Brigotti M, Carnicelli D, Alvergna P, Mazzaracchio R, Sperti S, et al. (1997) The RNA-N-glycosidase activity of Shiga-like toxin I: kinetic parameters of the native and activated toxin. Toxicon 35: 1431–1437.
13. Endo Y, Mitsui K, Motizuki M, Tsurugi K (1987) The mechanism of action of ricin and related toxic lectins on eukaryotic ribosomes. The site and the characteristics of the modification in 28 S ribosomal RNA caused by the toxins. J Biol Chem 262: 5908–5912.
14. Endo Y, Tsurugi K, Yutsudo T, Takeda Y, Ogasawara T, et al. (1988) Site of action of a Vero toxin (VT2) from Escherichia coli O157:H7 and of Shiga toxin on eukaryotic ribosomes. RNA N-glycosidase activity of the toxins. Eur J Biochem 171: 45–50.
15. McCluskey AJ, Poon GM, Bolewska-Pedyczak E, Srikumar T, Jeram SM, et al. (2008) The catalytic subunit of shiga-like toxin 1 interacts with ribosomal stalk proteins and is inhibited by their conserved C-terminal domain. J Mol Biol 378: 375–386.
16. Moazed D, Robertson JM, Noller HF (1988) Interaction of elongation factors EF-G and EF-Tu with a conserved loop in 23S RNA. Nature 334: 362–364.
17. Hausner TP, Atmadja J, Nierhaus KH (1987) Evidence that the G2661 region of 23S rRNA is located at the ribosomal binding sites of both elongation factors. Biochimie 69: 911–923.
18. Montanaro L, Sperti S, Mattioli A, Testoni G, Stirpe F (1975) Inhibition by ricin of protein synthesis in vitro. Inhibition of the binding of elongation factor 2 and of adenosine diphosphate-ribosylated elongation factor 2 to ribosomes. Biochem J 146: 127–131.
19. Endo Y, Tsurugi K (1988) The RNA N-glycosidase activity of ricin A-chain. The characteristics of the enzymatic activity of ricin A-chain with ribosomes and with rRNA. J Biol Chem 263: 8735–8739.
20. Chiou JC, Li XP, Remacha M, Ballesta JP, Tumer NE (2008) The ribosomal stalk is required for ribosome binding, depurination of the rRNA and cytotoxicity of ricin A chain in Saccharomyces cerevisiae. Mol Microbiol 70: 1441–1452.

21. Li XP, Grela P, Krokowski D, Tchorzewski M, Tumer NE (2010) Pentameric organization of the ribosomal stalk accelerates recruitment of ricin a chain to the ribosome for depurination. J Biol Chem 285: 41463–41471.

22. Chan DS, Chu LO, Lee KM, Too PH, Ma KW, et al. (2007) Interaction between trichosanthin, a ribosome-inactivating protein, and the ribosomal stalk protein P2 by chemical shift perturbation and mutagenesis analyses. Nucleic Acids Res 35: 1660–1672.

23. Chan SH, Hung FS, Chan DS, Shaw PC (2001) Trichosanthin interacts with acidic ribosomal proteins P0 and P1 and mitotic checkpoint protein MAD2B. Eur J Biochem 268: 2107–2112.

24. Bargis-Surgey P, Lavergne JP, Gonzalo P, Vard C, Filhol-Cochet O, et al. (1999) Interaction of elongation factor eEF-2 with ribosomal P proteins. Eur J Biochem 262: 606–611.

25. Datta PP, Sharma MR, Qi L, Frank J, Agrawal RK (2005) Interaction of the G′ domain of elongation factor G and the C-terminal domain of ribosomal protein L7/L12 during translocation as revealed by cryo-EM. Mol Cell 20: 723–731.

26. Helgstrand M, Mandava CS, Mulder FA, Liljas A, Sanyal S, et al. (2007) The ribosomal stalk binds to translation factors IF2, EF-Tu, EF-G and RF3 via a conserved region of the L12 C-terminal domain. J Mol Biol 365: 468–479.

27. Grela P, Helgstrand M, Krokowski D, Boguszewska A, Svergun D, et al. (2007) Structural characterization of the ribosomal P1A-P2B protein dimer by small-angle X-ray scattering and NMR spectroscopy. Biochemistry 46: 1988–1998.

28. Tchorzewski M, Krokowski D, Boguszewska A, Liljas A, Grankowski N (2003) Structural characterization of yeast acidic ribosomal P proteins forming the P1A-P2B heterocomplex. Biochemistry 42: 3399–3408.

29. Naganuma T, Shiogama K, Uchiumi T (2007) The N-terminal regions of eukaryotic acidic phosphoproteins P1 and P2 are crucial for heterodimerization and assembly into the ribosomal GTPase-associated center. Genes Cells 12: 501–510.

30. Krokowski D, Boguszewska A, Abramczyk D, Liljas A, Tchorzewski M, et al. (2006) Yeast ribosomal P0 protein has two separate binding sites for P1/P2 proteins. Mol Microbiol 60: 386–400.

31. Perez-Fernandez J, Remacha M, Ballesta JP (2005) The acidic protein binding site is partially hidden in the free Saccharomyces cerevisiae ribosomal stalk protein P0. Biochemistry 44: 5532–5540.

32. Hagiya A, Naganuma T, Maki Y, Ohta J, Tohkairin Y, et al. (2005) A mode of assembly of P0, P1, and P2 proteins at the GTPase-associated center in animal ribosome: in vitro analyses with P0 truncation mutants. J Biol Chem 280: 39193–39199.

33. Gonzalo P, Lavergne JP, Reboud JP (2001) Pivotal role of the P1 N-terminal domain in the assembly of the mammalian ribosomal stalk and in the proteosynthetic activity. J Biol Chem 276: 19762–19769.

34. Santos C, Ballesta JP (2005) Characterization of the 26S rRNA-binding domain in Saccharomyces cerevisiae ribosomal stalk phosphoprotein P0. Mol Microbiol 58: 217–226.

35. Yamasaki S, Furutani M, Ito K, Igarashi K, Nishibuchi M, et al. (1991) Importance of arginine at position 170 of the A subunit of Vero toxin 1 produced by enterohemorrhagic Escherichia coli for toxin activity. Microb Pathog 11: 1–9.

36. Hovde CJ, Calderwood SB, Mekalanos JJ, Collier RJ (1988) Evidence that glutamic acid 167 is an active-site residue of Shiga-like toxin I. Proc Natl Acad Sci U S A 85: 2568–2572.

37. Fields S, Song O (1989) A novel genetic system to detect protein-protein interactions. Nature 340: 245–246.

38. Bravman T, Bronner V, Lavie K, Notcovich A, Papalia GA, et al. (2006) Exploring "one-shot" kinetics and small molecule analysis using the ProteOn XPR36 array biosensor. Anal Biochem 358: 281–288.

39. Li XP, Chiou JC, Remacha M, Ballesta JP, Tumer NE (2009) A two-step binding model proposed for the electrostatic interactions of ricin a chain with ribosomes. Biochemistry 48: 3853–3863.

40. Qin S, Zhou HX (2009) Dissection of the high rate constant for the binding of a ribotoxin to the ribosome. Proc Natl Acad Sci U S A 106: 6974–6979.

41. Korennykh AV, Correll CC, Piccirilli JA (2007) Evidence for the importance of electrostatics in the function of two distinct families of ribosome inactivating toxins. Rna 13: 1391–1396.

42. Sturm MB, Schramm VL (2009) Detecting ricin: sensitive luminescent assay for ricin A-chain ribosome depurination kinetics. Anal Chem 81: 2847–2853.

43. Endo Y, Tsurugi K (1988) The RNA N-glycosidase activity of ricin A-chain. Nucleic Acids Symp Ser. pp 139–142.

44. Tam PJ, Lingwood CA (2007) Membrane cytosolic translocation of verotoxin A1 subunit in target cells. Microbiology 153: 2700–2710.

45. Hazes B, Read RJ (1997) Accumulating evidence suggests that several AB-toxins subvert the endoplasmic reticulum-associated protein degradation pathway to enter target cells. Biochemistry 36: 11051–11054.

46. Korennykh AV, Piccirilli JA, Correll CC (2006) The electrostatic character of the ribosomal surface enables extraordinarily rapid target location by ribotoxins. Nat Struct Mol Biol 13: 436–443.

47. Too PH, Ma MK, Mak AN, Wong YT, Tung CK, et al. (2009) The C-terminal fragment of the ribosomal P protein complexed to trichosanthin reveals the interaction between the ribosome-inactivating protein and the ribosome. Nucleic Acids Res 37: 602–610.

48. Yang Y, Mak AN, Shaw PC, Sze KH (2010) Solution structure of an active mutant of maize ribosome-inactivating protein (MOD) and its interaction with the ribosomal stalk protein P2. J Mol Biol 395: 897–907.

49. Di R, Kyu E, Shete V, Saidasan H, Kahn PC, et al. (2011) Identification of amino acids critical for the cytotoxicity of Shiga toxin 1 and 2 in Saccharomyces cerevisiae. Toxicon 57: 525–539.

50. Noris M, Remuzzi G (2005) Hemolytic uremic syndrome. J Am Soc Nephrol 16: 1035–1050.

51. Jackson MP, Neill RJ, O'Brien AD, Holmes RK, Newland JW (1987) Nucleotide sequence analysis and comparison of the structural genes for Shiga-like toxin I and Shiga-like toxin II encoded by bacteriophages from Escherichia coli 933. J FEMS Microbiology Letters 44: 109–114.

52. Ayub MJ, Smulski CR, Ma KW, Levin MJ, Shaw PC, et al. (2008) The C-terminal end of P proteins mediates ribosome inactivation by trichosanthin but does not affect the pokeweed antiviral protein activity. Biochem Biophys Res Commun 369: 314–319.

53. Frank C, Werber D, Cramer JP, Askar M, Faber M, et al. (2011) Epidemic Profile of Shiga-Toxin-Producing Escherichia coli O104:H4 Outbreak in Germany - Preliminary Report. N Engl J Med.

54. Kupferschmidt K (2011) Infectious diseases. As E. coli outbreak recedes, new questions come to the fore. Science 333: 27.

55. Voelker R (2011) Rare E. coli strain races through Europe; high rate of kidney failure reported. Jama 306: 29.

56. Bai Y, Watt B, Wahome PG, Mantis NJ, Robertus JD (2010) Identification of new classes of ricin toxin inhibitors by virtual screening. Toxicon 56: 526–534.

57. Pang YP, Park JG, Wang S, Vummenthala A, Mishra RK, et al. (2011) Small-molecule inhibitor leads of ribosome-inactivating proteins developed using the doorstop approach. PLoS One 6: e17883.

58. Roxas-Duncan V, Enyedy I, Montgomery VA, Eccard VS, Carrington MA, et al. (2009) Identification and biochemical characterization of small-molecule inhibitors of Clostridium botulinum neurotoxin serotype A. Antimicrob Agents Chemother 53: 3478–3486.

59. Tchorzewski M, Boldyreff B, Issinger O, Grankowski N (2000) Analysis of the protein-protein interactions between the human acidic ribosomal P-proteins: evaluation by the two hybrid system. Int J Biochem Cell Biol 32: 737–746.

60. McCluskey AJ, Poon GM, Gariepy J (2007) A rapid and universal tandem-purification strategy for recombinant proteins. Protein Sci 16: 2726–2732.

17

Beta-Strand Interfaces of Non-Dimeric Protein Oligomers Are Characterized by Scattered Charged Residue Patterns

Giovanni Feverati[1], Mounia Achoch[1,2], Jihad Zrimi[1,2], Laurent Vuillon[3], Claire Lesieur[1,4]*

1 Université de Savoie, Annecy le Vieux Cedex, France, **2** Laboratoire de Chimie Bioorganique et Macromoléculaire (LCBM), Faculté des Sciences et Techniques-Guéliz, Université Cadi Ayyad, Marrakech, Maroc, **3** LAMA, Université de Savoie, Le Bourget du Lac, France, **4** AGIM, Université Joseph Fourier, Archamps, France

Abstract

Protein oligomers are formed either permanently, transiently or even by default. The protein chains are associated through intermolecular interactions constituting the protein interface. The protein interfaces of 40 soluble protein oligomers of stœchiometries above two are investigated using a quantitative and qualitative methodology, which analyzes the x-ray structures of the protein oligomers and considers their interfaces as interaction networks. The protein oligomers of the dataset share the same geometry of interface, made by the association of two individual β-strands (β-interfaces), but are otherwise unrelated. The results show that the β-interfaces are made of two interdigitated interaction networks. One of them involves interactions between main chain atoms (backbone network) while the other involves interactions between side chain and backbone atoms or between only side chain atoms (side chain network). Each one has its own characteristics which can be associated to a distinct role. The secondary structure of the β-interfaces is implemented through the backbone networks which are enriched with the hydrophobic amino acids favored in intramolecular β-sheets (MCWIV). The intermolecular specificity is provided by the side chain networks via positioning different types of charged residues at the extremities (arginine) and in the middle (glutamic acid and histidine) of the interface. Such charge distribution helps discriminating between sequences of intermolecular β-strands, of intramolecular β-strands and of β-strands forming β-amyloid fibers. This might open new venues for drug designs and predictive tool developments. Moreover, the β-strands of the cholera toxin B subunit interface, when produced individually as synthetic peptides, are capable of inhibiting the assembly of the toxin into pentamers. Thus, their sequences contain the features necessary for a β-interface formation. Such β-strands could be considered as 'assemblons', independent associating units, by homology to the foldons (independent folding unit). Such property would be extremely valuable in term of assembly inhibitory drug development.

Editor: F. Gisou van der Goot, Ecole Polytechnique Federale de Lausanne, Switzerland

Funding: This work was supported by The system complex Rhone Alpes IXXI (5000 euros). Supported by the University of Savoie (4000 euros)The funders had no role in study design, data collection and analysis, decision to publish, or preparation of the manuscript.

Competing Interests: The authors have declared that no competing interests exist.

* E-mail: lesieur@lapp.in2p3.fr

Introduction

Most proteins are made of more than one polypeptide chain to carry out their biological function [1,2]. They are referred to as protein oligomers and have what is called a quaternary structure. In addition, numerous monomeric proteins associate transiently in binary or in higher stœchiometries (number of chains associated in a protein oligomer) during their life span. The formation of protein oligomer, known as protein assembly, is also a common reaction used by pathogens to produce killing "machineries". One good example is the pore forming toxins produced by pathogenic bacteria such as *Bacillus anthracis*, *Staphylococcus aurus* and *Aeromonas hydrohilae*. This mechanism is also responsible for protein misfolding diseases through the production of "amyloid" oligomers and fibers (e.g. Alzheimer, Parkinson, Creuzfeld Jacob) [3,4,5,6,7,8,9].

Intermolecular contacts (contacts between chains) exist only in multiple chain proteins. These contacts constitute what is called the protein interface and are formed through particular interaction patterns. Unfortunately, despite extensive analyses, the identification of the patterns responsible for permanent contacts remains difficult.

This is due to the broad diversity of the contact solutions [10,11]. The rationalization of known patterns of protein interfaces is also far from accomplished.

The patterns result from geometrical and chemical complementarities between the two partners. Numerous reports on protein interfaces, based on theoretical and experimental approaches, allow understanding some of the general rules underlying intermolecular contacts (for reviews see [2,10,12]).

First, one needs to distinguish within the interface, the amino acids involved in intermolecular contacts, the so called "hot spots", from those who are not. Several programs can identify theoretical hot spot residues at interfaces based on: (i) distance cuts-off combined or not with some chemical selection, (ii) solvent accessible surfaces, (iii) geometrical selection (e.g. Voronoi cells) or (iv) evolutionary conserved residues [2,13,14,15]. All require the atomic structure of the protein oligomer. Experimental evidences have also confirmed the presence of hot spot residues in interfaces (for review see [2]). One beautiful example is the selective effect of the mutation of only some of the residues of the interface on the protein assembly of the heptameric co-chaperone cpn10 [16].

Second, the interaction patterns of protein interfaces are related to their secondary and tertiary structures as it was initially described by Sir Francis Crick for α-coiled interfaces with the discovery of the heptaed sequences [17,18,19,20,21,22,23,24]. The importance of the structure of the interface in the implementation of a particular motif has been now generalized with high-throughput interaction discovery [25,26].

Third, at the amino acid level, a versatile solution has to be sought rather than a specific one. In fact, even for identical secondary structures, the geometry (triple helix, α-coiled, β-sandwich...) and/ or the symmetry of the protein interfaces also affect the patterns at the amino acid levels [11,17,18,20,27,28,29,30].

For a geometry of interface made of interacting β-strands (β-interfaces), dimers are the main stœchiometry studied, particularly when considering dataset analysis [21,31,32,33,34].

Here, we report the analysis of the β-interfaces of 40 soluble protein oligomers whose stoechiometries are from trimers to octamers. We used our tailor made program Gemini to select hot spots and to produce an interaction network -or a graph- of the subset of interactions that composes an interface [15]. Gemini quantitative and qualitative analyses reveal relatively long β-interfaces enriched with charged residues scattered within the interface. More precisely, arginine residues are preferred at N- and C- terminal extremities whereas histidine and glutamic acid residues are more frequent in the middle of the interfaces. Such a broad charge distribution has never been observed previously in dimeric β-interfaces or in intramolecular β-interactions.

Materials and Methods

Interfaces by Gemini

The computer programs (Gemini) relevant to the present paper have been described previously [15]. In summary, Gemini characterizes an interface as a subset of amino acids in interaction, or "hot spots". They emerge after a purely geometrical analysis of the 3D atomic structure of the protein, well described in the indicated publication. Gemini is equipped with an effective tool (GeminiGraph) that represents interfaces by (bipartite) graphs (Fig. 1). Throughout the paper, the graphs -and so the interfaces- are also referred to as 'interaction networks' or simply as 'networks'. Briefly, the two segments S1 and S2, of an interface are represented by two parallel rows. The interacting amino acids selected by Gemini are indicated by 'X' and the non interacting ones by dots '.' (Fig. 1C). The 'X' amino acids are the hot spots of the interface. The interactions (I) are illustrated by lines connecting two 'X'. The version used here includes the name of the amino acids at positions 'X', following the one-letter code. In few cases, the β-interface is so intimately close to a different interface geometry that Gemini keeps them together in the same interface region (see Table S2 and Dataset S1). In the present work only the β-interface part has been used; the corresponding graphs have therefore been manually annotated (supplementary material).

A supplementary feature has been added to Gemini, which describes the interfaces as two interaction sub-networks. One of them only includes interactions between backbone atoms (BB sub-network), the other interactions with at least one side chain atom (SC sub-network). The interactions of the BB sub-network (I_{BB}) are represented with dashed lines whereas those of the SC sub-network (I_{SC}) are represented with solid lines. X_{SC} and X_{BB} are the side chain and backbone hot spots, respectively.

Circular proteins

This is also a new addition to Gemini especially relevant to the present work. The goal of this part of the code is to recognize

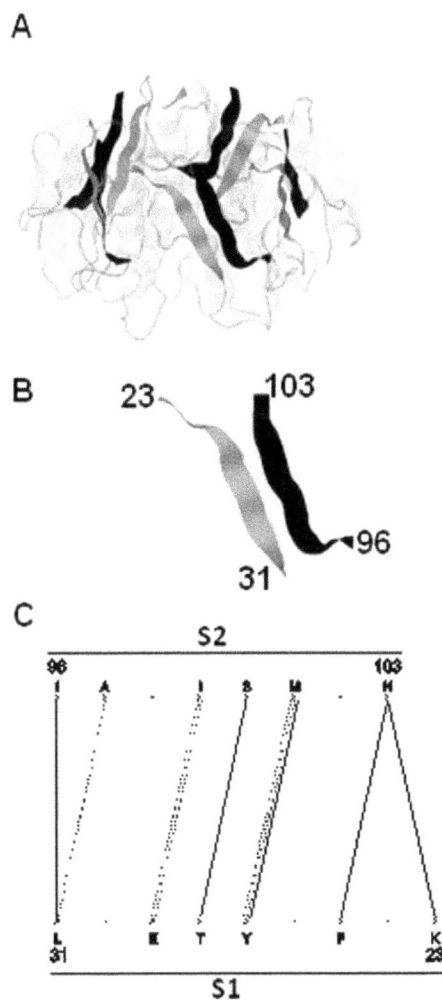

Figure 1. Example of one β-interface geometry. A. The x-ray structure of the whole cholera toxin B pentamer (CtxB$_5$) is shown in strands (PDB code: 1EEI) [66]. The two strands of the β-interface are highlighted in black and grey in ribbons. The image has been generated using Rasmol. **B.** The β-interface is made of the association of the segment composed of amino acids 23 to 31 on one chain (segment 1) and of the segment composed of the amino acids 96 to 103 on the adjacent chain (segment 2). **C.** Gemini graph of the CtxB β-interface. S1 and S2 stand for segments 1 and 2.

circular homo-oligomers (oligomers made of the same protein chain). The program classifies proteins into two classes: circular homo-oligomers and the rest that can contain hetero-oligomers and non circular homo-oligomers. For short, we call it non-circular (NC). The input information is the three-dimensional structure of PDB. No other database or author's annotation is used. The first step in the classification recognizes as NC those proteins whose chains are composed of different numbers of residues. Actually, given that in PDB files there can be additional or missing residues, an error of 25% is tolerated on the differences in the number of residues. The remaining proteins are therefore good candidates to be homo-oligomeric. In a second step, the program tries to find the first amino acid common to all the subunits. From it, five other common amino acids must be found, located at 15%, 30% and so on, of the sequence. If this step fails, the protein is NC. If it succeeds, the protein is very likely to be a homo-oligomer so a third step is needed to evaluate the spatial organization of the subunits. This is simply done by comparing the distances of the Cα of the six common amino acids already found. If the protein is a circular n-oligomer, there must be n identical distances (a tolerance of 5 Angstrom is used) otherwise the protein is NC. This algorithm is effective in finding circular homo-oligomers but is not enough to fully discriminate within the NC class. There are some false negatives, namely proteins that are circular homo-oligomers but are recognized as NC. This has the only effect of slightly reducing the size of our dataset. We did not observe false positives.

Cytoscape (http://www.cytoscape.org/)

It is an open source bioinformatics software platform for visualizing molecular interaction networks and biological pathways and integrating these networks with annotations, gene expression profiles and other data. Although Cytoscape was originally designed for biological research, now it is a general platform for complex network analysis and visualization. Among the several types of interaction data supported, the format SIF (simple interaction format) was used for the present paper.

RING (Residue Interaction Network Generator)

It is a web server with software for transforming a protein structure (in PDB format) into a network of interactions. Nodes represent single amino acids in the protein structure, while the edges represent the non-covalent bonding interactions that exist between them [35,36,37]. The interaction network and the edge attributes are stored in files with the SIF format. These files can then be easily loaded into CYTOSCAPE to visualize and manipulate the network [35,36,37]. In the present study, RING and CYTOSCAPE were used to produce and visualize the network of hydrogen bonds for the proteins of the dataset.

Statistics

Median, quartile- The median is the value that splits the dataset into two equally populated subsets (above and below the median). For example, for 40 cases and a median of 180 amino acids in size, there are 50% of the cases with a length above 180 and 50% with a length below 180 amino acids. The quartile is the value at which the dataset it divided into four parts, equally populated with the 25% of the samples. The lower separation point is the first quartile, the middle one is the median and the higher is the third quartile.

Global and Local propensy

The ratio between the amino acid frequency in a domain and the amino acid frequency in a database is called "*global propensy*". If the global propensy is above 1, the amino acid is "preferred" in the domain and if the propensy is below 1, the amino acid is "disfavored" in the domain. The "*local propensy*" is defined by the ratio between the amino acid frequency in a particular position (e.g. corner) of a sub-domain (e.g. β-interface) and its frequency in all the other positions in the sub-domain. A local propensy above 1 means the amino acid is preferred in that position than anywhere else in the sub-domain [38]. On the contrary, a local propensy below 1 means the amino acid is disfavored in that position compared to elsewhere in the sub-domain. The corner positions are the amino acids located at the four outer positions on a segment: two outer positions on each side of the segment. So each segment has four amino acids positioned on corners and two outer interactions. The central positions are anywhere else on the segment.

Secondary-structure prediction

GOR IV software was used to perform the secondary structure prediction of the segments of the proteins of the dataset. The secondary structure of each segment of the dataset was predicted (40×2 cases) considering all the wild-type amino acids of the segments and not only the -X-. Then, a residue was mutated and the secondary structure prediction was performed again. When a mutation affected the wild-type original secondary structure prediction, the mutated residue was considered important for the secondary structure of the segment. Hydrophobic residues of the BB or of the SC sub-networks, centrally located or at corners were mutated to charged residues (e.g. K, D, R, E, H). If one of the mutations affected the secondary structure prediction, mutation to other charged amino acids was not essayed. Polar and charged residues of the BB sub-networks centrally located in the full network, were also mutated to either polar or hydrophobic residues.

Probability

Let's call p_c the probability to find in an interface, a charged amino acid. We now evaluate p_{cc}, the probability to have at least one charged amino acid in (at least) one of the corners. This is evaluated as follows:

$$p_{cc} = 4 * p_c * (1 - p_c)^3 + 6 * p_c^2 * (1 - p_c)^2$$
$$+ 4 * p_c^3 * (1 - p_c) + p_c^4 = 1 - (1 - p_c)^4$$

where each addendum is respectively the probability to find: a charged amino acid in one corner only, a charged amino acid in two corners, a charger amino acid in three corners, a charged amino acid in all corners. Everything holds true for the corner probability within one of the sub-networks, provided p_c is the corresponding probability.

Reagents and buffers

Cholera toxin B pentamer (CtxB$_5$) and all other chemicals were obtained from Sigma. McIlvaine buffer (0.2 M disodium hydrogen phosphate, 0.1 M citric acid, pH 7.0), PBS and 0.1 M KCl/HCl at pH 1.0 were used. All buffers were filtered through sterile 0.22 μm filter before use. Synthetic peptides were ordered from proteogenix (www.proteogenix.fr).

SDS-PAGE analysis

SDS-PAGE (15% or 12%) were performed with a Bio-Rad mini-Protean 3 system using the Laemli method [39]. The gels were stained with Coomassie blue. 1 μg of sample was loaded on each lane of the gel.

Reassembly of CtxB into native pentamer

The conditions used for reassembly were adapted from elsewhere [40]. Briefly, native CxtB$_5$ was acidified in 0.1 M HCl/KCl at pH 1.0 for 15 min at a final toxin concentration of 86 μM, to induce the toxin dissociation into monomers (MW~11 600 kDa). The toxin was subsequently diluted to a final concentration of 8,6 μM, in McIlVaine buffers at pH 7.0 to promote reassembly. The samples were incubated for 15 min at 23°C before analysis by SDS-PAGE. The reassembly into native CtxB pentamer was inferred from SDS-PAGE analyses since CtxB$_5$ is stable in SDS-containing buffers and migrates in a gel, run on ice,with an apparent molecular weight characteristic of the B-subunit pentamer (MW~55 000 kDa). Only the native pentamer is SDS-resistant. The CtxB concentration for all experiments refers to the monomeric concentration.

Reassembly of CtxB in presence of peptides

The toxin reassembly was measured in presence of synthetic peptides whose sequences correspond to the toxin β-interfaces sequences (segments 1 and 2). The peptides were added in the neutralizing buffer at a molar ratio peptide to protein of 20. The reassembly conditions were identical to the one used for the toxin alone.

Results

The primary goal of the analysis is to seek protein interface features within a dataset of protein oligomers sharing only a common geometry of interfaces. This is inspired by the success obtained for α-coiled interfaces [17,18,19]. The second objective is to see if the features can be rationalized in term of assembly mechanisms. The interfaces are analyzed using our tailor made program Gemini, which considers interfaces as interaction networks and allows both quantitative and qualitative studies [15].

The dataset

The dataset was built by screening the Protein DataBank (PDB) [41]. First, cyclic protein oligomers were selected so all the cases had identical symmetry (circular, C_n). To this purpose a program called "Circular" (materials and methods) was made. In total 502 protein oligomers were identified with stœchiometries from 3 (trimer) to 8 (octamer) (Table 1). Stœchiometries above 8 contained too few cases to be considered. Second, the secondary structure of the protein interface was chosen as two interacting β-strands at least 4 amino acids apart on the individual chain. The two interacting β-strands had to be different in their amino acid sequences (Fig. 1). Each strand is called a segment. Segment 1 (S1) appears first (N-terminal side) followed by segment 2 (S2) (C-terminal side) on the primary sequence. This geometry is referred to as a β-interface throughout the paper. Third, dimers, hetero-oligomers, transient oligomers, viral and membrane proteins were discarded from the dataset as their interfaces are likely to be differently programmed. After selection, the dataset was made of 40 protein interfaces but the list is non exhaustive.

Properties of the whole chain proteins of the dataset

The protein oligomers are produced by organisms from the three super-kingdoms of life with 2% of archea, 75% of bacteria and 23% of eukaryotes (Table S1). For comparison, there are 8%, 54% and 38% of archea, bacteria and eukaryotic protein oligomers for the stœchiometries from 3 to 8 in the PDB. The atomic structures (PDB) of the protein oligomers of the dataset are shown in figure 2 to illustrate the diversity of their quaternary, tertiary (folds) and secondary structures. The folds are also represented by the SCOP superfamily codes in Table S1 [42].

The secondary structure content of the whole chains is also extensively variable with on average on the dataset 30 ± 20; 40 ± 20 and $30\pm10\%$ of α-, β- and random coiled structures. This is illustrated in figure 3 with the structures of the chaperone 1Q3S and of the oxidoreductase 1PVN which have a high content of α-structures (60 and 46%, respectively).

The distribution of the whole chain lengths is broad as can be seen on the histogram on figure 4. The median length is 160 amino acids for an interquartile of 148 amino acids. The average length is 203 ± 127 amino acids, value slightly smaller than the average length of monomeric proteins (~300 amino acids) (Tables S1) [1]. This might be due to the measurement of the protein lengths from the PDB sequences which contain gaps due to crystallization or diffraction issues.

The circular trimers are the most represented (67%) against an average of $7\pm4\%$ for the other stœchiometries (Table 1). The abundance of trimers might be related to the fact that the PDB over-represents low stœchiometries, dimer and trimer in particular, owing to the difficulties in crystallization. The β-interface geometry represents on average 8% of the circular protein oligomers (40/502) in good agreement with a previous measurement in dimers [21].

In summary, the protein oligomers of the dataset are produced by diverse organisms and cover a variety of functions, folds, amino acid lengths and stoichiometries (Table S1). Not surprisingly, the alignment of their amino acid sequences has no worthy of notice homology (not shown). Hence the dataset is characterized by a large heterogeneity.

Global beta interface characteristics

Gemini's interaction networks (or graphs) of the β-interfaces are in Dataset S1. The length and the number of hot spots (-X-) of each β-interface, are determined using the Gemini graphs (materials and methods). Both are counted considering the two segments, S1 and S2, of the interface (Table S2). The statistics on hot spots, interface length and number of interactions are summarized in Table 2. The average length and number of hot spots for the segment S1 or for the segment S2, are similar, indicative of indistinguishable characteristics of the two β-strands of the β-interfaces. The number of interactions between two hot spots (X) involved in the β-interfaces (I_β) is also provided by Gemini (Table S2 and Table 2).

The length, the hot spot number and the interaction number (I_β) have medians and interquartile ranges fairly similar to their respective average and standard deviation values indicative of a relative homogeneity of these features throughout the dataset (Table 2). Yet there is no visible common topological feature within the graphs of the β-interfaces or any specific chemical composition compared to the whole chains (Table 3). A slightly different chemical composition appears when the hot spots are considered instead of all the amino acids of the two segments S1 and S2 (Table 3). No particular sequence homology was observed upon alignments of the S1 and S2 segments (not shown).

It was then assumed that common features might be somehow diluted in a 'background' noise.

As the backbone atoms are identical for the twenty amino acids, it was possible that counting them in the chemical properties of the β-interfaces 'hid' some chemical specificity only distinguishable on the side chain atoms. Likewise, only the backbone atoms might

Table 1. Circular protein oligomers containing a β-interface.

Category	Trimer	Tetramer	Pentamer	Hexamer	Heptamer	Octamer	Total
Circular oligomers	339	39	54	43	22	5	502
β-interface	13	6	11	4	4	2	40
Circular oligomers (%)	67 (339/502)	8	11	9	4	1	100

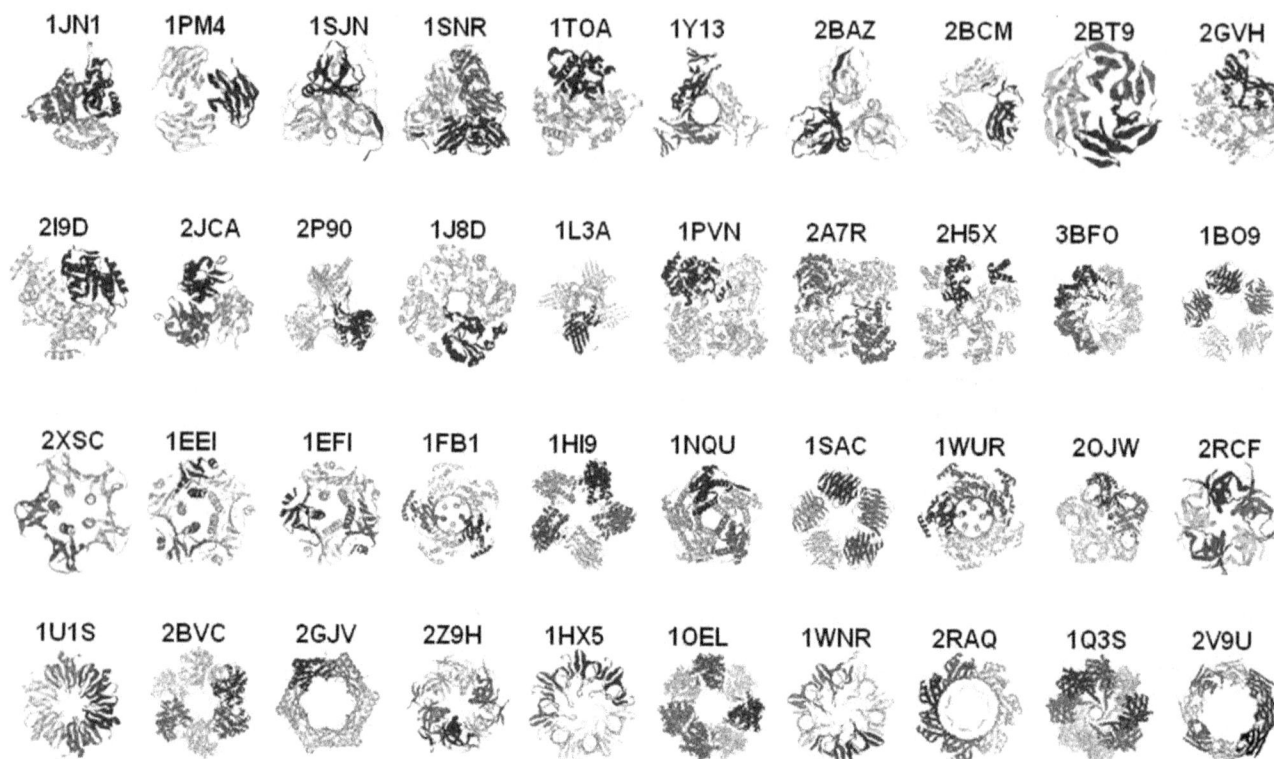

Figure 2. x-ray structures of the protein oligomers of the dataset. The respective PDB codes are indicated above the structures. The figure was made using RasMol. Each chain is shown in a different color.

carry topological information. Moreover, previous studies on protein interfaces had indicated the importance of distinguishing main chain (backbone atoms) contacts from side chain contacts [2,43,44].

Accordingly, the graphs of the β-interfaces were partitioned in two sub-graphs, one made of the backbone interactions (one atom of the backbone per segment, BB sub-networks) and one made of the side chain interactions (one atom of the side chain per segment or one atom of the side-chain on a segment and one atom of the backbone on the other segment, SC sub-network). They are shown in supplementary material 1 (Dataset S1). The interactions within the BB sub-networks are illustrated with dashed lines whereas the interactions within the SC sub-networks are illustrated with solid lines (see also materials and methods). It is important to note that the BB and SC sub-graphs can be considered individually (not considering the whole graphs) or within the whole graph. This nuance is important and when the two sub-networks are considered together, we will refer to as the "full" graph or the full network.

Characteristics of the BB sub-networks

The discrimination of the BB and SC sub-networks revealed significant features shared by the β-interfaces.

The BB sub-networks appeared characterized by common topological features but not by chemical specificities. First, different patterns of interactions show up in the BB sub-graphs. The first one, which appears in 19 graphs, is referred to as the "ladder" pattern because the BB interactions are running parallel to one another (Fig. 5). The second pattern which appears in 8 graphs is referred to as the "V-shape" pattern because it's a triplet interaction in the shape of a -V- (Fig. 6). The patterns are defined by elementary interaction blocks. One block "X.X" on one

segment interacts with one block "X.X" on the other segment in the ladder pattern. One "X" on one segment interacts with one block "X.X" on the other segment in the V-shape pattern. The elementary blocks appear singly or in multiple copies. Single versions of the ladder pattern appear in 1PVN, 2OJW, 1U1S and 1HX5 and in multiple copies in 1PM4, 1SNR, 1HI9, 1WUR, 2BCM, 2RCF, 2GJV, 2GVH, 2P90, 1J8D, 1WNR, 2RAQ, 1EEI and 1EFI . There are slightly altered versions of the ladder pattern. One graph (1FB1) is made of one block "X.X" on one segment interacting with one block "X . . X" on the other segment. Two graphs (2I9D and 2RCF) have one block of "XX" on one segment interacting with one block "XX" on the other segment.

Single version of the V-shape pattern can be observed in 2A7R and 2V9U and in multiple copies in 1SJN, 2BAZ, 1L3A, 1NQU, 1OEL and, 1Q3S.

There are also 5 graphs made of a mix of ladder and V-shape patterns (1Y13, 2I9D, 2H5X, 3BFO, 2Z9H).

The second topological information of the BB sub-networks is the fact that the ladder and the V-shape patterns appear related to the arrangement of the secondary structures of the β-interfaces. Indeed, they are observed mostly in anti-parallel and in parallel intermolecular β-strand interactions, respectively, and the pattern shapes' are reminiscent of the anti-parallel and parallel intramolecular main chain hydrogen bond networks found in β-sheets (Figs. 5B & 5C and 6B & 6C). To determine whether Gemini's BB networks were related to intermolecular hydrogen bonds, the program RING (materials and methods) was used, showing that out of the 100 atoms detected by RING as participating in hydrogen bonds, 98 are Gemini's backbone atoms. This is likely due to the selection process of Gemini which retains the closest atoms [15]. Gemini detects slightly more backbone atoms and bonds than RING (139 against 100) due to the fact that Gemini is

A

B

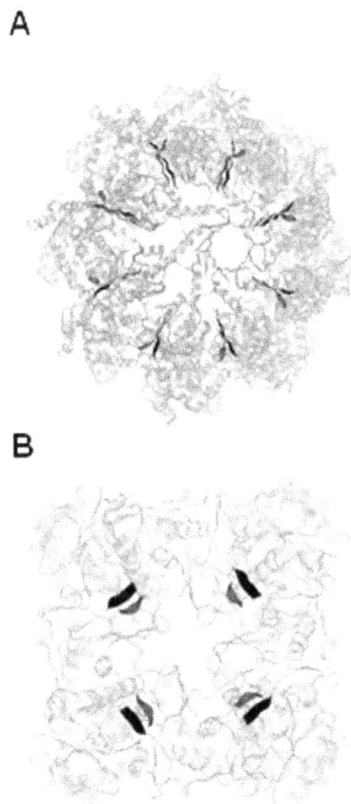

Figure 3. Protein oligomers containing a β-interface. A. The 1Q3S octameric bacterial chaperone [67] and **B.** The 1PVN tetrameric protozoa oxidoreductase [68]. Both structures are represented using RasMol. The chains are colored in light grey and the secondary-structures are represented by helices and strands. The β-strands of the interfaces are colored in black and dark grey in ribbons.

able to detect the double interactions per amino acids observed in the hydrogen bond network of intramolacular β-sheets (Fig. 5B & 5C and 6B & 6C). Thus, the BB sub-networks describe intermolecular β-sheets. This is confirmed by the observation that the graphs which have no BB interaction (1JN1, 1T0A, 2JCA, 1B09, 2XSC, 1SAC) or only one BB interaction (2BT9 and 2BVC)

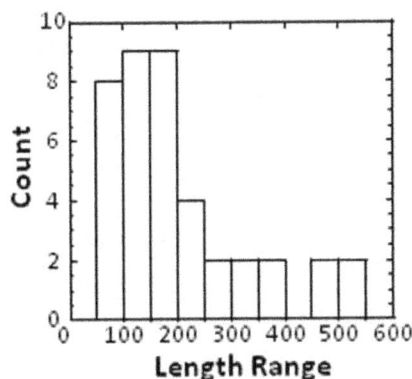

Figure 4. Histogram of the whole chain lengths. The length of the whole chain (range) is indicated on the x axis as the total number of amino acids.

Table 2. Statistics on the lengths of the dataset.

Sample	Average	SD[a]	Median	Q3-Q1	Q3[b]	Q1[b]
Length	17	6	17	7	19	12
Hot spot 'X'	12	4	12	5	14	9
Iβ	10	4	10	5	12	7

[a]SD stands for standard deviation.
[b]Q stands for quartile. The statistics are defined in materials and methods.

are not intermolecular β-sheets but are two rather perpendicular interacting β-strands, as can be seen on their respective PDB.

The BB sub-networks (X_{BB}) cannot be distinguished from the whole chains by a specific chemical composition (charged, polar and hydrophobic amino acids). Yet, they are dominated by hydrophobic properties: half of the amino acids of the BB sub-networks are hydrophobic and a third of the interactions are purely hydrophobic (Table 3 and table 4).

The global propensity (materials and methods) of the hydrophobic amino acids of the BB sub-networks was measured to evaluate which hydrophobic amino acids were over-represented in the β-interfaces compared to the whole chains (Table 5). A global propensity above 1 indicates a hydrophobic amino acid "preferred" in the BB sub-networks and on the contrary, a global propensity below 1, indicates a hydrophobic amino acid depleted in the BB sub-networks. Methionine (M), cysteine (C), tryptophane (W), isoleucine (I) and valine (V) are preferred in the BB sub-networks whereas proline (P), alanine (A), glycine (G) and leucine (L) residues are not favored in the BB sub-networks. The phenylalanine is equally present in the BB sub-networks and in the whole chains of the dataset (Global propensity around 1).

Characteristics of the SC sub-networks

In contrast to the BB sub-networks, the SC sub-networks have no topological information but some chemical specificity. In fact the SC sub-networks present an average chemical composition significantly different from the whole chains with a decrease of the percentage of hydrophobic amino acids in favor of an increase of the percentage of charged amino acids (Table 3). The percentage of polar residues remains similar for the SC sub-networks and the whole chains. This observation is even more obvious when the interactions (I_{SC}) are considered instead of the individual amino acids (X_{SC}), as the SC sub-networks have 5 times more purely charged interactions (Ch-Ch) than the BB sub-networks (Table 4). The SC sub-networks also have twice less purely hydrophobic interactions (F-F) than the BB sub-networks (Table 4).

Table 3. Average chemical composition, in percentage, of the amino acids of the whole chain of the protein dataset, of the two segments of the interface S1+S2) and of the hot spots of S1 and S2. SC and BB stand for side chain and backbone amino acids, respectively.

Interfaces	whole	S1+S2	S1+S2 'X'	X_{SC}	X_{BB}
Charged	24±17	24±10	28±14	30±17	23±16
Polar	23±15	26±14	29±16	29±17	27±24
Hydrophobic	53±34	50±12	45±15	41±15	50±14

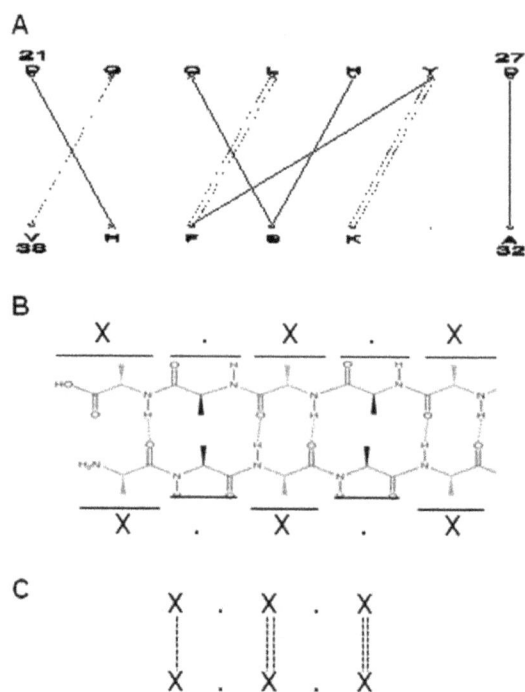

Figure 5. Anti-parallel BB sub-network and intramolecular hydrogen bond network. A. Gemini graph of an anti-parallel intermolecular β-interface **B.** Schematics of the hydrogen bond network of anti-parallel intramolecular β-sheet. **C.** Ladder pattern observed in BB sub-network and also visible in anti-parallel intramolecular β-sheet.

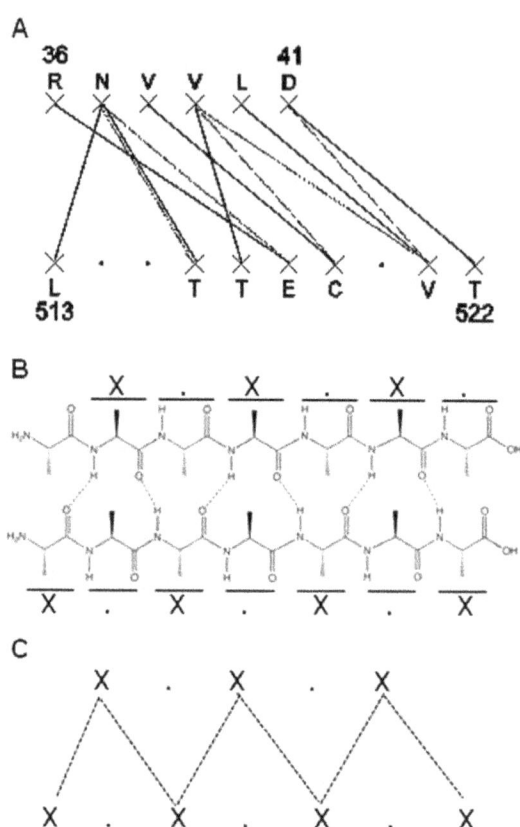

Figure 6. Parallel BB sub-network and intramolecular hydrogen bond network. A. Gemini graph of a parallel intermolecular β-interface **B.** Schematics of the hydrogen bond network of parallel intramolecular β-sheet. **C.** Ladder pattern observed in BB sub-network and also visible in parallel intramolecular β-sheet.

The global propensity (materials and methods) of the charged residues of the SC sub-networks compared to the whole chains is reported in Table 6. A charged amino acid with a global propensity above 1 is "preferred" in the SC sub-networks whereas a charged amino acid with a propensity below 1 is depleted. Apart from the histidine, which has a global propensity slightly above 1.0, all the charged residues of the SC sub-networks have a global propensity around 1.

The local propensity of the charged amino acids in the SC sub-networks was analyzed considering corner (the four outer SC amino acids) and central (non corner) positions (Table 7 and table 8, respectively). The local propensity (material and methods) is the ratio of the frequency of an amino acid in a particular position (e.g. corner) within a local structure (e.g. the β-interfaces) and of the frequency of the same amino acid in any other position within that local structure [38]. There are almost as much charged amino acids at corners than at central positions (44% in corner positions). But the two positions are made of different types of charged residues. Arginine (R) residues are more frequent at corners (local propensity above 1 in table 7) whereas it is glutamic acid and histidine residues which are favored centrally (local propensity above 1 in table 8). The lysine and aspartic acid residues have no local preferences (local propensity around 1 in both table 7 and table 8).

Comparison of BB and SC sub-networks

There exist several differences between the BB and the SC sub-networks (Table S3). There are 6 ± 3 I_{SC} interactions for only 4 ± 2 I_{BB} interactions. Additionally, there are 9 ± 4 X_{SC} amino acids for only 5 ± 3 X_{BB} amino acids. An amino acid with one atom involved in a BB interaction and one atom involved in a SC interaction is counted twice, one per network. But an amino acid having several atoms participating to the same network is counted only once. Thus, on average, the SC sub-network is bigger than the BB sub-network with roughly 60% of the interface amino acids and interactions devoted to it.

When considering the full graphs, it appears that the BB sub-networks are depleted of interactions and of hot spots at corners having only two graphs with two I_{BB} in the outer positions (1NQU and 2Z9H) and only 11 with one I_{BB} in the outer position (1Y13, 2BCM, 1PVN, 2A7R, 2H5X, 3BFO, 1EFI, 2OJW, 1U1S, 1WNR AND 1Q3S). In contrast, 28 graphs have two SC interactions in the outer positions and 39 (out of 40) have at least one. Likewise, the SC sub-networks are depleted of interactions and of hot spots

Table 4. Chemical composition of the interactions (amino acid -i- of segment 1 with amino acid -j- of segment 2 or vice-versa, data are added together) in the SC and in the BB (bracket) networks of the β-interfaces.

Chemical properties	Charged	Polar	Hydrophobic
Charged	17% (3,5%)		
Polar	13% (13%)	10% (9%)	
Hydrophobic	18% (23%)	25% (21%)	16% (30%)

Table 5. Global propensity of the hydrophobic residue in the BB sub-networks.

Hydrophobic	Number in the BB sub-networks	Percentage in the BB sub-networks	Number in the Whole chains	Percentage in the whole chains	Global propensity
I	26	0.20	577	0.13	1.6
L	17	0.13	700	0.16	0.8
V	27	0.21	714	0.16	1.3
A	13	0.10	756	0.17	0.6
C	4	0.03	95	0.02	1.5
M	11	0.09	180	0.04	2.1
F	9	0.07	295	0.07	1.1
G	15	0.12	714	0.16	0.7
P	4	0.03	351	0.08	0.4
W	3	0.02	82	0.02	1.3
Total	129		4464		

at central positions. There are 86 I_{SC} centrally located for a total of 240 I_{SC} (36%) and 143 X_{SC} centrally located for a total of 374 X_{SC} (38%). In the BB sub-networks, there are 86 I_{BB} centrally located for a total of 156 I_{BB} (55%) and 131 X_{BB} centrally located for a total of 219 X_{BB} (60%). This means that in a typical arrangement, the SC sub-network spatially contains and surrounds the BB one.

Consequently, the corners of the SC sub-networks are enriched with charged residues (32 graphs out of 40, 80%) while those of the BB sub-networks are depleted (10 graphs out 34: 29%). Similarly, the BB sub-networks are enriched centrally with hydrophobic residues (72 central hydrophobic residues for 110 in total: 65%) while the SC sub-networks are depleted (41 central hydrophobic residues for 101 in total: 41%).

Hence, the relative position of the sub-networks provides enrichment (or depletion) of a chemical property without having to vary the absolute number of amino acids of that property in the sub-networks. For example, there are 110 and 101 hydrophobic residues in the BB and SC sub-networks, respectively. Also, the probabilities of finding a charged residue in the corner of the SC or of the BB sub-networks, based on their respective chemical properties (Table 3), are indeed very similar 76% and 65%, respectively (materials and methods). Yet by positioning the X_{BB} centrally, the charged X_{SC} appear more frequently at corners.

Rationalization of the BB and SC features

Once common features are identified within the β-interfaces of the dataset, the next question is: can those features be rationalized in term of protein assembly or interface formation?

The first argument in that direction, is the weight of the β-interactions (Table S2). I_β are the interactions involved in the β-interface region of the protein oligomers of the dataset. Now, the total number of intermolecular interactions (I_{tot}) in a whole chain is the number of interactions in all the interface regions. I_{tot} is provided by Gemini. The average number of intermolecular interactions (I_{av}) per chain is the total number of interactions (I_{tot}) divided by the number of interface regions. The weight of the β-interactions is measured by the ratio -I_β/I_{av}- which gives the amount of interactions in a β-interface compared to the average number of interactions in the whole chains. On average, there are twice more interactions in the β-interfaces than in the whole interface (1.8±0.6). The high number of interactions due the beta geometry is consistent with a role of the β-interfaces in the assembly mechanism.

The data indicate that the BB sub-networks are related to the secondary structures of the interfaces and that they are enriched in hydrophobic residues and hydrophobic interactions. In order to test the involvement of the hydrophobic residues in the secondary structure of the interface, the effect of their mutation on secondary structure prediction was investigated.

The secondary structure of the segments (S1 and S2) with the wild-type (WT) sequence was predicted using GOR IV and compared to the prediction of the same segment after a point mutation of one hydrophobic residue. The mutation of centrally located hydrophobic residues to a charged residue (e.g. K, D, R, E, H) altered the secondary-structure prediction in 83% of the cases. The mutation of hydrophobic residues located at corners to

Table 6. Global propensity of the charged residue in the SC sub-networks.

Charged	Number in the SC sub-networks	Percentage in the SC sub-networks	Number in the whole chains	Percentage in the whole chains	Global propensity
R	19	0.16	403	0.18	0.9
E	31	0.27	584	0.27	1.0
K	26	0.22	497	0.23	1.0
D	24	0.21	507	0.24	0.9
H	13	0.11	188	0.09	1.3
Total	113		2179		

Table 7. Local propensity of the corner charged residue in the SC sub-networks.

Charged	Number in the corner position	Percentage in the corner position	Number in the SC sub-networks	Percentage in the SC sub-networks	Local propensity
R	12	0.24	19	0.17	1.4
E	11	0.22	31	0.27	0.8
K	12	0.24	26	0.23	1.0
D	11	0.22	24	0.21	1.0
H	4	0.08	13	0.12	0.7
Total	51		113		

charged residue, also disturbed the secondary-structure prediction but to a much lesser extent (44% of the cases). In the same way, the mutation of polar or of charged residues of the BB sub-networks centrally located, to hydrophobic, charge or polar amino acids affected the secondary-structure prediction in only 44% of the cases.

We then measured the local propensity of the hydrophobic residues located centrally in the BB sub-networks and affecting the 2D structure prediction (Table 9). It appears that among the secondary-influencing hydrophobic residues centrally located, the valine (V) and the phenylalanine (F) are preferred (local propensity above 1). The leucine (L), the isoleucine and the methionine (M) appear neutral in the central position (local propensity around 1). Tryptophan (W), proline (P), glycine (G), alanine (A) and cysteine (C) are not favored (local propensity below 1).

The local propensity results were tested using secondary-structure prediction again. Mutations of central hydrophobic amino acids of the BB sub-networks to hydrophobic amino acids which have a local propensity above 1 were expected to have a secondary-structure prediction identical to the wild-type one. This is referred to as the amino acid having a positive versatility (act as wild-type amino acid). On the contrary, mutations to amino acid with a local propensity below 1 were expected to alter the wild-type secondary-structure prediction. These amino acids are referred to as having a negative versatility. In total 331 mutations-predictions have been performed and on average 69% behave as expected (229/331). Both the versatilities are giving similar results with 67% (116/172) of the mutations to amino acids of positive versatility not affecting the secondary structure prediction and 71% (113/159) of the mutations to amino acids of negative versatility affecting it.

This is consistent with the involvement of the features of the BB sub-networks in the secondary structure formation of the β-interfaces.

The SC sub-networks have no topological information and therefore cannot be related to geometrical features. But they have enrichment in charged residues and more precisely a specific distribution of the type of charges along the interface. This suggests a chemical role of the SC sub-networks in the formation of the β-interfaces, via electrostatic interactions.

We have seen that the local positions of the hydrophobic and of the charged residues of the BB and SC sub-networks were connected to the relative position of the two sub-networks. Now, remarkably for the 11 graphs which have one outer BB interaction, 7 have one charged BB residue at a corner. Following the same drift, the graphs with a low content of SC interactions but made of a majority of BB interactions have a charged BB residue in a corner in 44% of the case (7 out 16 graphs) whereas this occurs only in 12% of the graphs made of a minority of BB interactions (3/24).

So even if having a charged residue in a corner appears a trademark of the SC sub-networks, a corner charged residue is maintained via the BB sub-networks if necessary. This looks like a compensatory or a substitutive mechanism.

A similar phenomenon can be observed for the hydrophobic property of the graphs. On average twice more SC hydrophobic residues are located centrally (1,1 central SC hydrophobic) in graphs made of a minority of BB interactions than in graphs made of a majority of BB interactions (0,45 central SC hydrophobic). More precisely, the number of centrally located hydrophobic residues is maintained at a value of 2,8±0,6 across the dataset with 2,2±0,5 of them affecting the secondary structure predictions (Fig. 7). This value is kept constant using either BB or SC residues, or a balance of both. The mutation of the centrally located hydrophobic residues of the SC sub-networks to charged residue affects the secondary prediction in 83% of the case, as for the BB sub-networks. Thus the regulation of the secondary structure

Table 8. Local propensity of the corner charged residue in the SC sub-networks.

Charged	Number in a NOT corner position	Percentage in a NOT corner position	Number in the SC sub-networks	Percentage in the SC sub-networks	Local propensity
R	7	0.11	19	0.17	0.7
E	20	0.32	31	0.27	1.2
K	14	0.22	26	0.23	1.0
D	13	0.21	24	0.21	1.0
H	9	0.14	13	0.12	1.2
Total	64		113		

Table 9. Local propensity of the central hydrophobic residue of the BB sub-networks affecting the 2D-structure prediction.

Hydrophobic	Number in central position	Percentage in the central position	Number in BB sub-networks	Percentage in BB sub-networks	Local propensity
I	11	0.20	26	0.20	1.0
L	8	0.15	17	0.13	1.1
V	18	0.33	27	0.21	1.6
A	3	0.06	13	0.10	0.6
C	1	0.02	4	0.03	0.6
M	4	0.07	11	0.09	0.9
F	5	0.09	9	0.07	1.3
G	3	0.06	15	0.12	0.5
P	1	0.02	4	0.03	0.6
W	0	0.00	3	0.02	0.0
Total	54		129		

through hydrophobic amino acids located centrally is organized by the BB sub-networks in most cases. But the BB sub-networks can be substituted by the SC sub-networks as an alternative.

Such compensatory or substitutive phenomenon is also in favor of the features being involved in the formation of the interface.

No distinction between the stœchiometries was found for any of the properties of the β-interfaces (not shown).

Autonomous β-interface segments

As mentioned earlier, the features describing the β-interfaces are rather homogeneous compared to the heterogeneity observed for their whole chains. In addition, it seems possible to associate the β-interface features to geometrical and chemical properties. This hinted the possibility that the β-interfaces had some autonomous capacity to associate in absence of the whole chain. This was further supported by the narrow distribution of the β-interface lengths and by the absence of proportion between the

lengths of the β-interface and the length of their respective whole chain (Fig. 8). To test that possibility, a simple experiment was carried out using the pentamer of the cholera toxin B ($CtxB_5$) as a prototype of the β-interfaces (Fig. 1). Conditions to follow the assembly of the $CtxB_5$ *in vitro* had been established previously and are indicated in material and methods [40]. Briefly, the native toxin (Fig. 9, lane 2) is acidified for 15 min at room temperature (RT) to lead to its dissociation into monomers (Fig. 9, lane 3). Subsequently, it is neutralized for 15 min at RT, time during which the reassembly into pentamer takes place (Fig. 9, lane 4). In subsequent experiments, 9mer (P1) or/and 8mer (P2) synthetic peptides with sequences corresponding to S1 ([23]KIFSYTESL[31]) and S2 ([96]IAAISMAN[103]), respectively, of the wild-type CtxB β-interface were added to the neutralizing buffer. The amounts of CtxB reassembled into pentamer under the different conditions,

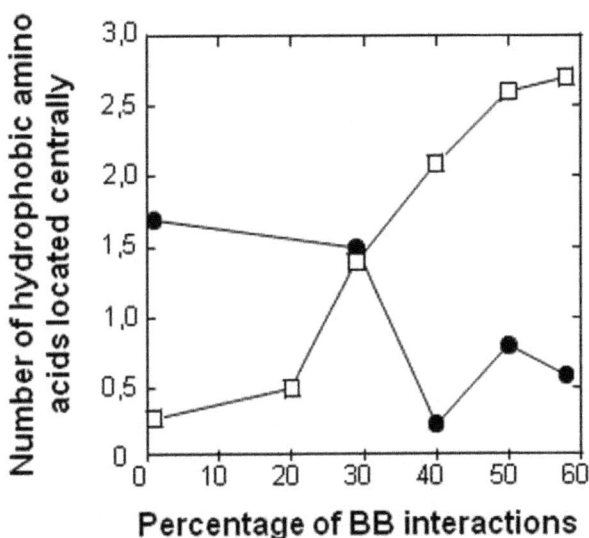

Figure 7. Central hydrophobic residues and percentage of BB interactions. The number of hydrophobic amino acids of the BB (β) or of the SC sub-networks (●) located centrally in the full networks are plotted against the percentage of BB interactions.

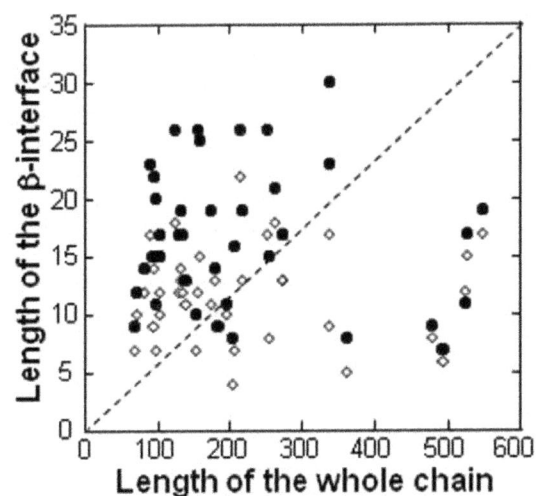

Figure 8. Absence of correlation between the lengths of the whole chains and of the β-interfaces. The length of the β-interface (sum of the amino acids of the two segments) of each protein of the dataset is plotted against the length of its respective whole chain (●, 'all amino acids' and ◇, 'X', respectively). If there was a correlation between the size of the whole chain and the size of its interface or the size of its hot spot numbers, the points would appear on the dashed line.

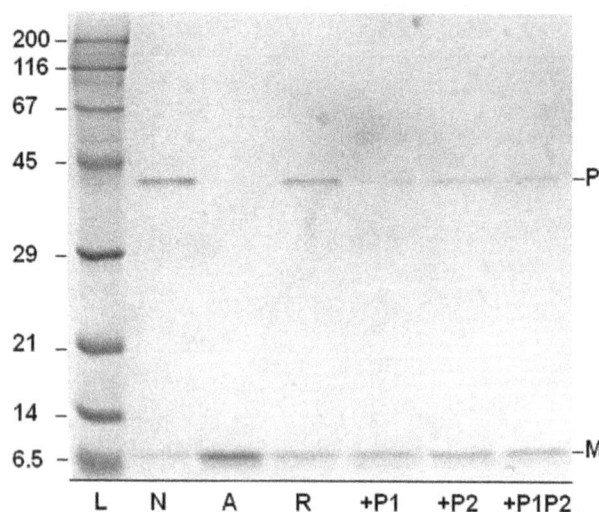

Figure 9. *In vitro* assembly of the cholera toxin B subunit into pentamer (CtxB$_5$). The formation of the CtxB β-interface is monitored by SDS-PAGE. The initial native CtxB$_5$ is indicated in lane 2 (N) whereas the acidified CtxB monomer is indicated in lane 3 (A). The toxin reassembly after 15 min in neutral condition is shown from lane 4 to 7 for the toxin alone (R, lane 4), or with a synthetic peptide of CtxB segment 1 sequence (+P1, lane 5) or with a synthetic peptide whose sequence corresponds to CtxB segment 2 (+P2, lane 6) or with a mixture of both peptides (+P1P2, lane 7). L stands for low molecular weight standard.

were then compared using SDS-PAGE (Fig. 9). The addition of P1 (Fig. 9, lane 5), of P2 (Fig. 9, lane 6) and of P1 and P2 together (Fig. 9, lane 7) strongly inhibited the reassembly of the toxin into pentamer. This indicates that P1 as well as P2 do interfere with the formation of CtxB-CtxB interfaces. P1 inhibited more than P2 and the mixture P1+P2 inhibited more than P2 but less than P1. Thus P1 and P2 must be reacting together.

Discussion

As for the α-coiled interfaces, the choice of a common geometry of interfaces proved to be successful in isolating characteristics among the β-interfaces of otherwise unrelated protein oligomers. The results are thus devoid of potential bias introduced when protein interfaces of proteins with similar folds or similar functions are compared. It was also possible to associate geometrical and chemical properties to the identified features. On one hand, this provides an evaluation of the features so their reliability improves. On the other hand, it also gives some rational about the 'mode of action' of the features in term of interface formation. Thus, using the CtxB model, the role of the hydrophobic and of the charged residues on the formation of the secondary structure and on the formation of the CtxB β-interface, respectively, can be tested. However, the study entirely focuses on the β-interfaces and as such the results are far from providing a full picture of the parameters involved in the assembly of the whole chains of the dataset. As an illustration, we have seen that the mutations of the central hydrophobic residues of the BB sub-networks have little effect on the secondary structure predictions of the whole length sequences (~25%) (not shown). The true essence of the results resides in the observation of interdigitated networks in which the interface features are made through strategic positioning of chemical characteristics rather than through drastic chemical modulation. Thus the search of a sequence of an interface cannot be done as the search of a sequence of a biological function (e.g. active site).

In summary, the β-interfaces are made of two interactions sub-networks. One is involving atoms of the main chain (BB sub-networks) and the other is involving atoms from the side chains (SC sub-networks). The characteristics of the BB sub-networks are related to the hydrophobic residues which seem particularly involved in the secondary structures of the β-interfaces. This is well supported by the fact that the hydrophobic residues favored in the β-interfaces (IVMWC) are also favored in intramolecular β-sheet (IVMCW) [34,45,46,47]. Likewise, the hydrophobic residues disfavored in the β-interfaces (AGP) are disfavored in intramolecular β-sheet (AGP) [34,45,46,47]. There are some discrepancies for the leucine and phenylalanine residues which are favored in intramolecular β-sheets but disfavored or neutral in the β-interfaces, respectively. Intriguingly, these two amino acids are enriched in amyloid β-fiber (LIF) [33]. The role of hydrophobic forces in interfaces (dimers) was previously reported but not in connection with the geometry of the interface [21,48,49] and for review see [2,12,33].

The hydrophobic amino acids of the BB sub-networks are thus devoid of 'intermolecular' specificity since they are shared with intramolecular interactions.

In contrast, the charged amino acids favored in the SC sub-networks present some specificity. First, intra-molecular β-interactions as well as dimeric β-interfaces are rather depleted in charged residues, apart from arginine for the dimeric interfaces ([21,32,33,45,46,50] and for review [2]). On the contrary, in the β-interface side chains, charged residues represent a third of the interfacial amino acids and have only a slight preference for histidine residues. It is interesting that the histidine residue stands out as it is the only amino acid charged under physiological conditions. It is also an amino acid already shown to take part in the assemblies of several protein oligomers [51,52,53]. Second, the β-interfaces of our dataset have an average net charge of −0.5 which differs from the one required for the formation of amyloid β-fiber (net charge of ±1), another type of β-interface [54,55,56].

The third and most practical information about the charge specificity, resides in the distribution of the charged residues. The arginine residues are frequent at both the corners (N- and C-terminal caps) of the β-interfaces whereas histidine and glutamic acid are favored centrally. Lysine and aspartic acid residues have no preferred position in the β-interfaces.

This is in contrast to parallel intramolecular β-sheet in which positively charged residues (KR) are located at the N-terminal extremities only and negatively charged residues (DE) are present at the C-terminal extremities only [47]. The presence of charges at the N- or C- terminal extremities is believed to act as β-breakers [45,47]. Additionally, the formation of amyloid β-fiber is promoted with positively charged residues (KR) located at the N-terminal extremities of the amyloid β-strands and negatively charged residues (DE) at both the N- or C-terminal extremities [54,55]. Finally, charged residues centrally located are observed in intra-molecular edge β-strands and are thought to prevent their aggregation [34]. Hence, the scattered distribution observed on the β-interfaces differentiates them from other types of intramolecular and intermolecular dimeric β-interactions (Fig. 10).

Altogether the data lead us to propose some hypothesis on the construction mechanism of the β-interfaces following two principles: (i) interfaces are built via geometrical and chemical recognition of the interacting domains and (ii) there are a recognition phase ('binding') and a stabilization phase. The BB sub-networks, via the hydrophobic residues, could provide the geometrical recognition whereas the side chain charged residues could provide the chemical one. It is tempting to speculate that the long arginine residue located at the extremities is employed as a

β-interface

| R | HE | R |

β-intramolecular

| RK | | DE |

Amyloid β-fiber

| RKDE | | DE |

β-edge

| RKDE |

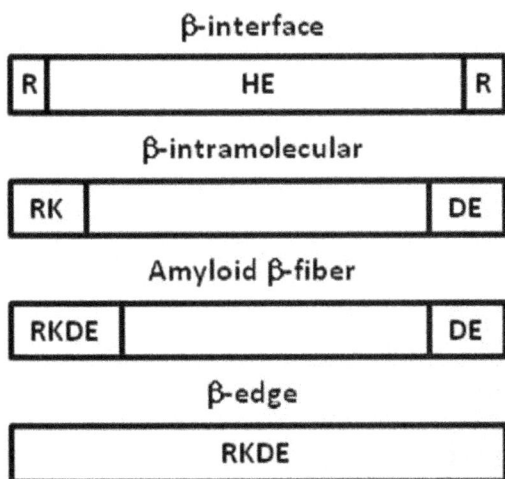

Figure 10. Schematic of the charge distribution in β-interactions. The amino acids are indicated using the single letter code.

hook to promote encounter. The central smaller histidine and glutamic acid residues could act as clips to stabilize the interface. Alternatively, they might, as proposed for the β-edge strands, maintain the two domains soluble prior the recognition.

Some experimental data are consistent with a relation between Gemini's hotspot residues and their involvement in the process of a β-interface formation. For example, the heat labile enterotoxin B (LTB$_5$) and the cholera toxin B (CtxB$_5$) pentamers, which shares 84% sequence identity and almost superimposable x-ray structures, have nevertheless different assembly mechanisms and different β-interface graphs (1EFI and 1EEI, respectively). The two toxin pentamers have only 14 different amino acids and one of them is in the β-interface (Leu 25 and Phe 25 in 1EFI and 1EEI, respectively). Residue 25 is involved in a I$_{BB}$ in both graphs but leucine and phenylalanine have been measured with different global propensities (Table 5). There are 6 I$_{BB}$ for 4 I$_{SC}$ in LTB$_5$ compatible with a geometry-regulated assembly as observed experimentally since only folded LTB chains associate [57]. On the other hand, there are 5 I$_{BB}$ for 5 I$_{SC}$ in CtxB$_5$ consistent with a more 'chemically'-regulated assembly also observed experimentally with partially folded CtxB chains capable of associating [40,52]. The presence of a I$_{SC}$ involving a lysine residue only in CtxB$_5$ (K23-N103) also supports a more 'chemically'-regulated assembly. Similarly, shiga-like toxin I and II have different stabilities and different graphs (2XSC and not shown) [58]. In the bacterial hexameric (1U1S) from *Pseudomonas aeruginosa* , the mutation of His 57, to alanine (Ala) or to threonine (Thr) destabilizes the hexamer by disturbing the side chain hydrogen

bond network of the His 57 with the side chains of Lys 56 and Ile 59 of the adjacent chain [59]. The His 57 side chain hydrogen bond network is properly seen on the Gemini graph of the β-interface of Hfq (Dataset 1, 1U1S). Disappearance of that network (or changes of that network) for mutant Ala 57 (or for mutant Thr 57) is also seen properly on the Gemini graphs of the mutated Hfq (not shown). Moreover, the conserved main chain hydrogen bond network made of the residues Met 53 and Tyr 55 of chain M with the residues Val 62 and Ser 60 of the adjacent chain is also identified by Gemini (not shown) [60]. However, cautious is necessary with interpreting the graph features. At this stage, they should be used as a tool to formulate hypothesizes for experimental tests.

There are several arguments, mentioned in the result section, supporting the idea that the β-interfaces are independent assembly unit. The most indicative one is the experimental observation that the CtxB β-interface peptides recognize the CtxB individual chains. Such peptides could be called "assemblons" by homology to the foldons [61,62]. Some peptides have been found to lead to the trimerization of proteins when genetically added to their sequence, supporting the 'assemblons' concept [63,64,65].

Supporting Information

Dataset S1   Gemini Graphs of the 40 β-interfaces. Each graph appears on a separate page. The stœchiometry and the PDB code of the concerned protein oligomer is indicated on the box in the left hand side of the image. The amino acid number is indicated with the type of amino acid at position X. Segments 1 and 2 appear on two parallel rows. X indicates amino acids involved in atomic interactions according to Gemini. SC and BB interactions are illustrated by solid and dashed lines, respectively [15]. The graphs which interfaces have been annotated manually are indicated with a straight line above the segments. A top (left) and a side view (right) of the x-ray structure of the protein oligomer is shown above its respective graph.

Table S1 Features of the protein oligomers of the dataset.

Table S2 Features of the β-interfaces.

Table S3 Properties of the two sub-graphs.

Author Contributions

Conceived and designed the experiments: CL GF LV. Performed the experiments: MA JZ. Analyzed the data: CL. Contributed reagents/materials/analysis tools: CL GF JZ LV. Wrote the paper: CL.

References

1. Goodsell DS, Olson AJ (2000) Structural symmetry and protein function. Annu Rev Biophys Biomol Struct 29: 105–153.
2. Janin J, Bahadur RP, Chakrabarti P (2008) Protein-protein interaction and quaternary structure. Q Rev Biophys 41: 133–180.
3. Iacovache I, van der Goot FG, Pernot L (2008) Pore formation: An ancient yet complex form of attack. Biochim Biophys Acta.
4. Lesieur C, Vecsey-Semjen B, Abrami L, Fivaz M, Gisou van der Goot F (1997) Membrane insertion: The strategies of toxins (review). Mol Membr Biol 14: 45–64.
5. Kirkitadze MD, Bitan G, Teplow DB (2002) Paradigm shifts in Alzheimer's disease and other neurodegenerative disorders: the emerging role of oligomeric assemblies. J Neurosci Res 69: 567–577.
6. Soto C (2003) Unfolding the role of protein misfolding in neurodegenerative diseases. Nature Reviews Neuroscience 4: 49–60.
7. Klein W, Stine W (2004) Small assemblies of unmodified amyloid [beta]-protein are the proximate neurotoxin in Alzheimer's disease. Neurobiology of aging 25: 569–580.
8. Harrison RS, Sharpe PC, Singh Y, Fairlie DP (2007) Amyloid peptides and proteins in review. Rev Physiol Biochem Pharmacol 159: 1–77.
9. Miller Y, Ma B, Nussinov R (2010) Polymorphism in Alzheimer A amyloid organization reflects conformational selection in a rugged energy landscape. Chemical reviews.
10. Larsen TA, Olson AJ, Goodsell DS (1998) Morphology of protein-protein interfaces. Structure 6: 421–427.
11. Grueninger D, Treiber N, Ziegler MOP, Koetter JWA, Schulze MS, et al. (2008) Designed protein-protein association. Science 319: 206.
12. Tuncbag N, Kar G, Keskin O, Gursoy A, Nussinov R (2009) A survey of available tools and web servers for analysis of protein–protein interactions and interfaces. Briefings in Bioinformatics 10: 217.

13. Cazals F, Proust F, Bahadur RP, Janin J (2006) Revisiting the Voronoi description of protein-protein interfaces. Protein Sci 15: 2082–2092.

14. Shulman-Peleg A, Shatsky M, Nussinov R, Wolfson HJ (2007) Spatial chemical conservation of hot spot interactions in protein-protein complexes. BMC Biol 5: 43.

15. Feverati Ga, Lesieur C (2010) Oligomeric Interfaces under the Lens: Gemini. Plos OneAvailable: http://dx.plos.org/10.1371/journal.pone.0009897.

16. Guidry JJ, Shewmaker F, Maskos K, Landry S, Wittung-Stafshede P (2003) Probing the interface in a human co-chaperonin heptamer: residues disrupting oligomeric unfolded state identified. BMC Biochem 4: 14.

17. Crick FHC (1953) The packing of alpha-helices: simple coiled-coils. Acta Crystallogr 6: 689–697.

18. Lupas A, Van Dyke M, Stock J (1991) Predicting coiled coils from protein sequences. Science 252: 1162–1164.

19. Lupas A (1996) Coiled coils: new structures and new functions. Trends Biochem Sci 21: 375–382.

20. Walshaw J, Woolfson DN (2003) Extended knobs-into-holes packing in classical and complex coiled-coil assemblies. J Struct Biol 144: 349–361.

21. Guharoy M, Chakrabarti P (2007) Secondary structure based analysis and classification of biological interfaces: identification of binding motifs in protein-protein interactions. Bioinformatics 23: 1909–1918.

22. Yan C, Wu F, Jernigan RL, Dobbs D, Honavar V (2008) Characterization of protein-protein interfaces. Protein J 27: 59–70.

23. Davis FP, Sali A (2005) PIBASE: a comprehensive database of structurally defined protein interfaces. Bioinformatics 21: 1901–1907.

24. Tsai CJ, Lin SL, Wolfson HJ, Nussinov R (1996) Protein-protein interfaces: architectures and interactions in protein-protein interfaces and in protein cores. Their similarities and differences. Crit Rev Biochem Mol Biol 31: 127–152.

25. Gao M, Skolnick J (2010) Structural space of protein–protein interfaces is degenerate, close to complete, and highly connected. Proceedings of the National Academy of Sciences 107: 22517.

26. Stein A, Mosca R, Aloy P (2011) Three-dimensional modeling of protein interactions and complexes is going omics. Current opinion in structural biology.

27. Tsai CJ, Lin SL, Wolfson HJ, Nussinov R (1996) A dataset of protein-protein interfaces generated with a sequence-order-independent comparison technique. J Mol Biol 260: 604–620.

28. Grigoryan G, Keating AE (2008) Structural specificity in coiled-coil interactions. Curr Opin Struct Biol 18: 477–483.

29. Calladine CR, Sharff A, Luisi B (2001) How to untwist an alpha-helix: structural principles of an alpha-helical barrel. J Mol Biol 305: 603–618.

30. Hadley EB, Testa OD, Woolfson DN, Gellman SH (2008) Preferred side-chain constellations at antiparallel coiled-coil interfaces. Proc Natl Acad Sci U S A 105: 530–535.

31. Tsai CJ, Lin SL, Wolfson HJ, Nussinov R (1997) Studies of protein-protein interfaces: a statistical analysis of the hydrophobic effect. Protein Sci 6: 53–64.

32. Ma B, Nussinov R (2007) Trp/Met/Phe hot spots in protein-protein interactions: potential targets in drug design. Current topics in medicinal chemistry 7: 999–1005.

33. Ma B, Elkayam T, Wolfson H, Nussinov R (2003) Protein–protein interactions: structurally conserved residues distinguish between binding sites and exposed protein surfaces. Proceedings of the National Academy of Sciences 100: 5772.

34. Richardson JS, Richardson DC (2002) Natural -sheet proteins use negative design to avoid edge-to-edge aggregation. Proceedings of the National Academy of Sciences 99: 2754.

35. Krishnan A, Zbilut JP, Tomita M, Giuliani A (2008) Proteins as networks: usefulness of graph theory in protein science. Curr Protein Pept Sci 9: 28–38.

36. Bode C, Kovacs IA, Szalay MS, Palotai R, Korcsmaros T, et al. (2007) Network analysis of protein dynamics. FEBS Lett 581: 2776–2782.

37. Martin AJ, Vidotto M, Boscariol F, Di Domenico T, Walsh I, et al. (2011) RING: networking interacting residues, evolutionary information and energetics in protein structures. Bioinformatics 27: 2003–2005.

38. Penel S, Hughes E, Doig AJ (1999) Side-chain structures in the first turn of the alpha-helix. J Mol Biol 287: 127–143.

39. Laemli U (1970) Cleavage of structural proteins during the assembly of the head of bacteriophage T4. Nature 227: 680–685.

40. Lesieur C, Cliff MJ, Carter R, James RF, Clarke AR, et al. (2002) A kinetic model of intermediate formation during assembly of cholera toxin B-subunit. J Biol Chem 277: 16697–16704.

41. Guex N, Peitsch MC (1997) SWISS-MODEL and the Swiss-PdbViewer: an environment for comparative protein modeling. Electrophoresis 18: 2714–2723.

42. Murzin AG, Brenner SE, Hubbard T, Chothia C (1995) SCOP: a structural classification of proteins database for the investigation of sequences and structures. J Mol Biol 247: 536–540.

43. Lo Conte L, Chothia C, Janin J (1999) The atomic structure of protein-protein recognition sites. J Mol Biol 285: 2177–2198.

44. Ma B, Nussinov R (2000) Molecular dynamics simulations of a [beta]-hairpin fragment of protein G: balance between side-chain and backbone forces1. Journal of molecular biology 296: 1091–1104.

45. Garratt RC, Thornton JM, Taylor WR (1991) An extension of secondary structure prediction towards the production of tertiary structure. FEBS letters 280: 141–146.

46. Minor DL, Jr., Kim P (1994) Measurement of the b-sheet-forming propensities of amino acids. Nature 367: 660–663.

47. FarzadFard F, Gharaei N, Pezeshk H, Marashi SA (2008) [beta]-Sheet capping: Signals that initiate and terminate [beta]-sheet formation. Journal of structural biology 161: 101–110.

48. Merkel JS, Sturtevant JM, Regan L (1999) Sidechain interactions in parallel [beta] sheets: the energetics of cross-strand pairings. Structure 7: 1333–1343.

49. Chakrabarti P, Janin J (2002) Dissecting protein-protein recognition sites. Proteins 47: 334–343.

50. Jones S, Thornton JM (1996) Principles of protein-protein interactions. Proc Natl Acad Sci U S A 93: 13–20.

51. Tacnet P, Cheong EC, Goeltz P, Ghebrehiwet B, Arlaud GJ, et al. (2008) Trimeric reassembly of the globular domain of human C1q. Biochim Biophys Acta 1784: 518–529.

52. Zrimi J, Ng Ling A, Giri-Rachman Arifin E, Feverati G, Lesieur C (2010) Cholera toxin B subunits assemble into pentamers - proposition of a fly-casting mechanism. PLoS One 5: e15347.

53. Dang LT, Purvis AR, Huang RH, Westfield LA, Sadler JE (2011) Phylogenetic and functional analysis of histidine residues essential for pH-dependent multimerization of von Willebrand factor. Journal of Biological Chemistry.

54. Lopez De La Paz M, Goldie K, Zurdo J, Lacroix E, Dobson CM, et al. (2002) De novo designed peptide-based amyloid fibrils. Proc Natl Acad Sci U S A 99: 16052–16057.

55. López De La Paz M, Serrano L (2004) Sequence determinants of amyloid fibril formation. Proceedings of the National Academy of Sciences of the United States of America 101: 87.

56. Marshall KE, Serpell LC (2009) Structural integrity of beta-sheet assembly. Biochem Soc Trans 37: 671–676.

57. Ruddock LW, Coen JJ, Cheesman C, Freedman RB, Hirst TR (1996b) Assembly of the B subunit pentamer of Escherichia coli heat-labile enterotoxin. Kinetics and molecular basis of rate-limiting steps in vitro. J Biol Chem 271: 19118–19123.

58. Conrady DG, Flagler MJ, Friedmann DR, Vander Wielen BD, Kovall RA, et al. (2010) Molecular basis of differential B-pentamer stability of Shiga toxins 1 and 2. PLoS One 5: e15153.

59. Moskaleva O, Melnik B, Gabdulkhakov A, Garber M, Nikonov S, et al. (2010) The structures of mutant forms of Hfq from Pseudomonas aeruginosa reveal the importance of the conserved His57 for the protein hexamer organization. Acta Crystallographica Section F: Structural Biology and Crystallization Communications 66: 760–764.

60. Nikulin A, Stolboushkina E, Perederina A, Vassilieva I, Blaesi U, et al. (2005) Structure of Pseudomonas aeruginosa Hfq protein. Acta Crystallographica Section D: Biological Crystallography 61: 141–146.

61. Panchenko AR, Luthey-Schulten Z, Wolynes PG (1996) Foldons, protein structural modules, and exons. Proc Natl Acad Sci U S A 93: 2008–2013.

62. Panchenko AR, Luthey-Schulten Z, Cole R, Wolynes PG (1997) The foldon universe: a survey of structural similarity and self-recognition of independently folding units. J Mol Biol 272: 95–105.

63. Mitraki A, van Raaij MJ (2005) Folding of beta-structured fibrous proteins and self-assembling peptides. Methods Mol Biol 300: 125–140.

64. Papanikolopoulou K, Teixeira S, Belrhali H, Forsyth VT, Mitraki A, et al. (2004a) Adenovirus fibre shaft sequences fold into the native triple beta-spiral fold when N-terminally fused to the bacteriophage T4 fibritin foldon trimerisation motif. J Mol Biol 342: 219–227.

65. Papanikolopoulou K, Forge V, Goeltz P, Mitraki A (2004b) Formation of highly stable chimeric trimers by fusion of an adenovirus fiber shaft fragment with the foldon domain of bacteriophage t4 fibritin. J Biol Chem 279: 8991–8998.

66. Zhang RG, Scott DL, Westbrook ML, Nance S, Spangler BD, et al. (1995) The three-dimensional crystal structure of cholera toxin. J Mol Biol 251: 563–573.

67. Shomura Y, Yoshida T, Iizuka R, Maruyama T, Yohda M, et al. (2004) Crystal structures of the group II chaperonin from Thermococcus strain KS-1: steric hindrance by the substituted amino acid, and inter-subunit rearrangement between two crystal forms. J Mol Biol 335: 1265–1278.

68. Gan L, Seyedsayamdost MR, Shuto S, Matsuda A, Petsko GA, et al. (2003) The immunosuppressive agent mizoribine monophosphate forms a transition state analogue complex with inosine monophosphate dehydrogenase. Biochemistry 42: 857–863.

18

Amyloid Core Formed of Full-Length Recombinant Mouse Prion Protein Involves Sequence 127–143 but Not Sequence 107–126

Biswanath Chatterjee[1], Chung-Yu Lee[1,2], Chen Lin[1], Eric H.-L. Chen[1], Chao-Li Huang[3], Chien-Chih Yang[3], Rita P.-Y. Chen[1,2]*

1 Institute of Biological Chemistry, Academia Sinica, Taipei, Taiwan, **2** Institute of Biochemical Sciences, National Taiwan University, Taipei, Taiwan, **3** Department of Biochemical Science and Technology, National Taiwan University, Taipei, Taiwan

Abstract

The principal event underlying the development of prion disease is the conversion of soluble cellular prion protein (PrP^C) into its disease-causing isoform, PrP^{Sc}. This conversion is associated with a marked change in secondary structure from predominantly α-helical to a high β-sheet content, ultimately leading to the formation of aggregates consisting of ordered fibrillar assemblies referred to as amyloid. *In vitro*, recombinant prion proteins and short prion peptides from various species have been shown to form amyloid under various conditions and it has been proposed that, theoretically, any protein and peptide could form amyloid under appropriate conditions. To identify the peptide segment involved in the amyloid core formed from recombinant full-length mouse prion protein mPrP(23–230), we carried out seed-induced amyloid formation from recombinant prion protein in the presence of seeds generated from the short prion peptides mPrP(107–143), mPrP(107–126), and mPrP(127–143). Our results showed that the amyloid fibrils formed from mPrP(107–143) and mPrP(127–143), but not those formed from mPrP(107–126), were able to seed the amyloidogenesis of mPrP(23–230), showing that the segment residing in sequence 127–143 was used to form the amyloid core in the fibrillization of mPrP(23–230).

Editor: Kaustubh Datta, University of Nebraska Medical Center, United States of America

Funding: This work was supported by the Academia Sinica and the National Science Council, Taiwan, R. O. C. (grant nos. NSC 97-2321-B-001-032, NSC 98-2311-B-001-019, NSC 99-2321-B-001-038, and NSC 100-2321-B-001-026). The funders had no role in study design, data collection and analysis, decision to publish, or preparation of the manuscript.

Competing Interests: The authors have declared that no competing interests exist.

* E-mail: pyc@gate.sinica.edu.tw

Introduction

Transmissible spongiform encephalopathies (TSEs) are a group of fatal neurodegenerative disorders which affect many mammalian species such as human, bovine, sheep, deer, mink, cat etc [1–4]. The principle pathogenic event in TSE is the formation of an abnormally folded isoform, known as PrP^{Sc} from a normal cellular protein called PrP^C [5]. PrP^C is an evolutionary conserved glycosylphosphatidylinositol (GPI)-anchored glycoprotein of ~254 amino acids and has been characterized structurally as having an unstructured N-terminal part and a C-terminal globular domain consisting of three α-helices and two short β-strands, together with one disulfide bond formed between helix 2 and 3 [6], [7]. Little is known about the exact structure of PrP^{Sc}, except that it has high cross-β structure content [8]. As estimated by circular dichroism spectroscopy and Fourier transformed infrared spectroscopy, PrP^{Sc} has a β-sheet content of 43% compared to the 4% β-sheet content of PrP^C, estimated from its NMR structure [9–11]. The conversion of PrP^C to PrP^{Sc} leads to a conformational change associated with an increase in β-sheet content [12], [13]. The characteristic features of PrP^{Sc} include its partial resistance to proteinase-K (PK) degradation, insolubility in non-ionic detergents, and a fibrillar structure typical of amyloid. After protease digestion, the protease-resistant core of PrP^{Sc} which consists of

residues 90–231, has a molecular mass of 27–30 kDa and is therefore denoted as PrP 27–30 [14]. PrP^{Sc} has the ability to induce misfolding of PrP^C, leading to the maintenance, propagation, and manifestation of disease phenotype in the host organism, with the result that, unlike other neurodegenerative disorders, prion diseases are transmissible.

The region of PrP^C, which participates in the formation of PrP^{Sc}, has long been a subject of interest, since it might provide information useful in designing inhibitors to interfere with the structural conversion of PrP^C and/or PrP^{Sc} propagation. Synthetic prion peptides of hamster origin covering residues 109–122 together with another overlapping 15-residue sequence 113–127 were found to form amyloids [15]. A chimeric mouse-hamster PrP of 106 amino acids (PrP106) with two deletions (Δ23–88 and Δ141–176) was found to retain the ability to support PrP^{Sc} formation in transgenic mice [16]. In addition, many short prion protein segments have been reported to have the ability to form amyloid fibrils. Synthetic prion peptides consisting of residues 106–126, 127–143, 106–147, 90–145, 171–193 (helix 2), or 199–226 (helix 3) has been found to form amyloid fibrils *in vitro* [17–21]. Human PrP(106–126) monomer has been reported to induce apoptosis in cultured neurons, and oligomers of this peptide have been shown to form cation channels and disrupt a model membrane non-specifically [22–27]. The importance of segment

106–147 was further established by the study of the crystal structure of the PrP globular domain, consisting of residues 123–230, which showed that the first β-strand (residues 128–131) of PrP^C forms an intramolecular β-sheet with residues 161–163 [28]. Molecular dynamics simulations showed that all three helices were conserved during the conversion of PrP^C to PrP^{Sc}, while an additional β-strand was predicted to form in the loop between β-strand 1 (residues 128–131) and helix 1(residues 144–154) [29], [30]. Moreover, it has been reported that a lipid-anchored 61–amino acid peptide from prion protein (lacking residues 23 to 88 and 141 to 221 of the mature prion protein), termed PrP61, can form β-sheet-rich protease-resistant fibrils at a physiological pH [31]. Therefore the region encompassing the first β-strand and the loop connecting the first β-strand to helix 1 might participate in the structural conversion and amyloid fibril formation.

Since amyloid formation is one of the hallmarks of prion pathogenesis, it is an excellent *in vitro* model for studying the critical region involved in the structural conversion. Considering the nucleation-dependent polymerization model of amyloidogenesis, the ability of synthetic peptide-derived amyloid fibrils to act as nuclei for the polymerization of full-length PrP would shed light upon the relative importance of different regions as cores for PrP amyloid formation. In this study, three synthetic peptides, mPrP(107–143), mPrP(107–126), and mPrP(127–143), were synthesized and the amyloid fibrils formed from these three peptides were used as seeds to determine the segment within sequence 107–143 which can act as the core region in prion protein amyloidogenesis *in vitro*, based on the ability of these peptides to cross-seed full-length prion protein mPrP(23–230).

Materials and Methods

Peptide Synthesis

The prion peptides used were synthesized using the Fmoc-polyamide method on a PS3 peptide synthesizer (Rainin, USA) [32]. The N- and C-terminal ends of the peptides were acetylated and amidated, respectively, in order to mimic the polypeptide bond in the full-length protein. The peptides were characterized by mass spectrometry after purification. After lyophilization, the peptides were stored at –30°C.

Expression and Purification of Full-length mPrP(23–230)

For expression and purification of mPrP(23–230), we followed the protocol described by Makarava et al [33]. Briefly, pET101/D-TOPO-mPrP(23–230), a kind gift from Dr. Ilia V. Baskakov (Center for Biomedical Engineering and Technology, University of Maryland Biotechnology Institute, USA), was transformed into *Escherichia coli* strain BL21 Star (DE3) (Invitrogen, Carlsbad, California, U.S.A). After induction in the presence of 1 mM IPTG for 5 hr, the cells were harvested, suspended in cell lysis buffer (50 mM Tris-HCl, 100 mM NaCl, 1 mM EDTA, pH 8.0), and subjected to repeated freeze-thaw cycles. Unless stated otherwise, all subsequent steps were conducted at room temperature. The

lysate was incubated with 0.2 mg/mL of lysozyme and 0.1 mM PMSF for 40 min with continuous stirring, then 1 mg/mL of deoxycholic acid was added and the mixture was incubated for 45 min, followed by addition of 5 μg/mL of DNase I and further 45 min incubation. Inclusion bodies were harvested by centrifugation of the lysate at 12,000g for 30 min at 4°C and solubilized in buffer A (8 M urea, 0.1 M Na$_2$HPO$_4$, 10 mM Tris-HCl, 10 mM reduced glutathione, pH 8.0). After centrifugation at 20000g for 30 minutes at 4°C, the soluble fraction was loaded onto a prepacked Ni column (HisTrapTM ^{FF} 1 mL, Amersham Biosciences) previously equilibrated with buffer A and non-bound proteins were removed by washing. Then mPrP(23–230) was eluted with buffer A at pH 4.5. The eluted protein was desalted on a HiPrep^{TM} 26/10 desalting column (Amersham Biosciences) at room temperature using 6 M urea, 0.1 M Tris-HCl, pH 7.5, as desalting buffer. Disulfide bond formation of the prion protein was induced by overnight oxidation at room temperature in the presence of 0.2 mM oxidized glutathione and 5 mM EDTA. The oxidized protein was purified at room temperature by reverse-phase chromatography on a C5 column (Discovery BIO Wide Pore C5, 10 μm, 25 cm×10.0 mm, Supelco, USA) with a 30 min linear gradient of 28–43% of buffer B (acetonitrile containing 0.1% trifluoroacetic acid). Oxidized mPrP(23–230) was eluted at about 33.3% of buffer B. The eluted protein was lyophilized and identified by ESI-TOF mass spectrometry and SDS-PAGE and stored at −30°C.

Thioflavin T (ThT) Binding Assay

Amyloid fibril formation of spontaneous and seeded amyloidogenesis of mPrP(23–230) was monitored using the Thioflavin T (ThT) binding assay [34]. Briefly, 30 μL of 200 μM ThT in 140 mM NaCl, 100 mM phosphate buffer (pH 8.5) was mixed with 30 μL of fibril solution for 1 min at room temperature and the fluorescence emission between 460 and 600 nm was measured in a 3-mm path-length rectangular cuvette in a FP-750 spectrofluorometer (JASCO, Japan) with excitation at 442 nm. Both excitation and emission slits were set at 5 nm.

Spontaneous Amyloid Fibril Formation by Mouse Prion Protein and Peptides

Purified recombinant mPrP(23–230) was dissolved in 6 M guanidine hydrochloride (GdnHCl) as a 132 μM stock solution. For fibrillization, 100 μL of the stock solution was diluted to 22 μM in 300 μL of fibril formation buffer (2x phosphate-buffered saline (PBS), 6 M urea, pH 6.0) and 200 μL of de-ionized water to give a final buffer composition of 1 M GdnHCl, 3 M urea in 1XPBS, pH 6.0 and the mixture was incubated 37°C with vigorous shaking (around 200–250 r.p.m.).

Peptides mPrP(107–143) and mPrP(127–143) were dissolved in deionized water as 100 μM stock solutions. The kinetics of amyloid formation was monitored in SpectraMax Gemini EM (Molecular Devices). Samples containing 50 μM of peptides in presence of 140 mM NaCl and 20 mM NaOAc, pH 3.7 and 10 μM ThT were incubated in 96 well assay plate (Corning, NY) at 25°C without shaking and kinetics was monitored by bottom reading of fluorescence intensity at every three hours interval using 445 nm excitation and 487 nm emission.

Peptide mPrP(107–126) was dissolved at a concentration of 754 μM in 20 mM HEPES buffer, pH 7.4, 100 mM NaCl, 0.01% NaN$_3$ and dissolution was assisted by sonication. The kinetics was measured at 37°C using SpectraMax Gemini EM as described above.

All set of experiments were measured in triplicate and subsequent results were expressed as average.

```
                  107                                                    143
mPrP(107-143)  NLKHVAGAAAAGAVVGGLGGYMLGSAMSRPMIHFGND

                  107                          126
mPrP(107-126)  NLKHVAGAAAAGAVVGGLGG

                                     127              143
mPrP(127-143)                     YMLGSAMSRPMIHFGND
```

Figure 1. Amino acid sequences of the prion peptides used.

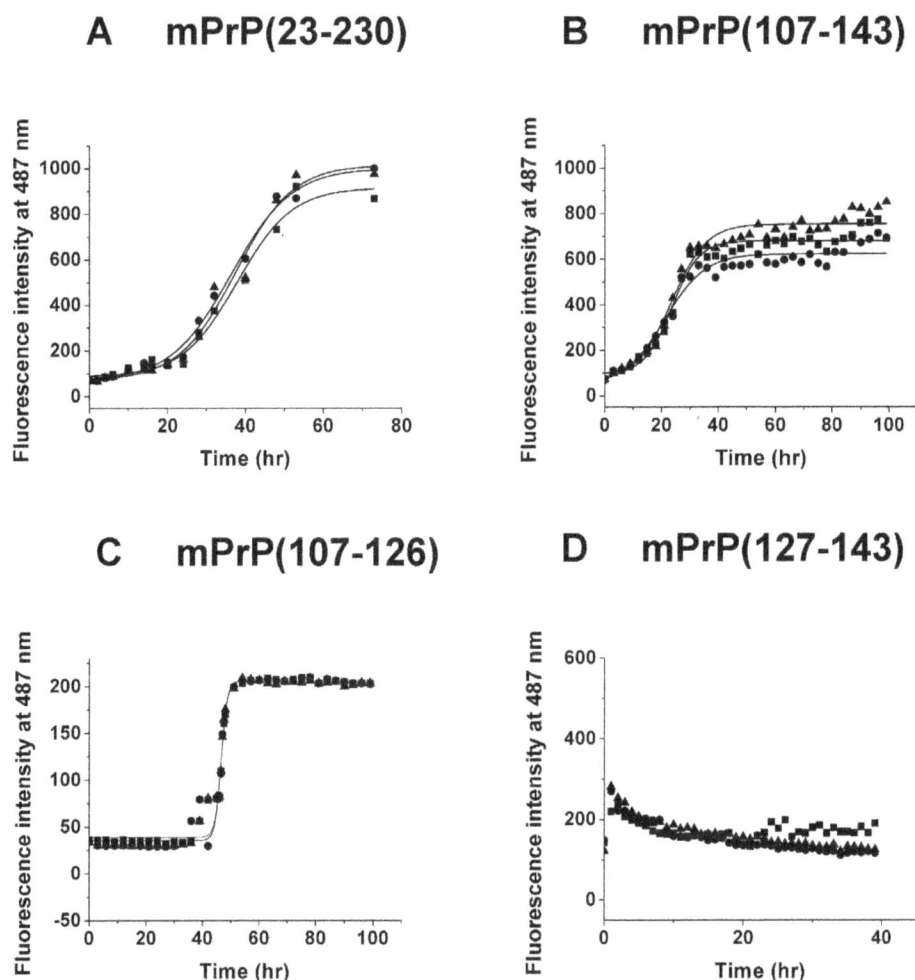

Figure 2. Spontaneous amyloid fibril formation of full length prion protein and prion peptides. Results from three independent measurements (denoted by closed square, ■; closed circle, ● and closed up triangle, ▲) are shown. (A) mPrP(23–230) (22 μM) in 1 M GdnHCl, 3 M urea in PBS, pH 6.0, was incubated at 37°C with shaking at 220 rpm. (B and D) mPrP(107–143) and mPrP(127–143) (50 μM), respectively in 140 mM NaCl and 20 mM NaOAc, pH 3.7, were incubated at 25°C without shaking. (C) mPrP(107–126) (754 μM) in 100 mM NaCl, 20 mM HEPES, pH 7.4, 0.01% NaN₃, was incubated at 37°C without shaking. The kinetics of amyloidogenesis was monitored by ThT binding assay.

Seed Preparation for the Seeding Assay

Amyloid fibrils, generated as described above, were spun down and re-suspended in de-ionized water. The concentration of monomer remaining in solution after the above centrifugation was determined by UV absorption, except for mPrP(107–126), which does not contain Tyr, where the monomer concentration in solution was determined by HPLC. Fibrils generated from peptides or from full-length protein were fragmented using, respectively, 20 or 60 cycles of intermittent pulses (one cycle consists of five pulses of 0.6 sec with a 5 sec interval between two consecutive cycles) with an ultrasonic processor (UP100H, Hielscher, USA) equipped with a 1 mm microtip immersed in the sample. The power during operation was set at 40%. The length of the fragmented fibrils was around 200 nm [34].

To prepare PK-digested mPrP(23–230) seed, the amyloid fibrils were spun down at 15,600 g for 30 minutes and digested for 1 h at 37°C with PK (enzyme to substrate ratio of 1:50), then the digestion was terminated by addition of 5 mM PMSF, and the PK-digested fibrils were spun down and used for preparing seed as described above.

Seeding Assay

For the seeding reaction, different amounts of sonicated seed were added to the monomer solution and amyloidogenesis was monitored by the ThT binding assay at different incubation times. The seed concentration could not be exactly determined because of the polymeric nature of the fibrils and was therefore expressed as the amount of protein or peptide monomer incorporated into amyloid fibrils. For example, in case of fibrillization of 1 mL of mPrP(23–230) solution (protein concentration 22.07 μM), 17.74 μM of monomer (17.74 nmoles of monomer) was recruited to amyloid fibrils whereas 4.33 μM of monomer (4.33 nmoles of monomer) remained in the supernatant after the amyloid fibrils were spun down. After centrifugation the fibrils were suspended in 200 μL of de-ionized water which gave rise to a concentration of 88.7 pmoles of mPrP(23–230) monomer per microliter fibril suspension. Likewise, in case of mPrP(107–143) and mPrP(127–143) after centrifugation the fibrils were suspended in 200 μL of de-ionized water which gave rise to 131.9 pmoles of mPrP(107–143) and 43.5 pmoles of mPrP(127–143) monomer per microliter fibril suspension. Lastly, 742.6 nmoles of mPrP(107–126) monomer was participated in fibril formation and the amyloid fibril was

A mPrP(23-230) fiber   **B mPrP(107-143) fiber**

C mPrP(107-126) fiber   **D mPrP(127-143) fiber**

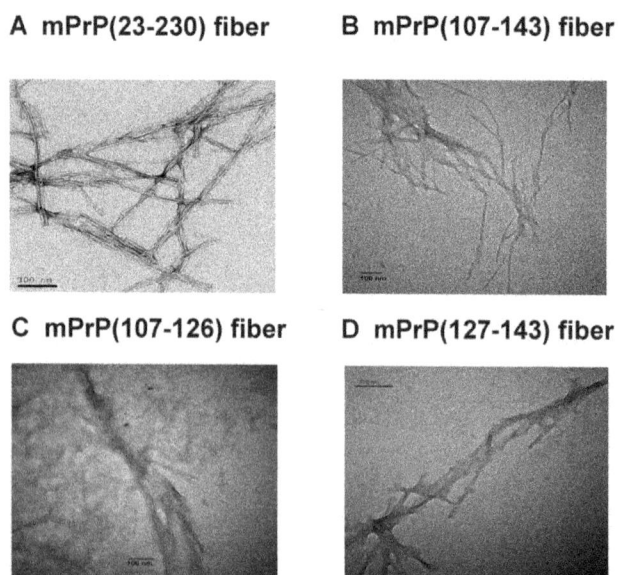

Figure 3. Morphology of amyloid fibrils. Transmission electron microscope images showing morphology of amyloid fibrils formed from (A) mPrP(23–230), (B) mPrP(107–143), (C) mPrP(107–126), or (D) mPrP(127–143). The bars represent 100 nm.

suspended in 200 μL of de-ionized water which gave rise to 1.8 nmoles of mPrP(107–126) monomer per microliter.

Data Analysis

Kinetic parameters for aggregation were obtained by monitoring the ThT-amyloid fluorescence signal over time and fitting fluorescence (F) data points to the sigmoidal curve of the following equation using Origin 8.0 graphical software [35].

$$F = A + B / \left(1 + e \wedge \left(k * \left(t_{1/2} - t\right)\right)\right)$$

The above equation fits very well with fibril formation data. In this equation, A is the fluorescence during the lag phase, B the difference in fluorescence between the lag phase and post-transition plateau, t the incubation time, $t_{1/2}$ the time required to reach half the maximum ThT emission and k the rate constant of fibril growth (h^{-1}). We defined the lag time as $t_{1/2} - 2/k$ for each fitted curve.

Transmission Electron Microscopy

Amyloid fibrils, generated as described above, were loaded onto carbon-coated 300-mesh copper grids, incubated for 4 min for absorption and then stained with 2% uranyl acetate for 4 min. After overnight drying in a desiccator, the samples were viewed on a Hitachi H-7000 electron microscope at 75 kV.

Results

Spontaneous Amyloid Fibril Formation by Mouse Full-length Prion Protein and Prion Peptides

The sequences of the three prion peptides used are shown in Figure 1. Spontaneous amyloid fibril formation by full-length mouse prion protein mPrP(23–230) in 1X PBS containing 1 M GdnHCl and 3 M urea, pH 6.0; by mouse prion peptides mPrP(107–143) and mPrP(127–143) in 140 mM NaCl in

20 mM NaOAc, pH 3.7, and by mPrP(107–126) in 20 mM HEPES buffer, pH 7.4, containing 100 mM NaCl and 0.01% NaN$_3$ was monitored by measuring the fluorescence emission of the amyloid fibril-ThT complex at 487 nm over time (Figure 2). For each experimental condition, three independent kinetic measurements were carried out. The typical sigmoid nature of the ThT-amyloid fluorescence plot indicates the generation of a molecular population with increased β-sheet content, an indicator of amyloid formation. In the case of the full-length protein mPrP(23–230), 1 M GdnHCl and 3 M urea were required to destabilize the native structure to facilitate its conversion into fibrils, and shaking was necessary to increase the chance of protein contact [36], [37]. The average lag time for spontaneous amyloidogenesis of mPrP(23–230) monomer in three independent experiments was 22.4 hours (Figure 2A). When peptides mPrP(107–143) and mPrP(127–143) were incubated in 140 mM NaCl and 20 mM NaOAc, pH 3.7, the average lag time for amyloidogenesis of mPrP(107–143) was 12.9 hours (Figure 2B), whereas mPrP(127–143) formed fibrils very rapidly and its lag time was hardly detectable (Figure 2D) suggesting mPrP(127–143) is much more amyloidogenic than mPrP(107–143). In contrast, mPrP(107–126) did not form fibrils under the same conditions (data not shown) and a very high peptide concentration (754 μM) and a different buffer (20 mM HEPES buffer, pH 7.4, 100 mM NaCl, 0.01% NaN$_3$) were required for amyloidogenesis (Figure 2C) [38]. Under this condition the average lag time for spontaneous amyloidogenesis of mPrP(107–126) was found to be 41.2 hours. Figure 3 shows the morphology of these four kinds of fibrils under transmission electron microscope and that no significant difference in morphology was apparent.

Cross-seeding Ability of mPrP(107–143) and mPrP(23–230)

If the amyloid fibrils of mPrP(107–143) and mPrP(23–230) use the same segment to form the cross-β structure, fibrils formed from mPrP(107–143) should be able to serve as template (seed) to initiate fibril formation by mPrP(23–230) monomer, and the fibrils formed from mPrP(23–230) should be able to initiate fibril formation by the mPrP(107–143) monomer. To prepare seed from a fibril suspension, centrifugation was first used to remove any remaining monomer, then ultrasonication was used to fragment long fibrils to generate seed [34]. In order to assess efficient fragmentation of fibril clump by sonication, we subjected the sonicated fibrils (seeds) to transmission electron microscopy as shown in Figure S1. Homologous seeding of the mPrP(23–230) monomer with the mPrP(23–230) seed showed a typical seeding effect in three independent experiments, as seen in Figure 4A where addition of 20 and 40 μL of mPrP(23–230) seed shortened the average lag time from 22.4 hours to 12.5 and 1.3 hours, respectively. On the other hand, heterologous seeding of mPrP(23–230) monomer with mPrP(107–143) seed showed immediate fibrillogenesis in all three independent measurements (Figure 4B).

We also examined the cross seeding effect of preformed mPrP(23–230) fibril on the amyloidogenesis of mPrP(107–143) monomer. As shown in Figure 5, addition of 20 μL of mPrP(23–230) seed shortened the average lag time from 12.9 to 5.2 hours. Because mPrP(23–230) is more than 5 times longer than mPrP(107–143), we wondered whether the presence of residues other than the amyloid core region might interfere the association between mPrP(107–143) monomer and mPrP(23–230) seed and affect seeding efficiency [39]. To test this, the same amount of mPrP(23–230) seed was digested with PK for 1 hr at 37°C to remove the non-amyloid part of the protein fibrils, then the PK-

A mPrP(23-230) monomer + mPrP(23-230) seed

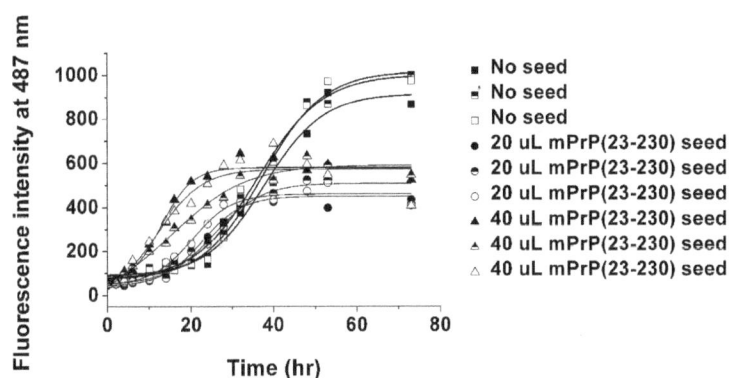

B mPrP(23-230) monomer + mPrP(107-143) seed

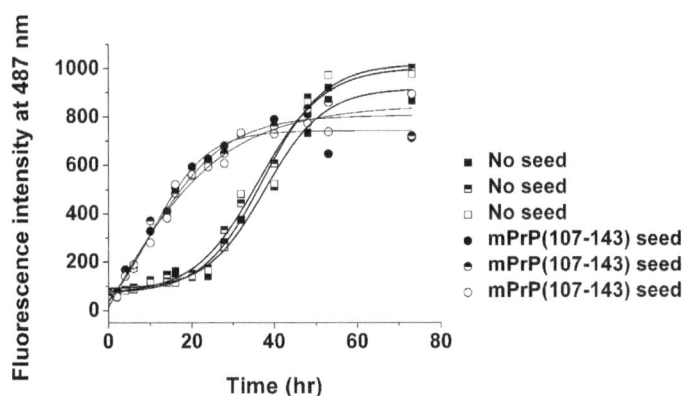

Figure 4. Amyloid fibril formation by mPrP(23–230) seeded with mPrP(23–230) seed or mPrP(107–143) seed. (A) Amyloidogenesis of mPrP(23–230) (22 µM) in 1 M GdnHCl, 3 M urea in PBS, pH 6.0, incubated at 37°C with vigorous shaking in the absence of seed (denoted by closed, half-filled and open squares for three independent measurements) or in the presence of 20 µL (denoted by closed, half-filled and open circles for three independent measurements) or 40 µL (denoted by closed, half-filled and open up triangles for three independent measurements) of sonicated mPrP(23–230) seed. In these three seeding experiments, mPrP(23–230) seed was prepared independently and three batches of seed contained 88.7, 88.3 and 96.2 pmoles of mPrP(23–230) monomer per microliter seed solution. (B) Amyloidogenesis of mPrP(23–230) under the same conditions as in (A) cross-seeded with 20 µL of sonicated mPrP(107–143) seed (denoted by closed, half-filled and open circles for three independent measurements). The seed contained 131.9, 129.1, and 119.7 pmoles of mPrP(107–143) per microliter solution in these three independent measurements. The data of spontaneous amyloidogenesis without seeding are also shown in the same plot for comparison.

digested mPrP(23–230) fibrils were collected by centrifugation of the reaction mixture, re-suspended in the same volume of water (shown in Figure S2), and its seeding ability was tested. As shown in Figure 5, the PK-digested seed was apparently more effective at initiating amyloidogenesis of mPrP(107–143), indicating that the effective association between the amyloid core of the seed and the monomer is an important determinant of seeding efficiency.

Comparison of the Cross-seeding Effect of mPrP(107–126) and mPrP(127–143) Fibrils on the mPrP(23–230) Monomer

The results above suggested that mPrP(23–230) and mPrP(107–143) might share the same amyloid core. Within mPrP(107–143), the N-terminal half contains the amyloidogenic sequence "AGAAAAGA" and is more hydrophobic than the C-terminal half. However, as shown in Figures 2C and 2D, mPrP(127–143)

formed amyloid rapidly and at a low concentration, whereas, in the case of mPrP(107–126), the amyloidogenesis was slow and required a very high peptide concentration. In the amyloidogenesis of the mPrP(23–230) monomer, shown in Figure 6, addition of 50 µL of mPrP(107–126) seed which contains as high as 1.8 nmoles mPrP(107–126) monomers per microliter seed solution still showed no seeding effect. In fact, the average lag time of amyloid formation of mPrP(23–230) was even slightly prolonged in the presence of mPrP(107–126) seed (Figure 6A), suggesting that mPrP(107–126) fibrils might interfere the self-association of mPrP(23–230). To test whether the lack of seeding effect of preformed mPrP(107–126) fibril was due to its less stability in the denaturing condition used for amyloidogenesis of mPrP(23–230) monomer throughout this study, we incubated all three fibrils generated from short prion peptides for 4 days in the fibrillization buffer and subjected to transmission electron microscopy to

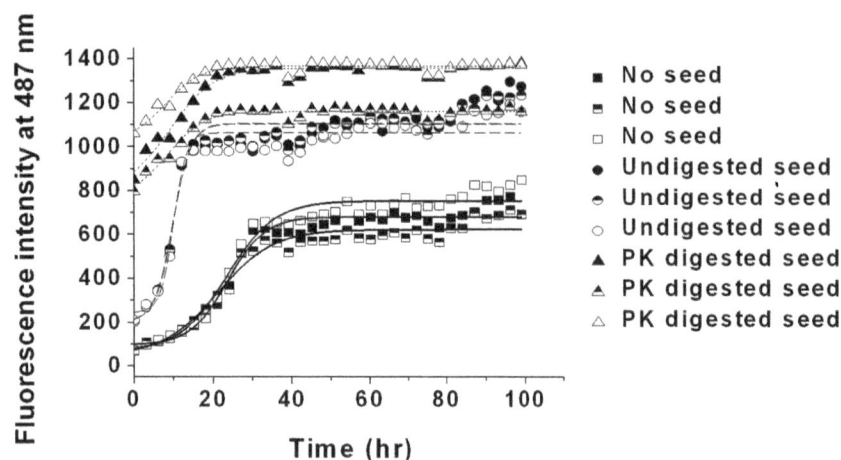

Figure 5. Amyloid fibril formation by mPrP(107–143) monomer in the absence of seed or in the presence of mPrP(23–230) seed with or without proteinase K digestion. mPrP(107–143) (50 μM) in 140 mM NaCl and 20 mM NaOAc, pH 3.7, was incubated at 25°C without shaking alone (denoted by closed, half-filled and open squares for three independent measurements), in the presence of 30 μL of mPrP(23–230) seed (denoted by closed, half-filled and open circles for three independent measurements), or in the presence of the same amount of mPrP(23–230) seed digested by proteinase K (denoted by closed, half-filled and open up triangles for three independent measurements).

analyze morphology. As shown in Figure S3, all three peptide-generated fibrils remained intact under denaturing condition, ruling out the possibility of poor stability. In contrast, amyloido-genesis of mPrP(23–230) was induced immediately on addition of only 20 μL of sonicated mPrP(127–143) seed containing only 44 pmoles of mPrP(127–143) per microliter seed solution (Figure 6B), a seeding effect similar to that seen with mPrP(107–143) seed in Figure 4B. Our results showed that, though peptide mPrP(107–143) can seed full-length recombinant prion protein, the seeding ability resides in the C-terminal segment of this peptide.

Discussion

The *in vitro* formation of amyloid fibril from soluble monomeric recombinant prion protein provides an insight into the structural conversion of prion protein, which ultimately leads to amyloido-genesis. With regard to the structure of soluble prion protein, it is important to locate the regions, which take part in the conversion process. According to several models, the process of β-aggregation starts when segments that possess high hydrophobicity, a high β-sheet propensity, and low net charge become exposed to the solvent and can associate [40–43]. Hydrophobicity analysis of the prion protein sequence revealed the existence of three hydropho-bic clusters, one in the region of amino acids 110–137 and the other two reside in helices 2 and 3 [21]. The N-terminal half of mPrP(107–143), i.e. mPrP(107–126), formed spontaneous amyloid fibrils, though with a considerable lag phase. This is in agreement with the findings reported by Gasset *et al* [15]. One interesting point is that this peptide needed a much higher monomer concentration (754 μM) for initiation of fibril formation, but the monomer concentration remained in solution after fibrillization was only 12.4, 11.1 and 6.6 μM in three independent experiments. In contrast, the C-terminal half of mPrP(107–143), i.e. mPrP(127–143), underwent fibrillization without any detectable lag time for nucleus formation at a peptide concentration of 50 μM but the monomer concentration remained in solution after fibrillization was 32.6, 35.6 and 27.2 μM in three independent experiments. Our data suggested that (1) mPrP(127–143) might contain an intrinsic structural element that drives nucleation; (2) mPrP(127–143) has a higher thermodynamic solubility than mPrP(107–126);

and (3) mPrP(107–126) might have a much higher energy barrier in the nucleation step. In this connection it is worth to mention that only having high hydrophobicity does not ensure a peptide segment of a protein to act as nucleation site where amyloidogen-esis can begin. This notion is supported by the fact that in spite of having high hydrophobicity mPrP(107–126) requires high mono-mer concentration probably to overcome a high energy barrier during nucleation. In order to locate potential sites of nucleation which can act as amyloidogenic hot-spots we utilized two bioinformatic prediction methods, namely, FoldAmyloid [44] and Aggrescan [45], which use amino acid composition of proteins as the basic approach for assigning amyloidogenic hot-spots. Prediction from both the methods revealed potential amyloido-genic hot-spots within the sequence 127–143. Thus immediate fibrillization of prion peptide mPrP(127–143) might happen due to the presence of amyloidogenic hot-spots within this region that can act as a nucleation site.

While investigating the cross-seeding ability of mPrP(23–230) seed in the fibrillization of mPrP(107–143) monomer, the lag time was found to be shortened compared to the unseeded reaction, but not eliminated. This is in agreement with the 'surface competition hypothesis' that we proposed earlier [39]. Based on this hypothesis the process of seed induced fibrillization is a two steps phenomenon which involves initial 'docking' - the association of monomer with the preexisting seed followed by 'locking' i.e., the formation of cross-β structure between incoming monomer and the seed. In presence of protein segments not involved in the formation of amyloid core, the probability of incorrect binding between incoming monomer and seed becomes fairly high enough that leads to a detectable lag time for the reason that incorrect binding cannot go through subsequent elongation step resulting in a delay of fibril growth. Pre-digestion of the fibrils with PK reduces the probability of incorrect binding and favors the association between the monomer and the seeding nucleus thus shortening the lag phase greatly. Hence our data infer that the residues not involved in the structural conversion process could contribute to the species barrier in prion transmission [39]. Bocharova et al. reported that the PK-digested mPrP(23–230) fibrils harbor an epitope of a monocolonal antibody D18, part of which spans amino acid residues 133–143. This result also supports our

A mPrP(23-230) monomer + mPrP(107-126) seed

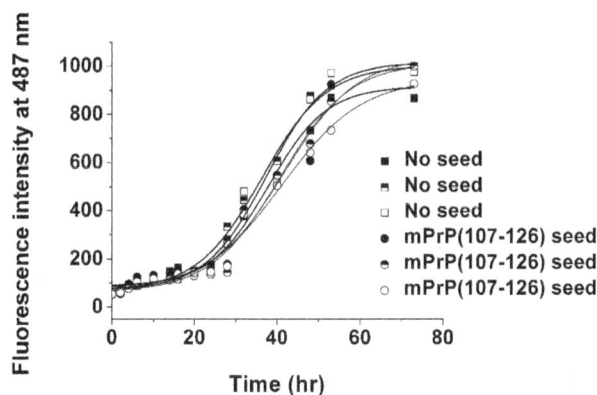

B mPrP(23-230) monomer + mPrP(127-143) seed

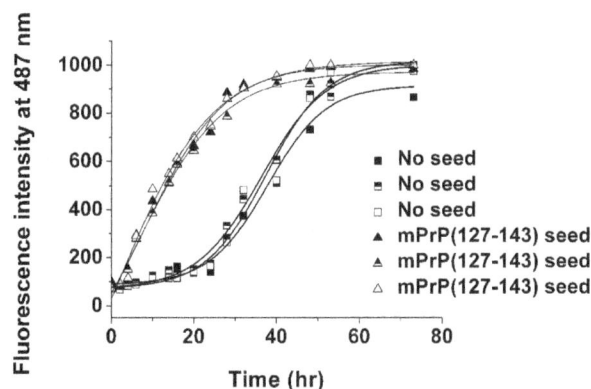

Figure 6. Amyloid fibril formation by mPrP(23–230) monomer cross-seeded with mPrP(107–126) seed or mPrP(127–143) seed. mPrP(23–230) (22 µM) in 1 M GdnHCl, 3 M urea in PBS, pH 6.0, was incubated at 37°C with vigorous shaking alone (in (A) as well as (B) denoted by closed, half-filled and open squares for three independent measurements) or in the presence of (A) 50 µL of mPrP(107–126) seed (denoted by closed, half-filled and open circles for three independent measurements) containing 1.8 nmoles of mPrP(107–126) per microliter or (B) 20 µL of mPrP(127–143) seed (denoted by closed, half-filled and open up triangles for three independent measurements) containing 43.5, 36 and 57 pmoles (for three independent measurements) of mPrP(127–143) monomer per microliter.

conclusion that sequence 127–143 is in the PK resistance core of amyloid fibrils derived from mPrP(23–230) [46].

Cross-seeding of mPrP(23–230) monomer with mPrP(107–126) seed failed to shorten the lag time, even though a very high amount of mPrP(107–126) seed was used, indicating the inability of the full-length monomer to interact with the seed. The recruitment of a new monomer to a preexisting nucleus/fibril is an essential step in amyloid propagation. Successful cross-seeding relies on conformational adaptability between monomer and seed. Although, theoretically, all peptides or proteins have the potential to form amyloid structure and many prion peptides have been reported to be able to form amyloid fibrils, the lack of conformational adaptability between a specific monomer/seed pair might impede the successful propagation of prion amyloid formed from that particular monomer. Amyloid generated from mPrP(107–126) appears to be inaccessible to mPrP(23–230), accounting for the inability of mPrP(107–126)-derived fibrils to seed amyloidogenesis of mPrP(23–230) monomer. This has also been demonstrated by Kundu et al [47], who showed that amyloid

fibril formed from synthetic PrP106–126 peptide (5%, wt/wt) was unable to seed 400 µM of huPrP23–144 monomer. In fact, in our cross-seeding experiment, the lag time for fibrillization with the mPrP(107–126) seed was even longer than in the unseeded reaction (Figure 6A). These results suggest that a longer time is required for the nucleation of the mPrP(23–230) monomer in the presence of mPrP(107–126) seed, i.e., the mPrP(107–126) seed inhibits the homologous association of mPrP(23–230) to form a nucleus. This is consistent with a previous report that hamster prion peptide P106–128 (corresponding to mouse sequence 105–127) has an inhibitory effect on the conversion of PrPsen (i.e. PrP^C) to PrPres (PrP^{Sc}) in a cell-free system [48].

The absence of a lag phase in the cross-seeding experiment using mPrP(23–230) monomer and fibrils derived from mPrP(127–143) suggests that the segment containing residues 127–143 probably forms the amyloid core of mPrP(23–230). Recently, Yamaguchi et al. [21] examined the amyloidogenic properties of several synthetic peptides and the ability of the peptide fibrils to induce amyloid formation of full-length mPrP and concluded that helix 2, and not helix 3, is important in initiating full-length mPrP fibrillization. Under their incubation conditions, only peptides corresponding to helix 2 and helix 3 could form amyloid fibrils, so only the fibrils derived from these two peptides were used in the cross-seeding reaction. In this regard it is essential to consider that a degree of structural cooperation might exist between sequence 127–143 and the C-terminal helices of prion protein in driving the structural conversion and the propagation of the converted structure. This argument is supported by a recent report showing that two sub-domains made up of [β-strand 1]-[helix 1]-[β-strand 2] and [helix 2]-[helix3] swap and separate from each other before dimerization [49]. Our results are also corroborated by the observation, made by Singh *et al.* [50], where they identified segments in prion protein, which exhibit multiple local conformations. They showed that the structural heterogeneity in the prion protein is not restricted to only the C-terminal domain rather it also manifests itself in the N-terminal segments like sequences 32–55 and 109–132 also. This study is highly pertinent to our work in a manner that both indicate to the contention that structural conversion of prion protein might be a collaborative phenomenon.

In conclusion, our study using seed-mediated *in vitro* amyloid formation shows that residues 127–143 harbor a critical region of prion protein, which, either independently or in conjunction with other critical regions, can initiate the seed-directed structural conversion of full-length prion protein.

Supporting Information

Figure S1   Fragmentation of preformed amyloid fibrils for seed preparation. Preformed amyloid fibrils made from full length prion protein and prion peptides were collected by centrifugation at 15,600g for 30 minutes. The fibrils were suspended in water and sonicated as described in Materials and Methods. The sonicated fibrils were used as seed in seeding experiments. The sonicated fibrils (A) mPrP(23–230), (B) mPrP(107–143), (C) mPrP(107–126) and (D) mPrP(127–143) were

viewed by transmission electron microscopy. The bars represent 100 nm.

Figure S2   Proteinase K (PK) digestion of preformed amyloid fibrils made from mPrP(23–230). Amyloid fibrils from spontaneous amyloidogenesis of mPrP(23–230) monomer were collected by centrifugation at 15,600g for 30 minutes at room temperature. The fibrils were then digested with PK as described in Materials and Methods. The PK-treated fibrils were viewed by transmission electron microscopy.

Figure S3   Assessment of stability of amyloid fibrils made from short prion peptides in denaturing condition. To assess the chemical stability of amyloid fibrils made from short prion peptides, preformed fibrils were incubated in the fibril formation buffer used for amyloidogenesis of mPrP(23–230) which contains 1 M GdnHCl and 3 M urea in PBS, pH 6.0 for 4 days at 37°C with shaking at 220 rpm. The integrity of fibril morphology (A) mPrP(107–143), (B) mPrP(107–126) and (C) mPrP(127–143) was analyzed by transmission electron microscopy.

Acknowledgments

We thank Dr. Ilia Baskakov for kindly providing the plasmid containing the mouse PrP gene. The ESI-TOF mass identification of proteins was performed by the Core Facilities at the Institute of Biological Chemistry, Academia Sinica, supported by the National Science Council and the Academia Sinica. We thank Mr. Tai-Lang Lin and the Core Facility of the Institute of Cellular and Organismic Biology, Academia Sinica, Taiwan for assistance in transmission electron microscopy.

Author Contributions

Conceived and designed the experiments: BC RPC. Performed the experiments: BC CYL CL CLH. Analyzed the data: BC EHC RPC. Contributed reagents/materials/analysis tools: CCY RPC. Wrote the paper: BC RPC.

References

1. Prusiner SB (1998) Prions. Proc Natl Acad Sci USA 95: 13363–13383.
2. Aguzzi A, Polymenidou M (2004) Mammalian prion biology: one century of evolving concepts. Cell 116: 313–327.
3. Chesebro B (2003) Introduction to the transmissible spongiform encephalopathies or prion diseases. Br Med Bull 66: 1–20.
4. Collinge J (2001) Prion diseases of humans and animals: their causes and molecular basis. Annu Rev Neurosci 24: 519–550.
5. Cohen FE, Prusiner SB (1998) Pathologic conformations of prion proteins. Annu Rev Biochem 67: 793–819.
6. Riek R, Hornemann S, Wider G, Billeter M, Glockshuber R, et al. (1996) NMR structure of the mouse prion protein domain PrP(121–231). Nature 382: 180–182.
7. Riek R, Hornemann S, Wider G, Glockshuber R, Wuthrich K (1997) NMR characterization of the full-length recombinant murine prion protein, mPrP(23–231). FEBS Lett 413: 282–288.
8. Riesner D (2003) Biochemistry and structure of PrP(C) and PrP(Sc). Br Med Bull 66: 21–33.
9. Pan K, Baldwin M, Nguyen J, Gasset M, Serban A, et al. (1993) Conversion of α-helices β-sheets features in the formation of the scrapie prion proteins. Proc Natl Acad Sci USA 90: 10962–10966.
10. Caughey BW, Dong A, Bhat KS, Ernst D, Hayes SF, et al. (1991) Secondary structure analysis of the scrapie-associated protein PrP 27–30 in water by infrared spectroscopy. Biochemistry 30: 7672–7680.
11. Donne DG, Viles JH, Groth D, Mehlhorn I, James TL, et al. (1997) Structure of the recombinant full-length hamster prion protein PrP(29–231): The N terminus is highly flexible. Proc Natl Acad Sci USA 94: 13452–13457.
12. Gasset M, Baldwin MA, Fletterick RJ, Prusiner SB (1993) Perturbation of the secondary structure of the scrapie prion protein under conditions that alter infectivity. Proc Natl Acad Sci USA 90: 1–5.
13. Safar J, Roller P, Gajdusek D, Gibbs C Jr (1993) Conformational transitions, dissociation, and unfolding of scrapie amyloid (prion) protein. J Biol Chem 268: 20276–20284.
14. McKinley MP, Bolton DC, Prusiner SB (1983) A protease-resistant protein is a structural component of the scrapie prion. Cell 35: 57–62.
15. Gasset M, Baldwin MA, Lloyd DH, Gabriel JM, Holtzman DM, et al. (1992) Predicted alpha-helical regions of the prion protein when synthesized as peptides form amyloid. Proc Natl Acad Sci U S A 89: 10940–10944.
16. Baskakov IV, Aagaard C, Mehlhorn I, Wille H, Groth D, et al. (2000) Self-assembly of recombinant prion protein of 106 residues. Biochemistry 39: 2792–2804.
17. Walsh P, Simonetti K, Sharpe S (2009) Core structure of amyloid fibrils formed by residues 106–126 of the human prion protein. Structure 17: 417–426.
18. Lin NS, Chao JC, Cheng HM, Chou FC, Chang CF, et al. (2010) Molecular structure of amyloid fibrils formed by residues 127 to 147 of the human prion protein. Chemistry 16: 5492–5499.
19. Tagliavini F, Prelli F, Verga L, Giaccone G, Sarma R, et al. (1993) Synthetic peptides homologous to prion protein residues 106–147 form amyloid-like fibrils in vitro. Proc Natl Acad Sci USA 90: 9678–9682.
20. Zhang H, Kaneko K, Nguyen JT, Livshits TL, Baldwin MA, et al. (1995) Conformational transitions in peptides containing two putative α-helices of the prion protein. J Mol Biol 250: 514–526.
21. Yamaguchi K, Matsumoto T, Kuwata K (2008) Critical region for amyloid fibril formation of mouse prion protein: unusual amyloidogenic properties of the helix 2 peptide. Biochemistry 47: 13242–13251.
22. Ettaiche M, Pichot R, Vincent JP, Chabry J (2000) In vivo cytotoxicity of the prion protein fragment 106–126. J Biol Chem 275: 36487–36490.
23. Forloni G, Angeretti N, Chiesa R, Monzani E, Salmona M, et al. (1993) Neurotoxicity of a prion protein fragment. Nature 362: 543–546.
24. Thellung S, Florio T, Corsaro A, Arena S, Merlino M, et al. (2000) Intracellular mechanisms mediating the neuronal death and astrogliosis induced by the prion protein fragment 106–126. Int J Dev Neurosci 18: 481–492.
25. Dupiereux I, Zorzi W, Lins L, Brasseur R, Colson P, et al. (2005) Interaction of the 106–126 prion peptide with lipid membranes and potential implication for neurotoxicity. Biochem Biophys Res Commun 331: 894–901.

26. Lin MC, Mirzabekov T, Kagan BL (1997) Channel formation by a neurotoxic prion protein fragment. J Biol Chem 272: 44–47.

27. Kourie JI, Culverson A (2000) Prion peptide fragment PrP[106–126] forms distinct cation channel types. J Neurosci Res 62: 120–133.

28. Haire LF, Whyte SM, Vasisht N, Gill AC, Verma C, et al. (2004) The crystal structure of the globular domain of sheep prion protein. J Mol Biol 336: 1175–1183.

29. DeMarco ML, Daggett V (2004) From conversion to aggregation: protofibril formation of the prion protein. Proc Natl Acad Sci USA 101: 2293–2298.

30. DeMarco ML, Silveira J, Caughey B, Daggett V (2006) Structural properties of prion protein protofibrils and fibrils: an experimental assessment of atomic models. Biochemistry 45: 15573–15582.

31. Supattapone S, Bouzamondo E, Ball HL, Wille H, Nguyen HO, et al. (2001) A protease-resistant 61-residue prion peptide causes neurodegeneration in transgenic mice. Mol Cell Biol 21: 2608–2616.

32. Hsu RL, Lee KT, Wang JH, Lee LY, Chen RP (2009) Amyloid-degrading ability of nattokinase from Bacillus subtilis natto. J Agri Food Chem 57: 503–508.

33. Makarava N, Baskakov IV (2008) Expression and purification of full-length recombinant PrP of high purity. Methods Mol Biol 459: 131–143.

34. Lee LY, Chen RP (2007) Quantifying the sequence-dependent species barrier between hamster and mouse prions. J Am Chem Soc 129: 1644–1652.

35. Alvarez-Martinez MT, Fontes P, Zomosa-Signoret V, Arnaud JD, Hingant E, et al. (2011) Dynamics of polymerization shed light on the mechanisms that lead to multiple amyloid structures of the prion protein. Biochim Biophys Acta 1814: 1305–1317.

36. Kocisko DA, Come JH, Priola SA, Chesebro B, Raymond GJ, et al. (1994) Cell-free formation of protease-resistant prion protein. Nature 370: 471–474.

37. Bocharova OV, Breydo L, Parfenov AS, Salnikov VV, Baskakov IV (2005) In vitro conversion of full-length mammalian prion protein produces amyloid form with physical properties of PrP^Sc. J Mol Biol 346: 645–659.

38. Lee SW, Mou Y, Lin SY, Chou FC, Tseng WH, et al. (2008) Steric zipper of the amyloid fibrils formed by residues 109–122 of the Syrian hamster prion protein. J Mol Biol 378: 1142–1154.

39. Liao TY, Lee LY, Chen RP (2011) Leu-138 in the bovine prion peptide fibrils is involved in the seeding discrimination related to codon-129 M/V polymorphism in the prion peptide seeding experiment. FEBS J 278: 4351–4361.

40. Serpell LC, Sunde M, Blake CC (1997) The molecular basis of amyloidosis. Cell Mol Life Sci 53: 871–887.

41. Blake C, Serpell L (1996) Synchrotron X-ray studies suggest that the core of the transthyretin amyloid fibril is a continuous beta-sheet helix. Structure 4: 989–998.

42. Serpell LC, Blake CC, Fraser PE (2000) Molecular structure of a fibrillar Alzheimer's A beta fragment. Biochemistry 39: 13269–13275.

43. Lopez De La Paz M, Goldie K, Zurdo J, Lacroix E, Dobson CM, et al. (2002) De novo designed peptide-based amyloid fibrils. Proc Natl Acad Sci USA 99: 16052–16057.

44. Garbuzynskiy SO, Lobanov MY, Galzitskaya OV (2010) FoldAmyloid: a method of prediction of amyloidogenic regions from protein sequence. Bioinformatics 26: 326–332.

45. Conchillo-Sole O, de Groot NS, Aviles FX, Vendrell J, Daura X, et al. (2007) AGGRESCAN: a server for the prediction and evaluation of "hot spots" of aggregation in polypeptides. BMC Bioinformatics 8: 65–81.

46. Bocharova OV, Breydo L, Salnikov VV, Gill AC, Baskakov IV (2005) Synthetic prions generated in vitro are similar to a newly identified subpopulation of PrPSc from sporadic Creutzfeldt-Jakob Disease. Protein Sci 14: 1222–1232.

47. Kundu B, Maiti NR, Jones EM, Surewicz KA, Vanik DL, et al. (2003) Nucleation-dependent conformational conversion of the Y145Stop variant of human prion protein: structural clues for prion propagation. Proc Natl Acad Sci USA 100: 12069–12074.

48. Chabry J, Caughey B, Chesebro B (1998) Specific inhibition of in vitro formation of protease-resistant prion protein by synthetic peptides. J Biol Chem 273: 13203–13207.

49. Hafner-Bratkovic I, Bester R, Pristovsek P, Gaedtke L, Veranic P, et al. (2011) Globular domain of the prion protein needs to be unlocked by domain swapping to support prion protein conversion. J Biol Chem 286: 12149–12156.

50. Singh J, Sabareesan AT, Mathew MK, Udgaonkar JB (2012) Development of the structural core and of conformational heterogeneity during the conversion of oligomers of the mouse prion protein to worm-like amyloid fibrils. J Mol Biol 423: 217–231.

19

Amyloid Precursor Protein (APP) Mediated Regulation of Ganglioside Homeostasis Linking Alzheimer's Disease Pathology with Ganglioside Metabolism

Marcus O. W. Grimm[1,2,3]*, Eva G. Zinser[3], Sven Grösgen[3], Benjamin Hundsdörfer[3], Tatjana L. Rothhaar[3], Verena K. Burg[3], Lars Kaestner[4], Thomas A. Bayer[5], Peter Lipp[4], Ulrike Müller[6], Heike S. Grimm[3], Tobias Hartmann[1,2,3]*

1 Deutsches Institut für DemenzPrävention (DIDP), Saarland University, Homburg/Saar, Germany, 2 Neurodegeneration and Neurobiology, Saarland University, Homburg/Saar, Germany, 3 Experimental Neurology, Saarland University, Homburg/Saar, Germany, 4 Molecular Cellbiology, Saarland University, Homburg/Saar, Germany, 5 Department for Psychiatry, University of Goettingen, Goettingen, Germany, 6 Institute for Pharmacy and Molecular Biotechnology (IPMB), University of Heidelberg, Heidelberg, Germany

Abstract

Gangliosides are important players for controlling neuronal function and are directly involved in AD pathology. They are among the most potent stimulators of Aβ production, are enriched in amyloid plaques and bind amyloid beta (Aβ). However, the molecular mechanisms linking gangliosides with AD are unknown. Here we identified the previously unknown function of the amyloid precursor protein (APP), specifically its cleavage products Aβ and the APP intracellular domain (AICD), of regulating GD3-synthase (GD3S). Since GD3S is the key enzyme converting a- to b-series gangliosides, it therefore plays a major role in controlling the levels of major brain gangliosides. This regulation occurs by two separate and additive mechanisms. The first mechanism directly targets the enzymatic activity of GD3S: Upon binding of Aβ to the ganglioside GM3, the immediate substrate of the GD3S, enzymatic turnover of GM3 by GD3S was strongly reduced. The second mechanism targets GD3S expression. APP cleavage results, in addition to Aβ release, in the release of AICD, a known candidate for gene transcriptional regulation. AICD strongly down regulated GD3S transcription and knock-in of an AICD deletion mutant of APP in vivo, or knock-down of Fe65 in neuroblastoma cells, was sufficient to abrogate normal GD3S functionality. Equally, knock-out of the presenilin genes, presenilin 1 and presenilin 2, essential for Aβ and AICD production, or of APP itself, increased GD3S activity and expression and consequently resulted in a major shift of a- to b-series gangliosides. In addition to GD3S regulation by APP processing, gangliosides in turn altered APP cleavage. GM3 decreased, whereas the ganglioside GD3, the GD3S product, increased Aβ production, resulting in a regulatory feedback cycle, directly linking ganglioside metabolism with APP processing and Aβ generation. A central aspect of this homeostatic control is the reduction of GD3S activity via an Aβ-GM3 complex and AICD-mediated repression of GD3S transcription.

Editor: Tsuneya Ikezu, Boston University School of Medicine, United States of America

Funding: The research leading to these results has received funding from the EU FP7 project LipiDiDiet, Grant Agreement No 211696 (TH), the Deutsche Forschungsgemeinschaft (TH), the Bundesministerium für Bildung, Forschung, Wissenschaft und Technologie via NGFNplus and Kompetenznetz degenerative Demenzen (TH), the HOMFOR 2008 (MG) and Homburger Forschungsförderungsprogramm 2009 (MG, TH) (Saarland University research grants). The funders had no role in study design, data collection and analysis, decision to publish, or preparation of the manuscript.

Competing Interests: The authors have declared that no competing interests exist.

* E-mail: marcus.grimm@uks.eu (MG); Tobias.Hartmann@uniklinikum-saarland.de (TH)

Introduction

Alzheimer's disease (AD) is a devastating neurodegenerative disorder, pathologically characterized by extracellular senile plaques and intracellular neurofibrillary tangles [1]. Major constituents of senile plaques are 40–42 amino acids (aa) long peptides termed β-amyloid (Aβ) [2]. Aβ is generated by sequential processing of the type-I transmembrane amyloid precursor protein (APP), involving β-secretase BACE1 [3] and γ-secretase. The β-secretase cleaves APP within the extracellular/luminal domain, generating the N-terminus of Aβ and a 99 aa long membrane-bound C-terminal fragment C99/β-CTF, which is further cleaved by γ-secretase to release Aβ and the APP intracellular domain (AICD). The γ-secretase consists of at least four proteins, presenilin 1 (PS1), presenilin 2 (PS2), nicastrin, anterior phar-ynx-defective 1 (Aph-1) and presenilin enhancer 2 (Pen-2) [4]. The polytopic transmembrane proteins presenilin (PS) constitute the active site of the protease [5] and mutations in the genes encoding PS1 and PS2 are responsible for most cases of familial early-onset Alzheimer's disease (EOAD) [6]. As the γ-secretase cleavage site is centered within the transmembrane domain, lipid composition of cellular membranes has been shown to influence proteolytic processing of APP [7,8,9,10,11,12]. Beside cholesterol and sphingomyelin (SM) [8,10], there are several strong indications which point towards an important role of gangliosides in AD pathogenesis [13,14,15,16]. Gangliosides are a family of sialic acid containing glycosphingolipids, highly enriched in neuronal and glial membranes, where they play important roles for development, proliferation, differentiation and maintenance of neuronal tissues and cells [17]. The ganglioside GM3 serves as a common precursor for the a- and b-series gangliosides. The GD3-synthase (GD3S) catalyzes the synthesis of GD3 by adding sialic acid to GM3, segregating the a- and b-series of gangliosides (Fig. 1A) [18] and therefore controlling the levels of the major brain gangliosides GM1, GD1a, GD1b and GT1b. Gangliosides are directly involved in AD pathology, e.g. the concentration and composition of gangliosides are altered in the brains of AD patients and in transgenic mouse models of AD [19,20,21]; ganglioside clusters in neuronal membranes take part in the formation of amyloid fibrils [13,15,16]; GM1 drastically increases Aβ production [22], binds to Aβ and is discussed to act as a seed for Aβ aggregation in amyloid plaques [23,24,25]; and Aβ aggregation as well as Aβ induced cell death are reduced in AD-model mice lacking GD3S [14]. The underlying mechanism of how APP processing and AD interfere with ganglioside homeostasis, however, remains unclear. Here we identified a regulatory cycle in which gangliosides regulate APP processing, and Aβ and AICD control the a/b-series ganglioside homeostasis. Central to this homeostatic control is the regulation of the GD3S activity via an Aβ-GM3 complex and AICD mediated repression of GD3S transcription.

Results

GD3-synthase is affected in PS- and APP-deficient cells

To investigate a potential role of PS as the active part of the γ-secretase complex in ganglioside metabolism, we analyzed PS1/PS2-deficient mouse embryonic fibroblasts (MEF PS1/2-/-) [26,27] and PS1 wild-type retransfected control cells (MEF PS1r) (Fig. S1). Lipid extraction of the major brain gangliosides [28] followed by thin-layer chromatography (TLC) analysis showed an increase in b-series gangliosides GD3, GD1b and GT1b in PS-deficient cells, whereas levels of the a-series gangliosides GM3, GM1 and GD1a were strongly decreased (Fig. 1B, Fig. S2). The conversion of a- to b-series gangliosides is catalyzed by the enzyme GD3S, suggesting that GD3S activity is blocked by γ-secretase activity. Indeed, quantitative real-time PCR of MEF PS1/2-/- cells showed that GD3S expression was strongly elevated and in accordance with increased GD3S expression, enzyme activity was also strongly increased in MEF PS1/2-/- cells (Fig. 1C), explaining the elevated b-series gangliosides and decreased a-series gangliosides observed. Beside γ-secretase cleavage of APP, γ-secretase is involved in processing numerous type-I transmembrane proteins [29]. To elucidate the mechanism how PS affects GD3S activity and to identify the γ-secretase substrate involved in GD3S activity regulation, we analyzed mouse embryonic fibroblasts devoid of APP and the APP-like protein APLP2 (MEF APP/APLP2-/-). Similar to the results obtained with PS-deficient cells, we observed an increase in GD3S activity and gene transcription in MEF APP/APLP2-/- cells (Fig. 1D), clearly suggesting that the cleavage of APP and/or APLP2 by PS/γ-secretase regulates ganglioside metabolism. Because PS knock-out mice are not viable, the role of PS could not be directly assessed *in vivo*. However, APP knock-out mice are viable, which allowed us to evaluate the role and *in vivo* relevance of APP in GD3S regulation directly. Indeed, the absence of APP alone increased GD3S expression in the brain (Fig. 1E), providing clear evidence for APP as a γ-secretase target involved in GD3S activity regulation.

Aβ peptides decrease GD3-synthase activity by mediating substrate availability

To identify the molecular mechanism of APP mediated GD3S regulation, we analyzed the potential role of the APP cleavage products Aβ and AICD, which are missing in PS- or APP-deficient cells. To identify whether Aβ peptides may influence GD3S activity, MEF PS1/2-/- and MEF APP/APLP2-/- cells were incubated with synthetic Aβ40 or Aβ42 peptides at physiological peptide concentrations. Silver stain analysis of the Aβ peptides revealed soluble monomeric Aβ, but no stable small oligomers (Fig. S3). Incubation with synthetic soluble Aβ40 or Aβ42 peptides resulted in decreased GD3S activity, both in cells lacking either PS1/2 or APP/APLP2 (Fig. 2A). To investigate whether the effect of Aβ on GD3S is due to a direct interaction, homogenates of MEF PS1/2-/- cells and a cell-free assay, which only contained purified GD3S and the GD3S substrate GM3, were incubated with Aβ and GD3S activity was determined. Both cell homogenates and the cell-free assay showed significantly decreased GD3S activity in presence of Aβ (Fig. 2B), suggesting that decreased GD3S activity is mediated directly by the interaction of Aβ with components of the cell-free assay. Inverted Aβ peptides and aggregated Aβ had no influence on GD3S activity (Fig. 2C), indicating a specific requirement for soluble Aβ peptides as expected for a physiological regulatory mechanism. Since Aβ peptides were able to decrease GD3S activity in a cell-free assay, consisting only of the substrate GM3 and the enzyme GD3S, two molecular mechanisms explaining how Aβ influences GD3S activity are conceivable: Firstly, Aβ could bind to the substrate, thus reducing substrate availability of the enzyme or secondly in principle, Aβ could directly bind to the enzyme, reducing the turnover of GM3 to GD3. Of interest in the first context is that Aβ is known to bind to GM1, another a-series ganglioside [24]. Indeed, Aβ co-immunoprecipitated GM3, the substrate for GD3S, but interestingly, it did not bind GD3, the product of GD3S (Fig. 2D, Fig. S4A, S4B). In agreement with our previous results we found, using synthetic Aβ40 and Aβ42 peptides and the Aβ40- and Aβ42-specific antibodies G2-10 and G2-11 for co-immunoprecipitation (co-IP), that both Aβ species bind GM3 (Fig. 2D; Fig. S4C). To rule out effects caused by the use of synthetic rather than naturally derived Aβ peptides, we analyzed conditioned media of APP expressing cells, secreting Aβ40 and Aβ42 in a ratio of approximately 10:1 [30]. The experiments with conditioned media for co-IP confirmed that naturally derived Aβ40 and Aβ42 bind GM3 at a similar 10:1 ratio (Fig. 2D; Fig. S4B). As a positive control we used GM1, which is known to bind Aβ [23,24,25]. GM1 bound to both Aβ species, Aβ40 and Aβ42 (Fig. S4D). These experiments suggest that these GM3-Aβ complexes are mechanistically involved in the reduced GD3S activity we observed in the APP or PS knock-out experiments. As already mentioned, a second mechanism could be that Aβ directly binds to GD3S and in this way reduces the turnover of GM3 to GD3. Co-IP studies revealed that Aβ was not able to bind to GD3S in the absence of GM3, ruling out that Aβ reduces GD3 production by direct binding to the enzyme (Fig. 2E). Importantly, under the same experimental conditions, but in presence of GM3, Aβ was bound to GD3S (Fig. 2E). This emphasizes that the inhibition of GD3S is mediated by binding of Aβ to GM3. When the GM3-Aβ complex binds to GD3S, as indicated by Fig. 2E, the conversion of GM3 to GD3 is reduced or prevented, consequently reducing substrate availability. Additionally, the GM3-Aβ complex might compete with unbound GM3 for GD3S binding and therefore acts as a competitive inhibitor, which would further reduce the amount of GM3 available for GD3S.

AICD decreases GD3-synthase gene expression

In addition to the Aβ-mediated reduction in GD3S activity, a further mechanism of action of influencing GD3S gene transcription, apparently exists. This is evidenced by the observation that

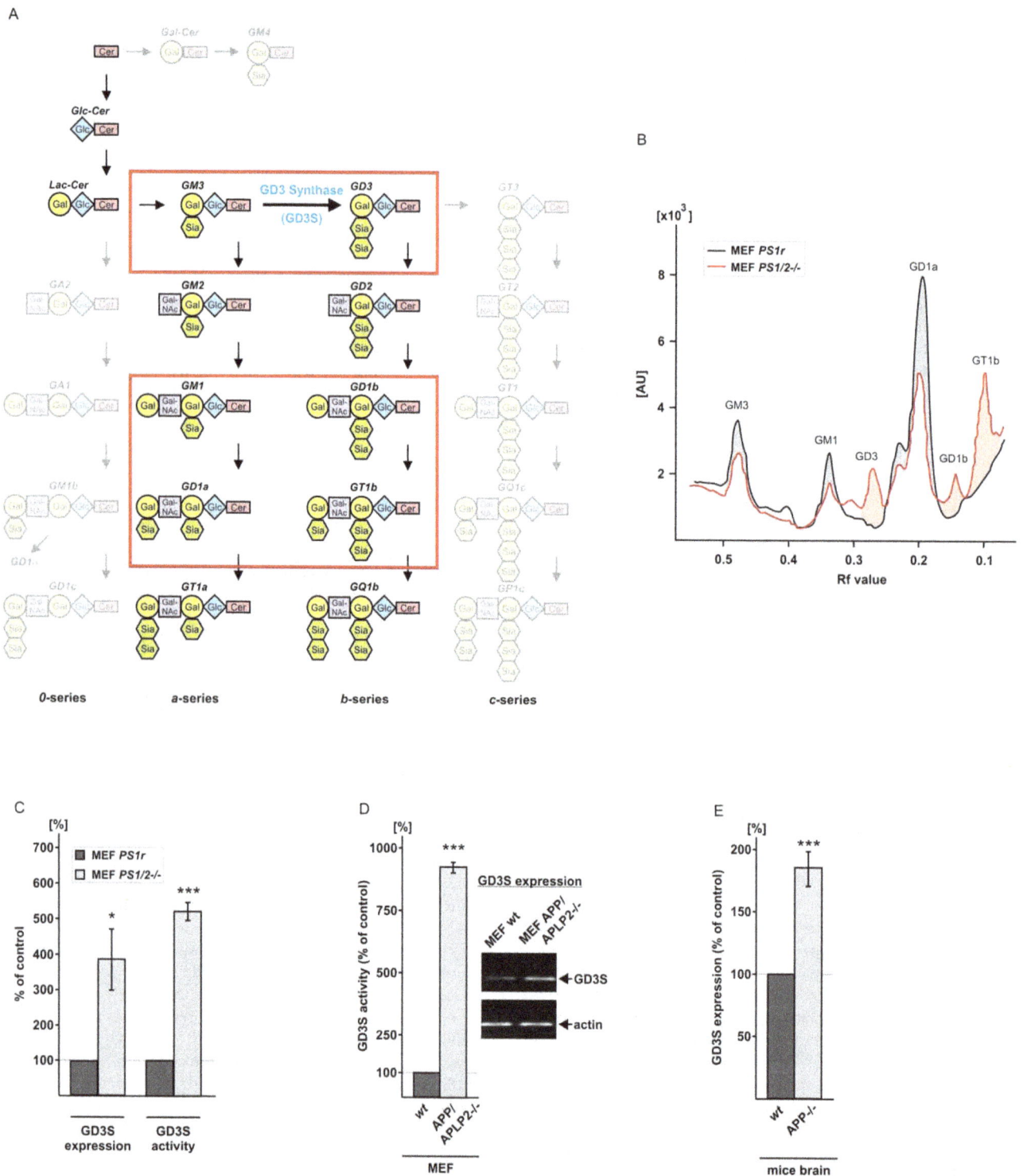

Figure 1. Ganglioside biosynthesis and effects on GD3s in dependence of APP-processing. (*A*) Biosynthetic pathway of ganglio-series gangliosides is shown. GD3-synthase (GD3S) converts GM3 to GD3 by adding sialic acid to GM3, generating the *b*-series of gangliosides. The major brain gangliosides GM1, GD1a, GD1b and GT1b are highlighted. (*B*) TLC analysis of *a*- and *b*-series gangliosides in PS-deficient cells. Representative densiometric quantification of TLC, analyzing the ganglioside pattern in PS1/2 deficient mouse embryonic fibroblasts (MEF PS1/2-/-) and MEF PS1/2-/- retranfected with PS1 wildtype (MEF PS1r). Peaks were identified by external standard and Rf-value. (*C*) GD3S expression and activity in PS1/2-deficient cells. In accordance with the altered *b:a* ratio of gangliosides, GD3S expression and activity is increased in PS-deficient cells (MEF PS1/2-/-) compared to MEF PS1r. (*D*) Analysis of GD3S activity and GD3S expression in APP/APLP2-deficient mouse embryonic fibroblasts (MEF APP/APLP2-/-) compared to wildtype mouse embryonic fibroblasts (MEF wt). MEF APP/APLP2-/- show increased GD3S activity and enhanced corresponding GD3S expression, analyzed by quantitative real-time PCR. (*E*) *In vivo* analysis of GD3S expression in brains of APP-deficient mice (APP-/-) show increased GD3S gene transcription compared to wt mice.

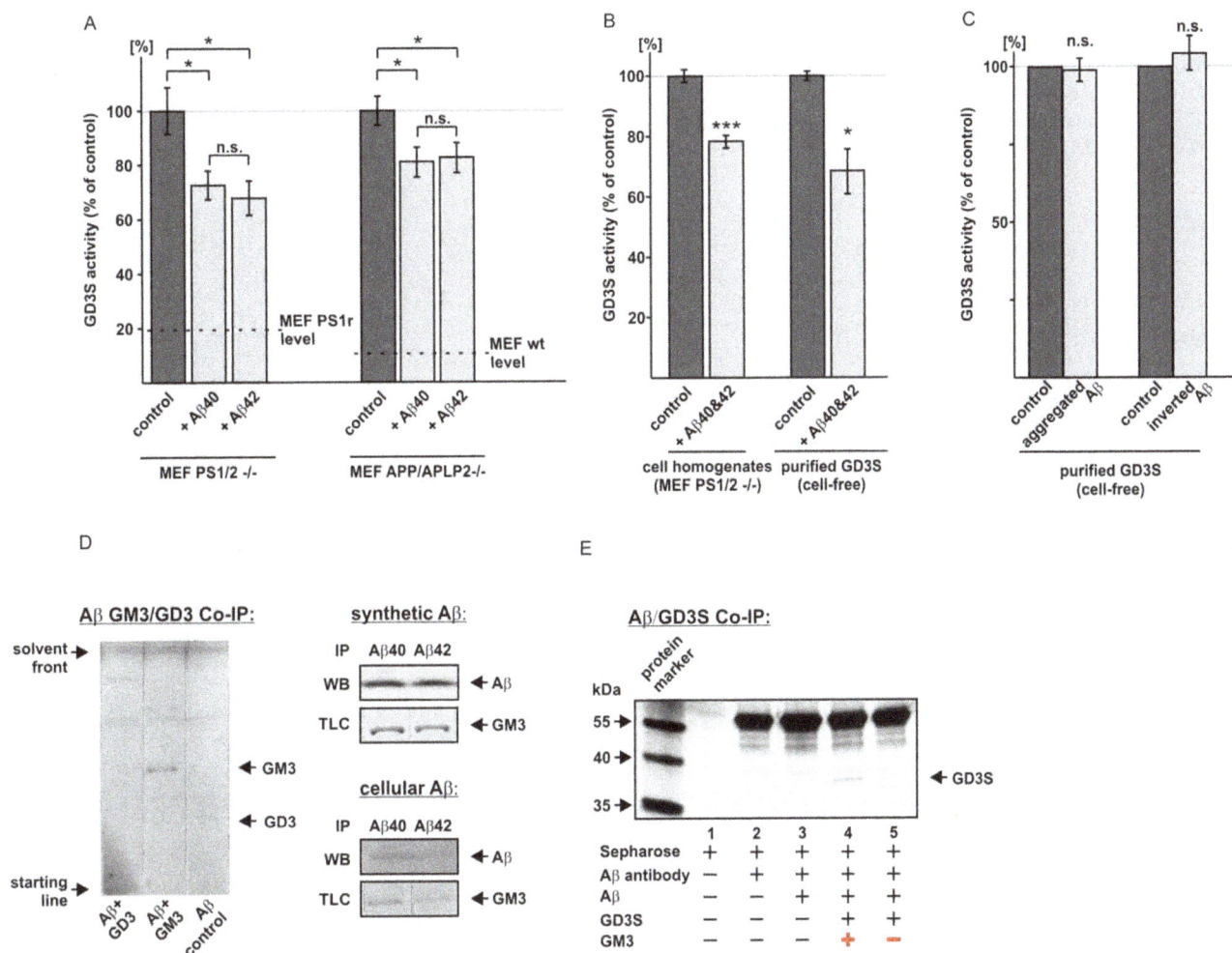

Figure 2. Influence of Aβ on GD3S activity. (A) Effect of Aβ on GD3S activity. MEF lacking Aβ because of PS- or APP-deficiency (MEF PS1/2-/- and MEF APP/APLP2-/-, respectively) were incubated with physiological Aβ40 or Aβ42 concentrations or the solvent used for the Aβ peptides (control). The level of the GD3S activity of the corresponding wildtype cells are indicated (horizontal dotted black line); corresponding wildtype/control cells for MEF PS1/2-/-: PS1/2 deficient MEF retransfected with PS1 (MEF PS1r); corresponding wildtype/control cells for MEF APP/APLP2: mouse embryonic fibroblasts of wildtype mice (MEF wt). Both Aβ species partially rescued the increased GD3S activity in PS- or APP-deficient MEF. No significant differences were observed between Aβ40 and Aβ42 peptides. (B) Direct effect of Aβ40 and Aβ42 on GD3S activity. Homogenates of PS1/2-/- cells incubated with Aβ40 and Aβ42 show decreased GD3S activity. Similar results are obtained with Aβ40 and Aβ42 in a cell-free assay containing only purified GD3S and the substrate GM3. (C) Influence of aggregated Aβ and inverted Aβ on GD3S activity. Inverted Aβ peptides and aggregated Aβ showed no influence on GD3S activity. (D) GM3, the substrate for GD3S, binds to Aβ: Physiological concentrations of cellular derived Aβ bind GM3 analyzed by co-immunoprecipitation (co-IP) of Aβ in presence of GD3 or GM3. GM3, but not GD3, binds to Aβ. After IP, Aβ bound gangliosides were detected via TLC. GM3 binds to synthetic Aβ40 and Aβ42 (shown for equimolar concentrations), and cellular derived Aβ40 and Aβ42 (shown for physiological concentrations, approx. 10:1 ratio). Thin black vertical lines (Fig. 2D left) indicate that the TLC plates were scraped to separate lines. (E) Co-IP of Aβ and GD3S in dependence of GM3. Aβ only binds to GD3S in presence of the substrate GM3. Data are represented as mean +/− SEM.

PS- and APP/APLP2-deficient cells as well as brains of APP-/- mice showed elevated GD3S gene expression (Fig. 1C, D, E). In accordance with this, Aβ only partially rescued the increased GD3S activity in PS- or APP-deficient MEF (Fig. 2A), indicating that besides Aβ other PS- and APP-dependent factors are involved in GD3S activity regulation. The intracellular domain of APP (AICD) is assumed to contribute to the regulation of gene expression [31,32] comparable to the function of the Notch intracellular domain (NICD), which is also released by γ-secretase [33]. To investigate the effect of AICD on GD3S gene expression we analyzed APP knock-in mouse embryonic fibroblasts deficient of full-length APP, expressing an APP construct, that lacks the last 15 aa from the C-terminus (MEF APPΔCT15) and hence a functional AICD domain [34]. The deleted last 15 aa include the YENPTY motif of APP, which apparently plays a crucial role in regulating gene transcription by binding to Fe65/X11 [31]. Indeed, MEF APPΔCT15 cells showed increased GD3S gene transcription and a nearly identical increase in GD3S activity (Fig. 3A). Supporting the *in vivo* relevance of these findings, brains of mice expressing APPΔCT15 also showed increased GD3S expression (Fig. 3B), which is in line with the result of APP knock-out mice (Fig. 1E). These results indicate that the presence of AICD strongly modulates GD3S gene transcription. To exclude that altered Aβ production, which might be caused by the truncated APP construct APPΔCT15 [35,36], could be responsible for the observed effect, we incubated MEF APPΔCT15 cells with a synthetic AICD peptide, corresponding to the last 20 aa of the C-terminus of APP, which was applied together with SAINT-

2:DOPE for efficient peptide uptake [37]. Indeed, the synthetic peptide significantly decreased GD3S expression in MEF APPΔCT15 cells (Fig. 3C), validating a function of AICD in regulating GD3S gene transcription. To further verify the role of AICD in regulating gene expression of GD3S, we generated Fe65 knock-down cells. As expected, a 58% reduction of Fe65

expression (Fig. S5), was sufficient to almost double GD3S expression levels (Fig. 3D).

As described above, Aβ directly causes a decrease in GD3S activity. To evaluate if, beside the observed AICD effect on gene expression, AICD has a similar direct effect on GD3S activity, homogenates of MEF APPΔCT15 cells were incubated with

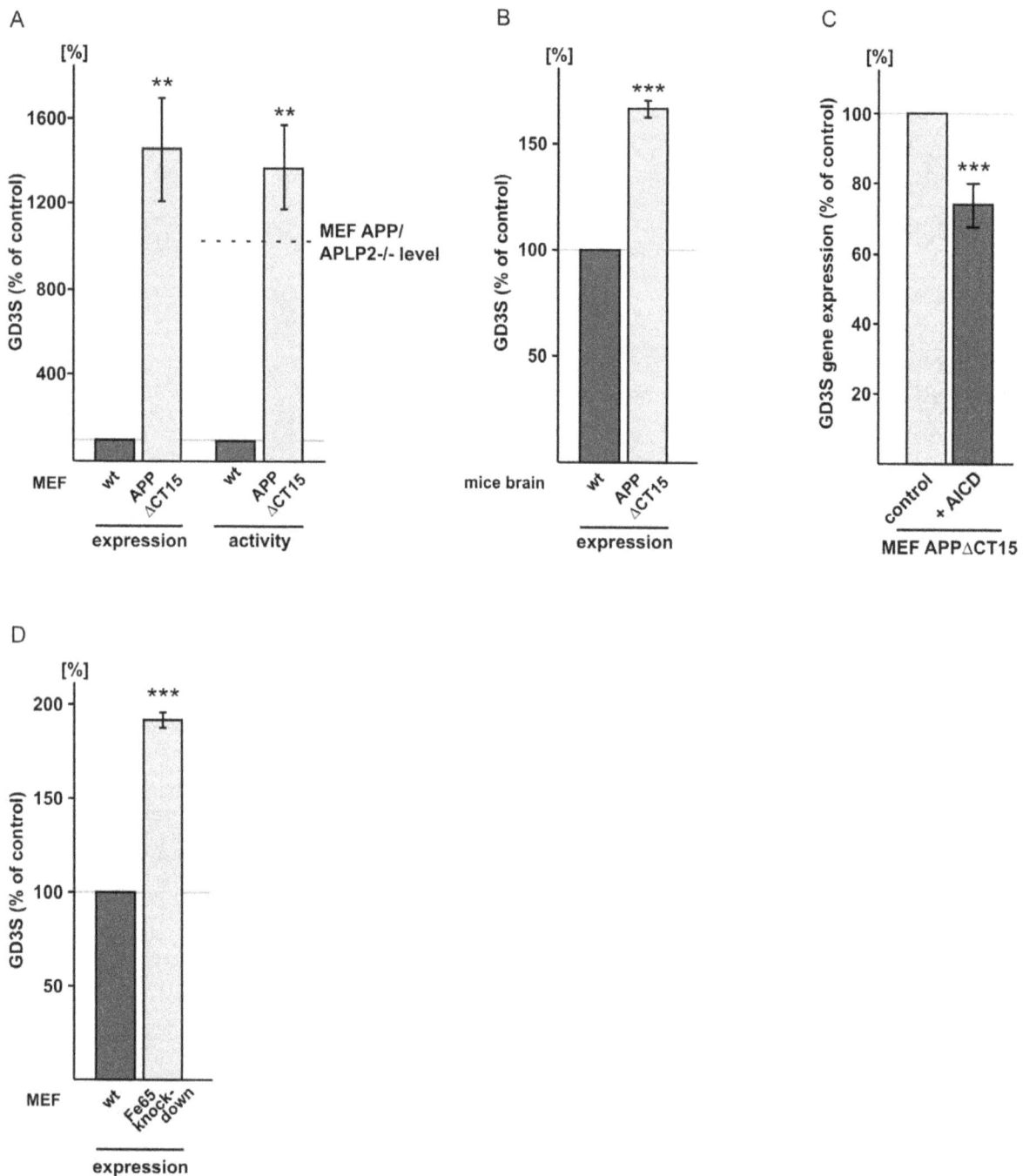

Figure 3. Effect of the APP intracellular domain (AICD) on GD3S. (A) MEF, deficient in full-length APP expressing a truncated APP construct lacking 15 aa from the C-terminus (APPΔCT15) show increased GD3S expression and activity. The level of the GD3S activity of the APP/APLP2 knock-out cells (MEF APP/APLP2-/-) are indicated (horizontal dotted black line). Interestingly the MEF APP/APLP2 knock-out level showed a slight less effect strength compared to the MEF APPΔCT15 cells. However this difference did not reach a significant level and might be due to clonal heterogeneity. (B) Increased GD3S expression in APPΔCT15 mouse brains. (C) GD3S gene expression: AICD peptide partially rescues elevated GD3S gene expression in MEF cells expressing the C-terminal truncated APP. An AICD peptide consisting of the last 20 aa from the APP C-terminus (AICD) is able to decrease GD3S expression. (D) shRNA generated Fe65 knock-down cells show increased GD3S gene transcription.

AICD peptides. Because homogenated cells were used for this experiment, alterations in GD3S activity cannot be caused by an AICD-induced change in GD3S gene expression. No effect of AICD peptide on GD3S activity was observed in these cell homogenates, indicating that AICD has no influence on GD3S enzymatic activity directly (Fig. S6A). Additionally, an influence of Aβ peptides on GD3S gene expression was ruled out by analyzing GD3S expression in PS-deficient cells supplemented with Aβ40 or Aβ42 (Fig. S6B).

Taken together, a dual mechanism of GD3S regulation mediated by PS-dependent APP processing can be postulated. Aβ peptides bind to GM3, thus decreasing GD3S activity by reducing substrate availability (Fig. 4A), whereas AICD inhibits gene expression of GD3S (Fig. 4B). Both mechanisms synergistically result in decreased cellular GD3S activity and thus in increased GM3 and reduced GD3 levels and an altered ratio of b:a series gangliosides.

GM3 decreases, GD3 increases Aβ production

It has been shown that the ganglioside GM1 increases Aβ production [22]. In order to analyze the effect of GM3 and GD3, which are themselves affected by APP processing, on Aβ generation and to evaluate the presence of a potential regulatory cycle in ganglioside homeostasis and APP processing, we incubated COS7 cells, stably transfected with the truncated APP construct SPC99, with GM3 or GD3. This shortened APP construct represents the C-terminal fragment of β-secretase cleaved APP, and allows the study of GM3 and GD3 influence on γ-secretase activity, independent of β-secretase activity [38,39].

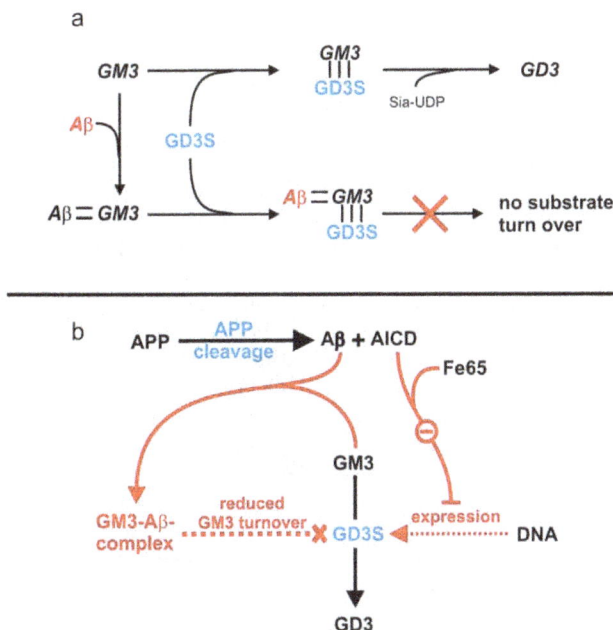

Figure 4. Molecular mechanisms of APP cleavage products in the regulation of GD3S enzyme activity. (A) In absence of Aβ peptides a-series ganglioside GM3 binds to GD3S and is converted to the b-series ganglioside GD3. In presence of Aβ, Aβ binds ganglioside GM3, forming an Aβ-GM3 complex. This complex still binds to GD3S, but cannot be converted to GD3. (B) Dual function of Aβ and AICD in GD3S regulation. Aβ reduces enzyme activity of GD3S by forming an Aβ-GM3 complex, resulting in reduced turnover of GM3 to GD3. AICD binds the adaptor protein Fe65 and reduces GD3S gene transcription, which also results in reduced turnover of GM3 to GD3.

Unexpectedly, the ganglioside GM3 decreased (Fig. 5A), but GD3 increased Aβ levels (Fig. 5B). Both effects were dose-dependent in a range from 10 to 100 μM. LDH-assay analysis revealed no signs of elevated cytotoxicity or reduced membrane integrity in the presence of gangliosides (Fig. S7). The effect of GM3 and GD3 is apparently cell line independent, since we obtained similar results in the human neuroblastoma cell line SH-SY5Y (Fig. S8).

Discussion

Gangliosides are sialic acid containing glycosphingolipids ubiquitously present in eukaryotic membranes with numerous cellular functions like signal transduction, cell adhesion, protein transport and brain development [17]. The numerous functions of gangliosides and the sequential anabolic pathway as described in Fig. 1A suggest that these enzymes must be tightly regulated. The major brain gangliosides belong either to the a- or b-series of gangliosides, emphasizing the importance of GD3S for brain ganglioside homeostasis. Although it is well established that brain ganglioside composition alters continuously during aging and in AD [19,28], the underlying cellular mechanisms remain unknown. To evaluate the link between ganglioside metabolism, APP processing and the resulting consequence for AD pathology, we analyzed ganglioside homeostasis for dependence on APP processing.

Absence of either APP or PS in cells or in the corresponding mouse models increased GD3S activity and expression, which, as a consequence, resulted in an increase in the b- to a-series ratio of gangliosides. Breaking down the molecular mechanism in detail, we identified that the regulation of GD3S by APP processing depends on two molecular factors, Aβ peptides and AICD. Aβ selectively binds GM3, the GD3S substrate, but not GD3, the GD3S product. The GM3-Aβ complex is still able to bind to GD3S but Aβ effectively prevents the conversion of GM3 to GD3, hence resulting in lowered cellular GD3S activity. Because Aβ is secreted and taken up by cells [40,41] and since there are, especially in neural cells, abundant amounts of Aβ present intracellularly [42,43,44], binding of Aβ to GM3 could in principle take place in any intracellular compartment, especially along the secretory pathway and endosomes, at the plasma membrane, where large amounts of Aβ are present, as well as in the extracellular space [45]. Selective reduction of substrate availability by binding of the substrate to an inhibitor, often described as substrate depletion, is a common molecular mechanism to modulate substrate turnover. A similar mechanism is described for phospholipids, which can be cleaved by phospholipases (PLA2). In the presence of the protein annexin, PLA2 activity is reduced by binding of annexin to the phospholipids [46]. Interestingly, Aβ has been observed to bind several other lipids, such as cholesterol [47] and GM1 [24]. GM1, another a-series ganglioside, is elevated in amyloid plaques further strengthening the link between GD3S regulation and AD [48]. With this conditions used here neither synthetic nor naturally derived soluble Aβ bound GD3. It has previously been reported that synthetic Aβ 1–40 as well as synthetic inverted Aβ 40-1 binds to all sialylated gangliosides [49]. This might indicate that conditions exist, such as amyloid formation as typical for AD, under which Aβ like peptides have increased affinity to all sialylated gangliosides. In such a case, binding of GD3 to Aβ, would likely further contribute to the disturbance of the physiological ganglioside homeostasis in the AD brain.

Reduced GM3 to GD3 turnover is further enhanced by another Aβ independent effect. Although Aβ directly decreased GD3S activity, this cannot account for the observed increase in GD3S

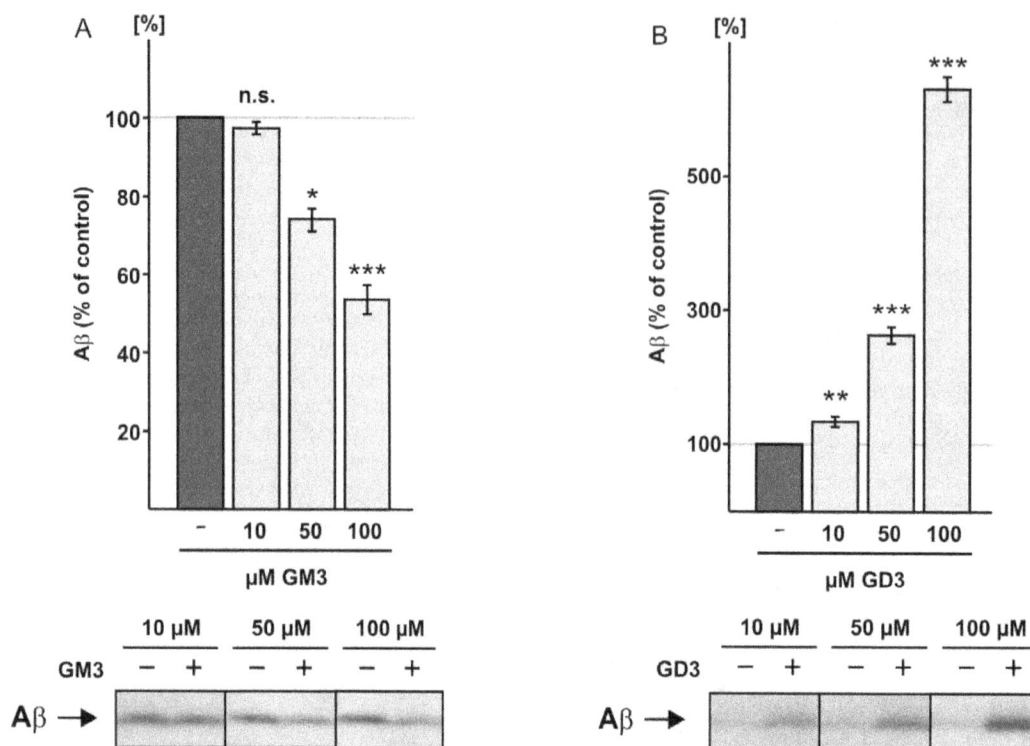

Figure 5. Effect of GM3 and GD3 on APP processing. (A) Dose-dependent effect of GM3 on Aβ production: COS7 cells stably transfected with SPC99, representing the β-secretase cleaved C-terminal fragment of APP, show a dose-dependent decrease in Aβ production in presence of GM3. Aβ levels were determined by IP and Western blot (WB) analysis. (B) Dose-dependent effect of GD3 on Aβ generation in SPC99 expressing COS7 cells. GD3 enhances Aβ generation dose-dependently. Aβ levels were determined by IP and WB analysis.

expression. Addition of Aβ to the knock-out cells had no influence on GD3S expression levels. Cao and others have found that AICD has structural and functional similarities with NICD, a well-established regulator of gene transcription [50]. However, the relevance of AICD for transcriptional regulation remains controversial [12,31,32,50]. To elucidate whether the increased GD3S expression is mediated *via* AICD, we used APPΔCT15 cells lacking the last 15 aa of the APP C-terminus and therefore a functional AICD. MEF APPΔCT15 cells showed drastically elevated GD3S gene transcription and similarly increased GD3S activity. Confirming the essential role of a functional AICD for GD3S regulation, the increased GD3S expression in APPΔCT15 cells could be partially rescued by addition of synthetic AICD peptide. To further verify our results that AICD triggers a cascade that results in decreased GD3S expression, we investigated whether a similar effect can be obtained by eliminating another protein from this cascade. Fe65 binds to the YENPTY motif of AICD and mediates nuclear targeting of AICD [51]. In line with the findings that absence of AICD causes an increase in GD3S expression, we found in Fe65 knock-down cells increased GD3S expression, validating the AICD/Fe65-mediated mechanism for regulating GD3S expression. Similar to the finding that Aβ altered GD3S enzymatic activity, but not GD3S expression, AICD suppressed GD3S expression, but had no influence on GD3S enzymatic activity.

These findings support an essential role of APP in ganglioside homeostasis in AD. The function of AICD, Aβ and γ-secretase in regulating GD3S would suggest that the alteration in ganglioside composition is a consequence of AD. Nevertheless, both the substrate and product of the GD3S, GM3 and GD3, themselves modulate APP processing making the distinction between cause

and consequence less clear. Like ganglioside GM1, GD3 increased Aβ generation whereas GM3 decreased Aβ release. Glycosphingolipids have been shown to be implicated in the regulation of the subcellular transport of APP in the secretory pathway [52], indicating that altered Aβ generation in presence of different gangliosides might be caused by altered APP transport to the cellular compartments where Aβ generation preferentially occurs, in post-Golgi secretory and endocytic compartments [45,53]. In addition changes in lipid raft composition might alter Aβ generation. For example it has been recently shown that docosahexaenoic acid (DHA) decreases Aβ generation by decreasing cholesterol in lipid raft membrane microdomains [54]. The hypothesis that lipids are also important for functional cellular protein transport as already reported for glycosphingolipids by Tamboli et al. [52] is further substantiated by the finding that BACE1 protein transport to the endosomal compartments, where β-secretase cleavage preferentially occurs, is impaired in presence of some lipids, e.g. DHA [54]. However, although it has been shown that membrane lipid composition influences APP cleavage [7,8,9,10,11], we do not exclude that lipids, such as GM3 and GD3, might directly affect secretase activities or other cellular mechanisms involved in Aβ generation. Reducing GD3S activity as potential therapeuthic target to treat or prevent AD would result in decreased GD3 and increased GM3 levels both resulting in decreased Aβ levels. However, one has to take into consideration that GM3 is further converted to ganglioside GM1, which is discussed to increase Aβ generation and to induce Aβ aggregation *in vitro*. In opposite to the *in vitro* findings concerning GM1 induced Aβ aggregation, *in vivo* studies revealed that GM1 is not required for Aβ aggregation [55] or even neuroprotective [14,56,57]. Disruption of the GM2 synthase gene in APP transgenic mice significantly increased Aβ aggregation, although GM1 is completely lacking in these mice [55]. Moreover, Bernardo et al. observed that deletion of GD3S, increasing neuronal expression of GM1, results in reduced soluble Aβ levels, decreased Aβ aggregation and plaque load in APP/PS transgenic mice paralleled by behavioural improvements [14]. From this line of evidence changes in the cerebral ganglioside composition might alter Aβ generation and aggregation and might be beneficial regarding AD pathology. Additionally, a recent study from Dhanushkodi et al. reports neuroprotective effects in vivo using sialidase from Vibrio cholerae producing a brain ganglioside profile similar to that of the GD3S knock-out, further implicating GD3S as a potential therapeutic target for AD [58]. However, further experiments are necessary to identify the role of specific gangliosides in AD pathology.

From our findings we propose a hypothetic model showing that GD3S and APP form a regulatory feedback cycle that links ganglioside metabolism with APP processing (Fig. 6). Since this regulatory cycle involves synergistic components the magnitude of the effect can be increased. Both Aβ and AICD decrease GD3S activity, resulting in strongly decreased GD3 levels. This in turn leads to reduced cleavage of APP. Additionally, decreased GD3S activity results in increased GM3 levels, further reducing APP cleavage to Aβ. It can therefore be proposed that interventions shifting the GD3/GM3 ratio towards increased GM3 and derceased GD3 levels could, by influencing this regulatory feedback cycle, cause reduced amyloidogenic processing of APP.

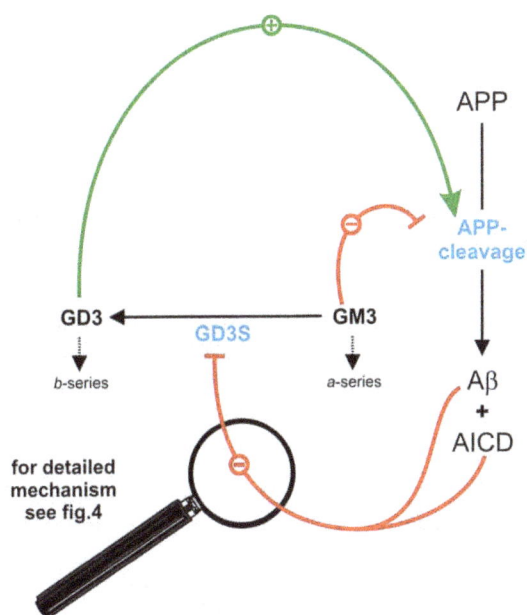

Figure 6. Hypothetic model of the physiological functions of Aβ and AICD in the regulation of GD3-synthase (GD3S) – the enzyme controlling major brain ganglioside composition. (A) Amyloidogenic proteolytic processing of the Alzheimer's amyloid precursor protein (APP) releases amyloid-beta peptides (Aβ) and the intracellular domain of APP (AICD). Aβ and AICD inhibit GD3S, resulting in reduced conversion of a- to b-series gangliosides. As a consequence of reduced conversion of a- to b-series gangliosides, GM3 increases whereas GD3 decreases. In return, both gangliosides, GM3 and GD3, themselves regulate the proteolytic cleavage of APP. The b-series ganglioside GD3 increases, whereas the a-series gangliosides GM3 decreases amyloidogenic proteolytic processing of APP.

This therefore may prove to be a potential strategy to treat or prevent AD.

Materials and Methods

Cell culture and biological material

COS7, SH-SY5Y and MEF cells were cultivated in DMEM (Sigma, Taufkirchen, Germany), 10% FCS (PAN Biotech, Aidenbach, Germany); for pCEP4/APP and pCEP4/SPC99 transfected cells additional HygromycinB (400 μg/ml) (PAN Biotech, Aidenbach, Germany) [44] was used. MEF PS1/2-/- cells were transfected with pCNA3.1/PS1 wildtype using Superfect as described by the manufacturer (Qiagen) to generate PS1 wildtype retransfected control cells (MEF PS1r). Stable transfectants were selected using 300 μg/ml Zeocin. MEF PS1r cells were cultivated in DMEM (Sigma, Taufkirchen, Germany), 10% FCS (PAN Biotech, Aidenbach, Germany), 300 μg/ml Zeocin (Invitrogen, Karlsruhe, Germany). Stable transfection and functionality of MEF PS1r cells were validated by PS1 western blot analysis (Fig. S1A), Aβ generation (Fig. S1B) and enzymatic measurement of γ-secretase activity (Fig. S1C). MEF PS1/2-/- cells [26,27], MEF PS1r cells [10], MEF APP/APLP2-/- cells [59] and MEF APPΔCT15 cells [34] have been described previously.

Ganglioside (Axxora, Lörrach, Germany) exposure was carried out in culture media without FCS for 16+6 hours. Notably the critical micelle concentration (CMC) of gangliosides examined in previous studies varies in dependence of the used methods. Whereas one study reports that for all gangliosides the CMC is between 70 to 100 μM [60], another study suggests for mono-sialogangliosides a CMC of $8.5*10^{-5}$ M and for disialogangliosides a CMC of $9.5*10^{-5}$ M [61]. In addition a more recent study suggests an even lower CMC for ganglioside GM3 of $3.4*10^{-9}$ M [62]. Therefore we used different ganglioside concentrations from 10 to 100 μM, which revealed a dose-dependent effect (Fig. 5).

Western Blot analysis

Aβ analysis of conditioned media of cultured cells was performed with antibody W02 as described earlier [63]. Control experiments using antibody WO2 for Aβ detection are shown in supplement Figures S9A, S9B and S9C. For the detection of PS1, cell-lysates were separated on 10–20% Tricine gels (Anamed, Groß-Bieberau, Germany). Cell-lysates were generated as previously described [44]. WB analysis was performed with the antibody sc-7860 (1:500; Santa Cruz, Heidelberg, Germany).

Densiometric quantification was performed using Image Gauge software. Samples were adjusted to equal protein amount before WB analysis [64].

Ganglioside determination and thin layer chromatography

After washing confluent grown cells three times with ice-cold phosphate-buffered saline (PBS), cells were scraped off and homogenized using a PotterS (B. Braun, Melsungen, Germany) at maximum speed (30 strokes). Protein amount was adjusted to 70 mg/ml according to Smith et al. [64]. After protein adjustments, a modified Bligh & Dyer method [65] was used to isolate gangliosides. After desalting the lipids via a reversed-phase cartridge (Waters Oasis, Eschborn, Germany) according to Whitfield et al. [66] a highly concentrated extract was applied to silica gel thin-layer chromatography (TLC) plates (Merck, Darmstadt, Germany). As a solvent system CHCl₃/MeOH/ H₂O-CaCl₂ 0.2% (60/35/8) was used [67]. The glycolipids were visualized by iodine and identified by their Rf-values and commercially available standards (AvantiPolarLipids, Alabaster,

USA). Densiometric quantification was performed using Image Gauge software. All solvents used were HPLC grade (VWR International GmbH, Darmstadt, Germany). To validate the use of iodine to stain gangliosides, we separated brain samples and ganglioside standard via TLC and stained either with iodine, orcinol or resorcinol (Fig. S10).

Determination of binding properties of Aβ to gangliosides and GD3S

Co-IP of Aβ with gangliosides was carried out with Aβ, produced by SH-SY5Y cells stably transfected with APP695 or with synthetic Aβ peptides (10 ng/ml) (B. Penke, Szeged, Hungary). For immunoprecipitation of Aβ peptides bound to gangliosides, monoclonal antibodies W02 (1 μg/ml), G2-10 (12.5 μg/ml), specific for the detection of Aβ40, and G2-11 (17.3 μg/ml), specific for the detection of Aβ42, were used [63]. Gangliosides bound to Aβ peptides were analyzed by TLC. Purified GD3S was purchased from Abnova (Taipei, Taiwan), used at a final concentration of 0.5 μg/ml and detected by silverstain [68].

Co-Immunoprecipitation of GD3S with Aβ peptide and Ganglioside GM3

13.65 nM of GD3S (Abnova, Taipei, Taiwan), 11.55 nM Aβ40, 1.155 nM Aβ42 were preincubated in PBS (pH 7.0, final volume 1 ml) in absence or in presence of 100 μM GM3 in a glass tube for 3 hours at 4°C with gentle shaking. After preincubation, samples were transferred in a microcentrifuge tube and 5 μg of antibody WO2 and 20 μl ProteinG-Sepharose were added. Immunoprecipitation of Enzyme-Peptide-Lipid-complex was performed over night at 4°C on an end over end shaker. At the end of incubation time, samples were centrifuged at 13.000 rpm for 1 min at 4°C. Supernatant was removed and Sepharose Beads were washed three times with 1 ml of 10 mM Tris (pH 7.4) Samples were boiled in SDS-sample buffer (187.5 mM Tris/HCl pH 6.8, 6% SDS, 30% Glycerol, 15% β-Mercaptoethanol 0.03% Bromphenolblau) and separated on a 10–20% Tris-Tricine Gel. Proteins were detected by silver staining according to the method of Switzer et al. [68].

Determination of monomeric Aβ40 and Aβ42 by SDS PAGE and Silver Staining

To determine the monomeric status of Aβ40 and Aβ42, used in *in vitro* and cell culture experiments, the synthetic peptides (200 ng Aβ40+20 ng Aβ42) were separated by SDS PAGE on a 12% NuPAGE Bis-Tris gel (Invitrogen), according to Dahlgren et al., which were also able to detect Aβ oligomers by utilizing this method [69]. Proteins were visualized by silver stain.

Enzymatic assays

For measuring **GD3S** activity confluent grown cells were washed three times with ice-cold sodium cacodylate 7.5 μM pH 5.8 including 1.5% Triton X-100, scraped off and homogenized using a PotterS (B. Braun, Melsungen, Germany) at maximum speed (100 strokes) [70]. After protein adjustment to 25 mg the reaction was started by addition of 5 nM CMP-sialic acid [^{3}H] (ARC, St. Louis, USA), 10 nM GM3 (Axxora, Lörrach, Germany) and 50 nM CMP-sialic acid (Sigma, Taufkirchen, Germany) in sodium cacodylate 7.5 μM pH 5.8 at 37°C. After 45 min the enzymatic reaction was stopped by addition of MeOH and CHCl₃, gangliosides were extracted as described above and separated via TLC. The GD3-band was scraped off and the including radioactivity was measured via scintillation counting in

Tri-Carb2800TR (Perkin Elmer, Rodgau-Jügesheim, Germany). To determine the effect of Aβ40 (10 ng/ml), Aβ42 (1 ng/ml) (B. Penke, Szeged, Hungary), inverted Aβ (10 ng/ml) or AICD (2 μM) (Genscript Corporation, Piscatway, USA) synthetic peptides were incubated for 1 h to cell-lysate and for 9 days in cell culture. After incubation the reaction was started by adding 5 nM CMP-sialic acid [^{3}H] (ARC, St. Louis, USA), 10 nM GM3 (Axxora, Lörrach, Germany) and 50 nM CMP-sialic acid (Sigma, Taufkirchen, Germany) in sodium cacodylate 7.5 μM pH 5.8 to the preincubated GM3-Aβ complex and stopped after another 45 min by addition of MeOH.

For the **GD3S in vitro assay** 25 nM Aβ and 5 nM GM3 were preincubated for 1 h to allow the formation of the complex GM3-Aβ. After that the reaction was started by adding 5 nM CMP-sialic acid [^{3}H] (ARC, St. Louis, USA), 10 nM GM3 (Axxora, Lörrach, Germany) and 50 nM CMP-sialic acid (Sigma, Taufkirchen, Germany) in sodium cacodylate 7.5 μM pH 5.8 to the preincubated GM3-Aβ complex and stopped after another 45 min by addition of MeOH. For cell-free enzymatic assay purified GD3S (1 μg/ml; 27.3 nM) (Abnova, Taipei City, Taiwan) was used instead of lysate. Aggregated Aβ was prepared as 22.2 μM solution by incubation >24 h at 37°C [10] and finally used as 22.2 nM. All solvents used were HPLC grade (VWR International GmbH, Darmstadt, Germany). GD3S assay unspecificity was determined using ganglioside GM2 (Fig. S11).

Determination of γ-secretase activity

Detection of γ-activity was performed as described before [38,54]. Briefly, cells were washed three times with ice-cold phosphate-buffered saline (PBS), scraped off in sucrose buffer (10 mM Tris/HCl pH 7.4 including 1 mM EDTA and 200 mM sucrose) and homogenized using a PotterS at maximum speed (25 strokes). After protein adjustment to 1 mg according to Smith et al. 1985 [64], the samples were centrifuged at 900rcf for 10 min at 4°C and the obtained post-nuclear fractions were ultracentrifuged at 55000 rpm for 75 min at 4°C. Pelleted membranes were resuspended using cannulaes with decreasing diameter in 300 μl sucrose buffer. The volume is partioned in 96well plates (γ: 100 μl, equates to 250 μg protein) and γ-secretase substrate [71] (Calbiochem, Darmstadt, Germany) (10 μM) was added. Fluorescence was measured for 3 h under light exclusion using excitation at 355±10 nm and fluorescence detection at 440±10 nm or 345±5 nm/500±2.5 nm respectively. Assay specificity is >90% for γ-secretase activity assay, and was validated using γ-secretase inhibitor L-658458 (Merck, Darmstadt, Germany).

Knock-down experiments

According to the manufacturer's protocol SureSilencing™ shRNA Plasmid (SABioscience, Frederick, USA) was used. The following insert sequences were used:
Fe65: 5′-TCC CTG GAC CAC TCT AAA CTT-3′
5′-CAA CCC AGG GAT CAA GTG TTT-3′
5′-AAG GCT TTG AGG ATG GAG AAT-3′
5′-TGT CCA CAC GTT TGC ATT CAT-3′.
Control: 5′-GGA ATC TCA TTC GAT GCA TAC-3′.
This control was used as randomized sequence in all knock-down experiments. The knock-down was verified by quantitative real-time experiments.

Quantitative real-time experiments

Total RNA was extracted from cells or tissue using RNeasyPlus Mini Kit (Qiagen, Hilden, Germany) or TRIzol reagent (Invitrogen, Karlsruhe, Germany), using manufacturer's protocols. 2 μg were reverse-transcribed using High Capacity cDNA Reverse

Transcription Kits and quantitative real-time PCR analysis was carried out using Fast SYBR Green Master Mix on 7500 Fast Real-Time PCR System (7500 Fast System SDS Software 1.3.1.; Applied Biosystems, Darmstadt, Germany). As control RNA samples were normalized to β-actin gene expression.

To determine the effect of Aβ40 (10 ng/ml), Aβ42 (1 ng/ml) (B. Penke, Szeged, Hungary) or AICD on gene transcription (2 μM) (Genscript Corporation, Piscataway, USA) synthetic peptides were incubated for 9 days in cell culture. Alternatively, the compound SAINT-2:DOPE was added together with the synthetic peptides for 12 h to the cells to achieve efficient delivery of the peptides [37].

Changes in gene expression was calculated using $2^{-(\Delta\Delta Ct)}$ method [72]. The $\Delta\Delta CT$ data are shown in table S1. To verify the results obtained by quantitative real-time experiments, samples were separated on 1.5% agarose gels in TBE buffer (90 mM Tris, 90 mM boric acid, 2 mM EDTA pH 8.0).

The following primer sequences were used:
Fe65 pair1: 5'-TCT TGC ACC AGC AGA CAG AG-3' and 5'-CAG CCA TGA TGA ATG CAA AC-3';
pair2: 5'- TTT GGA AGG ATG AAC CCA GT-3' and 5'-AAG CTT CTC CTC CTC TTG GG-3';
pair3: 5'- GCT CTA AGA TCA TGG CCG AA-3' and 5'-GGA ATT CCA CTT GGA AAG GG-3'.
β-Actin: 5'-CCT AGG CAC CAG GGT GTG AT-3' and 5'-TCT CCA TGT CGT CCC AGT TG-3'.

Cytotoxicity measurement

Cytotoxicity was measured using Lactate Dehydrogenase Cytotoxicity Assay Kit (LDH-Assay) and performed as described in the manufacturer's protocol (Cayman Chemical, Ann Arbor, USA). Briefly, conditioned media of incubated cells were centrifuged for 5 min at 400 g. 100 μl supernatant was used to determine lactate dehydrogenase cytotoxicity levels. For each cytotoxicity measurement lactate dehydrogenase standards were used as described in the manufacturer's protocol. After adding 100 μl reaction solution, the plate was agitated for 30 min at RT. Absorbance was measured at 490 nm using a MultiskanEX plate reader (Thermo Fisher Scientific, Waltham, USA).

Statistical analysis

All quantified data represent an average of at least three independent experiments. Error bars represent standard deviation of the mean. Statistical significance was determined by two-tailed Student's t-test; significance was set at *$P \leq 0.05$; **$P \leq 0.01$ and ***$P \leq 0.001$, n.d. = not detectable.

Supporting Information

Figure S1 Analysis of PS1 retransfected mouse embryonic fibroblasts devoid of PS1 and PS2. (*A*) PS expression in mouse embryonic fibroblasts (MEF) devoid of PS1 and PS2 (PS1/2-/-), corresponding wild-type embryonic fibroblasts (wt) and MEF PS1/2-/- cells retransfected with PS1 (PS1r). Western blot analysis of cell-lysates; PS was detected using the antibody sc-7860. (*B*) Determination of Aβ generation in PS1 retransfected MEF PS1/ 2-/-. MEF PS1/2-/- and MEF PS1r were infected with Semliki Forest Virus expressing APP695. Conditioned media were collected, immunoprecipitated with antibody W02 and detected via WB analysis using W02. (*C*) γ-secretase activity of MEF wild-type (wt) cells and MEF PS1r. γ-secretase activity of MEF wt and MEF PS1r was measured as described in Material and Methods.

Figure S2 TLC analysis of *a*- and *b*-series gangliosides in PS-deficient cells. (A) A representative TLC, analyzing the ganglioside pattern in PS1/2 deficient mouse embryonic fibroblasts (MEF PS1/2-/-) and MEF PS1/2-/- retranfected with PS1 wildtype (MEF PS1r) is shown. A ganglioside standard was loaded on the TLC. (B) Densiometric quantification of a representative TLC.

Figure S3 Determination of monomeric Aβ40 and Aβ42 by SDS Page and Silver staining. Synthetic Aβ40 and Aβ42 peptides (200 ng Aβ40, 20 ng Aβ42) used for the *in vitro* and cell culture experiments were loaded and separated on a 12% NuPAGE Bis-Tris gel Proteins were visualized by silver stain.

Figure S4 Co-immunoprecipitation of gangliosides with Aβ40 and Aβ42 peptides. (A) Co-immunoprecipitation studies using ganglioside GD3, cellular derived Aβ40 and Aβ42 peptides and the Aβ40 and Aβ42 specific antibodies G2-10 and G2-11. Cellular derived Aβ40 and Aβ42 peptides were obtained from conditioned media of APP wildtype expressing cells. Ganglioside GD3 does not co-immunoprecipitates with Aβ40 and Aβ42. (B) Co-immunoprecipitation studies using ganglioside GM3, cellular derived Aβ40 and Aβ42 peptides and the Aβ40 and Aβ42 specific antibodies G2-10 and G2-11. Cellular derived Aβ40 and Aβ42 peptides were obtained from conditioned media of APP wildtype expressing cells, secreting Aβ40/Aβ42 peptides in a ratio of approximately 10:1. Ganglioside GM3 co-immunoprecipitates with Aβ40 and Aβ42. In accordance to the 10:1 ratio of cellular derived Aβ40/Aβ42 peptides, the detected GM3 band shows a similar 10:1 ratio. (C) Co-immunoprecipitation studies using ganglioside GM3 and equimolar concentrations of synthetic Aβ40 and Aβ42 peptides. GM3 co-immunoprecipitates with synthetic Aβ40 and Aβ42 peptides. As expected using equimolar concentrations of Aβ40 and Aβ42 peptides, the stained (co-immunoprecipitated) GM3 bands have nearly identical intensity. (D) Co-immunoprecipitation studies using ganglioside GM1 and equimolar concentrations of synthetic Aβ40 and Aβ42 peptides. Black dotted vertical line: Samples were loaded on the same TLC but were not next to each other.

Figure S5 Fe65 knock-down in human neuroblastoma SH-SY5Y cells. The Fe65 expression level measured by RT-PCR analysis was reduced to 42%. Corresponding agarose gel is shown.

Figure S6 Influence of AICD on GD3S activity and of Aβ peptides on GD3S expression. (*A*) AICD peptides were incubated on APPΔCT15 cell homogenates and GD3S activity was determined as described. (*B*) Aβ peptides (Aβ40 and Aβ42) were incubated in cell culture on MEF PS1/2-/-. GD3S expression was determined by RT-PCR.

Figure S7 Lactate dehydrogenase assay in cells incubated with 100 μM GM3 or 100 μM GD3 versus corresponding solvent control (ddH$_2$O). No signs for increased cytotoxicity were observed in cells treated with gangliosides.

Figure S8 Aβ production in SPC99 expressing SH-SY5Y cells incubated with GM3 and GD3. Cells were incubated with 100 μM GM3 and 100 μM GD3, respectively, versus corresponding solvent control (ddH$_2$O). Conditioned media were collected and immunoprecipitated with the antibody W02. WB analysis was performed using the antibody W02.

Figure S9 Control experiments using antibody WO2 for Aβ detection. (A) 5 ng, 10 ng, 20 ng and 40 ng of synthetic Aβ40 and Aβ42 peptides, respectively, were loaded and separated on Tris-Tricine gels. Western blot (WB) analysis to detect Aβ was performed with antibody WO2. (B) Human neuroblastoma SH-SY5Y cells were incubated with 2 µM γ-secretase inhibitor X (Calbiochem, Darmstadt, Germany) and the solvent control. Conditioned media were immunoprecipitated with antibody WO2. Immunoprecipitated proteins were loaded and separated on a Tris-Tricine gel and detected via WB analysis using antibody WO2. In presence of γ-secretase inhibitor, Aβ cannot be detected (negative control). (C) Human neuroblastoma SH-SY5Y cells were incubated with 100 µM GM1 and the solvent control. Conditioned media were immunoprecipitated with antibody WO2. Immunoprecipitated proteins were loaded and separated on a Tris-Tricine gel. WB analysis was performed with antibody WO2. GM1 increases Aβ generation (positive control).

Figure S10 Staining of gangliosides separated via TLC using iodine, orcinol and resorcinol. Brain samples and ganglioside standard were separated via TLC as described in the material and method section and stained with iodine, orcinol and resorcinol. TLC1 was first stained with iodine and afterwards with orcinol. TLC2 was first stained with iodine and afterwards with resorcinol. This analysis shows that gangliosides (GM3, GM1, GD3, GD1a, GD1b and GT1b) can be stained with iodine, orcinol and resorcinol.

Figure S11 Determination of GD3S assay unspecificity. As negative control for the GD3S assay we used ganglioside GM2. Compared to the positive control using ganglioside GM3, GM2 was unable to stimulate GD3S. Only a background signal is obtained.

Table S1 Rowdata obtained from quantitative Real-Time PCR experiments. Tables display all ΔCt values, ΔΔCt values and $2^{-(\Delta\Delta Ct)}$ normalized to mRNA actin levels as described in Livak and Schmittgen [72]. At least three independent RNA preparations of at least three different brains or cell culture dishes were analyzed. Figure numbers refer to the original figures in the manuscript.

Acknowledgments

We gratefully thank Bart de Strooper for providing PS-deficient mouse embryonic fibroblasts and Inge Tomic for technical assistance.

Author Contributions

Conceived and designed the experiments: MG TH. Performed the experiments: EZ SG BH TR VB LK PL. Analyzed the data: MG EZ SG BH TR VB LK TB PL UM HG TH. Contributed reagents/materials/analysis tools: MG TB UM. Wrote the paper: MG HG TH.

References

1. Selkoe DJ (2004) Cell biology of protein misfolding: the examples of Alzheimer's and Parkinson's diseases. Nat Cell Biol 6: 1054–1061.
2. Masters CL, Simms G, Weinman NA, Multhaup G, McDonald BL, et al. (1985) Amyloid plaque core protein in Alzheimer disease and Down syndrome. Proc Natl Acad Sci U S A 82: 4245–4249.
3. Sinha S, Anderson JP, Barbour R, Basi GS, Caccavello R, et al. (1999) Purification and cloning of amyloid precursor protein beta-secretase from human brain. Nature 402: 537–540.
4. Steiner H, Fluhrer R, Haass C (2008) Intramembrane proteolysis by gamma-secretase. J Biol Chem 283: 29627–29631.
5. Wakabayashi T, De Strooper B (2008) Presenilins: members of the gamma-secretase quartets, but part-time soloists too. Physiology (Bethesda) 23: 194–204.
6. St George-Hyslop PH, Petit A (2005) Molecular biology and genetics of Alzheimer's disease. C R Biol 328: 119–130.
7. Fassbender K, Simons M, Bergmann C, Stroick M, Lutjohann D, et al. (2001) Simvastatin strongly reduces levels of Alzheimer's disease beta-amyloid peptides Abeta 42 and Abeta 40 in vitro and in vivo. Proc Natl Acad Sci U S A 98: 5856–5861.
8. Wolozin B (2004) Cholesterol and the biology of Alzheimer's disease. Neuron 41: 7–10.
9. Kovacs DM, Fausett HJ, Page KJ, Kim TW, Moir RD, et al. (1996) Alzheimer-associated presenilins 1 and 2: neuronal expression in brain and localization to intracellular membranes in mammalian cells. Nat Med 2: 224–229.
10. Grimm MO, Grimm HS, Patzold AJ, Zinser EG, Halonen R, et al. (2005) Regulation of cholesterol and sphingomyelin metabolism by amyloid-beta and presenilin. Nat Cell Biol 7: 1118–1123.
11. Osenkowski P, Ye W, Wang R, Wolfe MS, Selkoe DJ (2008) Direct and potent regulation of gamma-secretase by its lipid microenvironment. J Biol Chem 283: 22529–22540.
12. Liu Q, Zerbinatti CV, Zhang J, Hoe HS, Wang B, et al. (2007) Amyloid precursor protein regulates brain apolipoprotein E and cholesterol metabolism through lipoprotein receptor LRP1. Neuron 56: 66–78.
13. Yamamoto N, Matsubara T, Sato T, Yanagisawa K (2008) Age-dependent high-density clustering of GM1 ganglioside at presynaptic neuritic terminals promotes amyloid beta-protein fibrillogenesis. Biochim Biophys Acta 1778: 2717–2726.
14. Bernardo A, Harrison FE, McCord M, Zhao J, Bruchey A, et al. (2009) Elimination of GD3 synthase improves memory and reduces amyloid-beta plaque load in transgenic mice. Neurobiol Aging 30: 1777–1791.
15. Yanagisawa K (2011) Pathological significance of ganglioside clusters in Alzheimer's disease. J Neurochem.
16. Matsuzaki K (2011) Formation of Toxic Amyloid Fibrils by Amyloid beta-Protein on Ganglioside Clusters. Int J Alzheimers Dis 2011: 956104.
17. Degroote S, Wolthoorn J, van Meer G (2004) The cell biology of glycosphingolipids. Semin Cell Dev Biol 15: 375–387.
18. Lahiri S, Futerman AH (2007) The metabolism and function of sphingolipids and glycosphingolipids. Cell Mol Life Sci 64: 2270–2284.
19. Kracun I, Rosner H, Drnovsek V, Heffer-Lauc M, Cosovic C, et al. (1991) Human brain gangliosides in development, aging and disease. Int J Dev Biol 35: 289–295.
20. Molander-Melin M, Blennow K, Bogdanovic N, Dellheden B, Mansson JE, et al. (2005) Structural membrane alterations in Alzheimer brains found to be associated with regional disease development; increased density of gangliosides GM1 and GM2 and loss of cholesterol in detergent-resistant membrane domains. J Neurochem 92: 171–182.
21. Barrier L, Ingrand S, Damjanac M, Rioux Bilan A, Hugon J, et al. (2007) Genotype-related changes of ganglioside composition in brain regions of transgenic mouse models of Alzheimer's disease. Neurobiol Aging 28: 1863–1872.
22. Zha Q, Ruan Y, Hartmann T, Beyreuther K, Zhang D (2004) GM1 ganglioside regulates the proteolysis of amyloid precursor protein. Mol Psychiatry 9: 946–952.
23. Yanagisawa K, Odaka A, Suzuki N, Ihara Y (1995) GM1 ganglioside-bound amyloid beta-protein (A beta): a possible form of preamyloid in Alzheimer's disease. Nat Med 1: 1062–1066.
24. Wakabayashi M, Okada T, Kozutsumi Y, Matsuzaki K (2005) GM1 ganglioside-mediated accumulation of amyloid beta-protein on cell membranes. Biochem Biophys Res Commun 328: 1019–1023.
25. Okada T, Wakabayashi M, Ikeda K, Matsuzaki K (2007) Formation of toxic fibrils of Alzheimer's amyloid beta-protein-(1–40) by monosialoganglioside GM1, a neuronal membrane component. J Mol Biol 371: 481–489.
26. Herreman A, Hartmann D, Annaert W, Saftig P, Craessaerts K, et al. (1999) Presenilin 2 deficiency causes a mild pulmonary phenotype and no changes in amyloid precursor protein processing but enhances the embryonic lethal phenotype of presenilin 1 deficiency. Proc Natl Acad Sci U S A 96: 11872–11877.
27. Herreman A, Van Gassen G, Bentahir M, Nyabi O, Craessaerts K, et al. (2003) gamma-Secretase activity requires the presenilin-dependent trafficking of nicastrin through the Golgi apparatus but not its complex glycosylation. J Cell Sci 116: 1127–1136.
28. Svennerholm L, Bostrom K, Jungbjer B, Olsson L (1994) Membrane lipids of adult human brain: lipid composition of frontal and temporal lobe in subjects of age 20 to 100 years. J Neurochem 63: 1802–1811.
29. Kopan R, Ilagan MX (2004) Gamma-secretase: proteasome of the membrane? Nat Rev Mol Cell Biol 5: 499–504.
30. Scheuner D, Eckman C, Jensen M, Song X, Citron M, et al. (1996) Secreted amyloid beta-protein similar to that in the senile plaques of Alzheimer's disease is increased in vivo by the presenilin 1 and 2 and APP mutations linked to familial Alzheimer's disease. Nat Med 2: 864–870.

31. von Rotz RC, Kohli BM, Bosset J, Meier M, Suzuki T, et al. (2004) The APP intracellular domain forms nuclear multiprotein complexes and regulates the transcription of its own precursor. J Cell Sci 117: 4435–4448.

32. Hebert SS, Serneels L, Tolia A, Craessaerts K, Derks C, et al. (2006) Regulated intramembrane proteolysis of amyloid precursor protein and regulation of expression of putative target genes. EMBO Rep 7: 739–745.

33. Selkoe D, Kopan R (2003) Notch and Presenilin: regulated intramembrane proteolysis links development and degeneration. Annu Rev Neurosci 26: 565–597.

34. Ring S, Weyer SW, Kilian SB, Waldron E, Pietrzik CU, et al. (2007) The secreted beta-amyloid precursor protein ectodomain APPs alpha is sufficient to rescue the anatomical, behavioral, and electrophysiological abnormalities of APP-deficient mice. J Neurosci 27: 7817–7826.

35. Kouchi Z, Kinouchi T, Sorimachi H, Ishiura S, Suzuki K (1998) The deletion of the C-terminal tail and addition of an endoplasmic reticulum targeting signal to Alzheimer's amyloid precursor protein change its localization, secretion, and intracellular proteolysis. Eur J Biochem 258: 291–300.

36. Ono Y, Kinouchi T, Sorimachi H, Ishiura S, Suzuki K (1997) Deletion of an endosomal/lysosomal targeting signal promotes the secretion of Alzheimer's disease amyloid precursor protein (APP). J Biochem 121: 585–590.

37. van der Gun BT, Monami A, Laarmann S, Rasko T, Slaska-Kiss K, et al. (2007) Serum insensitive, intranuclear protein delivery by the multipurpose cationic lipid SAINT-2. J Control Release 123: 228–238.

38. Grimm MO, Grimm HS, Tomic I, Beyreuther K, Hartmann T, et al. (2008) Independent inhibition of Alzheimer disease beta- and gamma-secretase cleavage by lowered cholesterol levels. J Biol Chem 283: 11302–11311.

39. Lichtenthaler SF, Multhaup G, Masters CL, Beyreuther K (1999) A novel substrate for analyzing Alzheimer's disease gamma-secretase. FEBS Lett 453: 288–292.

40. Haass C, Koo EH, Teplow DB, Selkoe DJ (1994) Polarized secretion of beta-amyloid precursor protein and amyloid beta-peptide in MDCK cells. Proc Natl Acad Sci U S A 91: 1564–1568.

41. Hu X, Crick SL, Bu G, Frieden C, Pappu RV, et al. (2009) Amyloid seeds formed by cellular uptake, concentration, and aggregation of the amyloid-beta peptide. Proc Natl Acad Sci U S A 106: 20324–20329.

42. Hartmann T, Bieger SC, Bruhl B, Tienari PJ, Ida N, et al. (1997) Distinct sites of intracellular production for Alzheimer's disease A beta40/42 amyloid peptides. Nat Med 3: 1016–1020.

43. Cook DG, Forman MS, Sung JC, Leight S, Kolson DL, et al. (1997) Alzheimer's A beta(1–42) is generated in the endoplasmic reticulum/intermediate compartment of NT2N cells. Nat Med 3: 1021–1023.

44. Grimm HS, Beher D, Lichtenthaler SF, Shearman MS, Beyreuther K, et al. (2003) gamma-Secretase cleavage site specificity differs for intracellular and secretory amyloid beta. J Biol Chem 278: 13077–13085.

45. Zhang YW, Thompson R, Zhang H, Xu H (2011) APP processing in Alzheimer's disease. Mol Brain 4: 3.

46. Buckland AG, Wilton DC (1998) Inhibition of secreted phospholipases A2 by annexin V. Competition for anionic phospholipid interfaces allows an assessment of the relative interfacial affinities of secreted phospholipases A2. Biochim Biophys Acta 1391: 367–376.

47. Harris JR (2008) Cholesterol binding to amyloid-beta fibrils: a TEM study. Micron 39: 1192–1196.

48. Yanagisawa K (2005) GM1 ganglioside and the seeding of amyloid in Alzheimer's disease: endogenous seed for Alzheimer amyloid. Neuroscientist 11: 250–260.

49. Ariga T, Kobayashi K, Hasegawa A, Kiso M, Ishida H, et al. (2001) Characterization of high-affinity binding between gangliosides and amyloid beta-protein. Arch Biochem Biophys 388: 225–230.

50. Cao X, Sudhof TC (2001) A transcriptionally [correction of transcriptively] active complex of APP with Fe65 and histone acetyltransferase Tip60. Science 293: 115–120.

51. Radzimanowski J, Simon B, Sattler M, Beyreuther K, Sinning I, et al. (2008) Structure of the intracellular domain of the amyloid precursor protein in complex with Fe65-PTB2. EMBO Rep 9: 1134–1140.

52. Tamboli IY, Prager K, Barth E, Heneka M, Sandhoff K, et al. (2005) Inhibition of glycosphingolipid biosynthesis reduces secretion of the beta-amyloid precursor protein and amyloid beta-peptide. J Biol Chem 280: 28110–28117.

53. Cupers P, Bentahir M, Craessaerts K, Orlans I, Vanderstichele H, et al. (2001) The discrepancy between presenilin subcellular localization and gamma-secretase processing of amyloid precursor protein. J Cell Biol 154: 731–740.

54. Grimm MO, Kuchenbecker J, Grösgen S, Burg VK, Hundsdorfer B, et al. (2011) Docosahexaenoic Acid Reduces Amyloid {beta} Production via Multiple Pleiotropic Mechanisms. J Biol Chem 286: 14028–14039.

55. Oikawa N, Yamaguchi H, Ogino K, Taki T, Yuyama K, et al. (2009) Gangliosides determine the amyloid pathology of Alzheimer's disease. Neuroreport 20: 1043–1046.

56. Wu G, Lu ZH, Wang J, Wang Y, Xie X, et al. (2005) Enhanced susceptibility to kainate-induced seizures, neuronal apoptosis, and death in mice lacking gangliotetraose gangliosides: protection with LIGA 20, a membrane-permeant analog of GM1. J Neurosci 25: 11014–11022.

57. Kreutz F, Frozza RL, Breier AC, de Oliveira VA, Horn AP, et al. (2011) Amyloid-beta induced toxicity involves ganglioside expression and is sensitive to GM1 neuroprotective action. Neurochem Int 59: 648–655.

58. Dhanushkodi A, McDonald MP (2011) Intracranial V. cholerae Sialidase Protects against Excitotoxic Neurodegeneration. PLoS One 6: e29285.

59. Heber S, Herms J, Gajic V, Hainfellner J, Aguzzi A, et al. (2000) Mice with combined gene knock-outs reveal essential and partially redundant functions of amyloid precursor protein family members. J Neurosci 20: 7951–7963.

60. Oshima H, Soma G, Mizuno D (1993) Gangliosides can activate human alternative complement pathway. Int Immunol 5: 1349–1351.

61. Yohe HC, Rosenberg A (1972) Interaction of triiodide anion with gangliosides in aqueous iodine. Chem Phys Lipids 9: 279–294.

62. Palestini P, Pitto M, Sonnino S, Omodeo-Sale MF, Masserini M (1995) Spontaneous transfer of GM3 ganglioside between vesicles. Chem Phys Lipids 77: 253–260.

63. Ida N, Hartmann T, Pantel J, Schroder J, Zerfass R, et al. (1996) Analysis of heterogeneous A4 peptides in human cerebrospinal fluid and blood by a newly developed sensitive Western blot assay. J Biol Chem 271: 22908–22914.

64. Smith PK, Krohn RI, Hermanson GT, Mallia AK, Gartner FH, et al. (1985) Measurement of protein using bicinchoninic acid. Anal Biochem 150: 76–85.

65. Bligh EG, Dyer WJ (1959) A rapid method of total lipid extraction and purification. Can J Biochem Physiol 37: 911–917.

66. Whitfield P, Johnson AW, Dunn KA, Delauche AJ, Winchester BG, et al. (2000) GM1-gangliosidosis in a cross-bred dog confirmed by detection of GM1-ganglioside using electrospray ionisation-tandem mass spectrometry. Acta Neuropathol (Berl) 100: 409–414.

67. Christie WW (2003) Analysis of Phospholipids and Glycosyldiacylglycerols. Lipid Analysis. Bridgwater: The Oily Press. pp 137–180.

68. Switzer RC, 3rd, Merril CR, Shifrin S (1979) A highly sensitive silver stain for detecting proteins and peptides in polyacrylamide gels. Anal Biochem 98: 231–237.

69. Dahlgren KN, Manelli AM, Stine WB, Jr., Baker LK, Krafft GA, et al. (2002) Oligomeric and fibrillar species of amyloid-beta peptides differentially affect neuronal viability. J Biol Chem 277: 32046–32053.

70. Busam K, Decker K (1986) Ganglioside biosynthesis in rat liver. Characterization of three sialyltransferases. Eur J Biochem 160: 23–30.

71. Farmery MR, Tjernberg LO, Pursglove SE, Bergman A, Winblad B, et al. (2003) Partial purification and characterization of gamma-secretase from post-mortem human brain. J Biol Chem 278: 24277–24284.

72. Livak KJ, Schmittgen TD (2001) Analysis of relative gene expression data using real-time quantitative PCR and the 2(-Delta Delta C(T)) Method. Methods 25: 402–408.

20

Establishment of Real Time Allele Specific Locked Nucleic Acid Quantitative PCR for Detection of HBV YIDD (ATT) Mutation and Evaluation of Its Application

Yongbin Zeng[1,2♪], Dezhong Li[3♪], Wei Wang[1,2], Mingkuan Su[1,2], Jinpiao Lin[1,2], Huijuan Chen[1,2], Ling Jiang[1,2], Jing Chen[1,4], Bin Yang[1,2], Qishui Ou[1,2]*

1 First Clinical College, Fujian Medical University, Fuzhou, China, **2** Department of Laboratory Medicine, The First Affiliated Hospital of Fujian Medical University, Fuzhou, China, **3** Department of Laboratory Medicine, The Armed Police Hospital of Fujian, Fuzhou, China, **4** Center of Liver Diseases, The First Affiliated Hospital of Fujian Medical University, Fuzhou, China

Abstract

Background: Long-term use of nucleos(t)ide analogues can increase risk of HBV drug-resistance mutations. The rtM204I (ATT coding for isoleucine) is one of the most important resistance mutation sites. Establishing a simple, rapid, reliable and highly sensitive assay to detect the resistant mutants as early as possible is of great clinical significance.

Methods: Recombinant plasmids for HBV YMDD (tyrosine-methionine-aspartate-aspartate) and YIDD (tyrosine-isoleucine-aspartate-aspartate) were constructed by TA cloning. Real time allele specific locked nucleic acid quantitative PCR (RT-AS-LNA-qPCR) with SYBR Green I was established by LNA-modified primers and evaluated with standard recombinant plasmids, clinical templates (the clinical wild type and mutant HBV DNA mixture) and 102 serum samples from nucleos(t)ide analogues-experienced patients. The serum samples from a chronic hepatitis B (CHB) patient firstly received LMV mono therapy and then switched to LMV + ADV combined therapy were also dynamically analyzed for 10 times.

Results: The linear range of the assay was between 1×10^9 copies/μl and 1×10^2 copies/μl. The low detection limit was 1×10^1 copies/μl. Sensitivity of the assay were 10^{-6}, 10^{-4} and 10^{-2} in the wild-type background of 1×10^9 copies/μl, 1×10^7 copies/μl and 1×10^5 copies/μl, respectively. The sensitivity of the assay in detection of clinical samples was 0.03%. The complete coincidence rate between RT-AS-LNA-qPCR and direct sequencing was 91.2% (93/102), partial coincidence rate was 8.8% (9/102), and no complete discordance was observed. The two assays showed a high concordance (*Kappa* = 0.676, *P* = 0.000). Minor variants can be detected 18 weeks earlier than the rebound of HBV DNA load and alanine aminotransferase level.

Conclusions: A rapid, cost-effective, high sensitive, specific and reliable method of RT-AS-LNA-qPCR with SYBR Green I for early and absolute quantification of HBV YIDD (ATT coding for isoleucine) variants was established, which can provide valuable information for clinical antiretroviral regimens.

Editor: Bruno Verhasselt, Ghent University, Belgium

Funding: The study was supported by the grants from Key Project of Industry-university Cooperation of Fujian province (2013Y4002), Social Key Project of Fujian Technology Department (2011Y0023; http://www.fjkjt.gov.cn/) and National Natural Science Foundation of China (81371888). The funders had no role in study design, data collection and analysis, decision to publish, or preparation of the manuscript.

Competing Interests: The authors have declared that no competing interests exist.

* E-mail: ouqishui@163.com

♪ These authors contributed equally to this work.

Introduction

Nucleos(t)ide analogues (NAs), such as Lamivudine (LMV), Telbivudine (LdT), Adefovir Dipivoxil (ADV) and Entecavir (ETV), are widely used for the treatment of chronic hepatitis B [1]. Unfortunately, long-term use of these drugs can increase risk of HBV drug-resistance mutations. The main site associated with resistance is mainly in the highly conserved tyrosine-methionine-aspartate-aspartate (YMDD) motif of the C domain of HBV polymerase. Methionine substituted by valine or isoleucine in codon 204 (i.e., rtM204V/I) were confirmed to confer resistance to LMV. The rtM204I can not only cause the selection of

Lamivudine and Telbivudine resistance, but can also lead to Entecavir-resistance [2]. For many years, early detection of HBV drug-resistance mutants has received much attention due to its importance for the strategic treatment of chronic hepatitis B virus-infected patients.

Lots of highly sensitive detection technologies, such as next-generation sequencing, peptide nucleic acid-mediated PCR clamping, line probe assay (LiPA), liquid array, mass spectrometric analysis, have become available in recent years [3–7]. However, some limitations have also been found, i.e., special instruments may be present, testing costs may be high in some technologies, etc. Consequently, their wide application are somewhat restricted in clinical laboratories.

Although real time PCR such as allele-specific PCR (AS-PCR) for the detection of HBV mutants was reported years ago, little attention has been paid to AS-PCR using LNA-modified primers. Locked nucleic acid (LNA) is a nucleic acid analog with a 2-O, 4-C methylene bridge which generally increases its melting temperature. PCR primers modified with LNA nucleotides show very accurate mismatch discrimination and high specificity [8]. In the present study, we establish a novel assay named real time allele specific locked nucleic acid quantitative PCR (RT-AS-LNA-qPCR) with SYBR Green I for detection of HBV YIDD (ATT coding for isoleucine) mutation, and focus on its methodological evaluation and clinical application.

Materials and Methods

Patients

112 serum samples were obtained from chronic hepatitis B patients receiving nucleos(t)ide analogues (NAs) therapy in The First Affiliated Hospital of Fujian Medical University. 83 were male and 29 were female. The mean age was 38 ± 13 years. Clinical data such as alanine aminotransferase (ALT) was measured. The study was in accordance with the approval of the ethics committee of The First Affiliated Hospital of Fujian Medical University and the ethical principles of the 1975 Declaration of Helsinki. Written informed consent was also obtained from each patient.

Serum HBV DNA extraction

HBV DNA was extracted from serum samples using viral genomic DNA extraction kit (Beijing Xinnuo Company, China) according to the manufacturer's instruction and stored at $-20°C$ until used.

Primers for RT-AS-LNA-qPCR

Primers for RT-AS-LNA-qPCR were designed using Primer Premier 5.0 software (Premier Biosoft International, USA) according to the principle of AS-PCR technique. The design guidelines were provided by Exiqon (http://www.exiqon.com/oligo-tools). The polymorphism analysis of the most common genotypes (B and C) in China obtained from GenBank was performed on DNAMAN software (Lynnon BioSoft, Canada). The forward primer (KF) was at nucleotide positions 396~416 for the amplification of both YMDD and YIDD (tyrosine-isoleucine-aspartate-aspartate). The reverse primers modified with a LNA at the 3'-end terminal position (L1) or penultimate position (L2) were designed to detect YIDD and YMDD, respectively. Wild type primer with last position of the nucleic acids lock (L2$_l$) was synthesized. Non-degenerated primers including the forward (KF$_n$) and the reverse (L1$_n$ and L2$_n$) with exact match to mutant and wild type plasmids were also included in this study (Table 1).

Preparation of YMDD and YIDD plasmids for calibration curves

Two clinical samples harboring wild-type (YMDD, ATG coding for methionine) and mutant (YIDD, ATT coding for isoleucine) DNA sequences confirmed by direct sequencing were used for plasmids construction. Target DNA was amplified with primers (KF, L1 and L2) listed in Table 1. PCR was performed in a 25-µl reaction mixture containing 12.5 µl of 2× HotStart Taq PCR Mastermix (Tiangen, China), 0.7 µl of each primer (10 µM), 9.1 µl of ddH$_2$O and 2.0 µl of DNA template. Thermal cycling conditions were as follows: initial denaturation at 94°C for 3 min,

Table 1. Primers for RT-AS-LNA-qPCR.

Primer names	Sequences (5'-3')	Positions (nt)	Length (bp)
KF	TCATMTTCCTCTKCATCCTGC	396~416	21
L1	CCCCAAWACCACATCATC+A	741~759	19
L2	CCCAAWACCACATCATC+CA	740~758	19
L2$_l$	CCCAAWACCACATCATC+C	740~757	18
KF$_n$	TCATCTTCCTCTGCATCCTGC	396~416	21
L1$_n$	CCCCAATACCACATCATC+A	741~759	19
L2$_n$	CCCAATACCACATCATC+CA	740~758	19

KF/KF$_n$ indicate the common forward primers to all reactions. L1/L2/L2$_l$/L1$_n$/L2$_n$ indicate specific RT-AS-LNA-qPCR reverse primers. M indicates A/C; K indicates G/T; W indicates A/T; +A/+C indicate LNA nucleosides.

then 35 cycles with denaturation at 94°C for 30 s, annealing at 60°C for 30 s and extension at 72°C for 45 s, and a final elongation step at 72°C for 10 min. PCR products were electrophoresed on a 2.0% agarose gel and purified via PCR purification kit (Sangon, China), then cloned into pMD 18-T vector (Takara, Japan) and transformed into E. coli DH5α competent cell (Takara, Japan). The positive plasmids were sequenced and then isolated using Tiangen mini plasmid isolation kit (Tiangen, China). Plasmids were quantified by NanoDrop 2000 spectrophotometer (Thermo Fisher Scientific, USA). The corresponding copy number were calculated and 10-fold serially diluted from 1×10^{10} copies/µl to 1×10^{1} copies/µl using Easy Dilution Buffer (Takara, Japan) to generate standard concentrations.

Preparation of different proportions of clinical mutant template

DNA from clinical sample harboring rtM204I confirmed by sequencing were mixed with wild-type clinical sample in different proportions. Specifically, clinical template containing 50% mutant HBV DNA was obtained by mixing 10 µl 1×10^{5} copies/µl clinical mutant DNA sample with 10 µl 1×10^{5} copies/µl clinical wild-type DNA sample, and clinical template containing 25% mutant HBV DNA was obtained by mixing 5 µl mutant DNA with 15 µl wild-type DNA. Clinical templates containing different proportions of mutant DNA (i.e., 20%, 10%, 5%, 1%, 0.5%, 0.05%, 0.04%, 0.03%, 0.02% and 0.01% of mutants) were also prepared in a similar way.

RT-AS-LNA-qPCR

RT-AS-LNA-qPCR assay was performed on ABI 7500 Real-Time PCR system (Life Technologies, USA). The 25-µl PCR amplification reaction mixtures contained 12.5 µl of 2× SYBR Premix Ex Taq^{TM} (Takara, Japan), 0.7 µl of each primer (10 µM), 0.5 µl of 50× ROX II reference dye (Takara, Japan), 8.6 µl of ddH$_2$O and 2.0 µl of DNA template. Real time PCR conditions were: initial denaturation at 95°C for 30 s, followed by 40 cycles of denaturation at 95°C for 5 s, and annealing/extension at 60°C for 30 s. The post-amplification melting curve analysis was performed to confirm whether the nonspecific amplification was generated from primer-dimers.

Direct Sequencing

To evaluate the performance of RT-AS-LNA-qPCR, all clinical samples were subjected to direct sequencing. The HBV DNA fragments containing the polymerase RT (reverse-transcriptase) domain were amplified using HBV sequencing kit (Shenyou, Shanghai) according to the manufacturer's instruction. Purified PCR products were sequenced with ABI 3130 genetic analyzer (Life Technologies, USA).

Statistical analysis

Results obtained from RT-AS-LNA-qPCR were compared with those from sequencing. The statistical analysis was performed using statistical analysis software SPSS version 16.0 (SPSS Inc, USA) and GraphPad Prism software version 5.0 (GraphPad Software, USA). Complete concordance was considered if the results detected by RT-AS-LNA-qPCR and direct sequencing were identical. Partial concordance was considered if (1) RT-AS-LNA-qPCR provided additional information compared to that provided by sequencing, meaning that RT-AS-LNA-qPCR showed a mixture of wild-type and mutant sequences, whereas sequencing showed only one of the two results, or (2) sequencing showed a mixture of wild-type and mutant sequences but RT-AS-LNA-qPCR showed only a wild-type or a mutant sequence. Complete discordance was considered if one test showed a wild type and the other showed a mutant. Concordance was assessed by Cohen's *Kappa* test. A $P<0.05$ was considered statistically significant.

Results

Identification of recombinant plasmids by sequencing

The pMD-18-YMDD and pMD-18-YIDD recombinant clones were picked out and sequenced. The obtained DNA sequences were subjected to BLAST alignment against HBV genome database (http://www.ncbi.nlm.nih.gov/BLAST/). Blast results indicated the sequences were completely consistent with HBV reference genome. Therefore, the pMD-18-YMDD and pMD-18-YIDD recombinant plasmids were successfully constructed.

Linear range and detection limit

1×10^{10} copies/μl~1×10^{1} copies/μl recombinant plasmids were used to test the linear range and detection limit of RT-AS-LNA-qPCR (Figure 1 and Figure 2). There was an excellent linear correlation between the cycle number and the HBV DNA copy number from the concentration of 1×10^{9} copies/μl to 1×10^{2} copies/μl with correlation coefficients of 0.999 and 0.984 for the wild-type and mutant target sequences, respectively. The amplification efficiency were 94.9% and 96.3% for wild-type and mutant standard curves, respectively. Experiments using degenerated primers against each pure polymorphic templates (genotype B and C) were also done, and no differences were observed for the linear range and the detection limit (Figure S1)

Specificity test

The cross-reactivity test was carried out using a dilution series of the template i.e., different concentrations of pMD-18-YMDD recombinant plasmids (1×10^{7} copies/μl~1×10^{2} copies/μl) were amplified with mutant specific and wild-type specific primers, respectively, and so were the pMD-18-YIDD plasmids. As shown in Figure 3, no nonspecific amplification phenomenon was observed by mutant specific primers set (KF/L1), but, nonspecific amplification with the wild-type specific primers set (KF/L2) was detected at the concentration equal to or above 1×10^{6} copies/μl. However, the observed copy number was less than the expected copy number at least 4 logs. Interestingly, there was no nonspecific amplification observed when the mismatch template was below 1×10^{6} copies/μl. The priming efficiency of L2 and L2$_{l}$ was

compared and no difference was observed (Figure S2: The Ct values were found to be almost the same for primer L2 and primer L2$_{l}$ when the same plasmid concentration was included in the test). The specificity of L2 was also compared with L2$_{l}$; while 1×10^{8} copies/μl mutant DNA was added to the PCR reaction system, Ct (cycle threshold) values were 27.0 and 25.94 for L2 and L2$_{l}$, respectively; when the template was 1×10^{6} copies/μl, the Ct values were 34.00 and 32.77 respectively. The Ct values were 36.97 and undetectable respectively, when mutant DNA was 1×10^{5} copies/μl. These results indicated the specificity of L2 was higher than L2$_{l}$ (Figure S2).

Accuracy test

To assess the accuracy, a mixture of known quantity of wild-type and mutant standard plasmids were used to generate different proportions of mutant DNA, and both wild-type and mutant DNA were quantitatively analyzed by RT-AS-LNA-qPCR. As indicated in Figure 4, there was a significant linear correlation ($R^{2}=0.9907$, $P<0.0001$) between actual proportion of mutant and calculated proportion of mutant for HBV standard plasmid at the concentration of 1×10^{7} copies/μl.

Reproducibility test

High (1×10^{7} copies/μl), medium (1×10^{5} copies/μl) and low (1×10^{3} copies/μl) concentrations of YMDD and YIDD recombinant plasmids were used as templates for 20 separate, simultaneous measurements by RT-AS-LNA-qPCR. The intra-run coefficient of variation (CV) was 0.74%, 0.91% and 1.07% for YMDD plasmids and 0.42%, 0.75% and 0.29% for YIDD plasmids. The quantitative assay was also done for 20 days consecutively. The inter-run CV was 1.12%, 1.79% and 1.93% for YMDD plasmids and 2.72%, 1.89%, 2.10% for YIDD plasmids.

Sensitivity test

A mixture of the dilution series of mutant plasmids with different fixed concentrations of wild-type DNA (1×10^{9} copies/μl, 1×10^{7} copies/μl and 1×10^{5} copies/μl, respectively) were tested with mutant primer to determine the minimum mutant DNA concentration at which RT-AS-LNA-qPCR could accurately and steadily quantify. The 100% pure wild-type DNA and no-template control were also tested. Results (Figure 5 and Figure S3) showed the assay could accurately and steadily detect 1×10^{3} copies/μl of mutant in the three different fixed concentrations of wild-type DNA, meaning that the sensitivity were 10^{-6}, 10^{-4} and 10^{-2} in the wild-type background of 1×10^{9} copies/μl, 1×10^{7} copies/μl and 1×10^{5} copies/μl, respectively. Figure 5B showed it began to deviate from linearity when the concentration of mutant DNA was below 1×10^{3} copies/μl. The sensitivity test was also carried out using non-degenerated primers (KF$_{n}$ and L1$_{n}$) with exact match to mutant plasmids and there was no difference in detection sensitivity using the non-degenerated primers and the degenerated primers (Figure S4).

Clinical utility and reproducibility

The sensitivity was estimated using the different proportions of mutant above-mentioned (5%, 1%, 0.5%, 0.05%, 0.04%, 0.03%, 0.02% and 0.01%, respectively) by RT-AS-LNA-qPCR. We found that 0.03% of mutants could be accurately and steadily quantified by RT-AS-LNA-qPCR. Whereas the amplification curves of both 0.02% and 0.01% of mutants could not be discriminated from that of 0.03% of mutants (Figure 6). So 0.03% was regarded as the sensitivity of the assay of clinical samples with the most common polymorphisms.

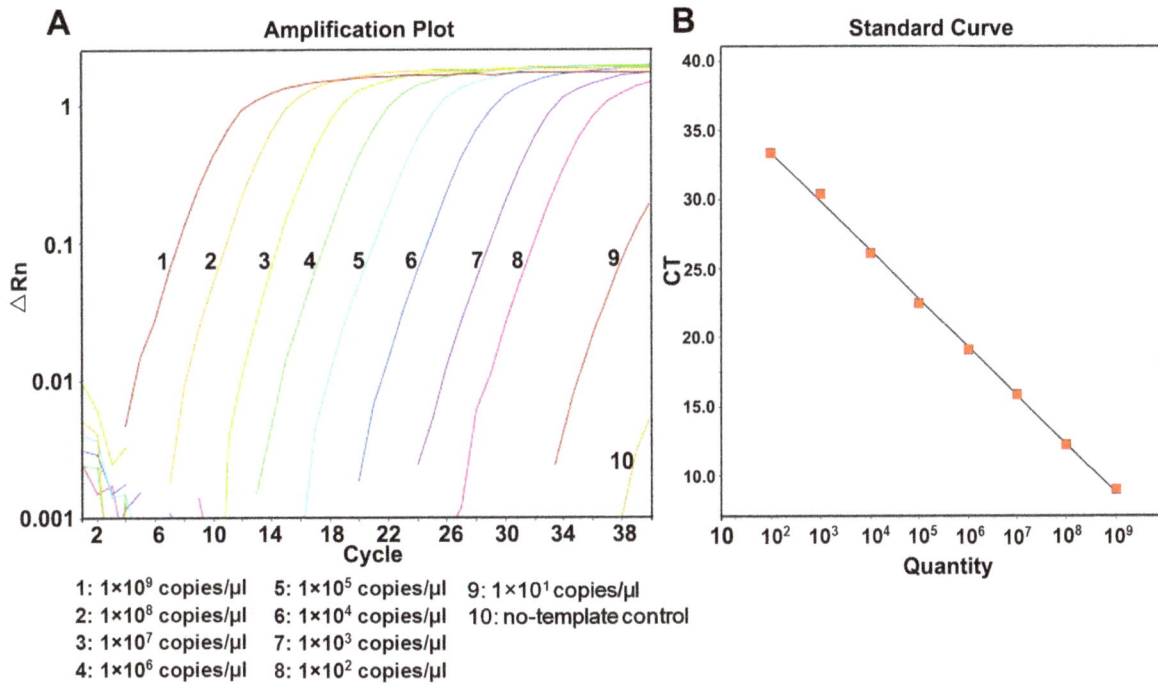

Figure 1. Amplification plot and standard curve of RT-AS-LNA-qPCR for pMD-18-YMDD. (A) Amplification plot with different colors represented different concentrations of pMD-18-YMDD plasmids which were illustrated in the figure. (B) Standard curve of pMD-18-YMDD: $Y = -3.478X+41.654$ ($R^2 = 0.999$).

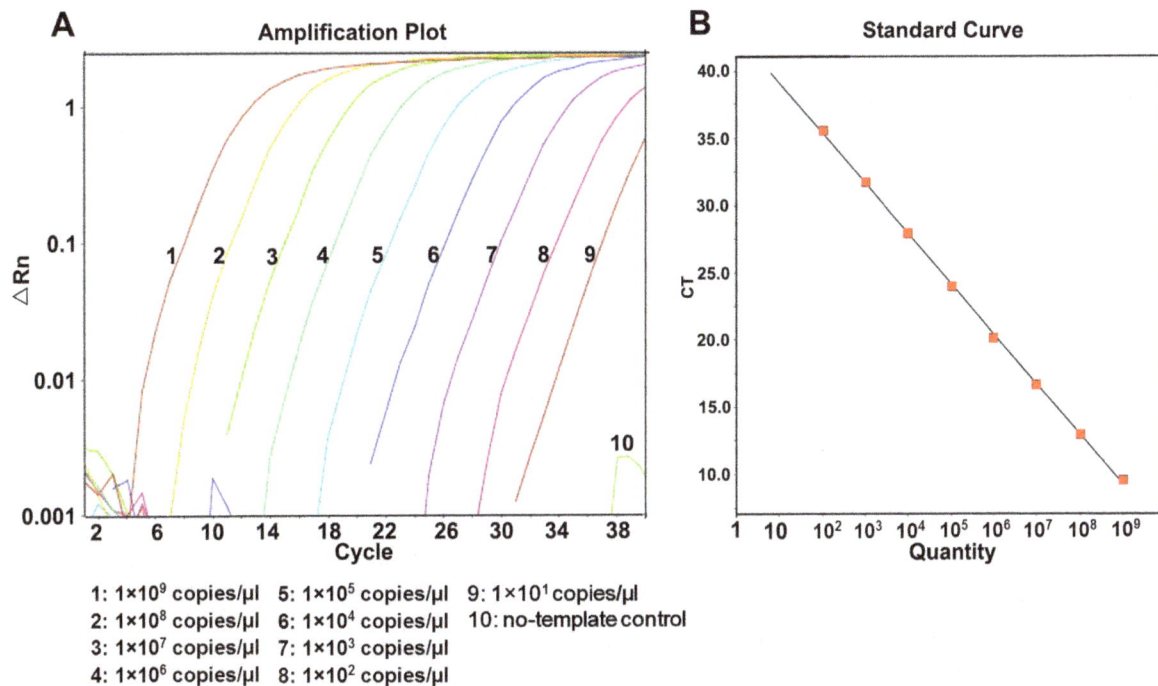

Figure 2. Amplification plot and standard curve of RT-AS-LNA-qPCR for pMD-18-YIDD. (A) Amplification plot with different colors represented different concentrations of pMD-18-YIDD plasmids which were illustrated in the figure. (B) Standard curve of pMD-18-YIDD: $Y = -3.413X+37.793$ ($R^2 = 0.984$).

A

B

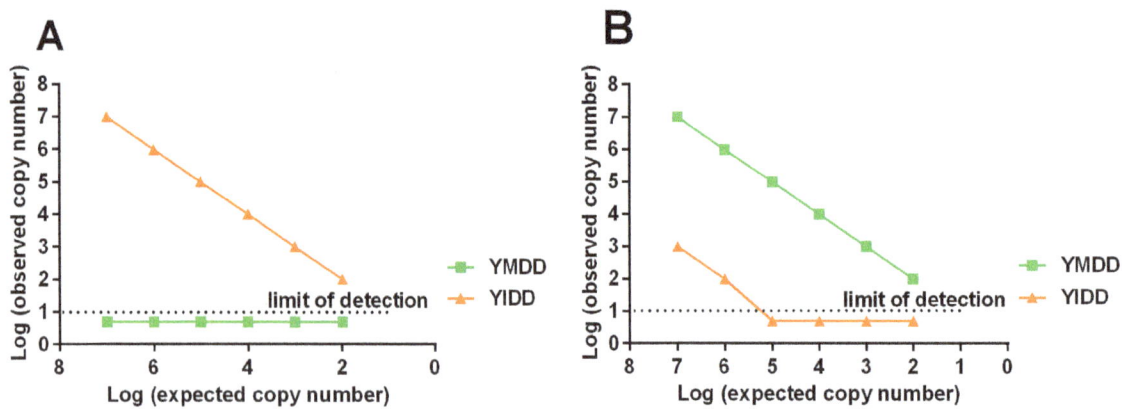

Figure 3. Cross-reactivity test of RT-AS-LNA-qPCR. (A) Specificity of mutant primer for detection of YMDD and YIDD plasmids. No nonspecific amplification was observed. (B) Specificity of wild-type primer for detection of YMDD and YIDD plasmids. Nonspecific amplification was detected at the concentration equal to or above 1×10^6 copies/μl, but the corresponding mismatch products decrease significantly.

To demonstrate reproducibility of quantification, the intra-run and inter-run reproducibility tests using clinical samples were also done. Repeated analysis demonstrated good reproducibility when the proportions of mutants was 0.03%.

Sensitivity comparison of RT-AS-LNA-qPCR with sequencing analysis on clinical samples

Clinical samples containing different proportions (50%, 25%, 20%, 10%, 5%, 1%, 0.5%, 0.05%~0.01%) of rtM204I were used to compare the sensitivity of RT-AS-LNA-qPCR with sequencing. As indicated in Table 2, RT-AS-LNA-qPCR could detect rtM204I at a proportion as low as 0.01% (detection limit) of the total population, whereas sequencing analysis only detect at a proportion of 10%.

Comparison of RT-AS-LNA-qPCR with direct sequencing

102 NAs-experienced patients' sera were parallel analyzed by RT-AS-LNA-qPCR and direct sequencing. As shown in Table 3, among the 102 samples analyzed, 85 YMDD, 4 YIDD and 13

Figure 4. The correlation between actual and calculated proportion of mutant plasmid DNA at the concentration of 1×10^7 copies/μl. $R^2 = 0.9907$, $P < 0.0001$.

mixtures of YMDD + YIDD were detected by RT-AS-LNA-qPCR. By comparing the results obtained from RT-AS-LNA-qPCR with those from sequencing, the complete coincidence rate was 91.2% (93/102), partial coincidence rate was 8.8% (9/102), and no complete discordance was observed. Thus the two assays showed a high concordance ($Kappa = 0.676$, $P = 0.000$). Among the 6 partial concordant results (Only YIDD was detected by direct sequencing), the proportion of variants ranged from 80.8% to 99.8% of the total viral population was calculated by RT-AS-LNA-qPCR (Table 4). Among another 3 samples collected from CHB patients who were highly suspicious of drug resistance, only wild-type DNA was detected at rt204 by direct sequencing, however, both wild-type and mutant were detected by RT-AS-LNA-qPCR, and the proportion of mutant was only 8.4%, 9.8% and 9.5%, respectively, which was below the sensitivity of direct sequencing. 3 typical samples whose detected results were partially concordant between direct sequencing and RT-AS-LNA-qPCR were selected for cloning sequencing; and the established method is consistent with the cloning sequencing finding. The typical cloning sequencing chromatograms of HBV rt204 were showed in Figure 7.

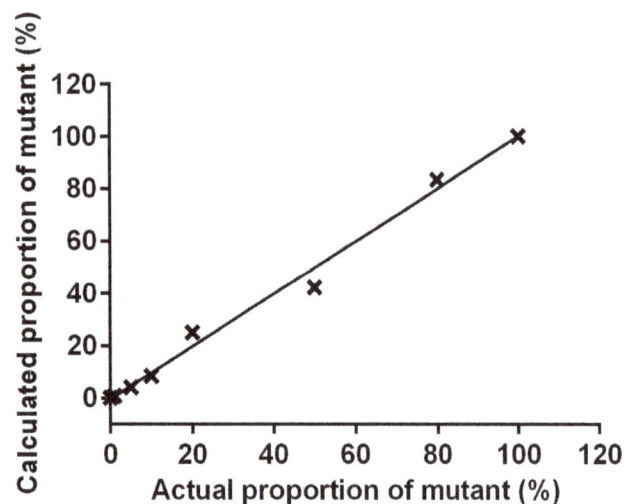

Dynamic analysis of wild-type and mutant HBV DNA in a CHB patient by RT-AS-LNA-qPCR

A 48-week-follow-up study of dynamic of YMDD and YIDD HBV serum DNA levels and ALT levels during first LMV mono therapy and then LMV + ADV combined therapy was illustrated in Figure 8. As can be seen, HBV DNA level and ALT level declined substantially at the first 7 weeks and no YIDD variants were detectable at that time. However, at week 17, a minor fraction of YIDD variants had already been detected, accounting for 13.7%, 18 weeks before the rebound of HBV DNA load and alanine aminotransferase level. It was indicated that variants had become the predominant population with a percentage of 58.3% 11 weeks later. At week 36, YIDD variants population increased to 100%. ADV was added to ongoing lamivudine treatment at that time point. The bi-therapy decreased HBV DNA, however, very slowly; and variants remained the predominant population.

Discussion

HBV drug resistant mutants are confirmed to exist in CHB patients prior to the initiating antiviral therapy with NAs. Previous researches have documented that the minor preexistence mutants

A

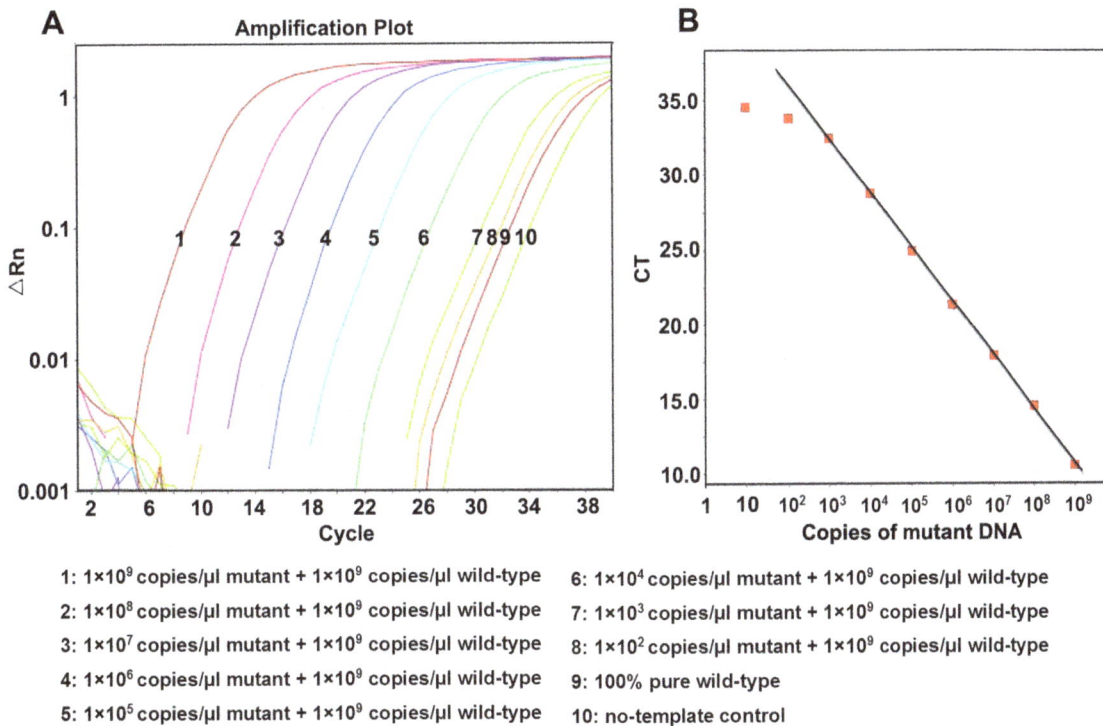

Amplification Plot

1: 1×10⁹ copies/µl mutant + 1×10⁹ copies/µl wild-type
2: 1×10⁸ copies/µl mutant + 1×10⁹ copies/µl wild-type
3: 1×10⁷ copies/µl mutant + 1×10⁹ copies/µl wild-type
4: 1×10⁶ copies/µl mutant + 1×10⁹ copies/µl wild-type
5: 1×10⁵ copies/µl mutant + 1×10⁹ copies/µl wild-type

6: 1×10⁴ copies/µl mutant + 1×10⁹ copies/µl wild-type
7: 1×10³ copies/µl mutant + 1×10⁹ copies/µl wild-type
8: 1×10² copies/µl mutant + 1×10⁹ copies/µl wild-type
9: 100% pure wild-type
10: no-template control

Figure 5. Sensitivity of RT-AS-LNA-qPCR in 1×10^9 copies/µl wild-type DNA background. (A) Amplification plot with different colors represented different copies of mutant DNA balanced mixing with 1×10^9 copies/µl wild-type DNA which were indicated in the figure. (B) Linear correlation diagram of sensitivity of RT-AS-LNA-qPCR in 1×10^9 copies/µl wild-type DNA background.

can be gradually selected to become the dominant species under dual pressures of NAs and host immunity, and finally precede the occurrence of viral breakthrough or biochemical breakthrough [9]. Therefore, establishing a simple, rapid, reliable and highly sensitive assay to detect the resistant mutants as early as possible

and analysis of their evolution during NAs-experienced are of great clinical significance.

A number of techniques regarding HBV mutants detection have been described. Direct sequencing, a qualitative assay, is considered to be the golden standard for HBV drug-resistance mutations detection, which is applied widely in clinical laborato-

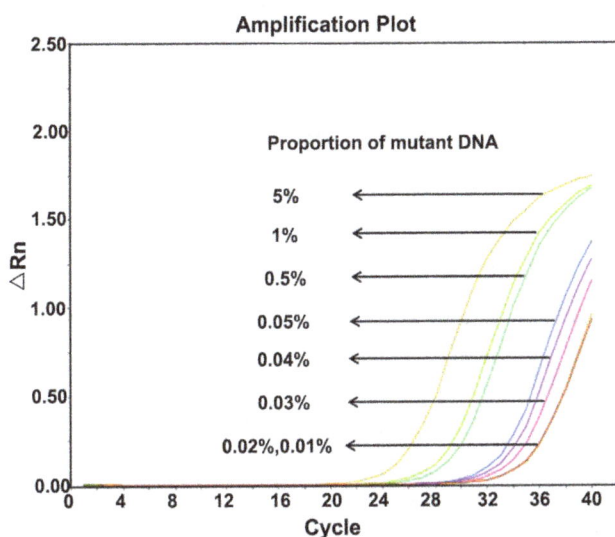

Figure 6. Clinical templates with different proportions of mutant DNA were tested by mutant specific primer set. Amplification plot with different colors represented different proportions of mutant DNA which were illustrated in the figure. The Ct of no-template control was undetectable.

Table 2. Detection sensitivity comparison of RT-AS-LNA-qPCR and sequencing on clinical samples containing different proportions of rtM204I variant.

Proportions of rtM204I	RT-AS-LNA-qPCR	Sequencing
50%	M/I	M/I
25%	M/I	M/I
20%	M/I	M/I
10%	M/I	M/I
5%	M/I	M
1%	M/I	-
0.5%	M/I	-
0.05%	M/I	-
0.04%	M/I	-
0.03%	M/I	-
0.02%	M/I	-
0.01%	M/I	-
NTC	N	N

M: Methionine; I: isoleucine; -: not performed; NTC: no-template control; N: undetectable.

Table 3. Concordance between RT-AS-LNA-qPCR and direct sequencing.

RT-AS-LNA-qPCR	Direct sequencing			Total
	YMDD	**YMDD+YIDD**	**YIDD**	
YMDD	**85**	0	0	85
YMDD+YIDD	*3*	**4**	*6*	13
YIDD	0	0	**4**	4
Total	88	4	10	102

Complete concordant results are shown in bold. Partial concordant results are shown in italics.

ries. But only mutations in circulating quasispecies pool present at>20% can be detectable [9]. Thus, it is not suitable for early detection of drug-resistance mutations. INNO-LiPA HBV DR, a commercial assay kit, has an analytical sensitivity between 5% and 10% and is much easier to perform and to automate [4], however, variants below 5% may be missed. In addition, it requires high cost and is only applied in developed countries. DNA chip is a high throughput, parallel and automatic technique and has become available in many fields, but it was not widely accepted in clinical laboratories due to its high cost. Ultradeep pyrosequencing (UDPS) can detect mutation of at least 1% of the total viral population, nonetheless, disadvantages have also been observed, including the short read length, high cost, complex data analysis and high error rate [3,10]. Therefore, so far it has been used only for research purposes and not suitable for clinical use.

Allele-specific PCR, has also been termed amplification refractory mutation system PCR, selective PCR or mismatch PCR, which is a convenient and easily performed assay for mutation detection. However, it would be very important to reduce or to avoid false positive results by the optimization conditions. Basically speaking, primer design is the most important factor that would determine whether AS-PCR works specifically. There have been several studies with respect to the detection of HBV variants by AS-PCR [11–13], however, to our knowledge, few AS-PCR assays have established taking advantage of LNA-modified primers and SYBR Green I.

Locked nucleic acid is a special nucleic acid analog with a 2-oxygen and 4-carbon atoms methylene bridge that locks the ribose group into the C3-endo conformation, thus, decreases ribosome structure flexibility and increases the phosphate backbone stability. It is generally accepted that LNA increases the melting DNA heteroduplex temperature between $1\sim8°C$ per LNA nucleotide [8,14].

To date, many reports have documented that primers at or near 3'-end modified with LNA nucleotides can be applied to improve the mismatch discrimination ability of allele-specific PCR assay [8,14,15]. But data about this application in HBV drug-resistance mutations detection are still scarce. Bhattacharya D et al. successfully developed an allele-specific quantitative PCR with combination of locked nucleic acid primers and a minor groove binder probe for the quantitative determination of minor viral quasispecies of the triple combination mutation rtV173L+rtL180M+rtM204V within one HBV genome. The assay could accurately detected 3×10^2 copies of the triple mutant in the 3×10^8 copies background of wild-type DNA [16].

Though genotypic detection of HBV resistance by AS-PCR using primers containing LNA bases has also been reported by Fang J et al. [17], several improvements were included in our study. There were some differences in primers. Firstly, ambiguity codes were introduced into the primers design because of the polymorphisms of HBV genomes, thereby allowing the primers to hybridize to different genotypes which are common in China. Different polymorphisms in the primer binding region may affect the sensitivity of mutant detection in the presence of the dominant wild type population. To validate detection sensitivity, experiments using non-degenerated primers with exact match to wild type and mutant plasmids were performed. The results indicated that there were no difference in sensitivity between the non-degenerated primers and degenerated primers. Degenerated primers against each pure polymorphic templates (genotype B and C) were also evaluated, whereas the detection sensitivity showed no difference as well. The likely explanations were as follows: On one hand, only two ambiguity codes were included in the forward primer design and only one in the reverse, consequently, the degeneracy is low. On the other hand, the positions of ambiguity codes were in the middle rather than 3'-end of the primers which may have limited influence on sensitivity test. Undoubtedly, the primers meet the degenerated primer design criteria. But the polymorphisms in HBV strains required the use of degenerated primer pairs according to a similar successful primer design described in

Table 4. Partial concordant results between RT-AS-LNA-qPCR and direct sequencing.

Patient number	Copy number of HBV DNA(copies/μl)		Mutant/Wild-type ratio	%Mutant/(Mutant+Wild-type)	Sequencing result	Motif
	Wild-type	**Mutant**				
2	9.32E+02	8.03E+04	86.2	98.9	ATT	YIDD
56	8.24E+05	3.47E+06	4.2	80.8	ATT	YIDD
11	5.25E+04	2.17E+07	413.3	99.8	ATT	YIDD
32	2.26E+04	1.13E+05	5.0	83.3	ATT	YIDD
7	3.31E+05	3.21E+06	9.7	90.7	ATT	YIDD
86	2.03E+04	1.13E+06	55.7	98.2	ATT	YIDD
22	3.47E+03	3.18E+02	0.092	8.4	ATG	YMDD
16	7.02E+04	7.63E+03	0.11	9.8	ATG	YMDD
65	1.22E+04	1.28E+03	0.10	9.5	ATG	YMDD

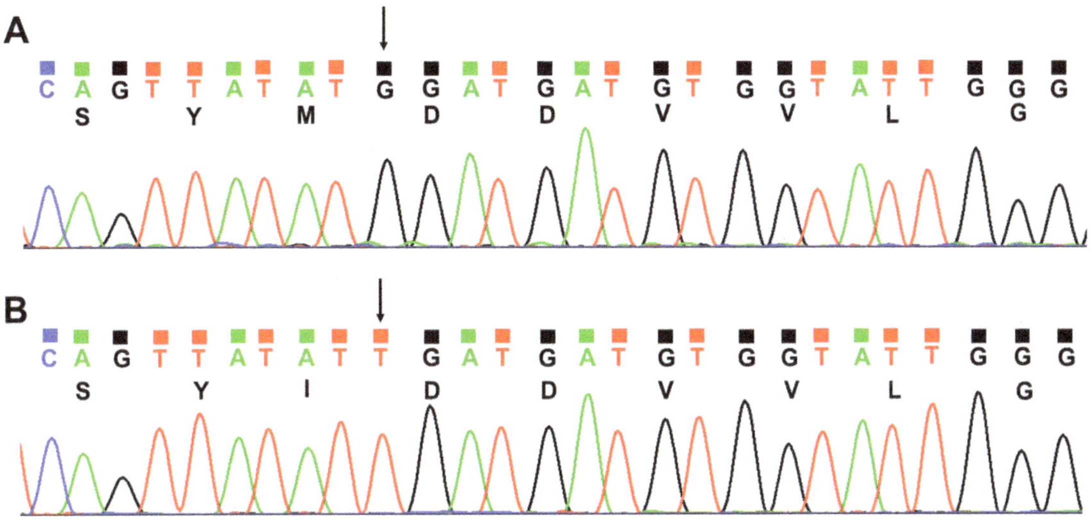

Figure 7. Typical chromatograms of the cloning sequencing of HBV rtM204 and rtM204I. (A) Chromatograms of rtM204 (wild-type). The bases in 204 site were ATG, indicated by the arrow. (B) Chromatograms of rtM204I (mutant). The bases in 204 site were ATT, indicated by the arrow.

Ntziora's research with respect to quantitative detection of the M204V hepatitis B virus minor variants [18]. Secondly, the LNA position of reverse primer specific for the wild-type sequence was at the penultimate position instead of 3′ termini which made it work more specifically. There were different opinions about positions of the nucleic acids locks in the literatures. And there was not a specific guideline on choosing the last position versus the penultimate position for incorporation of the LNA base so far. To a great extent, the position of LNA was determined empirically [16]. Most studies had shown a 3′ LNA residue in the primer at the SNP site (i.e., the mismatch site) could improve mismatch discrimination excellently. Therefore, the LNA nucleotide position in our primers was also chosen at the mismatch site. In a study of effect of locked nucleic acid (LNA) modification position upon representative DNA polymerase and exonuclease activities, Di Giusto et al. reported single LNA at the penultimate (L-2) nucleotide position generated nuclease resistance activity and provided improved discrimination [19]. And a similar successful primer design had also been adopted by Bhattacharya D et al. to quantify minority resistance variants in hepatitis B infection [16].

Inspired by their researches, we introduced an adenine (A) base into Primer L2 to allow LNA to locate at the penultimate to improve the specificity of Primer L2. And we compared the performance of L2 with $L2_l$. It could be seen that the priming efficiency of L2 and $L2_l$ were almost the same. When cross-reactivity tests were carried out, the specificity of L2 the was higher than $L2_l$, as evidenced by an increase of the Ct value ($\Delta Ct = CtL2 - CtL2_l > 1$) while $1 \times 10^8 \sim 1 \times 10^6$ copies/µl mutant DNA was added to the PCR reaction system. Consequently, L2 exhibited an advantage over $L2_l$ in this study. Beyond that, we also systematically evaluated its performance and clinical application. Besides plasmids, in the present work, clinical materials were used to estimate detection sensitivity and reproducibility of RT-AS-LNA-qPCR. The value of 0.03% was regarded as the sensitivity performed in the clinic. The results from RT-AS-LNA-qPCR and sequencing clearly demonstrated that the two assays showed a high concordance (*Kappa* = 0.676, *P* = 0.000). The complete concordance rate between the two assays was 91.2% (93/102). Among 6 partial results (YIDD was detectable by direct sequencing), the quantity of variants were represented 80.8% to 99.8%, a

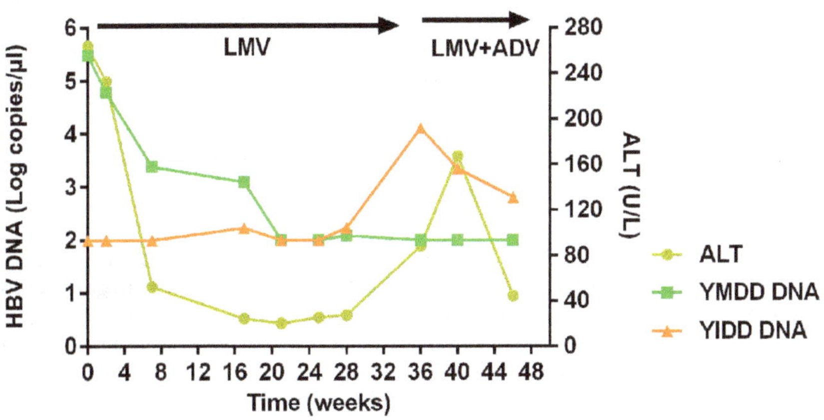

Figure 8. Dynamics of HBV YMDD, YIDD DNA levels and serum ALT level during LMV and LMV + ADV therapy in a CHB patient with breakthrough hepatitis.

considerable amount of the total viral population. Among another 3 samples collected from CHB patients who were highly suspicious of drug resistance, only wild-type was detected at rt204 by direct sequencing, but both YMDD and YIDD were detectable by RT-AS-LNA-qPCR. The mutant ratio were 8.4%, 9.8% and 9.5%, respectively, below the sensitivity of direct sequencing. The results from cloning sequencing performed on 3 typical samples whose detected results were partially concordant between direct sequencing and RT-AS-LNA-qPCR were concordant with our established assay. These results indicated that the assay was reliable and sensitive enough to detect minor HBV variants.

Previous studies demonstrated that YMDD variants can be detected about 7 months before clinical breakthrough [7], although different reports differed [20]. In the dynamic observation of virological and biochemical characteristics of a LMV-resistant CHB infection patient, we found that the minor variants can be detected 18 weeks earlier than the rebound of HBV DNA and alanine aminotransferase level. So it may provide useful information for predicting breakthrough hepatitis and adjusting treatment strategies, however, further work focusing on larger-scale sample analysis of its evolution regularity are necessary during follow-up study.

Results obtained from our experimental data showed RT-AS-LNA-qPCR had a much higher analytical sensitivity for detecting minor variants in high wild type background and had several advantages over direct sequencing. Moreover, the assay is SYBR Green I -based quantitative real-time PCR, and there is no need to design target-specific probes, which is an inexpensive and easily extended method. In addition, instead of $\Delta\Delta Ct$ calculation method, the assay uses standard curves in each run to absolute quantification wild type and mutant DNA amount, therefore, the accuracy of our assay is much more trustworthy.

Though RT-AS-LNA-qPCR showed excellent characteristics, some limitations were observed as well. Firstly, the specificity of wild-type primer is a little poor, the amplification of the mutant template with the wild-type primer occurred at a mutant viral concentration above 1×10^6 copies/μl, but, the observed copy number was less than the expected copy number at least 4 logs. There were evidences that nonspecific priming phenomena caused by AS-PCR cannot be completely avoided [11–13].The different degrees of specificity between wild-type and mutant primers may be due to different primer-template mismatch types [21]. However, no nonspecific products occurred when mutant DNA level was below 1×10^6 copies/μl. As the quantity of HBV DNA level is relatively low (below 1×10^6 copies/μl) in patients with NAs treatment and two separate PCR including wild-type and mutant reaction tubes were run in parallel, the limitation can be ignored to some extent. Secondly, mutant primer was only designed for ATT, however, there are three codons coding for isoleucine (ATT, ATC and ATA). It should be noted that ATC and ATA were omitted in the present study due to their low frequency in China [5]. In a survey taken in our laboratory, we investigated 57 serum samples whose YIDD sequence were confirmed by DNA sequencing. Results showed that ATT accounted for 94.7% (54/57), while non-ATT only accounted for 5.3% (3/57) of total and appeared concurrently with ATT. The results were consistent with what has been previously reported by Sun [22] and Pyo Hong S [6]. So, the false negative rate was obviously very low. Thirdly, the limitation is that validation data was restricted to rtM204I. It is known that rtM204I is one of the most important and common resistance mutation sites, which can not only cause the selection of LMV and LdT resistance, but also lead to ETV resistance. Therefore, establishing a simple, rapid, reliable and highly sensitive assay to detect the rtM204I mutant as early as possible is of great clinical

significance. In this study, we developed a new assay for the detection of the variant rtM204I with high-sensitivity. It provided a specific and reliable assay for detection of HBV YIDD variants, which is important for the management of CHB patients. Moreover, this assay may provide valuable reference for the detection of any other critical variants in the future study.

It remains unclear how quickly the minor variants will become the dominant and what mutant-to-wild-type ratio may cause treatment failure. In the follow-up study, we will combine with RT-ARMS-qPCR assay previously established in our laboratory to dynamic monitoring the evolution of HBV during NAs-experienced, so as to provide direct virological evidence for clinic.

In conclusion, a rapid, cost-effective, sensitive, specific and reliable method was developed for the quantification of the dynamic changes of HBV YIDD variants, which was a practical tool for HBV drug-resistant management. Further studies on methodological comparison including ultra-deep pyrosequencing, etc. will be included in our next study.

Supporting Information

Figure S1   Amplification plot of wild-type degenerated primers against each pure polymorphic templates (genotype B and C) for pMD-18-YMDD. (A) Amplification plot with different colors represented different concentrations of pMD-18-YMDD plasmids of genotype B amplified with wild-type degenerated primers which were illustrated in the figure. (B) Amplification plot with different colors represented different concentrations of pMD-18-YMDD plasmids of genotype C amplified with wild-type degenerated primers which were illustrated in the figure.

Figure S2   Efficiency and specificity test of L2 and L2₁. (A) Amplification plot with different colors represented different concentrations of pMD-18-YMDD plasmids amplified with L2 and L2₁ primers, respectively, which were illustrated in the figure (curve 1 and 2: 1×10^5 copies/μl; curve 3 and 4: 1×10^4 copies/μl; curve 5 and 6: 1×10^3 copies/μl; curve 7 and 8: 1×10^2 copies/μl.). (B) Amplification plot with different colors represented different concentrations of pMD-18-YIDD plasmids amplified with L2 and L2₁ primers, respectively, which were illustrated in the figure (curve 1 and 2: 1×10^8 copies/μl; curve 3 and 4: 1×10^7 copies/μl; curve 5 and 6: 1×10^6 copies/μl; curve 7 and 8: 1×10^5 copies/μl.).

Figure S3   Sensitivity of RT-AS-LNA-qPCR in 1×10^7 copies/μl and 1×10^5 copies/μl wild-type DNA background. (A) Amplification plot with different colors represented different copies of mutant DNA balanced mixing with 1×10^7 copies/μl wild-type DNA which were indicated in the figure. (B) Amplification plot with different colors represented different copies of mutant DNA balanced mixing with 1×10^5 copies/μl wild-type DNA which were indicated in the figure.

Figure S4   Sensitivity detection of mutants in 1×10^9 copies/μl wild-type DNA background using degenerated primer and non-degenerated primer. (A) Amplification plot with different colors represented different copies of mutant DNA balanced mixing with 1×10^9 copies/μl wild-type DNA, 100% pure wild-type DNA and no-template control respectively amplified with degenerated primer with exact match to mutant plasmid. (B) Amplification plot with different colors represented different copies of mutant DNA balanced mixing with 1×10^9 copies/μl wild-type DNA, 100% pure wild-type DNA and no-template control respectively amplified with non-degenerated primer with exact match to mutant plasmid.

Author Contributions

Conceived and designed the experiments: YZ QO BY. Performed the experiments: YZ DL WW JL. Analyzed the data: YZ DL MS JC. Contributed reagents/materials/analysis tools: LJ BY QO JL. Wrote the paper: YZ HC JL.

References

1. European Association For The Study Of The Liver (2012) EASL clinical practice guidelines: Management of chronic hepatitis B virus infection. J Hepatol 57: 167–185.

2. Kim SS, Cho SW, Kim SO, Hong SP, Cheong JY (2013) Multidrug-resistant hepatitis B virus resulting from sequential monotherapy with lamivudine, adefovir, and entecavir: clonal evolution during lamivudine plus adefovir therapy. J Med Virol 85: 55–64.

3. Chevaliez S, Rodriguez C, Pawlotsky JM (2012) New virologic tools for management of chronic hepatitis B and C. Gastroenterology 142: 1303–1313 e1301.

4. Degertekin B, Hussain M, Tan J, Oberhelman K, Lok AS (2009) Sensitivity and accuracy of an updated line probe assay (HBV DR v.3) in detecting mutations associated with hepatitis B antiviral resistance. J Hepatol 50: 42–48.

5. Liu H, Mao R, Fan L, Xia J, Li Y, et al. (2011) Detection of lamivudine- or adefovir-resistant hepatitis B virus mutations by a liquid array. J Virol Methods 175: 1–6.

6. Hong SP, Kim NK, Hwang SG, Chung HJ, Kim S, et al. (2004) Detection of hepatitis B virus YMDD variants using mass spectrometric analysis of oligonucleotide fragments. J Hepatol 40: 837–844.

7. Kirishima T, Okanoue T, Daimon Y, Itoh Y, Nakamura H, et al. (2002) Detection of YMDD mutant using a novel sensitive method in chronic liver disease type B patients before and during lamivudine treatment. J Hepatol 37: 259–265.

8. Latorra D, Campbell K, Wolter A, Hurley JM (2003) Enhanced allele-specific PCR discrimination in SNP genotyping using 3' locked nucleic acid (LNA) primers. Hum Mutat 22: 79–85.

9. Pallier C, Castera L, Soulier A, Hezode C, Nordmann P, et al. (2006) Dynamics of hepatitis B virus resistance to lamivudine. J Virol 80: 643–653.

10. Solmone M, Vincenti D, Prosperi MC, Bruselles A, Ippolito G, et al. (2009) Use of massively parallel ultradeep pyrosequencing to characterize the genetic diversity of hepatitis B virus in drug-resistant and drug-naive patients and to detect minor variants in reverse transcriptase and hepatitis B S antigen. J Virol 83: 1718–1726.

11. Lupo J, Larrat S, Hilleret MN, Germi R, Boyer V, et al. (2009) Assessment of selective real-time PCR for quantitation of lamivudine and adefovir hepatitis B virus-resistant strains and comparison with direct sequencing and line probe assays. J Virol Methods 156: 52–58.

12. Punia P, Cane P, Teo CG, Saunders N (2004) Quantitation of hepatitis B lamivudine resistant mutants by real-time amplification refractory mutation system PCR. J Hepatol 40: 986–992.

13. Wightman F, Walters T, Ayres A, Bowden S, Bartholomeusz A, et al. (2004) Comparison of sequence analysis and a novel discriminatory real-time PCR assay for detection and quantification of Lamivudine-resistant hepatitis B virus strains. J Clin Microbiol 42: 3809–3812.

14. Morandi L, de Biase D, Visani M, Cesari V, De Maglio G, et al. (2012) Allele specific locked nucleic acid quantitative PCR (ASLNAqPCR): an accurate and cost-effective assay to diagnose and quantify KRAS and BRAF mutation. PloS one 7: e36084.

15. Strand H, Ingebretsen OC, Nilssen O (2008) Real-time detection and quantification of mitochondrial mutations with oligonucleotide primers containing locked nucleic acid. Clin Chim Acta 390: 126–133.

16. Bhattacharya D, Lewis MJ, Lassmann B, Phan T, Knecht G, et al. (2013) Combination of allele-specific detection techniques to quantify minority resistance variants in hepatitis B infection: a novel approach. J Virol Methods 190: 34–40.

17. Fang J, Wichroski MJ, Levine SM, Baldick CJ, Mazzucco CE, et al. (2009) Ultrasensitive genotypic detection of antiviral resistance in hepatitis B virus clinical isolates. Antimicrob Agents Chemother 53: 2762–2772.

18. Ntziora F, Paraskevis D, Haida C, Magiorkinis E, Manesis E, et al. (2009) Quantitative detection of the M204V hepatitis B virus minor variants by amplification refractory mutation system real-time PCR combined with molecular beacon technology. J Clin Microbiol 47: 2544–2550.

19. Di Giusto DA, King GC (2004) Strong positional preference in the interaction of LNA oligonucleotides with DNA polymerase and proofreading exonuclease activities: implications for genotyping assays. Nucleic Acids Res 32: e32.

20. Umeoka F, Iwasaki Y, Matsumura M, Takaki A, Kobashi H, et al. (2006) Early detection and quantification of lamivudine-resistant hepatitis B virus mutants by fluorescent biprobe hybridization assay in lamivudine-treated patients. J Gastroenterol 41: 693–701.

21. Kwok S, Kellogg DE, McKinney N, Spasic D, Goda L, et al. (1990) Effects of primer-template mismatches on the polymerase chain reaction: human immunodeficiency virus type 1 model studies. Nucleic Acids Res 18: 999–1005.

22. Sun Y, Shen Z, Wu Y, Gao RT, Li Y, Yang Z (2007) Real-time PCR method for monitoring YMDD Motif mutations associated with Lamivudine resistance in patients with HBV infection. Chinese journal of laboratory medicine 30: 533–537.

21

Analysis of Clonal Type-Specific Antibody Reactions in *Toxoplasma gondii* Seropositive Humans from Germany by Peptide-Microarray

Pavlo Maksimov[1]*, **Johannes Zerweck**[2], **Aline Maksimov**[1], **Andrea Hotop**[3], **Uwe Groß**[3], **Katrin Spekker**[4], **Walter Däubener**[4], **Sandra Werdermann**[5], **Olaf Niederstrasser**[6], **Eckhardt Petri**[7], **Marc Mertens**[8], **Rainer G. Ulrich**[8], **Franz J. Conraths**[1], **Gereon Schares**[1]*

1 Federal Research Institute for Animal Health, Institute of Epidemiology, Friedrich-Loeffler-Institut, Wusterhausen, Germany, 2 JPT Peptide Technologies GmbH, Berlin, Germany, 3 German National Consulting Laboratory for Toxoplasmosis, Department of Medical Microbiology, University Medical Center Göttingen, Göttingen, Germany, 4 Institute of Medical Microbiology and Hospital Hygiene, Heinrich-Heine-University Düsseldorf, Düsseldorf, Germany, 5 Institut für Arbeits und Sozialhygiene Stiftung, Kyritz, Germany, 6 BG Kliniken Bergmannstrost, Halle, Germany, 7 Novartis Vaccines and Diagnostics, Marburg, Germany, 8 Federal Research Institute for Animal Health, Institute for Novel and Emerging Infectious Diseases, Friedrich-Loeffler-Institut, Greifswald - Insel Riems, Germany

Abstract

Background: Different clonal types of *Toxoplasma gondii* are thought to be associated with distinct clinical manifestations of infections. Serotyping is a novel technique which may allow to determine the clonal type of *T. gondii* humans are infected with and to extend typing studies to larger populations which include infected but non-diseased individuals.

Methodology: A peptide-microarray test for *T. gondii* serotyping was established with 54 previously published synthetic peptides, which mimic clonal type-specific epitopes. The test was applied to human sera (n = 174) collected from individuals with an acute *T. gondii* infection (n = 21), a latent *T. gondii* infection (n = 53) and from *T. gondii*-seropositive forest workers (n = 100).

Findings: The majority (n = 124; 71%) of all *T. gondii* seropositive human sera showed reactions against synthetic peptides with sequences specific for clonal type II (type II peptides). Type I and type III peptides were recognized by 42% (n = 73) or 16% (n = 28) of the human sera, respectively, while type II–III, type I–III or type I–II peptides were recognized by 49% (n = 85), 36% (n = 62) or 14% (n = 25) of the sera, respectively. Highest reaction intensities were observed with synthetic peptides mimicking type II-specific epitopes. A proportion of the sera (n = 22; 13%) showed no reaction with type-specific peptides. Individuals with acute toxoplasmosis reacted with a statistically significantly higher number of peptides as compared to individuals with latent *T. gondii* infection or seropositive forest workers.

Conclusions: Type II-specific reactions were overrepresented and higher in intensity in the study population, which was in accord with genotyping studies on *T. gondii* oocysts previously conducted in the same area. There were also individuals with type I- or type III-specific reactions. Well-characterized reference sera and further specific peptide markers are needed to establish and to perform future serotyping approaches with higher resolution.

Editor: Mauricio Martins Rodrigues, Federal University of São Paulo, Brazil

Funding: The study was supported by the German Federal Ministry of Education and Research (Toxonet01 and Toxonet02) by funds to Dr. Schares (01KI0765; 01KI1002F), Dr. Däubener (01KI0764; 01KI1002E) and Dr. Groß (01KI0766; 01KI1002B). The funders had no role in study design, data collection and analysis, decision to publish, or preparation of the manuscript.

Competing Interests: Dr. Zerweck is laboratory head for production of peptide microarray slides at JPT peptide Technologies. Dr. Petri is employed by Novartis Vaccines and Diagnostics. There are no patents, products in development or marketed products to declare.

* E-mail: pavlo.maksimov@fli.bund.de (PM); gereon.schares@fli.bund.de (GS)

Introduction

Infection with the intracellular protozoan parasite *Toxoplasma gondii* is often asymptomatic or causes flu-like symptoms in immunocompetent individuals. Primary maternal infection with the parasite during pregnancy may lead to abortion or induce disease in the transplacentally infected fetus. Toxoplasmosis is often fatal in immunocompromised patients [1,2,3].

T. gondii has a clonal population structure. North America and Europe are dominated by three clonal lineages of *T. gondii*, i.e. the clonal types I, II and III. Type II is most abundant in infected humans and domestic animals [4,5,6,7,8]. While type III strains are abundant in animals, they are rarely seen in humans [4,5,6,7,9], but this distribution may be impaired by a sampling bias. Previous studies suggested that type I strains are relatively rare in animals and humans and they have been predominantly found in immunocompromised patients who had experienced a reactivation of *T. gondii* infection, which frequently occurs in HIV-infected toxoplasmosis patients [4,10]. However, Ajzenberg and colleagues (2009) [11] demonstrated that most European immunocompromised patients with reactivated toxoplasmosis were infected with *T. gondii* clonal type II, whereas clonal type I and non-archetypal *T. gondii* types were isolated from African and South American patients. This suggests that the occurrence of particular *T. gondii* clonal types is influenced by the geographic origin of the patients. Most *T. gondii* isolates obtained in South America, Asia and Africa are genetically distinct from the clonal types I, II and III [12,13].

T. gondii of clonal types I, II and III show different virulence patterns in outbred mice inoculated intraperitoneally (i.p.) with tachyzoites [14,15]. In this experimental system, *T. gondii* of the clonal types II and III are characterized by LD50 values of $\geq 10^3$ tachyzoites, i.e. low virulence in mice. By contrast, *T. gondii* isolates of type I are highly virulent for mice with LD100 values of ≤ 10 tachyzoites [14,15]. It is not yet clear, whether these differences also imply differences in the pathogenicity of *T. gondii* in humans [15]. There is evidence, however, suggesting that host-genetic factors also contribute to the severity of toxoplasmosis [16,17,18,19,20,21].

Several serological assays have been reported that aim at predicting the clonal type of *T. gondii* by which animals or humans are infected [22,23,24,25,26]. Serotyping is based on the observation that the clonal lineages of *T. gondii* which dominate in North America and Europe differ not only genetically but also in the amino acid sequences of several parasite proteins, leading to polymorphic sites. Antibody responses against these polymorphic sites can thus be allele-specific [22,27]. Since the three clonal types may have arisen from common ancestors of two closely related but genetically different lineages [8,28], many of the polymorphic sites are specific for more than one of the three clonal types I, II or III. The pioneering work of Kong et al. (2003) [22] showed that short synthetic peptides derived from polymorphic regions could be used to serologically predict the clonal type of *T. gondii* humans or mice were infected with.

The aim of the present study was to test a panel of sera from *T. gondii* seropositive patients and volunteers (forest workers) from Germany against polymorphic, type-specific sites of 14 *T. gondii* antigens to obtain insights into the clonal types of *T. gondii* these persons were infected with and to explore potential differences in the peptide spectra recognized by patients and seropositive but non-diseased volunteers.

Materials and Methods

Patient sera from clinics

In total, 74 *T. gondii* positive human sera were provided by the Institute of Medical Microbiology and Hospital Hygiene, Heinrich-Heine-University, Düsseldorf and the Department of Medical Microbiology and the National Reference Center for Systemic Mycoses, University Medical Center, Göttingen. Out of these, 21 originated from individuals with acute toxoplasmosis, and 53 from individuals with chronic *T. gondii* infection. In addition, these institutions provided 65 samples from serologically *T. gondii*-negative individuals.

Screening of human sera for *T. gondii*-specific immunoglobulin G (IgG) was performed at the institutions providing the sera using an immunofluorescence test (IFT; bioMérieux, Nürtingen, Germany), the LIAISON IgG immunoassay (DiaSorin, Dietzenbach, Germany) or the Mini VIDAS immunoassay system (bioMérieux SA, Marcy l'Etoile, France). *T. gondii*-specific IgM was detected using the Mini VIDAS immunoassay system (bioMérieux SA, Marcy l'Etoile, France), the LIAISON IgM immunoassay (DiaSorin) or the ISAGA IgM immunoassay (bioMerieux). Detailed information about the serological results for each patient

serum is shown as supporting information (Table S1). Transient detection of *T. gondii*-specific IgM and eventually IgA was regarded as an indication of an acute infection. In a few patients (n = 7), a persistent IgM response was demonstrated by repeated testing. For these patients, a persistent but inactive (latent) infection was assumed. Presence of IgG and absence of IgM/IgA was regarded as an indication for persistent but inactive (latent) infection.

Sera from volunteers

A total number of 563 sera were collected from forest workers at all forest offices in the German Federal State Brandenburg [29].

Ethical considerations

The study reported in our manuscript was a collaborative work of the Toxonet01 project of the National Research Platform for Zoonoses and was approved by the respective ethical committees of the Medical Faculties of the Universities of Düsseldorf (3174, 20/01/09) and Göttingen (8/6/09) and by the State Medical Association of Brandenburg (19/04/10). Serum samples were collected under approved protocols.

For the anonymized patient sera provided by the Institute of Medical Microbiology and Hospital Hygiene, Heinrich-Heine-University, Düsseldorf and the Department of Medical Microbiology and the National Reference Center for Systemic Mycoses, University Medical Center, Göttingen informed consent was obtained verbally, which was in agreement with the ethical committee's approval. All volunteers (forest workers) were included in the study on the basis of written informed consent as described in detail by Mertens et al. (2011) [29].

Latex agglutination test

A latex agglutination test (LAT, TOXOREAGENT, MAST Diagnostica GmbH, Reinfeld, Germany) was performed according to the instructions of the manufacturer. Results were expressed as reciprocal antibody titres. Sera with reciprocal LAT titres of ≥ 16 were regarded as seropositive. Reciprocal LAT titres of <16 were considered as seronegative.

T. gondii surface antigen 1 (TgSAG1) immunoblot

Native *T. gondii* surface antigen 1 (TgSAG1) was affinity-purified as previously described [30]. The identity of the purified protein was confirmed using monoclonal antibodies against TgSAG1 (IgG2a P30/3 [ISL, Paignton, UK]). Detection of antibodies against TgSAG1 was performed essentially as described for animal sera [30] with a few modifications. Briefly, human sera were diluted 1:10 and the conjugate (horse radish peroxidase [HRP] AffiniPure rabbit anti-human IgA+IgG+IgM [H+L], Jackson ImmunoResearch, West Grove, PA, USA) was diluted 1:500. Reactivity with a protein of a relative molecular mass of 30 kDa was regarded as a *T. gondii* positive reaction. Sera obtained from a LAT positive and a LAT negative volunteer were used as controls.

Peptides

A total of 54 *T. gondii* synthetic peptides based on amino acid sequences representing polymorphic epitopes of the three archetypal lineages of *T. gondii* were used to detect type-specific antibodies in sera of *T. gondii* seropositive humans from Germany. The respective peptide sequences [22] were derived from 14 *T. gondii* immunogenic proteins, including dense granule proteins, surface antigens and rhoptry proteins (Table S2). Peptide sequences were based on information available for representative *T. gondii* strains of the clonal types I (RH), II (Me49 and Prugniaud) and III (VEG and CEP) (previously described by Kong et al. (2003) [22]). Some of the peptides had sequences specific for more than one of the three clonal lineages. These peptides are referred to as type I–II, type I–III or type II–III.

Preparation of peptide-microarray slides

Peptides were synthesized and printed on peptide-microarray slides by JPT Peptide Technologies GmbH, Berlin. First, aminooxy-acetylated peptides were synthesized on cellulose membranes in parallel using the SPOT synthesis technology [31,32]. After side chain de-protection, the solid phase-bound peptides were transferred into 96-well microtitre filtration plates (Millipore, Bedford, USA) and treated with 200 μl of aqueous triethylamine (0.5% v/v) to cleave the peptides from the cellulose support. Peptide-containing triethylamine solution was filtered off and the solvent removed by evaporation under reduced pressure. The resulting peptide derivatives (50 nmol) were re-dissolved in 25 μl printing solution (70% DMSO, 25% 0.2 M sodium acetate pH 4.5, 5% v/v glycerol) and transferred into 384-well microtitre plates. Two droplets of 0.5 nl peptide solution (1 mM) were deposited per spot on epoxy-functionalized glass slides (Corning Epoxy # 40042; Corning, Lowell, USA) using the non-contact printer Nanoplotter (GESIM, Großerkmannsdorf, Germany) equipped with a piezoelectric NanoTip (GESIM). The method for chemoselective immobilization on peptide-microarrays was originally described by Panse and colleagues (2004) [33]. This procedure was further optimized for peptide arrays for serum antibody detection and reviewed by Andresen and Grötzinger (2009) [34]. Chicken IgY, cat, human, mouse and pig IgG (Sigma, Munich, Germany and Diatec, Oslo, Norway) were also printed on the slides as antibody controls at a concentration of 500 μg/ml in 100 mM PBS buffer, pH 8.0.

The peptide library was spotted on each slide in triplicate. The slide layout consisted therefore of three identical sub-arrays; peptide and control spots were printed in 21 identical blocks. Printed peptide-microarrays were kept at room temperature for 5 h, washed with de-ionised water, quenched for 1 h with 0.1 mg/ml bovine serum albumin (BSA) in 75 mM saline sodium citrate (SSC) buffer, pH 7.0, containing 0.1% SDS and 750 mM NaCl, at 42°C, washed extensively with 1.5 mM SSC buffer, pH 7.0, followed by washings with de-ionised water and dried using a chip centrifuge (UNIEQUIP Laborgerätebau und Vertriebs GmbH, Planegg, Germany). Resulting peptide-microarrays were stored at 4°C until used.

Examination of sera by peptide-microarray

Array slides were first incubated with blocking solution (PBS, 0.05% Tween 20, 0.2% I-Block [Applied Biosystems, Bedford, MA, USA]) for 30 min. The slides were then placed into a Microplate Microarray Hardware (Arrayit Corporation, Sunnyvale, CA, USA), which allows to examine arrays separately in a 96 well ELISA format.

Human serum samples (150 μl/well), diluted 1:200 in blocking solution, were incubated at 37°C for 1 h and washed seven times for 3 min with PBS-T (PBS, pH 7.2; 0.5% Tween 20) on a shaker at room temperature. Conjugate (Cy5-AffiniPure donkey anti-human IgG, Fcγ fragment specific [min X Bov,Hrs,Ms Sr Prot], Jackson ImmunoResearch Laboratories, West Grove, USA) diluted 1:1000 (1 μg/ml) was added to the wells (150 μl/well), incubated at 37°C for 30 min, and washed as indicated above, followed by three additional washing steps, 1 min each, with sterile-filtered MilliQ water. Afterwards, the slides were spun dry for 10 s using a slide spinner (DW-41MA-230, Qualitron Inc/Eppendorf, Berzdorf, Germany).

Scanning and measurement of spot signal intensities and data extraction

Peptide-microarray slides were scanned at a wavelength of 635 nm using a GenePix 4000B microarray scanner (Axon Instruments, Concord, Canada) in a low-noise, high-sensitivity photomultiplier tube (PMT) at a level of 100% and a resolution of 10 μm. Images were saved electronically in TIFF and JPG formats.

Image analysis was performed using the circular feature alignment of the GenePix Pro 6.0 software (Axon Instruments) and GenePix Array List (GAL) files. Each circular feature consisted of the peptide spot to determine the foreground and a surrounding area to detect the background reaction. The signals from pixels of each circular feature were used to calculate median net fluorescence intensities of both, the foreground and background of each peptide spot [35,36].

Peptide-microarray data analysis

To analyze the raw data (median of signal intensity) in GPR (GenePix Results) files, index values (IVs) were recovered for each peptide-spot as log2 of the quotient of the medians of foreground and background [35,36]. Each serum was analyzed on a single block with the peptides printed in triplicate in each block. To obtain the serum-specific reaction against each peptide, the means of the IVs for each peptide spot per block (mean sample index value, MSIV) were calculated using the "corrected mean" formula (Microsoft Office EXCEL 2003) to exclude artefacts, i.e. false-positive and -negative signals within the replicas in each block. Application of the "corrected mean" formula had the following effect: If one out of three IVs per sample deviated more than 1.5-fold from the mean of all three IVs, the value was discarded and MSIV was calculated from the two remaining IVs. The peptide-microarrays used in this study failed to meet the criteria required for submission under MIAME based public databases [37,38]. Therefore MSIV for all sera and peptides are presented as supporting information (Table S3).

To ensure the specificity, we established an individual cut-off for each peptide to classify a reaction with this particular peptide as positive or negative using receiver-operating characteristic (ROC) analysis and the serological status of each serum (Table S3) as a reference standard. The cut-off was selected for each peptide separately using the MSIVs obtained for all *T. gondii* seronegative and seropositive human sera and accepting a maximum of 4% false-positive reactions. The results of ROC analysis (specificity, area under ROC curve, sensitivity and cut-off) for each peptide are shown as supporting information (Table S2). Table S4 shows the results of the application of these cut-offs to the MSIV for each serum and each peptide.

Statistical analysis

Fisher's exact test and logistic regression were computed with R, version 2.8.1 (R Foundation for Statistical Computing, Vienna, Austria, ISBN 3-900051-07-0, URL http://www.R-project.org) using packages "Stats" and "Epicalc" respectively [39]. Linear regression and the Wilcoxon rank test were performed using STATISTICA 8 (StatSoft, Tulsa, USA). P-values<0.05 were regarded as statistically significant. Kappa values were calculated using a web-based program (http://www.graphpad.com/quickcalcs/kappa1.cfm). To adjust p-values in multiple testing scenarios, Bonferroni correction was used [40].

To establish cut-offs for each peptide, ROC analysis was applied using the R-package "DiagnosisMed".

The R-package "*vcd*" was used for computing and visualizing log-linear independence models to examine whether reactions with specific peptide cohorts occurred more frequently than with others, i.e. whether the hypothesis of independence had to be rejected. Mosaic plots were used to visualize resulting contingency tables and Pearson residuals. Residuals displayed in mosaic plots represent standardized deviations of observed from expected values calculated by Pearson chi-square. The size of each box within the plot corresponds to the observed frequencies of positive and negative peptide reactions as well as the number of tested peptides within a peptide cohort specific of a clonal type. To present Pearson residuals in mosaic plots, the shading introduced by Friendly et al. (1994) [41] was used. Blue scale shading with a solid blue line (Pearson residuals: >2) or red scale shading with a dashed red line (Pearson residuals: <-2) indicate statistically significant (Pearson chi-squared p-value<0.05) over-, or under-representation of certain clonal type-specific peptide reactions within analyzed groups of sera, respectively (rejection of independence hypothesis). Pearson residuals from -2 to 2 are presented by filling the boxes in white colour, presenting the homogeneous distribution of peptide reaction within certain groups.

To perform multiple comparisons of the MSIVs for the tested peptides as well as for clonal type-specific peptide groups, a Post-Hoc-Test (LSD[Least Significant Difference]-Test) on ANOVA results was applied using the R package "*agricolae*". The differences between the means of positive peptide reactions in the analysis of peptide reactivities within tested groups were regarded as significant if the differences were equal to or higher than the LSD values.

Results

Examination for antibodies against *T. gondii* in sera from individuals and seropositive volunteers

All sera of *T. gondii*-infected individuals showed reciprocal LAT titres of 16 to >2048 (Table 1; Fig. 1 [A]). The majority of sera from individuals with an acute *T. gondii* infection (15 of 21; 71%) had reciprocal LAT titres of >256, while the majority of sera from individuals with a latent *T. gondii* infection (31 of 53; 59%) showed reciprocal LAT titres of 16 to ≤256. All sera from seronegative patients (n = 65) had reciprocal LAT titres of <16.

The initial LAT screening of a total of 563 sera of forest workers revealed the presence of antibodies against *T. gondii* in the LAT in 476 (84%) (Table 1). The majority of these sera (351 of 476; 74%) had LAT titres ranging between 16 and ≤256 (Fig. 1 [A]). The

remaining sera (n = 87) were regarded as seronegative (reciprocal LAT titres <16). Volunteers had significantly lower LAT titres than latently or acutely infected patients (Wilcoxon rank test, p-value<0.001). To confirm the LAT results, all volunteer sera were also tested by TgSAG1 immunoblot. Antibodies to TgSAG1 were detected in 485 of 563 (86%) sera. The agreement between the TgSAG1 immunoblot and the LAT was characterized by a kappa value of 0.913.

Logistic regression analysis revealed that seropositivity in volunteers (forest workers) was positively associated with age in both the LAT and the in-house TgSAG1 immunoblot (LAT: OR 1.09 [95% CI: 1.06–1.12], p_{Wald}-value<0.001; TgSAG1 immunoblot: OR 1.07 [95% CI: 1.04–1.1], p_{Wald}-value<0.001).

For further examination in the peptide-microarray, 100 volunteer sera which had tested *T. gondii* positive in both assays and 75 volunteer sera with negative results in both *T. gondii* tests were selected randomly. Seropositive and seronegative volunteers were interviewed during sampling to obtain information about their health status. None of the volunteers included in this study reported signs of acute toxoplasmosis.

Diagnostic specificity and sensitivity of peptide-microarray testing in seronegative and seropositive sera

All sera of seronegative patients and volunteers (n = 140) as well as sera from seropositive patients and volunteers (n = 174) were used to establish peptide-specific cut-offs by ROC analysis. The serological status of patients and volunteers was based on LAT results as a reference standard (Table S3). Application of these cut-offs revealed peptide-dependent diagnostic specificities for the *T. gondii*-negative sera which ranged between 96% and 97% (Table S2). A total of 174 sera, including all seropositive sera from patients (21 with acute and 53 with latent *T. gondii* infection) and 100 randomly selected sera from seropositive volunteers were tested on the peptide-microarray (Table S5). Twenty-two of these 174 (12.6%) seropositive sera failed to recognize any of the 54 peptides. All non-reactive sera had low LAT titres, i.e. showed reciprocal LAT titres between 16 and ≤256.

Sera of patients with an acute *T. gondii* infection recognized a significantly higher number of peptides than the sera of seropositive volunteers (Wilcoxon Rank Test, p-value = 0.012). Also sera of patients with a latent *T. gondii* infection reacted with a statistically significantly higher number of peptides than the sera of seropositive volunteers (Wilcoxon Rank test, p-value = 0.017). The differences between individuals with acute and latent *T. gondii* infections were not statistically significant (Wilcoxon rank test, p-value = 0.256). Linear regression analysis revealed that the number

Table 1. Results in the Latex-Agglutination-Test (LAT) for sera from groups of seropositive and seronegative individuals with acute or latent toxoplasmosis including patients from clinics and volunteers (forest workers).

Group	Infection status	LAT titre									
		<16	16	32	64	128	256	512	1024	2048	>2048
Patients (n = 21)	Acute*		1		1	1	3	4	3	4	4
Patients (n = 53)	Latent#		1	1	7	10	12	3	8	1	10
Volunteers (n = 476)	Positive$^{\Omega}$		22	44	79	111	95	76	35	9	5
Patients/Volunteers (n = 152)	Negative	152									

*Transient detection of *T. gondii* specific IgM and eventually IgA was regarded as an indication of acute infection.
#Presence of IgG and absence of IgM/IgA was regarded as an indication for persistent but inactive (latent) infection. In a few patients a persistent IgM response was demonstrated by repeated testing. For these patients, a persistent but inactive (latent) infection was assumed.
$^{\Omega}$Antibody isotypes not specified.

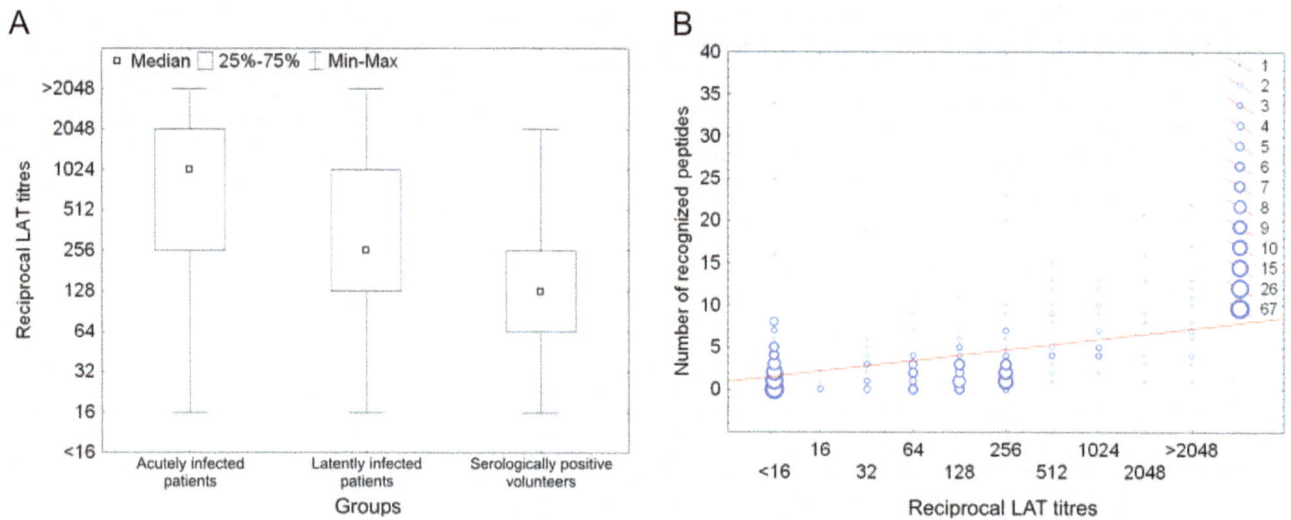

Figure 1. The number of recognized peptides is associated with the titre in LAT (Latex Agglutination Test). Wilcoxon rank test analysis of the *T. gondii* LAT titre distribution within groups of human sera revealed significantly lower LAT titres in sera from volunteers than in sera from latently or in acutely infected individuals (p-value<0.001). No significant differences were observed between sera from acutely and latently infected individuals (A). The association between LAT titre and number of recognized peptides was characterized by an R^2 value of 0.16 (p-value<0.001) (B).

of recognized peptides was statistically significantly associated with the log2-transformed reciprocal LAT titres (p-value<0.001). This association was characterized by an R^2 value of 0.16 (Figure 1[B]).

Type-specificity of peptide reactions

In total 9396 (54 peptides×174 sera) peptide reactions were possible: 2436 (14 peptides×174 sera) type I-specific, 2436 (14 peptides×174 sera) type II-specific, 1044 (6 peptides×174 sera) type III-specific, 696 (4 peptides×174 sera) type I–II-, 1740 (10 peptides×174 sera) type I–III-, and 1044 (6 peptides×174 sera) type II–III-specific. In total, 731 of 9396 (8%) possible peptide reactions were observed. Positive reactions were predominantly directed against type II-specific and the type II–III-specific peptides (proportions of peptide reactions 14% [336/2436]) and 10% [106/1044], respectively). The positive reactions against type I- (116/2436 [5%]), type III- (41/1044 [4%]), type I–II- (30/696 [4%]) or type I–III-specific (102/1740 [6%]) peptides were underrepresented.

Reactions within clonal type-specific peptide groups were statistically analyzed using a log-linear model. Contingency tables and deviations from independence hypothesis were visualized in mosaic plots (Fig. 2 [A, B, C, D]). Pearson residuals >4 indicated a statistically significant (Chi-squared p-value<0.001) overrepresentation of positive clonal type II-specific peptide reactions within all groups of sera (Fig. 2 [A, B, C, D]). In latently infected patients, type II–III-specific positive peptide reactions were also statistically significantly overrepresented (Pearson residuals: 2–4; Chi-squared p-value<0.05) (Fig. 2 [B]). In patients with an acute infection and in volunteers reactions with type II–III-specific peptides were also overrepresented (white rectangle with solid blue borderline), however, the hypothesis of independence could statistically not be rejected (Fig. 2 [A, C]). Reactions with peptides of type I, III, I–III and I–II were underrepresented in all tested groups of sera (Fig. 2 [A, B, C, D]).

Of 35 dense granule-derived peptides, 7 were recognized by 16%–42% of the sera (Table S5). The majority of these peptides (n = 5) had amino acid sequences specific for type II (dGRA6-II-216(9), GRA3-II-28, GRA6-II-214, dGRA6-II-214, GRA7-II-225; Table S5). The amino acid sequences of the remaining two

peptides were specific for both, clonal types I and III (GRA3-I/III-28) or had a sequence specific for clonal type I (NTP3-I-99). Two other dense granule peptides were recognized by 10–11% of the sera. One of these peptides had a type II-specific (dGRA6-II-214(9)) and the other peptide a type I–II-specific (GRA7-I/II-215) amino acid sequence.

Only one of 15 surface antigen-derived (SAG3-II-49) and one of six rhoptry-derived (ROP1-II/III-181) peptides with type II and type II–III specificity were recognized by more than 15% of the sera (Table S5). One rhoptry (ROP1-II/III-359) and none of the remaining surface-derived peptides were among those recognized by 10–15% of the sera.

Differences in intensity of type-specific peptide reactions

To detect differences between reaction intensities (MSIVs) for each tested peptide and for peptide groups presenting clonal type specificity in seropositive patients and volunteers, ANOVA and the LSD-Post-Hoc-Test were performed (Fig. 3). The analyses revealed that for the groups of acutely and latently infected patients, those peptide groups mimicking clonal type II and II–III specificities were recognized by the highest MSIVs as compared to the remaining peptide groups. These differences were statistically significant (LSD>0.36, p-value<0.05 [for acutely infected patients]; LSD>0.16, p-value<0.05 [for latently infected patients]) (Fig. 3 [A, C]). In volunteers, the clonal type II-specific peptide group was also recognized by the highest MSIVs as compared to remaining peptide groups. The difference was statistically significant (LSD>0.102, p-value<0.05 [for seropositive volunteers]) (Fig. 3 [E]).

The intensity of index values was also analysed for each peptide in patient groups with acute or latent *T. gondii* infection and in seropositive volunteers.

Peptides derived from dense granule antigens mimicking type II specificity (GRA6-II-214, dGRA6-II-214, dGRA6-II-216(9)) and one type II–III rhoptry derived peptide (ROP-II/III-181) were detected by the highest MSIVs in all patient and volunteer groups. The differences were statistically significant (LSD>0.67, p-value<0.05 [for acutely infected patients]; LSD>0.36, p-value<0.05 [for latently infected patients]; LSD>0.24, p-value<0.05 [for seropositive volunteers]) (Fig. 3 [B, D, F]).

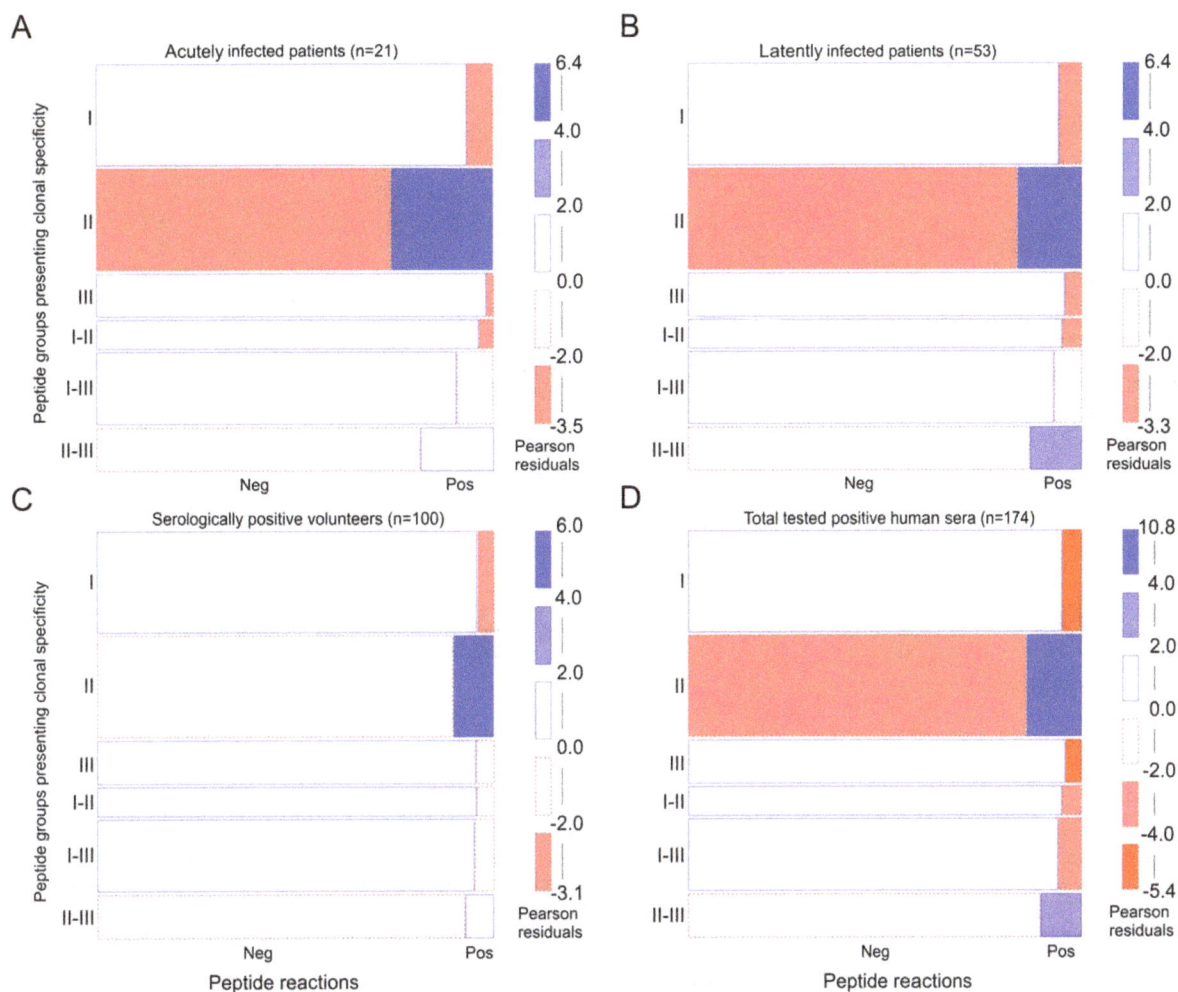

Figure 2. Statistically significant overrepresentation of reactions against clonal type II specific peptides. To determine whether reactions against certain clonal type-specific peptide cohorts (I, II, III, I–II, I–III, or II–III) were over- or underrepresented in various groups of *T. gondii* positive human sera a log-linear model analysis was performed and visualized by mosaic plot: acutely infected individuals (A), latently infected individuals (B) serologically positive volunteers (forest workers) (C) and all tested positive human sera (D). The size of each box corresponds to the observed frequencies of positive (Pos) and negative (Neg) peptide reactions as well as the number of tested peptides within each clonal type-specific peptide cohort. Pearson residuals represent standardized deviations of observed from expected values. The solid blue line indicates that the number of positive or negative reactions is higher than expected but not statistically significant. Blue scale shadings suggest the statistically significant rejection of the null hypothesis, i.e. overrepresentation of certain type-specific peptide reactions (Pearson chi-squared p-value<0.05). Dashed red lines indicate an underrepresentation of positive or negative peptide reactions which is not statistically significant. Red scale shadings suggest a statistically significant rejection of the null hypothesis, i.e. underrepresentation of peptide reactions within tested peptide and human groups (Pearson chi-squared p-value<0.05).

In patients with acute and latent toxoplasmosis, two further type II-specific peptides (GRA3-II-28, GRA7-II-225) were also detected by the highest MSIVs (LSD>0.67, p-value<0.05 [acutely infected patients]; LSD>0.36, p-value<0.05 [Latently infected patients]) (Fig. 3 [B, D]).

GRA3-I/III-28 and SRS-I-53 peptides also belong to the peptide group recognized by the highest MSIVs in acutely *T. gondii*-infected patients and in seropositive volunteers (LSD>0.67, p-value<0.05 [acutely infected patients]; LSD>0.24, p-value<0.05 [for seropositive volunteers]) (Fig. 3 [B, F]).

Differences in number of anti-peptide reactions between groups of sera

In all sera from patients and volunteers, reactions against type II dense granule based peptides (GRA3-II-28, GRA6-II-214, GRA7-II-225) and a type II–III rhoptry peptide (ROP1-II/III-181)

dominated, i.e. these peptides reacted with more than 20% of the sera (Table S5).

In addition, further peptides were recognized by more than 20% of the sera of patients with acute toxoplasmosis (dGRA6-II-216(9)) but not by those of seropositive volunteers and patients with latent infection. More than 20% of the sera from patients with latent infection recognized a type II surface antigen-derived peptide (SAG3-II-49), but those of volunteers and individuals with acute toxoplasmosis failed to react with this peptide in a proportion of >20%. Two dense granule-based peptides (dGRA6-II-214, GRA3-I/III-28) were recognized by more than 20% of the patients sera (acute, latent), but not by those of seropositive volunteers.

No major differences within each peptide category were observed between the different groups of persons (Table S6), with three exceptions. Individuals with acute toxoplasmosis recognized

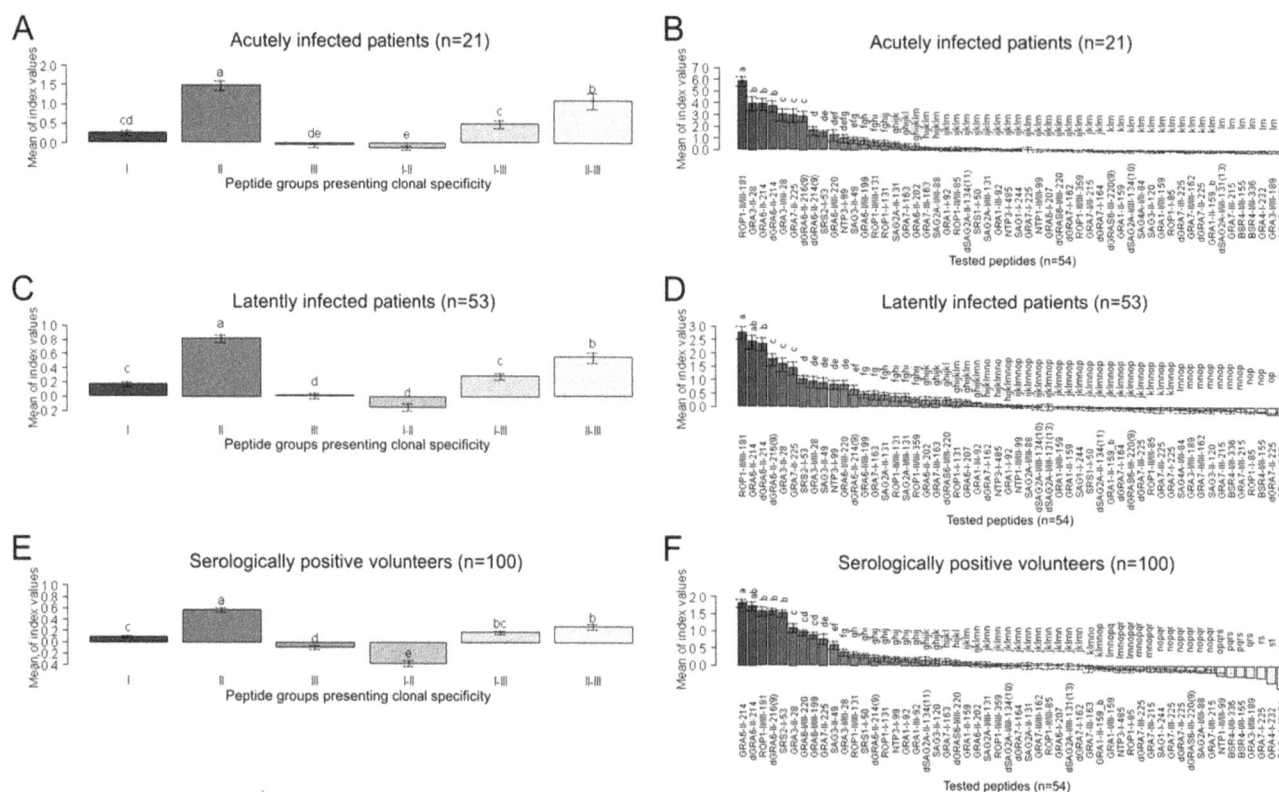

Figure 3. Strongest reaction intensities were recorded for clonal type II specific peptides. To evaluate the intensities (MSIVs) by which single peptides as well as peptide cohorts (I, II, III, I–II, I–III, or II–III) were recognized by *T. gondii* seropositive patient and volunteer groups, ANOVA and the Least Significant Difference (LSD)-Post-Hoc-Test were performed. Whiskers in barplots represent 95% confidence intervals of the means of MSIVs. The differences between the means of MSIVs for single peptides or peptide cohorts within tested groups were regarded as statistically significant, when the differences were equal or higher than the LSD values. Different letters above the whiskers indicate significant differences between the mean intensities in the Post-Hoc-LSD test. Means of MSIVs for each peptide cohort are presented in (A) for the acutely infected patient group (LSD>0.36, p-value<0.05); in (C) for the latently infected patient group (LSD>0.16, p-value<0.05); and in (E) for the seropositive volunteer group (LSD>0.10, p-value<0.05). Means of MSIVs for each single peptide are presented in (B) within the acutely infected patient group (LSD>0.67, p-value<0.05); in (D) within the latently infected patient group (LSD>0.36, p-value<0.05); and in (F) within the seropositive volunteer group (LSD>0.24, p-value<0.05).

a significantly higher proportion of peptides with type II-specific sequences than volunteers (p-value = 1.87×10^{-11}; Fisher's exact test; Table S6) and latently infected patients, (p-value = 8.24×10^{-11}). Latently infected patients recognized a significantly higher proportion of peptides with type II-specific sequences than seropositive volunteers (p-value = 2.98×10^{-5}, Fisher's exact test; Table S6). Acutely and latently infected patients recognized a significantly higher proportion of peptides with type II–III specificity as compared to volunteers (p-value = 0.00021, p-value = 0.0036; Fisher's exact test; Table S6).

Peptide-microarray-based serotyping

Data analysis was carried out to identify the clonal types responsible for the infection of the tested persons and to analyse potential type-specific differences in the peptide spectra recognized by individuals presenting acute or latent *T. gondii* infection and seropositive volunteers.

The majority (n = 124; 71%) of sera showed reactions against synthetic peptides with sequences specific for clonal type II (type II peptides) (Table 2). Forty-two percent (n = 73) or 16% (n = 28) of the sera reacted with type I and type III peptides, respectively, while type II–III, type I–III or I–II peptides were recognized by 49% (n = 85), 36% (n = 62) or 14% (n = 25) sera, respectively.

Based on the anti-peptide reactions, only a fraction of sera could be clearly attributed to either of the three clonal types I, II, or III (Table 3). Among sera that reacted with peptides containing type II specific sequences, 35% (50/142) showed reactions exclusively compatible with clonal type II (Table 3). The remaining 65% (92/142) reacted not only with type II peptides but also with peptides with sequences specific for other clonal types. Most sera reacting with type I and type III peptides could not be clearly assigned to one of the three clonal lineages as many of them also recognized peptides with sequences specific for the other clonal types (Table 3).

Discussion

A number of polymorphic peptides has been described in *T. gondii* antigens which might be suitable to indirectly determine the clonal type of *T. gondii*, humans or mice are infected with [22]. Using such peptides, we tested sera from seropositive volunteers and patients from Germany, to obtain insights into the clonal types of *T. gondii* by which these humans were infected and to examine potential differences in the spectra of peptides recognized by sera of various subgroups.

Several attempts have been made to type *T. gondii* infections by serological techniques using ELISA formats in which synthetic peptides were coupled via keyhole limpet hemocyanin [22,42] or

Table 2. Clonal type-specific anti-peptide reactivity of *T. gondii* positive humans.

Peptide specificity	Acute patients (n = 21)		Non-acute patients (n = 53)		Volunteers (forest workers) (n = 100)		Total (n = 174)	
	n	%	n	%	n	%	n	%
I	9	43	26	49	38	38	73	42
II	19	91	38	72	67	67	124	71
III	2	10	8	15	18	18	28	16
I–II	2	10	9	17	14	14	25	14
I–III	15	71	19	36	28	28	62	36
II–III	19	91	30	57	36	3	85	49

Reactions are sorted according to the specificities of peptides.

directly to the solid phase [24,25,43,44]. Others used recombinant antigens for serotyping [23,26]. We applied a synthetic peptide-microarray format to test a panel of sera simultaneously with all peptides that had previously been used in an ELISA by Kong et al. (2003) [22]. In studies on other infectious diseases, Melnyk et al. (2002) [45] and Mezzasoma et al. (2002) [46] compared peptide-ELISAs with peptide-microarrays and found that peptide-micro-arrays were much more sensitive than peptide-ELISAs. We therefore expected for the serological typing of *T. gondii* infections that the microarray format should have at least the same sensitivity as the previously reported ELISA format. To ensure a minimum diagnostic specificity of 96% for each peptide, i.e. to make sure that it is unlikely that *T. gondii*-negative humans react with any of these peptides, an individual cut-off was selected for each peptide based on the foreground-background ratio obtained for each peptide and the sera of 140 *T. gondii* seronegative humans.

By conventional techniques, i.e. by PCR-RFLP mediated genotyping using polymorphic loci, we have previously shown that almost all *T. gondii* parasites isolated from cats in Germany showed an allele combination resembling that of clonal type II [47,48]. Only a single clonal type III isolate and a few isolates with allele combinations different from those of clonal type I, II or III were observed [47]. We therefore expected that the majority of sera from seropositive humans from Germany would recognize peptides with type II-specific amino acid sequences. This turned out to be true since reactions with type II peptides were superior compared to reactions with other peptides in number as well as in intensity. Thus our results are in accord with the results of serotyping studies performed in France and Poland with a limited number (i.e. 8 or 2, respectively) of those 54 peptides we applied (GRA6-II-214, GRA6-I/III-220, dGRA6-II-214, dGRAS6-I/III-220, GRA7-II-225, GRA7-III-225, dGRA7-II-225, dGRA7-III-

225) [42,43]. In addition, our findings confirm the results of studies from Peyron et al. (2006) and Morisset et al. (2009) with recombinant polypeptides mimicking polymorphic clonal type-specific sites of *T. gondii* GRA5 and GRA6 which revealed a significantly dominant clonal type II-specific serological response in patients from France, Italy and Denmark [23,26].

Although reactions with type II-specific peptides dominated in number and intensity in our study, the sera of many of these humans reacted also with a few peptides with sequences specific for other clonal types. As it is unlikely that all these individuals experienced mixed infections or infections with atypical *T. gondii*, these conflicting results were probably due to the limited specificity of some of the peptides used in serotyping. In these cases, the clonal type of *T. gondii* the affected persons were infected with could not be unambiguously determined. One reason for a low discriminatory power of individual peptides might be the presence of at least one further epitope in the non-polymorphic part of the peptide in addition to the type-specific epitope in the polymorphic site [22,23]. Therefore, our results suggest that a large panel of well characterized human sera is needed to determine the specificity of each polymorphic peptide. The peptides that are finally used to differentiate clonal type-specific antibody reactions in individuals must be selected extremely carefully. Unfortunately, well-characterized human sera suitable for the evaluation of peptides are rare.

The results of this study also show that the sensitivity by which peptides were recognized varied considerably between the examined groups of patients or volunteers, respectively. For instance, individual type II-specific peptides were recognized by 1% to 42% of the sera.

Each individual serum recognized an almost unique spectrum of peptides. This may reflect an individual maturation of particular

Table 3. Proportion of human sera showing peptide reactions compatible with *T. gondii* infections by clonal types I, II, or III.

	Sera with anti-peptide reactions exclusively compatible with the respective clonal type				Sera with anti-peptide reactions not exclusively compatible with the respective clonal type			
	Total (%)	AP (%)*	LP (%)*	V (%)*	Total (%)	AP (%)*	LP (%)*	V (%)*
Clonal type I	11 (11)	0 (0)	5 (15)	6 (12)	89 (89)	15 (100)	28 (85)	46 (88)
Clonal type II	50 (35)	5 (25)	16 (36)	29 (37)	92 (65)	15 (75)	28 (64)	49 (63)
Clonal type III	12 (13)	1 (6)	5 (17)	6 (13)	80 (87)	15(94)	25 (83)	40 (87)

Reactions are sorted according to their compatibility with infections of *T. gondii* of the clonal type I, II, or III.
*Data resolved for seropositive patients with acute toxoplasmosis (AP), patients with latent toxoplasmosis (LP) and seropositive volunteers (V).

plasma cells leading to an increased affinity of the antibodies they produce against different antigens or epitopes of *T. gondii*. The variation in the sensitivity of peptide recognition by different groups of infected persons may be further influenced by a variety of variables, e.g. host-genetic factors, route of infection, secondary infections and time of primary infection [17,18,49].

We also found a statistically significant association between the LAT titre and the number of recognized peptides, with the LAT titres explaining 16% of the variability in the number of recognized peptides. Consequently, groups of humans showing differences in mean LAT titres showed similar differences in the number of peptides recognized in the microarray analysis. For instance, sera of patients with an acute *T. gondii* infection had a significantly higher mean LAT titre (Fig. 1 [A]) and recognized a significantly higher number of peptides than sera of seropositive volunteers.

In this study, 13% of the LAT positive sera did not react with any of the 54 peptides used. A previous study, in which two ELISAs with peptides presenting clonal type II and I–III specificity (GRA6-II-214, GRA6-I/III-220) were used, revealed that more than 30% of seropositive sera from Europe (France and Portugal) failed to react in these peptide ELISAs [43]. Sousa and colleagues (2009) suggested that the use of single peptides for serotyping could lead to mistyping. To overcome this problem, a large pool of polymorphic peptides from different antigens should be used [24]. Although we applied a much higher number of peptides as compared to these previous studies, we also observed a high proportion of sera that reacted only with a low number of peptides.

Individual peptides were only recognized by a limited number of sera. GRA6-II-214, for example, has previously been used in a number of other typing studies [22,24,25,41]. This peptide was recognized only by 31% of all tested *T. gondii* antibody positive sera. The truncated variations of this peptide (dGRA6-II-214; dGRA6-II-214(9); dGRA6-II-216(9)) were recognized by even lower proportions of *T. gondii* antibody-positive sera (19%, 10% and 18%, respectively). Therefore, the results of our study show that the sensitivity of individual peptides might be low and, consequently, allow to conclude that serotyping with synthetic peptides requires a large number of highly specific polymorphic peptides.

Our results clearly showed that peptides derived from dense granule proteins, i.e. GRA3, GRA6, and GRA7, were the most reactive ones when tested with human sera. Of 35 dense granule-derived peptides, 7 were recognized by more than 15% of the examined human sera. None of the 15 surface antigen-derived peptides and only 1 of 6 rhoptry antigen-derived peptides rendered a similar result. This finding is in accord with the high potential of dense granule proteins as diagnostic antigens [50,51,52,53].

In our study, a higher proportion of acutely infected patients recognized GRA6 and GRA3 derived peptides as compared to individuals with latent *T. gondii* infection (Table S5). This is in accord with previous results of others who showed that it is possible to discriminate between acute and chronic *T. gondii* infections by using recombinant GRA6 or GRA7 [54,55,56].

In conclusion, the results of this study demonstrate that a peptide-microarray assay can be used to detect *T. gondii* clonal type-specific antibody responses in seropositive humans. A previous study suggested that individuals in the study area were mainly exposed to clonal type II *T. gondii* [47,48]. Indeed, positive peptide reactions presenting clonal type II specificity were statistically significantly overrepresented in the tested human population and the intensity by which type II peptides were recognized was significantly higher than the intensity by which peptides with other specificities were detected. However, to establish serotyping assays with higher resolution, well-characterized reference sera and further specific peptide markers are needed.

Supporting Information

Table S1 Detailed serological results for patient sera.

Table S2 Peptides with clonal type specific amino-acid sequences used for typing the anti-*T. gondii* IgG response in humans.

Table S3 Corrected mean sample index values (signal intensity) listed for all peptides and sera.

Table S4 Serum-peptide reactions as determined by peptide specific cut-offs for all tested peptides and sera. Positive reactions are signed as original index value and negative reactions signed as "0".

Table S5 Number of sera from seropositive patients and volunteers (forest workers) recognizing peptides with clonal type-specific amino acid sequences.

Table S6 Statistical analysis (Fisher's exact test) of differences in the proportion of peptides recognized by different groups of toxoplasmosis patients (acute, latent) or seropositive volunteers (forest workers).

Acknowledgments

We acknowledge the excellent technical assistance of Andrea Bärwald, Lieselotte Minke and Robert Carus. We would like to thank D. C. Herrmann for his critical comments on the manuscript.

Author Contributions

Conceived and designed the experiments: PM JZ UG WD FJC GS. Performed the experiments: PM JZ AM AH KS WD SW ON EP MM. Analyzed the data: PM GS. Contributed reagents/materials/analysis tools: JZ AH UG KS WD SW ON EP MM RGU FJC GS. Wrote the paper: PM FJC GS.

References

1. Janitschke K, Held T, Kruiger D, Schwerdtfeger R, Schlier G, et al. (2003) Diagnostic value of tests for *Toxoplasma gondii*-specific antibodies in patients undergoing bone marrow transplantation. Clin Lab 49: 239–242.
2. Montoya JG, Liesenfeld O (2004) Toxoplasmosis. Lancet 363: 1965–1976.
3. Jones JL, Lopez A, Wilson M, Schulkin J, Gibbs R (2001) Congenital toxoplasmosis: a review. Obstet Gynecol Surv 56: 296–305.
4. Howe DK, Sibley LD (1995) *Toxoplasma gondii* comprises three clonal lineages: correlation of parasite genotype with human disease. J Infect Dis 172: 1561–1566.
5. Howe DK, Honor' S, Derouin F, Sibley LD (1997) Determination of genotypes of *Toxoplasma gondii* strains isolated from patients with toxoplasmosis. J Clin Microbiol 35: 1411–1414.

6. Ajzenberg D, Banuls AL, Tibayrenc M, Darde ML (2002) Microsatellite analysis of *Toxoplasma gondii* shows considerable polymorphism structured into two main clonal groups. Int J Parasitol 32: 27–38.

7. Sibley LD, Khan A, Ajioka JW, Rosenthal BM (2009) Genetic diversity of *Toxoplasma gondii* in animals and humans. Philos Trans R Soc Lond B Biol Sci 364: 2749–2761.

8. Khan A, Dubey JP, Su C, Ajioka JW, Rosenthal BM, et al. (2011) Genetic analyses of atypical *Toxoplasma gondii* strains reveal a fourth clonal lineage in North America. Int J Parasitol 41: 645–55.

9. Dubey JP, Graham DH, Dahl E, Sreekumar C, Lehmann T, et al. (2003) *Toxoplasma gondii* isolates from free-ranging chickens from the United States. J Parasitol 89: 1060–1062.

10. Khan A, Su C, German M, Storch GA, Clifford DB, et al. (2005) Genotyping of *Toxoplasma gondii* strains from immunocompromised patients reveals high prevalence of type I strains. J Clin Microbiol 43: 5881–5887.

11. Ajzenberg D, Yera H, Marty P, Paris L, Dalle F, et al. (2009) Genotype of 88 *Toxoplasma gondii* isolates associated with toxoplasmosis in immunocompromised patients and correlation with clinical findings. J Infect Dis 199: 1155–67.

12. Carme B, Demar M, Ajzenberg D, Darde ML (2009) Severe acquired toxoplasmosis caused by wild cycle of *Toxoplasma gondii*, French Guiana. Emerg Infect Dis 15: 656–658.

13. Grigg ME, Ganatra J, Boothroyd JC, Margolis TP (2001) Unusual abundance of atypical strains associated with human ocular toxoplasmosis. J Infect Dis 184: 633–639.

14. Sibley LD, Boothroyd JC (1992) Virulent strains of *Toxoplasma gondii* comprise a single clonal lineage. Nature 359: 82–85.

15. Su C, Howe DK, Dubey JP, Ajioka JW, Sibley LD (2002) Identification of quantitative trait loci controlling acute virulence in *Toxoplasma gondii*. Proc Natl Acad Sci U S A 99: 10753–10758.

16. Boothroyd JC, Grigg ME (2002) Population biology of *Toxoplasma gondii* and its relevance to human infection: do different strains cause different disease? Curr Opin Microbiol 5: 438–442.

17. Suzuki Y (2002) Host resistance in the brain against *Toxoplasma gondii*. J Infect Dis 185 Suppl 1: S58–S65.

18. Suzuki Y (1999) Genes, cells and cytokines in resistance against development of toxoplasmic encephalitis. Immunobiology 201: 255–271.

19. Holland GN (2004) Ocular toxoplasmosis: a global reassessment. Part II: disease manifestations and management. Am J Ophthalmol 137: 1–17.

20. Saeij JP, Boyle JP, Boothroyd JC (2005) Differences among the three major strains of *Toxoplasma gondii* and their specific interactions with the infected host. Trends Parasitol 21: 476–481.

21. Jamieson SE, Cordell H, Petersen E, McLeod R, Gilbert RE, et al. (2009) Host genetic and epigenetic factors in toxoplasmosis. Mem Inst Oswaldo Cruz 104: 162–169.

22. Kong JT, Grigg ME, Uyetake L, Parmley S, Boothroyd JC (2003) Serotyping of *Toxoplasma gondii* infections in humans using synthetic peptides. J Infect Dis 187: 1484–1495.

23. Peyron F, Lobry JR, Musset K, Ferrandiz J, Gomez-Marin JE, et al. (2006) Serotyping of *Toxoplasma gondii* in chronically infected pregnant women: predominance of type II in Europe and types I and III in Colombia (South America). Microbes Infect 8: 2333–2340.

24. Sousa S, Ajzenberg D, Marle M, Aubert D, Villena I, et al. (2009) Selection of polymorphic peptides from GRA6 and GRA7 sequences of *Toxoplasma gondii* strains to be used in serotyping. Clin Vaccine Immunol 16: 1158–1169.

25. Sousa S, Canada N, Correia da Costa JM, Darde ML (2010) Serotyping of naturally *Toxoplasma gondii* infected meat-producing animals. Vet Parasitol 169: 24–28.

26. Morisset S, Peyron F, Lobry JR, Garweg J, Ferrandiz J, et al. (2008) Serotyping of *Toxoplasma gondii*: striking homogeneous pattern between symptomatic and asymptomatic infections within Europe and South America. Microbes Infect 10: 742–747.

27. Parmley SF, Gross U, Sucharczuk A, Windeck T, Sgarlato GD, et al. (1994) 2 Alleles of the Gene Encoding Surface-Antigen P22 in 25 Strains of *Toxoplasma gondii*. J Parasitol 80: 293–301.

28. Grigg ME, Suzuki Y (2003) Sexual recombination and clonal evolution of virulence in *Toxoplasma*. Microbes Infect 5: 685–690.

29. Mertens M, Hofmann J, Petraityte-Burneikiene R, Ziller M, Sasnauskas K, et al. (2011) Seroprevalence study in forestry workers of a non-endemic region in eastern Germany reveals infections by Tula and Dobrava-Belgrade hantaviruses. Med Microbiol Immunol 200: 263–8.

30. Hosseininejad M, Azizi HR, Hosseini F, Schares G (2009) Development of an indirect ELISA test using a purified tachyzoite surface antigen SAG1 for sero-diagnosis of canine *Toxoplasma gondii* infection. Vet Parasitol 164: 315–319.

31. Wenschuh H, Volkmer-Engert R, Schmidt M, Schulz M, Schneider-Mergener J, et al. (2000) Coherent membrane supports for parallel microsynthesis and screening of bioactive peptides. Biopolymers 55: 188–206.

32. Frank R, Overwin H (1996) SPOT synthesis. Epitope analysis with arrays of synthetic peptides prepared on cellulose membranes. Methods Mol Biol 66: 149–169.

33. Panse S, Dong L, Burian A, Carus R, Schutkowski M, et al. (2004) Profiling of generic anti-phosphopeptide antibodies and kinases with peptide microarrays using radioactive and fluorescence-based assays. Mol Divers 8: 291–9.

34. Andresen H, Grötzinger C (2009) Deciphering the antibodyome - peptide arrays for serum antibody biomarker diagnostics. Curr Proteomics 6: 1–12.

35. Ngo Y, Advani R, Valentini D, Gaseitsiwe S, Mahdavifar S, et al. (2009) Identification and testing of control peptides for antigen microarrays. J Immunol Methods 343: 68–78.

36. Nahtman T, Jernberg A, Mahdavifar S, Zerweck J, Schutkowski M, et al. (2007) Validation of peptide epitope microarray experiments and extraction of quality data. J Immunol Methods 328: 1–13.

37. Vigil A, Ortega R, Jain A, Nakajima-Sasaki R, Tan X, et al. (2010) Identification of the feline humoral immune response to *Bartonella henselae* infection by protein microarray. PLoS one 5: e11447.

38. Pamelard F, Even G, Apostol C, Preda C, Dhaenens C, et al. (2009) PASE: a web-based platform for peptide/protein microarray experiments. Methods Mol Biol 570: 413–430.

39. Dominguez-Almendros S, Benitez-Parejo N, Gonzalez-Ramirez AR (2011) Logistic Regression Models. Allergol Immunopath (Madr.) 39: 295–305.

40. Bland JM, Altman DG (1995) Multiple significance tests: the Bonferroni method. B M J 310: 170.

41. Friendly M (1994) Mosaic Displays for Multiway Contingency-Tables. J A S A 89: 190–200.

42. Nowakowska D, Colon I, Remington JS, Grigg M, Golab E, et al. (2006) Genotyping of *Toxoplasma gondii* by multiplex PCR and peptide-based serological testing of samples from infants in Poland diagnosed with congenital toxoplasmosis. J Clin Microbiol 44: 1382–1389.

43. Sousa S, Ajzenberg D, Vilanova M, Costa J, Darde ML (2008) Use of GRA6-derived synthetic polymorphic peptides in an immunoenzymatic assay to serotype *Toxoplasma gondii* in human serum samples collected from three continents. Clin Vaccine Immunol 15: 1380–1386.

44. Xiao J, Buka SL, Cannon TD, Suzuki Y, Viscidi RP, et al. (2009) Serological pattern consistent with infection with type I *Toxoplasma gondii* in mothers and risk of psychosis among adult offspring. Microbes Infect 11: 1011–1018.

45. Melnyk O, Duburcq X, Olivier C, Urbes F, Auriault C, et al. (2002) Peptide arrays for highly sensitive and specific antibody-binding fluorescence assays. Bioconjug Chem 13: 713–720.

46. Mezzasoma L, Bacarese-Hamilton T, Di Cristina M, Rossi R, Bistoni F, et al. (2002) Antigen microarrays for serodiagnosis of infectious diseases. Clin Chem 48: 121–130.

47. Herrmann DC, Pantchev N, Globokar-Vrhovec M, Barutzki D, Wilking H, et al. (2010) Atypical *Toxoplasma gondii* genotypes identified in oocysts shed by cats in Germany. Int J Parasitol 40: 285–292.

48. Schares G, Globokar-Vrhovec M, Pantchev N, Herrmann DC, Conraths FJ (2008) Occurrence of *Toxoplasma gondii* and *Hammondia hammondi* oocysts in the faeces of cats from Germany and other European countries. Vet Parasitol 152: 34–45.

49. Liesenfeld O, Kosek J, Remington JS, Suzuki Y (1996) Association of CD4(+) T cell-dependent, interferon-gamma- mediated necrosis of the small intestine with genetic susceptibility of mice to peroral infection with *Toxoplasma gondii*. J Exp Med 184: 597–607.

50. Cesbron-Delauw MF (1994) Dense-granule organelles of *Toxoplasma gondii*: their role in the host-parasite relationship. Parasitol Today 10: 293–296.

51. Jacobs D, Vercammen M, Saman E (1999) Evaluation of recombinant dense granule antigen 7 (GRA7) of *Toxoplasma gondii* for detection of immunoglobulin G antibodies and analysis of a major antigenic domain. Clin Diagn Lab Immunol 6: 24–29.

52. Lecordier L, Fourmaux MP, Mercier C, Dehecq E, Masy E, et al. (2000) Enzyme-linked immunosorbent assays using the recombinant dense granule antigens GRA6 and GRA1 of *Toxoplasma gondii* for detection of immunoglobulin G antibodies. Clin Diagn Lab Immunol 7: 607–611.

53. Beghetto E, Spadoni A, Buffolano W, Del Pezzo M, Minenkova O, et al. (2003) Molecular dissection of the human B-cell response against *Toxoplasma gondii* infection by lambda display of cDNA libraries. Int J Parasitol 33: 163–173.

54. Redlich A, Müller WA (1998) Serodiagnosis of acute toxoplasmosis using a recombinant form of the dense granule antigen GRA6 in an enzyme-linked immunosorbent assay. Parasitol Res 84: 700–706.

55. Hiszczynska-Sawicka E, Brillowska-Dabrowska A, Dabrowski S, Pietkiewicz H, Myjak P, et al. (2003) High yield expression and single-step purification of *Toxoplasma gondii* SAG1, GRA1, and GRA7 antigens in *Escherichia coli*. Protein Expr Purif 27: 150–157.

56. Golkar M, Azadmanesh K, Khoshkholgh-Sima B, Babie J, Mercier C, et al. (2008) Serodiagnosis of recently acquired *Toxoplasma gondii* infection in pregnant women using enzyme-linked immunosorbent assays with a recombinant dense granule GRA6 protein. Diagn Microbiol Infect Dis 61: 31–9.

Permissions

All chapters in this book were first published in PLOS ONE, by The Public Library of Science; hereby published with permission under the Creative Commons Attribution License or equivalent. Every chapter published in this book has been scrutinized by our experts. Their significance has been extensively debated. The topics covered herein carry significant findings which will fuel the growth of the discipline. They may even be implemented as practical applications or may be referred to as a beginning point for another development.

The contributors of this book come from diverse backgrounds, making this book a truly international effort. This book will bring forth new frontiers with its revolutionizing research information and detailed analysis of the nascent developments around the world.

We would like to thank all the contributing authors for lending their expertise to make the book truly unique. They have played a crucial role in the development of this book. Without their invaluable contributions this book wouldn't have been possible. They have made vital efforts to compile up to date information on the varied aspects of this subject to make this book a valuable addition to the collection of many professionals and students.

This book was conceptualized with the vision of imparting up-to-date information and advanced data in this field. To ensure the same, a matchless editorial board was set up. Every individual on the board went through rigorous rounds of assessment to prove their worth. After which they invested a large part of their time researching and compiling the most relevant data for our readers.

The editorial board has been involved in producing this book since its inception. They have spent rigorous hours researching and exploring the diverse topics which have resulted in the successful publishing of this book. They have passed on their knowledge of decades through this book. To expedite this challenging task, the publisher supported the team at every step. A small team of assistant editors was also appointed to further simplify the editing procedure and attain best results for the readers.

Apart from the editorial board, the designing team has also invested a significant amount of their time in understanding the subject and creating the most relevant covers. They scrutinized every image to scout for the most suitable representation of the subject and create an appropriate cover for the book.

The publishing team has been an ardent support to the editorial, designing and production team. Their endless efforts to recruit the best for this project, has resulted in the accomplishment of this book. They are a veteran in the field of academics and their pool of knowledge is as vast as their experience in printing. Their expertise and guidance has proved useful at every step. Their uncompromising quality standards have made this book an exceptional effort. Their encouragement from time to time has been an inspiration for everyone.

The publisher and the editorial board hope that this book will prove to be a valuable piece of knowledge for researchers, students, practitioners and scholars across the globe.

List of Contributors

Masanori Yagi, Takahiro Tougan, Nirianne M. Q. Palacpac, Nobuko Arisue and Toshihiro Horii
Department of Molecular Protozoology, Research Institute for Microbial Diseases, Osaka University, Suita, Osaka, Japan

Gilles Bang and Pierre Druilhe
Laboratoire de Parasitologie Bio-Médicale, Institut Pasteur, Paris, France

Taiki Aoshi and Ken J. Ishii
Laboratoire de Parasitologie Bio-Médicale, Institut Pasteur, Paris, France
Laboratory of Vaccine Science, Immunology Frontier Research Center, Osaka University, Suita, Osaka, Japan

Yoshitsugu Matsumoto
Laboratory of Molecular Immunology, School of Agriculture and Life Sciences, The
University of Tokyo, Tokyo, Japan

Thomas G. Egwang
Med Biotech Laboratories, Kampala, Uganda

Susanne Elfert, Andreas Weise, Katja Bruser and Martin L. Biniossek
Institute of Molecular Medicine and Cell Research, Albert-Ludwigs-University Freiburg, Freiburg, Germany

Sabine Jägle and Niklas Senghaas
Institute of Molecular Medicine and Cell Research, Albert-Ludwigs-University Freiburg, Freiburg, Germany
Faculty of Biology, Albert-Ludwigs-University Freiburg, Freiburg, Germany

Andreas Hecht
Institute of Molecular Medicine and Cell Research, Albert-Ludwigs-University Freiburg, Freiburg, Germany
Faculty of Biology, Albert-Ludwigs-University Freiburg, Freiburg, Germany
BIOSS Centre for Biological Signalling Studies, Albert-Ludwigs-University Freiburg, Freiburg, Germany

Cristina Manguan-Garcia, Rosario Machado-Pinilla, Leandro Sastre and Rosario Perona
Instituto de Investigaciones Biomédicas CSIC/UAM, Madrid, Spain
CIBER de Enfermedades Raras, Valencia, Spain

Laura Pintado-Berninches and Jaime Carrillo
Instituto de Investigaciones Biomédicas CSIC/UAM, Madrid, Spain

Carme Pérez-Quilis, Isabel Esmoris and Amparo Gimeno
Biomedical Research Institute INCLIVA, Valencia, Spain
Department of Physiology, Faculty of Medicine and Dentistry, University of Valencia, Valencia, Spain

Jose Luis García-Giménez and Federico V. Pallardó
CIBER de Enfermedades Raras, Valencia, Spain
Biomedical Research Institute INCLIVA, Valencia, Spain
Department of Physiology, Faculty of Medicine and Dentistry, University of Valencia, Valencia, Spain

Yi Piao, Ayano Kimura, Satomi Urano, Yuhki Saito, Tohru Yamamoto, Saori Hata and Toshiharu Suzuki
Laboratory of Neuroscience, Graduate School of Pharmaceutical Sciences, Hokkaido University, Sapporo, Japan

Hidenori Taru
Laboratory of Neural Cell Biology, Graduate School of Pharmaceutical Sciences, Hokkaido University, Sapporo, Japan
Creative Research Institute, Hokkaido University, Sapporo, Japan

Benno Wölk
Department of Medicine II, University of Freiburg, Freiburg, Germany
Institute of Virology, Hannover Medical School, Hannover, Germany

Claudia Trautwein, Hans-Georg Rammensee and Stefan Stevanovic
Department of Immunology, University of Tübingen, Tübingen, Germany

Benjamin Büchele, Nadine Kersting, Hubert E. Blum and Volker Brass
Department of Medicine II, University of Freiburg, Freiburg, Germany

Andreas Cerny
Clinical Pharmacology and Clinical Immunology/Allergology, Inselspital, University of Bern, Bern, Switzerland

Darius Moradpour
Department of Medicine II, University of Freiburg, Freiburg, Germany, Division of Gastroenterology and Hepatology, Centre Hospitalier Universitaire Vaudois, University of Lausanne, Lausanne, Switzerland

Kim Henriksen, Mette G. Sørensen, Natasha Barascuk, Monika Pajak, Claus Christiansen and Morten A. Karsdal
Nordic Bioscience Biomarkers and Research, Herlev, Denmark

Yaguo Wang and Qinlong Zheng
Nordic Bioscience, Beijing, China

Joyce Suhy
CCBR-Synarc, San Francisco, California, United States of America

Jan T. Pedersen
Neurodegeneration, H. Lundbeck A/S, Copenhagen, Denmark

Kevin L. Duffin and Robert A. Dean
Translational Sciences Department, Eli Lilly and Company, Indianapolis, Indiana, United States of America

Alexis Rodríguez
Departamento de Medicina Molecular y Bioprocesos, Instituto de Biotecnología, Universidad Nacional Autónoma de México, Cuernavaca Morelos, México
Centro de Investigación en Biotecnología, Universidad Autónoma del Estado de Morelos, Cuernavaca, Morelos, México

Elba Villegas
Centro de Investigación en Biotecnología, Universidad Autónoma del Estado de Morelos, Cuernavaca, Morelos, México

Alejandra Montoya-Rosales and Bruno Rivas-Santiago
Medical Research Unit-Zacatecas, Mexican Institute of Social Security, UIMZ-IMSS, Zacatecas, Mexico

Gerardo Corzo
Departamento de Medicina Molecular y Bioprocesos, Instituto de Biotecnología, Universidad Nacional Autónoma de México, Cuernavaca Morelos, México

Hema Bora, Rupesh Kumar Tyagi and Yagya Dutta Sharma
Department of Biotechnology, All India Institute of Medical Sciences, New Delhi, India

Alessandro Mosco, Antonella Campanella and Piero G. Giulianini
Department of Life Sciences, University of Trieste, Trieste, Italy

Vientsislav Zlatev, Corrado Guarnaccia, Sándor Pongor and Sotir Zahariev
International Centre for Genetic Engineering and Biotechnology, AREA Science Park, Trieste, Italy

Betty Fleuchot, Alain Guillot, Christine Mézange, Colette Besset, Emilie Chambellon, Véronique Monnet and Rozenn Gardan
INRA, UMR1319 MICALIS, Jouy en Josas, France
AgroParistech, UMR MICALIS, Jouy en Josas, France

Yanhua Fan
Biotechnology Research Center, Southwest University, Beibei, Chongqing, People's Republic of China
Department of Microbiology and Cell Science, University of Florida, Gainesville, Florida, United States of America

Roberto M. Pereira
Department of Entomology and Nematology, University of Florida, Gainesville, Florida, United States of America

Engin Kilic, George Casella and Nemat O. Keyhani
Department of Microbiology and Cell Science, University of Florida, Gainesville, Florida, United States of America

Diana Hildebrand, Philipp Merkel, Lars Florian Eggers and Hartmut Schlüter
University Medical Centre Hamburg-Eppendorf, Institute of Clinical Chemistry, Mass Spectrometric Proteomics, Hamburg, Germany

Paula S. Santos, Rafael Nascimento, Fabiana A. A. Santos and Luiz R. Goulart
Laboratório de Nanobiotecnologia, Instituto de Genética e Bioquímica, Universidade Federal de Uberlândia, Uberlândia, Brazil

Luciano P. Rodrigues, Ana G. Brito-Madurro and João M. Madurro
Laboratório de Filmes Poliméricos e Nanotecnologia, Instituto de Química, Universidade Federal de Uberlândia, Uberlândia, Brazil

Paula C. B. Faria
Laboratório de Leishmanioses, Universidade Federal de Minas Gerais, Belo Horizonte, Brazil

João R. S. Martins
Laboratório rio de Parasitologia, Instituto de Pesquisas Veterinárias Desidério Finamor, Eldorado do Sul, Brazil

Rafael Diego Rosa, Agnès Vergnes, Julien de Lorgeril, Laure Sauné, Bernard Romestand, Julie Fievet, Yannick Gueguen, Evelyne Bachère and Delphine Destoumieux-Garzón
Ecologie des Systèmes Marins Côtiers, UMR5119, Centre National de la Recherche Scientifique, Institut Français de Recherche pour l'Exploitation de la Mer, Institut de la Recherche pour le Développement, Université Montpellier 1, Université Montpellier 2, Montpellier, France,

Priscila Goncalves and Luciane Maria Perazzolo
Laboratory of Immunology Applied to Aquaculture, Department of Cell Biology, Embryology and Genetics, Federal University of Santa Catarina, Florianópolis SC, Brazil

Andrew J. McCluskey
Department of Pharmaceutical Sciences, University of Toronto, Toronto, Ontario, Canada

Eleonora Bolewska-Pedyczak
Sunnybrook Research Institute, Toronto, Ontario, Canada

Nick Jarvik, Gang Chen and Sachdev S. Sidhu
Banting and Best Department of Medical Research, Terrence Donnelly Center for Cellular and Biomolecular Research, University of Toronto, Toronto, Ontario, Canada

Jean Gariépy
Department of Pharmaceutical Sciences, University of Toronto, Toronto, Ontario, Canada
Sunnybrook Research Institute, Toronto, Ontario, Canada,
Department of Medical Biophysics, University of Toronto, Toronto, Ontario, Canada

Giovanni Feverati
Université de Savoie, Annecy le Vieux Cedex, France

Mounia Achoch and Jihad Zrimi
Université de Savoie, Annecy le Vieux Cedex, France
Laboratoire de Chimie Bioorganique et Macromoléculaire (LCBM), Faculté des Sciences et Techniques-Guéliz, Université Cadi Ayyad, Marrakech, Maroc

Laurent Vuillon
LAMA, Université de Savoie, Le Bourget du Lac, France

Claire Lesieur
Université de Savoie, Annecy le Vieux Cedex, France
AGIM, Université Joseph Fourier, Archamps, France

Biswanath Chatterjee, Chen Lin and Eric H.-L. Chen
Institute of Biological Chemistry, Academia Sinica, Taipei, Taiwan

Chung-Yu Lee and Rita P.-Y. Chen
Institute of Biological Chemistry, Academia Sinica, Taipei, Taiwan
Institute of Biochemical Sciences, National Taiwan University, Taipei, Taiwan

Chao-Li Huang and Chien-Chih Yang
Department of Biochemical Science and Technology, National Taiwan University, Taipei, Taiwan

Marcus O. W. Grimm and Tobias Hartmann
Deutsches Institut für DemenzPrävention (DIDP), Saarland University, Homburg/Saar, Germany
Neurodegeneration and Neurobiology, Saarland University, Homburg/Saar, Germany
Experimental Neurology, Saarland University, Homburg/Saar, Germany

Eva G. Zinser, Sven Grösgen, Benjamin Hundsdörfer, Tatjana L. Rothhaar, Verena K. Burg and Heike S. Grimm
Experimental Neurology, Saarland University, Homburg/Saar, Germany

Lars Kaestner and Peter Lipp
Molecular Cellbiology, Saarland University, Homburg/Saar, Germany

Thomas A. Bayer
Department for Psychiatry, University of Goettingen, Goettingen, Germany

Ulrike Müller
Institute for Pharmacy and Molecular Biotechnology (IPMB), University of Heidelberg, Heidelberg, Germany

Yongbin Zeng, Wei Wang, Mingkuan Su, Jinpiao Lin, Huijuan Chen, Bin Yang, Qishui Ou and Ling Jiang
First Clinical College, Fujian Medical University, Fuzhou, China
Department of Laboratory Medicine, The First Affiliated Hospital of Fujian Medical University, Fuzhou, China

Dezhong Li
Department of Laboratory Medicine, The Armed Police Hospital of Fujian, Fuzhou, China

Jing Chen
First Clinical College, Fujian Medical University, Fuzhou, China
Center of Liver Diseases, The First Affiliated Hospital of Fujian Medical University, Fuzhou, China

Pavlo Maksimov, Aline Maksimov, Franz J. Conraths and Gereon Schares
Federal Research Institute for Animal Health, Institute of Epidemiology, Friedrich-Loeffler-Institut, Wusterhausen, Germany

Johannes Zerweck
JPT Peptide Technologies GmbH, Berlin, Germany

Andrea Hotop and Uwe Groß
German National Consulting Laboratory for Toxoplasmosis, Department of Medical Microbiology, University Medical Center Göttingen, Göttingen, Germany

Katrin Spekker and Walter Däubener
Institute of Medical Microbiology and Hospital Hygiene, Heinrich-Heine-University Düsseldorf, Düsseldorf, Germany

Sandra Werdermann
Institut für Arbeits und Sozialhygiene Stiftung, Kyritz, Germany

Olaf Niederstrasser
BG Kliniken Bergmannstrost, Halle, Germany

Eckhardt Petri
Novartis Vaccines and Diagnostics, Marburg, Germany

Marc Mertens and Rainer G. Ulrich
Federal Research Institute for Animal Health, Institute for Novel and Emerging Infectious Diseases, Friedrich-Loeffler-Institut, Greifswald - Insel Riems, Germany

Index